D0164396

**Third Edition**

# Convective Heat Transfer

# Third Edition

# Convective Heat Transfer

**Sadık Kakaç**
TOBB University of Economics and Technology, Ankara, Turkey

**Yaman Yener**
Northeastern University, USA

**Anchasa Pramuanjaroenkij**
Kasetsart University, Thailand

CRC Press
Taylor & Francis Group
Boca Raton  London  New York

CRC Press is an imprint of the
Taylor & Francis Group, an **informa** business

CRC Press
Taylor & Francis Group
6000 Broken Sound Parkway NW, Suite 300
Boca Raton, FL 33487-2742

© 2014 by Taylor & Francis Group, LLC
CRC Press is an imprint of Taylor & Francis Group, an Informa business

No claim to original U.S. Government works

Printed and bound in the United States of America by Edwards Brothers Malloy
Version Date: 20130820

International Standard Book Number-13: 978-1-4665-8344-3 (Hardback)

---

### Library of Congress Cataloging-in-Publication Data

---

Kakaç, S. (Sadik)
    Convective heat transfer, third edition / Sadik Kakac, Yaman Yener, Anchasa Pramuanjaroenkij. -- Third edition.
        pages cm
    Includes bibliographical references and index.
    ISBN 978-1-4665-8344-3 (hbk. : alk. paper)
    1. Heat--Convection. 2. Heat--Transmission. I. Yener, Yaman, 1946- II. Pramuanjaroenkij, A. (Anchasa) III. Title.

TJ260.K25 2014
621.402'2--dc23                                                                                    2013033728

---

**Visit the Taylor & Francis Web site at**
**http://www.taylorandfrancis.com**

**and the CRC Press Web site at**
**http://www.crcpress.com**

*To Filiz, Demet, and Supakorn*

*This book is dedicated to the memory of Yaman Yener*

True enlightenment in life is science-technology.

**Mustafa Kemal Atatürk**

Heat, like gravity, penetrates every substance of the universe; its rays occupy all parts of space. The theory of heat will hereafter form one of the most important branches of general physics.

**J. B. Joseph Fourier, 1824**

# Contents

# *Preface*

The second edition of this book evolved from a series of lecture notes that we developed over many years of teaching a graduate-level course on convective heat transfer in the Mechanical Engineering Department at the Middle East Technical University, Ankara, Turkey, and the University of Miami and Northeastern University, United States. Experiences gained from the use of the second edition and very valuable suggestions by the faculties who used it as a textbook led us to prepare the third edition of this book. Since the publication of the second edition, there have been emerging issues in science and technology, such as in the field of micro/nanotechnology, which have resulted in new research areas of heat transfer in microchannels and the enhancement of convective heat transfer with nanofluids, which in turn have led to valuable research contributions in the literature. We thus decided to revise the popular second edition with minor changes in some chapters and added two new chapters on heat transfer in microchannels and convective heat transfer with nanofluids with new correlations and solved examples. On the other hand, our general philosophy in writing a textbook on convective heat transfer for graduate students has been retained. In order to provide an exposition of an applied nature, we have followed the philosophy that we thought would be the most suitable to the combination of physical reasoning and theoretical analysis with solved examples. The topics covered in the chapters have been chosen carefully, and the whole book has been methodically organized for the best presentation of the subject matter.

This textbook is intended for first-year mechanical engineering graduate students. The chapter on boiling and condensation has been omitted in this edition due to space and time considerations in a one-semester course. As there are numerous edited volumes and handbooks on various topics of convective heat transfer, it is difficult to select the key topics for a one-semester textbook. We hope that the readers will appreciate our selection of fundamental topics and their clear presentation.

We have tried to cover topics that students should know about, including the physical mechanisms of convective heat transfer phenomena, exact or approximate solution methods, and solutions under various conditions. As a textbook it should contain essential and sufficient materials for a one-semester graduate course in convective heat transfer. Therefore, a rigorous proof or explanation is given wherever necessary for the benefit of the readers. Special attention has been given to the derivation of the basic equations of convective heat transfer and their solutions. It is assumed that the readers have a basic knowledge of thermodynamics, heat transfer, fluid mechanics, and differential equations at the undergraduate level.

Convective heat transfer is of great importance in industry. For practical use, important correlations are also listed whenever appropriate. Solved examples are given, and the problems at the end of each chapter are designed to clarify the physical and/or theoretically important points, as well as to supplement and extend the text. The Système International (SI) is used as the system of unit in the solution of the problems. We have made every effort to minimize typographical errors. We welcome feedback from readers to improve subsequent editions of this book.

**Sadik Kakaç**
*Ankara, Turkey*

**Anchasa Pramuanjaroenkij**
*Sakon Nakhon, Thailand*

# Acknowledgments

We express our sincere appreciation to our colleagues for their valuable comments on the first and second editions and to our graduate students for asking questions about our solution methods and solutions, which contributed in the improvement of the text. Special thanks go to our colleagues at the TOBB University of Economy and Technology, Northeastern University, Kasetsart University, and Middle East Technical University for their contributions and assistance during the preparation of the revised manuscript for this third edition, especially Amarin Tonkratoke and Nilgün F. Ünver.

We are indebted to the contribution of Ali Akgüneş, Middle East Technical University, Ankara, Turkey, in preparing the line drawings in this book. The first author is grateful to Prof. Warner Rohsennow and Prof. Nickelson at MIT for teaching him heat transfer at an advanced level and to Prof. W.B. Hall and Dr. Philipe Price for constant encouragement and support during his PhD work at the Victoria University of Manchester.

We have made every effort to minimize typographical errors. We welcome feedback from readers to improve subsequent editions of this book. Thanks are also due to Arlene Kopeloff, editorial assistant; Jessica Vakili, project coordinator; and Jonathan W. Plant, executive editor; as well as to other individuals at CRC Press who contributed their talents and energy to the third edition of this book.

Finally, we acknowledge the encouragement and support of our lovely families who made many sacrifices during the preparation of this text. We always remember Yaman Yener for his invaluable contributions to the field of heat transfer and to the first and second editions of this book.

# 1

# Foundations of Heat Transfer

## Nomenclature

| | |
|---|---|
| $A$ | surface area; heat transfer area, $m^2$ |
| $E$ | energy, J |
| $e$ | energy per unit mass, J/kg |
| $F$, $\mathbf{F}$ | force, N |
| $F_{ij}$ | radiation shape factor |
| $\mathfrak{I}_{12}$ | radiation factor defined by Equation 1.67 |
| $\mathbf{f}$ | body force per unit mass, N/kg |
| $g$ | gravitational acceleration, $m/s^2$ |
| $h$ | heat transfer coefficient, W/($m^2$ K) |
| $i$ | enthalpy per unit mass, J/kg |
| $\hat{\imath}$ | unit vector in $x$-direction |
| $\hat{\jmath}$ | unit vector in $y$-direction |
| $k$ | thermal conductivity, W/(m K) |
| $\hat{k}$ | unit vector in $z$-direction |
| $\mathbf{M}$ | linear momentum, kg m/s |
| $m$ | mass, kg |
| $\dot{m}$ | mass flow rate, kg/s |
| $n$ | distance; normal direction, m |
| $\hat{\mathbf{n}}$ | unit vector in $n$-direction |
| $P$ | pressure, $N/m^2$ |
| $Q$ | heat, J |
| $q$ | heat transfer rate, W |
| $q''$, $\mathbf{q}''$ | heat flux, $W/m^2$ |
| $\dot{q}_e$ | volumetric heat generation rate due to electrical sources, $W/m^3$ |
| $S$ | entropy, J/K |
| $s$ | entropy per unit mass, J/(kg K) |
| $T$ | temperature, °C, K |
| $t$ | time, s |
| $\mathcal{U}$ | internal energy per unit mass, J/kg |
| $u$ | velocity component in $x$-direction, m/s |
| $V$, $\mathbf{V}$ | velocity, m/s |
| $\upsilon$ | volume, $m^3$ |
| $v$ | velocity component in $y$-direction, m/s |
| $W$ | work, J |
| $\dot{W}$ | rate of work, W |

| $w$ | velocity component in $z$-direction, m/s |
|---|---|
| $x$ | rectangular coordinate; distance parallel to surface, m |
| $y$ | rectangular coordinate; distance normal to surface, m |
| $z$ | rectangular coordinate, m |

## Greek Symbols

| $\alpha$ | absorptivity |
|---|---|
| $\Delta$ | finite increment |
| $\delta$ | velocity boundary-layer thickness, m; inexact differential increment |
| $\delta_T$ | thermal boundary-layer thickness, m |
| $\varepsilon$ | emissivity |
| $\rho$ | density, kg/m$^3$; reflectivity |
| $\sigma$ | Stefan–Boltzmann constant, $5.6697 \times 10^{-8}$ W/(m$^2 \cdot$K$^4$) |
| $\tau$ | transmissivity |
| $\phi$ | gravitational potential, m$^2$/s$^2$ |

## Subscripts

| $b$ | blackbody |
|---|---|
| $c.s.$ | control surface |
| $c.v.$ | control volume |
| $f$ | fluid |
| $irr.$ | irreversible |
| $n$ | normal direction |
| $rev.$ | reversible |
| $s$ | solid |
| $w$ | wall condition |
| $x$ | $x$-direction |
| $\infty$ | free-stream conditions |

## Superscripts

| $\wedge$ | unit vector |
|---|---|

---

## 1.1 Introductory Remarks

Heat transfer is the study of energy transfer processes between material bodies solely as a result of temperature differences. Heat transfer problems confront the engineers and researchers in nearly every branch of engineering and science. Although it is generally regarded as most closely related to mechanical engineering, much work in this field has also been done in chemical, nuclear, metallurgical, and electrical engineering, where heat transfer problems are equally important. It is probably this fundamental and widespread influence that has helped heat transfer develop as an engineering science.

In thermodynamics, *heat* is defined as the form of energy that crosses the boundary of a thermodynamic system by virtue of a temperature difference existing between the system and its surrounding. That is, heat is the energy in transition across the system boundary and temperature difference is the driving potential for its propagation. *Heat flow* is vectorial in the sense that it is in the direction of a negative temperature gradient, that is, from higher toward lower temperatures.

Thermodynamics is that branch of science that deals with the study of heat and work interactions of a system with its surroundings. The laws of thermodynamics may be used to predict the gross amount of heat transferred to or from a system during a process in which the system goes from one thermodynamic state (i.e., mechanical and chemical, as well as thermal equilibrium) to another. In most instances, however, the overriding consideration may be the length of time over which the transfer of heat occurs or, simply, the time rate at which it takes place. The laws of thermodynamics alone are not sufficient to provide such information; neither can they explain the mechanisms of heat transfer, which is not strictly restricted to equilibrium states. The science of heat transfer, on the other hand, studies the mechanisms of heat transfer and extends thermodynamic analysis, through the development of necessary empirical and analytical relations, to calculate heat transfer rates.

The science of heat transfer is based upon foundations comprising both theory and experiment. As in other engineering disciplines, the theoretical part is constructed from one or more *physical* (*or natural*) *laws*. The physical laws are statements, in terms of various concepts, which have been found to be true through many years of experimental observations. A physical law is called a *general law* if its application is independent of the medium under consideration. Otherwise, it is called a *particular law*. There are, in fact, four general laws among others upon which all the analyses concerning heat transfer, either directly or indirectly, depend [1]. These are

1. The law of conservation of mass
2. Newton's second law of motion
3. The first law of thermodynamics
4. The second law of thermodynamics

In addition to the general laws, it is usually necessary to bring certain particular laws into analysis. Examples are Fourier's law of heat conduction, Newton's law of cooling, the Stefan–Boltzmann law of radiation, Newton's law of viscosity, and the ideal gas law.

## 1.2 Modes of Heat Transfer

The mechanism by which heat is transferred in a heat exchange or an energy conversion system is, in fact, quite complex. There appear, however, to be three rather basic and distinct *modes* of heat transfer. These are *conduction*, *convection*, and *radiation*.

*Conduction* is the process of heat transfer by molecular motion, supplemented in some cases by the flow of free electrons, through a body (solid, liquid, or gaseous) from a region of high temperature to a region of low temperature. Heat transfer by conduction also takes place across the interface between two bodies in contact when they are at different temperatures.

The mechanism of heat conduction in liquids and gases has been postulated as the transfer of kinetic energy of the molecular movement. Transfer of thermal energy to a fluid increases its internal energy by increasing the kinetic energy of its vibrating molecules and is measured by the increase of its temperature. Heat conduction is thus the transfer of kinetic energy of the more energetic molecules in the high-temperature region by successive collisions to the molecules in the low-temperature region.

Heat conduction in solids with crystalline structure, such as quartz, depends on energy transfer by molecular and lattice vibrations and free-electron drift. In the case of amorphous solids, such as glass, heat conduction depends only on the molecular transport of energy.

*Thermal radiation*, or simply *radiation*, is heat transfer in the form of electromagnetic waves. All substances, solid bodies as well as liquids and gases, emit radiation as a result of their temperature, and they are also capable of absorbing such energy. Furthermore, radiation can pass through certain types of substances (called *transparent* and *semitransparent* materials) as well as through vacuum, whereas for heat conduction to take place, a material medium is absolutely necessary.

Conduction is the only mechanism by which heat can flow in *opaque* solids. Through certain transparent or semitransparent solids, such as glass and quartz, energy flow can be by radiation as well as by conduction. With gases and liquids, if there is no observable fluid motion, the heat transfer mechanism will be conduction (and, if not negligible, radiation). However, if there is macroscopic fluid motion, energy can also be transported in the form of internal energy by the movement of the fluid itself. The process of energy transport by the combined effect of heat conduction (and radiation) and the movement of fluid is referred to as *convection* or *convective heat transfer*.

Although in the foregoing classification we have classified convection to be a mode of heat transfer, it is actually conduction (and radiation) in moving fluids. An analysis of convective heat transfer is, therefore, more involved than that of heat transfer by conduction alone because the motion of the fluid must be studied simultaneously with the energy transfer process.

In reality, temperature distribution in a medium is controlled by the combined effect of these three modes of heat transfer. Therefore, it is not actually possible to entirely isolate one mode from interactions with other modes. For simplicity in the analysis, however, these three modes of heat transfer are almost always studied separately. In this book we study convection only.

## 1.3 Continuum Concept

In an analysis of convective heat transfer in a fluid, the motion of the fluid must be studied simultaneously with the heat transport process. In its most fundamental form, the description of the motion of a fluid involves a study of the behavior of all discrete particles (molecules, atoms, electrons, etc.) that make up the fluid. The most fundamental approach in analyzing convective heat transfer would be, therefore, to apply the laws of mechanics and thermodynamics to each individual particle, or a statistical group of particles, subsequent to some initial state of affairs. Such an approach would give an insight into the details of the energy transfer processes; however, it is too cumbersome for most situations arising in engineering.

In most engineering problems, our primary interest lies not in the molecular behavior of the fluid but rather in how it behaves as a continuous medium. In our study of convective heat transfer, we shall, therefore, neglect the molecular structure of the fluid and consider it to be a continuous medium—*continuum*—which is fortunately a valid approach to many practical problems where only macroscopic information is of interest. Such a model may be used provided that the size and the mean free path of molecules are small enough compared with other dimensions existing in the medium and that a statistical average is meaningful. This approach, which is also known as *phenomenological* approach to convective heat transfer, is simpler than microscopic approaches and usually gives the answers required in engineering. On the other hand, in order to make up for the information lost by the neglect of the molecular structure, certain parameters such as thermodynamic state and transport properties have to be introduced empirically. Parallel to the study of convective heat transfer by continuum approach, however, molecular considerations can also be used to obtain information on thermodynamic and transport properties. In this book, we restrict our discussions to the phenomenological convective heat transfer only.

## 1.4  Some Definitions and Concepts of Thermodynamics

In this section, we review some of the definitions and concepts of thermodynamics that are needed for the study of convective heat transfer. The reader is, however, advised to refer to textbooks on thermodynamics, such as References 2,3, for an in-depth discussion of these definitions and concepts, as well as the laws of thermodynamics.

A *system* is any arbitrary collection of matter of fixed identity bounded by a closed surface, which can be a real or an imaginary one. All other systems that interact with the system under consideration are known as its *surroundings*. In the absence of any mass–energy conversion, the mass of a system not only remains constant, but the system must be made up of exactly the same submolecular particles. The four general laws listed in Section 1.1 are always stated in terms of a system. In fact, one cannot meaningfully apply a general law until a definite system is identified.

A *control volume* is any defined region in space across the boundaries of which matter, energy, and momentum may flow; within which matter, energy, and momentum storage may take place; and on which external forces may act. Its position, shape, and/or size may change with time. However, most often we deal with control volumes that are fixed in space and of fixed size and shape.

The dimensions of a system or a control volume may be finite in extent, or all of them may be infinitesimal. The complete definition of a system or a control volume must include at least the implicit definition of a coordinate system, since they may be moving or stationary.

The characteristic of a system we are most interested in is its thermodynamic state, which is described by a list of the values for all its *properties*. A property of a system is either a directly or an indirectly observable characteristic of the system, which can in principle be quantitatively evaluated. Volume, mass, pressure, temperature, etc., are all properties. If all the properties of a system remain unchanged, then the system is said to be in an equilibrium state. A *process* is a change of state and is described in part by the series of states passed through by the system. A *cycle* is a process wherein the initial and final states of a system are equal.

If no energy transfer as heat takes place between any two systems when they are placed in contact with each other, they are said to be in *thermal equilibrium*. Any two systems are said to have the same *temperature* if they are in thermal equilibrium with each other. Two systems that are not in thermal equilibrium have different temperatures, and energy transfer as heat may take place from one of the systems to the other. Temperature is, therefore, that property of a system that measures the thermal level of the system.

The laws of thermodynamics deal with interactions between a system and its surroundings as they pass through equilibrium states. These interactions may be divided into two as (1) *work* and (2) *heat* interactions. Heat has already been defined as the form of energy that crosses the boundary of a system due to a temperature difference existing between the system and its surroundings. Work, on the other hand, is a form of energy that is characterized as follows: When an energy form of one system (such as kinetic energy, potential energy, chemical energy) is transformed into an energy form of another system or surroundings without the transfer of mass from the system and not by means of a temperature difference, the energy is said to have been transferred through the performance of work.

## 1.5 General Laws

In the following sections, we first give the statements of the four general laws referred to in Section 1.1 in terms of a system and then develop their control-volume forms.

### 1.5.1 Law of Conservation of Mass

The law of conservation of mass, when referred to a system, simply states that, in the absence of any mass–energy conversion, the mass of the system remains constant. Thus, for a system,

$$\frac{dm}{dt} = 0 \quad \text{or} \quad m = \text{constant}, \tag{1.1}$$

where $m$ is the mass of the system.

We now proceed to develop the form of this law as it applies to a control volume. Consider an arbitrary control volume fixed in space and of fixed shape and size through which a fluid streams as illustrated in Figure 1.1. Define a system whose boundary at some time $t$ happens to correspond exactly to that of the control volume. By definition, the control volume remains fixed in space, but the system moves and at some later time $t + \Delta t$ occupies a different volume in space. The two positions of the system are shown in Figure 1.1 by dashed lines. Since the mass of the system is conserved, we can write

$$m_1(t) = m_1(t + \Delta t) + m_2(t + \Delta t) - m_3(t + \Delta t) \tag{1.2a}$$

or

$$m_1(t + \Delta t) - m_1(t) = m_3(t + \Delta t) - m_2(t + \Delta t), \tag{1.2b}$$

where $m_1$, $m_2$, and $m_3$ represent the instantaneous values of the mass contained in the three regions of space shown in Figure 1.1. Dividing Equation 1.2b by $\Delta t$, we get

$$\frac{m_1(t + \Delta t) - m_1(t)}{\Delta t} = \frac{m_3(t + \Delta t)}{\Delta t} - \frac{m_2(t + \Delta t)}{\Delta t}. \tag{1.3}$$

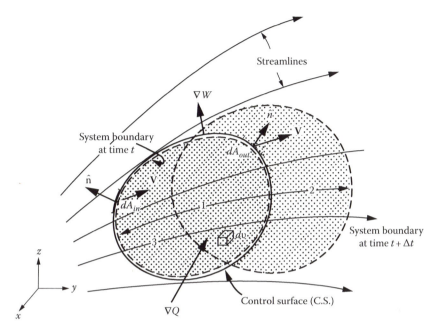

**FIGURE 1.1**
Flow through control volume.

As $\Delta t \to 0$, the left-hand side of Equation 1.3 becomes

$$\lim_{\Delta t \to 0} \frac{m_1(t + \Delta t) - m_1(t)}{\Delta t} = \frac{\partial m_{c.v.}}{\partial t} = \frac{\partial}{\partial t} \int_{c.v.} \rho d\upsilon, \qquad (1.4)$$

where
  $m_{c.v.}$ is the instantaneous mass within region 1, that is, the control volume
  $d\upsilon$ is an element of the control volume
  $\rho$ is the local density of that element
  $c.v.$ designates the control volume bounded by the control surface ($c.s.$)

Equation 1.4 represents the time rate of change of the mass within the control volume. Further, as $\Delta t \to 0$, the first term on the right-hand side of Equation 1.3 becomes

$$\lim_{\Delta t \to 0} \frac{m_3(t + \Delta t)}{\Delta t} = \dot{m}_{in} = - \int_{A_{in}} \rho \mathbf{V} \cdot \hat{\mathbf{n}} \, dA_{in}, \qquad (1.5)$$

and the second term reduces to

$$\lim_{\Delta t \to 0} \frac{m_2(t + \Delta t)}{\Delta t} = \dot{m}_{out} = \int_{A_{out}} \rho \mathbf{V} \cdot \hat{\mathbf{n}} \, dA_{out}, \qquad (1.6)$$

where
  $\dot{m}_{in}$ and $\dot{m}_{out}$ represent the mass flow rates in and out of the control volume
  $\mathbf{V}$ is the velocity vector
  $\hat{\mathbf{n}}$ is the outward-pointing unit vector normal to the control surface

Thus, as $\Delta t \to 0$, Equation 1.3 becomes

$$\frac{\partial}{\partial t} \int_{c.v.} \rho d\upsilon = - \int_{A_{in}} \rho \mathbf{V} \cdot \hat{\mathbf{n}} \, dA_{in} - \int_{A_{out}} \rho \mathbf{V} \cdot \hat{\mathbf{n}} \, dA_{out},$$

(1.7a)

which can be rewritten as

$$\frac{\partial}{\partial t} \int_{c.v.} \rho \, d\upsilon = - \int_{c.s.} \rho \mathbf{V} \cdot \hat{\mathbf{n}} \, dA.$$

(1.7b)

This result states that the net rate of mass flow into a control volume is equal to the time rate of increase of mass within the control volume. Since the control volume is fixed, Equation 1.7b can also be written as

$$\int_{c.v.} \frac{\partial \rho}{\partial t} \, d\upsilon = - \int_{c.s} \rho \mathbf{V} \cdot \hat{\mathbf{n}} \, dA.$$

(1.7c)

The surface integral on the right-hand side of Equation 1.7c can be transformed into a volume integral by the Gauss divergence theorem [4], that is,

$$\int_{c.s.} \rho \mathbf{V} \cdot \hat{\mathbf{n}} dA = \int_{c.v.} \nabla \cdot (\rho \mathbf{V}) d\upsilon.$$

(1.8)

Hence, Equation 1.7c may be rewritten as

$$\int_{c.v.} \frac{\partial \rho}{\partial t} \, d\upsilon = - \int_{c.v.} \nabla \cdot (\rho \mathbf{V}) d\upsilon,$$

(1.9a)

or

$$\int_{c.v.} \left[ \frac{\partial \rho}{\partial t} + \nabla \cdot (\rho \mathbf{V}) \right] d\upsilon = 0.$$

(1.9b)

Since this result would be valid for all arbitrary control volumes, the integrand must be zero everywhere, thus yielding

$$\frac{\partial \rho}{\partial t} + \nabla \cdot (\rho \mathbf{V}) = 0.$$

(1.10)

This equation is called the *continuity* equation. Further discussions on the continuity equation will be given in Section 2.2.

### 1.5.2 Newton's Second Law of Motion

Newton's second law of motion states that the net force $\mathbf{F}$ acting on a system in an inertial coordinate system is equal to the time rate of change of the total linear momentum $\mathbf{M}$ of the system; that is,

$$\mathbf{F} = \lim_{\Delta t \to 0} \frac{\Delta \mathbf{M}}{\Delta t} = \frac{d\mathbf{M}}{dt}. \tag{1.11}$$

Following the same approach we have used in the previous section, we now proceed to develop the form of this law as it applies to a control volume. Referring to Figure 1.1, the change in the total linear momentum of the system during the time interval $\Delta t$ can be written as

$$\Delta \mathbf{M} = \mathbf{M}_1(t + \Delta t) + \mathbf{M}_2(t + \Delta t) - \mathbf{M}_3(t + \Delta t) - \mathbf{M}_1(t), \tag{1.12}$$

where $\mathbf{M}_1$, $\mathbf{M}_2$, and $\mathbf{M}_3$ represent the instantaneous values of linear momentum of the masses contained in the three regions of space shown in Figure 1.1. Rearranging Equation 1.12 and dividing by $\Delta t$, we get

$$\frac{\Delta \mathbf{M}}{\Delta t} = \frac{\mathbf{M}_1(t + \Delta t) - \mathbf{M}_1(t)}{\Delta t} + \frac{\mathbf{M}_2(t + \Delta t)}{\Delta t} - \frac{\mathbf{M}_3(t + \Delta t)}{\Delta t}. \tag{1.13}$$

As $\Delta t \to 0$, the first term on the right-hand side of Equation 1.13 reduces to

$$\lim_{\Delta t \to 0} \frac{\mathbf{M}_1(t + \Delta t) - \mathbf{M}_1(t)}{\Delta t} = \frac{\partial}{\partial t} \mathbf{M}_{c.v.} = \frac{\partial}{\partial t} \int_{c.v.} \rho \mathbf{V} d\upsilon, \tag{1.14}$$

where

$\mathbf{M}_{c.v.}$ is the instantaneous value of the linear momentum within the control volume
$\rho$ and $\mathbf{V}$ are the local density and the velocity vector of the fluid element of volume $d\upsilon$

Equation 1.14 represents the time rate of change of linear momentum within the control volume. Similarly, the second and the third terms on the right-hand side of Equation 1.13 become

$$\lim_{\Delta t \to 0} \frac{\mathbf{M}_2(t + \Delta t)}{\Delta t} = \int_{A_{out}} \mathbf{V} \rho \mathbf{V} \cdot \hat{\mathbf{n}} dA_{out} \tag{1.15}$$

and

$$\lim_{\Delta t \to 0} \frac{\mathbf{M}_3(t + \Delta t)}{\Delta t} = \int_{A_{in}} \mathbf{V} \rho \mathbf{V} \cdot \hat{\mathbf{n}} dA_{in}. \tag{1.16}$$

Equations 1.15 and 1.16 represent the rates of linear momentum leaving and entering the control volume, respectively. Thus, Equation 1.11 can now be written as

$$\mathbf{F} = \frac{\partial}{\partial t} \int_{c.v.} \mathbf{V}\rho d\upsilon + \int_{A_{out}} \mathbf{V}\rho \mathbf{V} \cdot \hat{\mathbf{n}} dA_{out} + \int_{A_{out}} \mathbf{V}\rho \mathbf{V} \cdot \hat{\mathbf{n}} dA_{in} \tag{1.17a}$$

or

$$\mathbf{F} = \frac{\partial}{\partial t} \int_{c.v.} \mathbf{V}\rho d\upsilon + \int_{c.s.} \mathbf{V}\rho \mathbf{V} \cdot \hat{\mathbf{n}} dA. \tag{1.17b}$$

This result, which is usually called the *momentum theorem* or the *law of conservation of linear momentum,* states that the net force acting on the fluid instantaneously occupying a control volume is equal to the time rate of change of linear momentum within the control volume plus the net rate of linear momentum flow out of the control volume. It should be emphasized that Equation 1.17b is valid only when referred to axes moving without acceleration, since the usual form of Newton's second law of motion, Equation 1.11, holds under this condition.

Equation 1.17b is a vector relation. Referred to the rectangular coordinates $x$, $y$, and $z$, we can write the component in the $x$-direction as

$$F_x = \frac{\partial}{\partial t} \int_{c.v.} u\rho d\upsilon + \int_{c.s.} u\rho \mathbf{V} \cdot \hat{\mathbf{n}} dA, \tag{1.18}$$

where

$u$ is the $x$-component of the velocity vector, $\mathbf{V} = \hat{\mathbf{i}}u + \hat{\mathbf{J}}v + \hat{\mathbf{k}}w$
$F_x$ is the $x$-component of the net force, $\mathbf{F} = \hat{\mathbf{i}}F_x + \hat{\mathbf{J}}F_y + \hat{\mathbf{k}}F_z$, acting on the fluid instantaneously occupying the control volume

Equation 1.18 states, therefore, that the net force acting in the $x$-direction instantaneously on the fluid within the control volume is equal to the time rate of change of linear $x$-momentum within the control volume plus net flow rate of linear $x$-momentum out of the control volume. Of course, two similar relations also apply to the $y$- and $z$-directions.

The force $\mathbf{F}$ in Equation 1.17 is the resultant of all forces exerted by the surroundings on the fluid instantaneously occupying the control volume. A considerable number of such forces may be present in a specific problem, but they may, in general, be divided into two classes as follows:

1. Body (volume) forces
2. Surface forces

*Body forces* are those that are proportional to either the volume or mass of the fluid and comprise the forces involving action at a distance, such as gravitational, electrostatic, electromagnetic, centrifugal, and Coriolis forces. Body forces are usually expressed as per unit mass or volume of the fluid acted upon. *Surface forces* are, on the other hand, those acting on an element of fluid through its bounding surfaces and describe the influence of the surrounding fluid on the fluid element under consideration. Surface forces referred to a unit area are called *stresses*. In a viscous fluid, it is useful to divide surface stresses into two as follows:

1. Normal stresses
2. Shear stresses

Surface stresses in a viscous fluid will be discussed in more detail in Section 2.3.

### 1.5.3 First Law of Thermodynamics

When a system undergoes a cyclic process, the first law of thermodynamics can be expressed as

$$\oint \delta Q = \oint \delta W, \tag{1.19}$$

where the cyclic integral $\oint \delta Q$ represents the net heat transfer *to* the system and the cyclic integral $\oint \delta W$ is the net work done *by* the system during the process. Both heat and work are *path* functions; that is, the amount of heat transferred and the amount of work done when a system undergoes a change of state depend on the path the system follows during the change of state. This is why the differentials of heat and work are inexact differentials, denoted by the symbols $\delta Q$ and $\delta W$. For a process that involves an infinitesimal change of state during a time interval $dt$, the first law of thermodynamics is given by

$$dE = \delta Q - \delta W, \tag{1.20}$$

where
   $\delta Q$ and $\delta W$ are the differential amounts of heat added *to* the system and the work done *by* the system, respectively
   $dE$ is the corresponding increase in the total energy of the system during the time interval $dt$

The *energy E* is a property of the system and, like all other properties, is a *point* function. That is, $dE$ depends upon the initial and final states only and not on the path followed between the two states. For a more complete discussion of point and path functions, the reader is referred to References 2,3. Equation 1.20 can also be written as a rate equation:

$$\frac{dE}{dt} = \frac{\delta Q}{dt} - \frac{\delta W}{dt} \tag{1.21a}$$

or

$$\frac{dE}{dt} = q - \dot{W}, \tag{1.21b}$$

where $q = \delta Q/dt$ represents the rate of heat transfer to the system and $\dot{W} = \delta W/dt$ the rate of work done (power) by the system.

We now proceed to develop the control-volume form of the first law of thermodynamics. Referring to Figure 1.1, the first law for the system under consideration can be written as

$$\Delta E = \nabla Q - \nabla W, \tag{1.22}$$

where
  $\nabla Q$ is the amount of heat transferred to the system
  $\nabla W$ is the work done by the system
  $\Delta E$ is the corresponding increase in the energy of the system during time interval $\Delta t$

Dividing Equation 1.22 by $\Delta t$, one obtains

$$\frac{\Delta E}{\Delta t} = \frac{\nabla Q}{\Delta t} - \frac{\nabla W}{\Delta t}. \tag{1.23}$$

The increase in the energy of the system during the time interval $\Delta t$ may be written as

$$\Delta E = E_1(t + \Delta t) + E_2(t + \Delta t) - E_3(t + \Delta t) - E_1(t), \tag{1.24}$$

where $E_1$, $E_2$, and $E_3$ are the instantaneous values of energy of the masses contained in the three regions of space shown in Figure 1.1. Rearranging Equation 1.24 and dividing by $\Delta t$, we get

$$\frac{\Delta E}{\Delta t} = \frac{E_1(t + \Delta t) - E_1(t)}{\Delta t} + \frac{E_2(t + \Delta t)}{\Delta t} - \frac{E_3(t + \Delta t)}{\Delta t}. \tag{1.25}$$

As $\Delta t \to 0$, the first term on the right-hand side of Equation 1.25 becomes

$$\lim_{\Delta t \to 0} \frac{E_1(t + \Delta t) - E_1(t)}{\Delta t} = \frac{\partial E_{c.v.}}{\partial t} = \frac{\partial}{\partial t} \int_{c.v.} e\rho \, d\upsilon, \tag{1.26}$$

where
  $E_{c.v.}$ is the instantaneous value of energy of the mass occupying the control volume at time $t$
  $e$ is the *local specific energy*, that is, energy per unit mass

Equation 1.26 represents the time rate of change of energy within the control volume. Moreover,

$$\lim_{\Delta t \to t} \frac{E_2(t + \Delta t)}{\Delta t} = \int_{A_{out}} e\rho \mathbf{V} \cdot \hat{\mathbf{n}} \, dA_{out} \tag{1.27}$$

and

$$\lim_{\Delta t \to 0} \frac{E_3(t + \Delta t)}{\Delta t} = - \int_{A_{in}} e\rho \mathbf{V} \cdot \hat{\mathbf{n}} \, dA_{in}, \tag{1.28}$$

which represent, respectively, the rates of energy leaving and entering the control volume at time $t$. Thus, as $\Delta t \to 0$, Equation 1.25 reduces to

$$\frac{dE}{dt} = \lim_{\Delta t \to 0} \frac{\Delta E}{\Delta t} = \frac{\partial}{\partial t} \int_{c.v.} e\rho\, d\upsilon + \int_{A_{out}} e\rho\mathbf{V}\cdot\hat{\mathbf{n}}\, dA_{out} + \int_{A_{in}} e\rho\mathbf{V}\cdot\hat{\mathbf{n}}\, dA_{in}, \tag{1.29a}$$

which can also be written as

$$\frac{dE}{dt} = \frac{\partial}{\partial t} \int_{c.v.} e\rho\, d\upsilon + \int_{c.s.} e\rho\mathbf{V}\cdot\hat{\mathbf{n}}\, dA. \tag{1.29b}$$

The first term on the right-hand side of Equation 1.23 represents, as $\Delta t \to 0$, the rate of heat transfer across the control surface; that is,

$$\lim_{\Delta t \to 0} \frac{\nabla Q}{\Delta t} = \left[\frac{\delta Q}{dt}\right]_{c.s.} = q_{c.s.}. \tag{1.30}$$

Similarly, the second term becomes

$$\lim_{\Delta t \to 0} \frac{\nabla W}{\Delta t} = \frac{\delta W}{dt} = \dot{W}, \tag{1.31}$$

which is the rate of work done (power) by the fluid in the control volume (i.e., the system) on its surroundings at any time $t$. Hence, as $\Delta t \to 0$, Equation 1.23 becomes

$$\frac{\partial}{\partial t} \int_{c.v.} e\rho\, d\upsilon + \int_{c.s.} e\rho\mathbf{V}\cdot\hat{\mathbf{n}}\, dA = q_{c.s.} - \dot{W}, \tag{1.32}$$

which is the control-volume form of the first law of thermodynamics. However, a final form of this expression can be obtained after further consideration of the power term $\dot{W}$.

The work done by the fluid element in the control volume against its surroundings at any time $t$ may consist of the following four parts:

1. Work done against the body forces (i.e., gravitational, electrostatic, electromagnetic)
2. Work done against the surface forces (i.e., normal and shear forces)
3. Work done by the system on its surroundings that could cause a shaft to rotate (i.e., shaft work)
4. Work done on the system due to a power drawn from an external electric circuit

Let $\mathbf{f}$ be the net body force per unit mass acting on the fluid particles. Since

$$\text{Rate of work} = \text{Force} \times \text{Velocity in the direction of the force,}$$

the rate of work done by the fluid element in the control volume at time $t$ against the body forces may be expressed as

$$-\int_{c.v.}\rho\mathbf{f}\cdot\mathbf{V}\,d\upsilon,\tag{1.33}$$

where we have introduced the minus sign because work is done against the body forces when $\mathbf{f}$ and $\mathbf{V}$ are opposed.

Thus, Equation 1.32 may now be written as

$$\frac{\partial}{\partial t}\int_{c.v.}e\rho\,d\upsilon+\int_{c.s.}e\rho\mathbf{V}\cdot\hat{\mathbf{n}}\,dA=q_{c.s.}-\dot{W}_{normal}-\dot{W}_{shear}-\dot{W}_{shaft}+\int_{c.v.}\rho\mathbf{f}\cdot\mathbf{V}\,d\upsilon+\int_{c.v.}\dot{q}_e d\upsilon,\tag{1.34}$$

where

$\dot{W}_{normal}$, $\dot{W}_{shear}$, $\dot{W}_{shaft}$ represent the rates of work done against the normal forces, shear forces, and the rate of shaft work, respectively

$\dot{q}_e$ is the rate of internal energy generation per unit volume due to the power drawn to the system from an external electric circuit

So far, we have made no mention of the specific energy $e$. Since we have already included the *potential energy* of the fluid in the work term, that is, the work done against the gravity, the specific energy $e$ may conveniently be separated into two as bulk *kinetic energy* per unit mass and specific *internal energy*, that is,

$$e=\frac{1}{2}V^2+\mathcal{U},\tag{1.35}$$

where $V=|V|$. In Equation 1.35 kinetic energy is interpreted as the bulk energy associated with the observable fluid motion on a per unit mass basis. By internal energy we understand all forms of energy, other than bulk kinetic and potential, energies associated with molecular and atomic structure and behavior of the fluid.

If the body force $\mathbf{f}$ is only due to the gravity, then it can be expressed in terms of the negative gradient of the gravitational potential $\phi$, that is,

$$f=-\nabla\phi.\tag{1.36}$$

As a specific case, if $\mathbf{f}=-g\hat{k}$, where $g$ is the gravitational acceleration, then $\phi=gz$. Thus, the term involving the net body force per unit volume in Equation 1.34 becomes

$$\int_{c.v.}\rho\mathbf{f}\cdot\mathbf{V}\,d\upsilon=-\int_{c.v.}\rho\nabla\phi\cdot\mathbf{V}d\upsilon.\tag{1.37}$$

Since,

$$\rho\nabla\phi\cdot\mathbf{V}=\nabla\cdot(\rho\phi\mathbf{V})-\phi\nabla\cdot(\rho\mathbf{V}),\tag{1.38}$$

Equation 1.37 may be written as

$$\int_{c.v.} \rho \mathbf{f} \cdot \mathbf{V}\, d\upsilon = -\int_{c.v.} \nabla \cdot (\rho \phi \mathbf{V}) d\upsilon + \int_{c.v.} \phi \nabla \cdot (\rho \mathbf{V}) d\upsilon. \tag{1.39}$$

The first volume integral on the right-hand side of Equation 1.39 can be converted into a surface integral by employing the divergence theorem [4] as

$$\int_{c.v.} \nabla \cdot (\rho \phi \mathbf{V}) d\upsilon = \int_{c.s.} \rho \phi \mathbf{V} \cdot \hat{\mathbf{n}}\, dA. \tag{1.40}$$

In view of the continuity Equation 1.10, the second volume integral on the right-hand side of Equation 1.30 can be rewritten as

$$\int_{c.v.} \phi \nabla \cdot (\rho \mathbf{V}) d\upsilon = -\int_{c.v.} \phi \frac{\partial \rho}{\partial t}\, d\upsilon. \tag{1.41a}$$

If $\phi$ is time-independent, Equation 1.41a may also be written as

$$\int_{c.v.} \phi \nabla \cdot (\rho \mathbf{V}) d\upsilon = -\frac{\partial}{\partial t} \int_{c.v.} \rho \phi d\upsilon. \tag{1.41b}$$

Hence, Equation 1.39 becomes

$$\int_{c.v.} \rho \mathbf{f} \cdot \mathbf{V} d\upsilon = -\int_{c.s.} \rho \phi \mathbf{V} \cdot \hat{\mathbf{n}}\, dA - \frac{\partial}{\partial t} \int_{c.v.} \rho \phi d\upsilon. \tag{1.42}$$

Substitution of Equation 1.42 into Equation 1.34, together with the use of the relation (1.35), gives

$$\frac{\partial}{\partial t} \int_{c.v.} \rho \left( \mathcal{U} + \frac{1}{2} V^2 + \phi \right) d\upsilon + \int_{c.s.} \rho \left( \mathcal{U} + \frac{1}{2} V^2 + \phi \right) \mathbf{V} \cdot \hat{\mathbf{n}} dA$$

$$= q_{c.s.} - \dot{W}_{normal} - \dot{W}_{shear} - \dot{W}_{shaft} + \int_{c.v.} \dot{q}_e d\upsilon. \tag{1.43}$$

Equation 1.43 is the control-volume form of the first law of thermodynamics usually encountered in a thermodynamics text, and it is an equation of change for

$$\mathcal{U} + \frac{1}{2} V^2 + \phi, \tag{1.44}$$

which is known as the *specific total energy.*

The work term $\dot{W}_{normal}$, which involves normal stresses, can also be presented in a more usable form. As we shall discuss in Section 2.3, normal stresses can be separated into two as *pressure* and *normal viscous stresses*. Consider now the work done by the system against pressure, which is also called *flow work:* The rate of work done by the system against pressure acting at a surface element $dA_{out}$ will be given by $pdA_{out}\mathbf{V}\cdot\hat{\mathbf{n}}$, where $p$ is the pressure acting on the surface element. Therefore, the rate of work done against pressure over $A_{out}$ is $\int_{A_{out}} p\mathbf{V}\cdot\hat{\mathbf{n}}dA_{out}$. Similarly, the rate of work done on the system by pressure acting on $A_{in}$ would be given by $-\int_{A_{in}} p\mathbf{V}\cdot\hat{\mathbf{n}}dA_{in}$. Thus, the net rate of work done by the system against pressure will be

$$\int_{A_{out}} p\mathbf{V}\cdot\hat{\mathbf{n}}dA_{out} - \left[-\int_{A_{in}} p\mathbf{V}\cdot\hat{\mathbf{n}}dA_{in}\right] = \int_{c.s.} p\mathbf{V}\cdot\hat{\mathbf{n}}dA. \tag{1.45}$$

The work term $\dot{W}_{normal}$ can now be written as

$$\dot{W}_{normal} = \int_{c.s.} p\mathbf{V}\cdot\hat{\mathbf{n}}\,dA + \dot{W}_{normal\ viscous}, \tag{1.46}$$

where $\dot{W}_{normal\ viscous}$ represents the rate of work done by the system against normal viscous forces. Thus, Equation 1.43 may be rewritten as

$$\frac{\partial}{\partial t}\int_{c.v.} \rho\left(\mathcal{U}+\frac{1}{2}V^2+\phi\right)d\upsilon + \int_{c.s.} \rho\left(\mathcal{U}+\frac{1}{2}V^2+\phi\right)\mathbf{V}\cdot\hat{\mathbf{n}}dA$$

$$= q_{c.s.} - \int_{c.s.} p\mathbf{V}\cdot\hat{\mathbf{n}}dA - \dot{W}_{viscous} - \dot{W}_{shaft} + \int_{c.v.} \dot{q}_e d\upsilon, \tag{1.47}$$

where we have substituted

$$\dot{W}_{viscous} = \dot{W}_{normal\ viscous} + \dot{W}_{shear}. \tag{1.48}$$

Equation 1.47 can also be written as

$$\frac{\partial}{\partial t}\int_{c.v.} \rho\left(\mathcal{U}+\frac{1}{2}V^2+\phi\right)d\upsilon + \int_{c.s.} \rho\left(i+\frac{1}{2}V^2+\phi\right)\mathbf{V}\cdot\hat{\mathbf{n}}dA$$

$$= q_{c.s.} - \dot{W}_{viscous} - \dot{W}_{shaft} + \int_{c.v.} \dot{q}_e d\upsilon, \tag{1.49}$$

where $i$ is the *enthalpy* of the fluid per unit mass defined as

$$i = \mathcal{U}+\frac{p}{\rho}. \tag{1.50}$$

Equation 1.49 is an alternative expression for the control-volume form of the first law of thermodynamics.

### 1.5.4 Second Law of Thermodynamics

The first law of thermodynamics, which embodies the idea of conservation of energy, gives means for quantitative calculation of changes in the state of a system due to interactions between the system and its surroundings but tells nothing about the direction a process might take. In other words, physical observations such as the following cannot be explained by the first law: A cup of hot coffee placed in a cool room will always tend to cool to the temperature of the room, and once it is at room temperature, it will never return spontaneously to its original hot state; air will rush into a vacuum chamber spontaneously; the conversion of heat into work cannot be carried out on a continuous basis with a conversion efficiency of 100%; water and salt will mix to form a solution, but separation of such a solution cannot be made without some external means; a vibrating spring will eventually come to rest all by itself; etc. These observations concerning unidirectionality of naturally occurring processes have led to the formulation of the second law of thermodynamics. Over the years, many statements of the second law have been made. Here we shall give the following *Clausius* statement: It is impossible for a self-acting system unaided by an external agency to move heat from one system to another at a higher temperature.

The second law leads to a thermodynamic property—*entropy*. For any *reversible* process that a system undergoes during a time interval $dt$, the change in the entropy $S$ of the system is given by

$$dS = \left(\frac{\delta Q}{T}\right)_{rev}. \tag{1.51a}$$

For an *irreversible process*, the change, however, is

$$dS > \left(\frac{\delta Q}{T}\right)_{irr}, \tag{1.51b}$$

where
  $\delta Q$ is the small amount of heat added to the system during the time interval $dt$
  $T$ is the temperature of the system at the time of heat transfer

Equations 1.51a,b may be taken as the mathematical statement of the second law, which can also be written in rate form as

$$\frac{dS}{dt} \geq \frac{1}{T}\frac{\delta Q}{dt}. \tag{1.52}$$

The control-volume form of the second law can also be developed by following a procedure we have used in the previous sections. Rather than going through the entire development, here we shall give the result [3]

$$\frac{\partial}{\partial t}\int_{c.v.} s\rho\,dv + \int_{c.s.} s\rho\mathbf{V}\cdot\hat{\mathbf{n}}\,dA \geq \int_{c.s.} \frac{1}{T}\frac{\delta Q}{dt}, \tag{1.53}$$

where $s$ is the entropy per unit mass and the equality applies to reversible processes and the inequality to irreversible processes.

**TABLE 1.1**

Summary of General Laws

| Law | For a System | For a Control Volume |
|---|---|---|
| Law of conservation of mass | $\dfrac{dm}{dt} = 0$ | $\dfrac{\partial}{\partial t} \int\limits_{c.v.} \rho \, d\upsilon = - \int\limits_{c.s.} \rho \mathbf{V} \cdot \hat{\mathbf{n}} \, dA$ |
| Newton's second law of motion | $\mathbf{F} = \dfrac{d\mathbf{M}}{dt}$ | $\mathbf{F} = \dfrac{\partial}{\partial t} \int\limits_{c.v.} \rho \mathbf{V} d\upsilon + \int\limits_{c.s.} \mathbf{V} \rho \mathbf{V} \cdot \hat{\mathbf{n}} \, dA$ |
| First law of thermodynamics | $\dfrac{dE}{dt} = q - \dot{W}$ | $\dfrac{\partial}{\partial} \int\limits_{c.v.} \left( u + \dfrac{1}{2} V^2 + \phi \right) \rho \, d\upsilon$ |
| | | $+ \int\limits_{c.s.} \left( i + \dfrac{1}{2} V^2 + \phi \right) \rho \mathbf{V} \cdot \hat{\mathbf{n}} \, dA$ |
| | | $= q_{c.s.} - \dot{W}_{shaft} - \dot{W}_{viscous} + \int\limits_{c.v.} \dot{q}_e d\upsilon$ |
| Second law of thermodynamics | $\dfrac{dS}{dt} \geq \dfrac{1}{T} \dfrac{\delta Q}{dt}$ | $\dfrac{\partial}{\partial t} \int\limits_{c.v.} s\rho \, d\upsilon + \int\limits_{c.s.} s\rho \mathbf{V} \cdot \hat{\mathbf{n}} \, dA \geq \int\limits_{c.s.} \dfrac{1}{T} \dfrac{\delta Q}{dt}$ |

The irreversibility of a system instantaneously occupying a control volume is defined as

$$\Delta S_{c.v.} = \frac{\partial}{\partial t} \int\limits_{c.v.} s\rho \, d\upsilon + \int\limits_{c.s.} s\rho \mathbf{V} \cdot \hat{\mathbf{n}} \, dA - \int\limits_{c.s.} \frac{1}{T} \frac{\delta Q}{dt}, \tag{1.54}$$

which is a measure of the *entropy production* within the control volume. Moreover,

$$LW = T_o \Delta S_{c.v.} \tag{1.55}$$

gives the *lost work* in an energy producing system due to irreversibility, where $T_o$ is the absolute temperature of the system's so-called dead state at which the system is in thermodynamic equilibrium with its environment at $(p_o, T_o)$.

The most effective performance of systems in industrial applications involving heat transfer processes corresponds to the least generation of entropy; that is, the rate of loss of useful work in a process is directly proportional to the rate of entropy production during that process.

Equations 1.53 through 1.55 are important in the analysis of conventional power plants and in assessing thermal performance of heat exchangers. For more information, readers are referred to Reference 5.

Thus, we have completed the discussion of the general laws for systems and control volumes (i.e., closed and open systems), which are summarized in Table 1.1. We shall refer to these results in later chapters of this book.

## 1.6 Particular Laws

As mentioned in Section 1.1, the general principles or laws alone are not sufficient to solve heat transfer problems. We have to bring into an analysis certain particular laws. In the following sections, we will discuss the three particular laws of heat transfer.

### 1.6.1 Fourier's Law of Heat Conduction

Fourier's law of heat conduction, which is the basic law governing heat conduction based on the continuum concept, originated from experimental observations by J. B. Biot. But it was named after the well-known French scientist J. B. J. Fourier, who used it in his remarkable work "Theorie Analytique de la Chaleur," which was published in Paris in 1822 [6]. In this book he gave a very complete exposition of the theory of heat conduction. This law states that heat flux due to conduction in a given direction (i.e., the rate of heat transfer per unit area across an infinitesimal area element whose normal is in this direction) at a point within a medium (solid, liquid, or gaseous) is proportional to the temperature gradient in the same direction at the same point. For heat conduction in any direction $n$, this law is given by [7]

$$q_n'' = -k \frac{\partial T}{\partial n}, \tag{1.56}$$

where
$q_n''$ is the magnitude of the heat flux in the $n$-direction
$\partial T/\partial n$ is the temperature gradient in the same direction

Here, $k$ is a proportionality constant known as the *thermal conductivity* of the material of the medium under consideration and is a positive quantity. The minus sign is included in Equation 1.56 so that heat flow is defined to be positive when it is in the direction of a negative temperature gradient.

Thermal conductivity is a thermophysical property and has the units W/(m K) in the SI system. A medium is said to be *homogeneous* if its thermal conductivity does not vary from point to point within the medium, and *heterogeneous* if there is such a variation. Further, a medium is said to be *isotropic* if its thermal conductivity at any point in the medium is the same in all directions, and *anisotropic* if it exhibits directional variation. Materials having porous structure, such as cork and glass wool, are examples of heterogeneous media, and those having fibrous structure, such as wood or asbestos, are examples of anisotropic media. In anisotropic media the heat flux due to heat conduction in a given direction may also be proportional to the temperature gradients in other directions, and therefore Equation 1.56 may not be valid [8].

In an isotropic medium, an equation like Equation 1.56 can be written in each coordinate direction. For example, in rectangular coordinates, the heat-flux relations in the $x$-, $y$-, and $z$-directions can be written as

$$q_x'' = -k \frac{\partial T}{\partial x}, \quad q_y'' = -k \frac{\partial T}{\partial y}, \quad \text{and} \quad q_z'' = -k \frac{\partial T}{\partial z}. \tag{1.57a,b,c}$$

These are, in fact, the three components in the $x$-, $y$-, and $z$-directions of the *heat-flux vector*

$$q'' = -k\nabla T, \tag{1.58}$$

which is the vector form of Fourier's law in isotropic media. This equation is well established for heat conduction in isotropic solids, and practical applications of it for various problems require the laboratory measurement of thermal conductivities of representative specimens.

The magnitude of thermal conductivity of various substances varies over wide ranges, for example, from 0.0146 W/(m K) for $CO_2$ at 0°C to 410 W/(m K) for pure silver at 0°C. Thermal conductivity of a material depends on its chemical compositions, physical structure, and the state of it. It also depends on the temperature and pressure to which the material is subjected. In most cases thermal conductivity is much less dependent on pressure than on temperature, so that the dependence on pressure may be neglected and thermal conductivity is tabulated as a function of temperature only.

Heat conduction in gases and vapors depends mainly on the molecular transfer of kinetic energy of the molecular movements, that is, heat conduction is the transmission of kinetic energy by the more active molecules in the higher temperature regions by successive collisions with lower-energy molecules. Clearly, the faster the molecules move, the faster they will transfer energy. According to the kinetic theory of gases, the temperature of an element of gas is proportional to the mean kinetic energy of its constituent molecules. This implies that thermal conductivity of a gas should be dependent on its temperature.

In liquids, molecules are more closely spaced, and therefore, the molecular force fields exert a strong influence on the energy exchange in the collisions. Because of this, liquids usually have much higher values for thermal conductivity than gases. The thermal conductivities of various solids, liquids, and gases are listed along with other thermophysical properties in Appendices A and B.

As already discussed in Section 1.2, solid materials may have solely crystalline structures, such as quartz; may be in amorphous solid state, such as glass; mixture of the two materials; or may be somewhat porous in structure with air or other gases in the pores. Heat conduction in solids with crystalline structures depends on the energy transfer by molecular and lattice vibrations and free-electron drift. In general, energy transfer by molecular and lattice vibrations is not as large as the transfer by the electron transport, and it is for this reason that good electrical conductors are always good heat conductors, while electrical insulators are usually good heat insulators. In the case of amorphous solids, heat conduction depends only on the molecular energy transport, and this is the reason why the thermal conductivities of amorphous solids are usually smaller than the thermal conductivities of solids with crystalline structures.

### 1.6.2 Newton's Law of Cooling

Convection has already been defined in Section 1.2 as the process of heat transport in a fluid by the combined action of heat conduction (and radiation) and macroscopic fluid motion. As a mechanism of heat transfer, it is important not only between the layers of a fluid but also between a fluid and a solid surface when they are in contact.

When a fluid flows over a solid surface as illustrated in Figure 1.2, it is an experimentally observed fact that the fluid particles adjacent to the surface stick to it and therefore have zero velocity relative to the surface. Other fluid particles attempting to slide over the stationary ones at the surface are retarded as a result of viscous forces between the fluid particles. The velocity of the fluid particles thus asymptotically approaches that of the undisturbed free stream over a distance $\delta$ (*velocity boundary-layer thickness*) from the surface, with the resulting velocity distribution shown in Figure 1.2.

As illustrated in Figure 1.2, if $T_w > T_\infty$, then heat will flow from the solid to the fluid particles at the surface. The energy thus transmitted increases the internal energy of the fluid particles (*sensible* heat storage) and is carried away by the motion of the fluid. The temperature distribution in the fluid adjacent to the surface will then appear as shown in Figure 1.2, asymptotically approaching the free-stream value $T_\infty$ in a short distance $\delta_T$ (*thermal boundary-layer thickness*) from the surface.

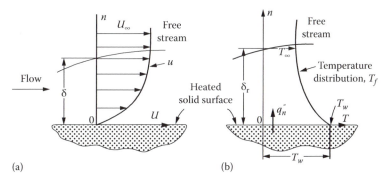

**FIGURE 1.2**
Boundary layers along a solid surface: (a) Velocity boundary layer and (b) thermal boundary layer.

Since the fluid particles at the surface are stationary, the heat flux from the surface to the fluid will be

$$q_n'' = -k_f \left( \frac{\partial T_f}{\partial n} \right)_w, \tag{1.59}$$

where
$k_f$ is the thermal conductivity of the fluid
$T_f$ is the temperature distribution in the fluid
the subscript $w$ means the derivative is evaluated at the surface
$n$ denotes the normal direction from the surface

In 1701, Newton expressed the heat flux from a solid surface to a fluid by the equation

$$q_n'' = h(T_w - T_\infty), \tag{1.60}$$

where $h$ is called *heat transfer coefficient, film conductance,* or *film coefficient.* In the literature, Equation 1.60 is known as *Newton's law of cooling.* In fact, it is not a law but the equation defining the heat transfer coefficient; that is,

$$h = \frac{q_n''}{T_w - T_\infty} = \frac{-k_f (\partial T_f / \partial n)_w}{T_w - T_\infty}. \tag{1.61}$$

The heat transfer coefficient has the units W/(m$^2$ K) in the SI system. It should be noted that $h$ is also given by

$$h = \frac{-k_s (\partial T_s / \partial n)_w}{T_w - T_\infty}, \tag{1.62}$$

where
$k_s$ is the thermal conductivity of the solid
$T_s$ is the temperature distribution in the solid

**TABLE 1.2**

Approximate Values of the Heat Transfer Coefficient $h$, W/(m² K)

| Fluid | Free Convection | Forced Convection |
|-------|-----------------|-------------------|
| Gases | 5–30 | 30–300 |
| Water | 30–300 | 300–10,000 |
| Viscous | 5–100 | 30–3,000 |
| Liquid metals | 50–500 | 500–20,000 |
| Boling water | 2,000–20,000 | 3,000–100,000 |
| Condensing water vapor | 3,000–30,000 | 3,000–200,000 |

If the fluid motion involved in the process is induced by some external means such as a pump, blower, or fan, then the process is referred to as *forced convection*. If the fluid motion is caused by any body force within the system, such as those resulting from the density gradients near the surface, then the process is called *natural* (or *free*) *convection*.

Certain convective heat transfer processes, in addition to sensible heat storage, may also involve *latent* heat storage (or release) due to phase change. Boiling and condensation are two such cases.

The heat transfer coefficient is actually a complicated function of the flow conditions, thermophysical properties (viscosity, thermal conductivity, specific heat, density) of the fluid, and geometry and dimensions of the surface. Its numerical value, in general, is not uniform over the surface. Table 1.2 gives the order of magnitude of the range of value of the heat transfer coefficient under various conditions [9].

### 1.6.3 Stefan–Boltzmann Law of Radiation

As mentioned in Section 1.2, all substances emit energy in the form of electromagnetic waves (i.e., thermal radiation) as a result of their temperature and are also capable of absorbing such energy. When thermal radiation is incident on a body, part of it is reflected by the surface as illustrated in Figure 1.3. The remainder may be absorbed as it travels

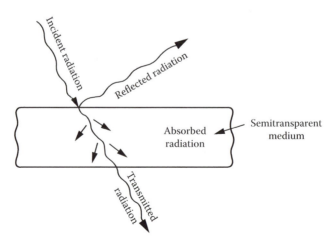

**FIGURE 1.3**
Absorption, reflection, and transmission of incident radiation.

through the body. If the material of the body is a strong absorber of thermal radiation, the energy that penetrates into the body will be absorbed and converted into internal energy within a very thin layer adjacent to the surface. Such a body is called *opaque*. If the material thickness required to substantially absorb radiation is large compared to the thickness of the body, then most of the radiation will pass through the body without being absorbed, and such a body is called *transparent*.

When radiation impinges on a surface, the fraction that is reflected is defined as the *reflectivity* ρ, the fraction absorbed as the *absorptivity a*, and the fraction transmitted as the *transmissivity* τ. Thus,

$$\rho + \alpha + \tau = 1. \tag{1.63}$$

For opaque substances, τ = 0, and therefore Equation 1.63 reduces to

$$\rho + \alpha = 1. \tag{1.64}$$

An ideal body that absorbs all the impinging radiation energy without reflection and transmission is called a *blackbody*. Therefore, for a blackbody Equation 1.63 reduces to α = 1. Only a few materials, such as carbon black and platinum black, approach the blackbody in their ability to absorb radiation energy. A blackbody also emits the maximum possible amount of thermal radiation [10]. The total emission of radiation per unit surface area and per unit time from a blackbody is related to the fourth power of the absolute temperature $T$ of the surface by the *Stefan–Boltzmann law of radiation*, which is

$$q_{r.b}'' = \sigma T^4, \tag{1.65}$$

where σ is the *Stefan–Boltzmann constant* with the value 5.6697 × 10⁻⁸ W/(m² K⁴) in the SI system.

Real bodies (surfaces) do not meet the specifications of a blackbody but emit radiation at a lower rate than a blackbody of the same size and shape and at the same temperature. If $q_r''$ is the radiative flux (i.e., radiation emitted per unit surface area and per unit time) from a real surface maintained at the absolute temperature $T$, then the *emissivity* of the surface is defined as

$$\varepsilon = \frac{q_r''}{\sigma T^4}. \tag{1.66}$$

Thus, for a blackbody ε = 1. For a real body exchanging radiation only with other bodies at the same temperature (i.e., for thermal equilibrium), it can be shown that α = ε, which is a statement of *Kirchhoff's law* in thermal radiation [10]. The magnitude of emissivity depends upon the material, its state, temperature, and the surface conditions.

If two isothermal surfaces $A_1$ and $A_2$, having emissivities $\varepsilon_1$ and $\varepsilon_2$ and absolute temperatures $T_1$ and $T_2$, respectively, exchange heat by radiation only, then the net rate of heat exchange between these two surfaces is given by

$$q_r = \sigma A_1 \Im_{12} \left( T_1^4 - T_2^4 \right), \tag{1.67}$$

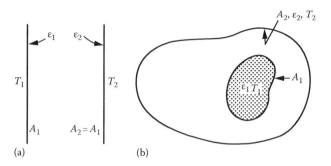

**FIGURE 1.4**
Two isothermal surfaces $A_1$ and $A_2$ exchange radiation energy: (a) Two large parallel surfaces and (b) enclosed surfaces.

where Kirchhoff's law is assumed to be valid. If $A_1$ and $A_2$ are two large parallel surfaces with negligible heat losses from the edges as shown in Figure 1.4a, then the factor $\mathfrak{F}_{12}$ in Equation 1.67 is given by

$$\frac{1}{\mathfrak{F}_{12}} = \frac{1}{\varepsilon_1} + \frac{1}{\varepsilon_2} - 1. \tag{1.68}$$

If $A_1$ is completely enclosed by the surface $A_2$ as shown in Figure 1.4b, then

$$\frac{1}{\mathfrak{F}_{12}} = \frac{1}{F_{12}} + \frac{1}{\varepsilon_1} - 1 + \frac{A_1}{A_2}\left[\frac{1}{\varepsilon_2} - 1\right], \tag{1.69}$$

where $F_{12}$ is a purely geometric factor called *radiation shape factor* or *view factor* between the surfaces $A_1$ and $A_2$ and is equal to the fraction of the radiation leaving surface $A_1$ that directly reaches surface $A_2$. Radiation shape factors are given in the form of equations and/or charts for several configurations in the literatures [10–12]. For surfaces $A_1$ and $A_2$, it is obvious that

$$\sum_{j=1}^{2} F_{ij} = 1, \quad i = 1, 2. \tag{1.70}$$

Obviously, if $A_1$ is a completely convex or a plane surface, then $F_{11} = 0$.

In certain applications it may be convenient to define a *radiation heat transfer coefficient* $h_r$ as

$$q_r = h_r A_1 (T_1 - T_2). \tag{1.71}$$

When this is applied to Equation 1.67, $h_r$ is given by

$$h_r = \sigma \mathfrak{F}_{12}(T_1 + T_2)\left(T_1^2 + T_2^2\right) \tag{1.72}$$

The particular laws of heat transfer are summarized in Table 1.3.

**TABLE 1.3**

Summary of the Particular Laws of Heat Transfer

| Mode | Mechanism | Particular Law |
|---|---|---|
| Conduction | Diffusion of thermal energy | $q_n'' = -k\partial T/\partial n$ |
| Convection | Diffusion and transport of thermal energy | $q'' = h(T_w - T_\infty)$ |
| Radiation | Heat transfer by electromagnetic waves | $q_r'' = h_r(T_1 - T_2)$ |

## Problems

**1.1** In many practical engineering problems involving steady fluid flows, the inlet and outlet flows to and from various devices are usually regarded as 1D. Develop the first law of thermodynamics (i.e., the steady-flow energy equation) for such a device as illustrated in Figure 1.5.

**1.2** Calculate the reaction force on the pipe bend illustrated in Figure 1.6, resulting from a steady flow of fluid in it. $P_1$ and $P_2$ are the gauge pressures at sections 1 and 2, respectively, and $R_x$ and $R_y$ are the components of the resultant force exerted by the pipe on the fluid.

Obtain the continuity equation by writing a mass balance over a stationary volume element $\Delta x \cdot \Delta y \cdot \Delta z$ located at any point $(x,y,z)$ through which a fluid streams.

The temperature profile at a location in water flowing over a flat surface is experimentally measured to be

$$T = 20 + 80e^{-800y},$$

where $T$ is in °C and $y$ in m is the distance measured normal to the surface with $y = 0$ corresponding to the surface. What is the value of heat transfer coefficient at this location? Assume that the thermal conductivity of water is $k = 0.62$ W/(m K).

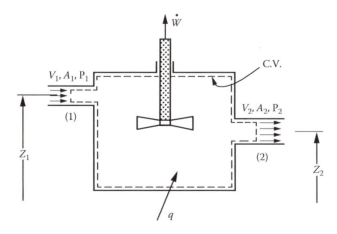

**FIGURE 1.5**
Problem 1.1.

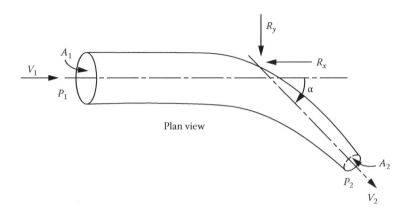

Plan view

**FIGURE 1.6**
Problem 1.2.

Estimate the equilibrium temperature of a long rotating cylinder of diameter $D$ and oriented in space with its axis normal to the sun rays. The cylinder is at a location in space where the irradiation from the sun (i.e., energy incident on a surface perpendicular to the sun rays per unit time per unit area) is 1500 W/m². Assume that the absorptivity of the surface of the cylinder to solar radiation, $\alpha_s$, is equal to its emissivity, $\varepsilon$, and the outer space is a blackbody at 0 K.

Show that Equation 1.17b can be expressed as

$$\mathbf{F} = \int_{c.v.} \rho \frac{D\mathbf{V}}{Dt} d\upsilon,$$

where $D\mathbf{V}/Dt$ represents the substantial derivative of $\mathbf{V}$ defined as

$$\frac{D\mathbf{V}}{Dt} = \frac{\partial \mathbf{V}}{\partial t} + (\mathbf{V} \cdot \nabla)\mathbf{V}.$$

Show that Equation 1.32 can be expressed as

$$\int_{c.v.} \rho \frac{De}{Dt} d\upsilon = q_{c.s.} - \dot{W}.$$

# References

1. Shapiro, A. H., *The Dynamics and Thermodynamics of Compressible Fluid Flow*, Vol. 1, The Ronald Press, New York, 1953.
2. Hatsopoulos, G. N. and Keenan, J. H., *Principles of General Thermodynamics*, Wiley, New York, 1965.
3. Van Wylen, G. J. and Sonntag, R. E., *Fundamentals of Classical Thermodynamics*, 3rd edn., John Wiley & Sons, New York, 1985.
4. Hildebrand, F. B., *Advanced Calculus for Applications*, 2nd edn., Prentice Hall, Upper Saddle River, NJ, 1976.

5. Bejan, A., *Entropy Generation through Heat and Fluid Flow*, John Wiley & Sons, New York, 1982.
6. Fourier, J., *The Analytical Theory of Heat*, Dover Publications, New York, 1955.
7. Yener, Y. and Kakaç, S., *Heat Conduction*, 4th edn., Taylor & Francis Group, Boca Raton, FL, 2008.
8. Özisik, M. N., *Heat Conduction*, John Wiley & Sons, New York, 1980.
9. Arpaci, V. S., *Conduction Heat Transfer*, Addison-Wesley, Reading, MA, 1966.
10. Siegel, R., Howell, J. R., and Mengüç, M. P., *Thermal Radiation Heat Transfer*, 5th edn., CRC Press, Boca Raton, FL, 2010.
11. Özisik, M. N., Radiative transfer and interactions with conduction and convection, John Wiley & Sons, New York, 1978.
12. Bejan, A., *Convection Heat Transfer*, 2nd edn., John Wiley & Sons, New York, 1995.

## Suggested Reading

In addition to the references listed previously, the following short list, selected from a large list of material available in the literature, may serve as a guide to newcomers to the convective heat transfer area.

### Basic Heat Transfer and General Principles

1. Eckert, E. R. G. and Drake, R. M., *Analysis of Heat and Mass Transfer*, McGraw-Hill, New York, 1972.
2. Incropera, F. P., DeWitt, D. P., Bergman, T. L., and Lavine, A. S., *Fundamentals of Heat and Mass Transfer*, 6th edn., John Wiley & Sons, New York, 2007.
3. Özisik, M. N., *Heat Transfer—A Basic Approach*, McGraw-Hill, New York, 1985.
4. Rohsenow, W. M. and Choi, H., *Heat, Mass and Momentum Transfer*, Prentice Hall, Upper Saddle River, NJ, 1961.

### Convection

1. Arpaci, V. S. and Larsen, P. S., *Convection Heat Transfer*, Prentice Hall, Upper Saddle River, NJ, 1984.
2. Bejan, A., *Convection Heat Transfer*, 2nd edn., John Wiley & Sons, New York, 1995.
3. Burmeister, L. C., *Convective Heat Transfer*, 2nd edn., Wiley-Interscience Publication, New York, 1993.
4. Cebeci, T. and Bradshaw, P., *Physical and Computational Aspects of Convective Heat Transfer*, Springer, New York, 1984.
5. Eckert, E. R. G., Pioneering contributions to our knowledge of convective heat transfer, *J. Heat Transfer*, 103, 409–414, 1981.
6. Kakaç, S., Shah, R. K., and Aung, W. (Eds.), *Handbook of Single-Phase Convective Heat Transfer*, John Wiley & Sons, New York, 1987.
7. Kays, W. M. and Crawford, M. E., *Convective Heat and Mass Transfer*, 3rd edn., McGraw-Hill, New York, 1993.
8. Cebeci, T., *Convective Heat Transfer*, Horizons Pub., Long Beach, CA and Springer-Verlag, Heidelberg, Germany, 2002.

### Fluid Mechanics and Boundary-Layer Theory

1. Schlichting, H., *Boundary-Layer Theory*, Translated into English by J. Kestin, 7th edn., McGraw-Hill, New York, 1979.
2. White, F. M., *Viscous Fluid Flow*, McGraw-Hill, New York, 1974.

# 2

## Governing Equations of Convective Heat Transfer

### Nomenclature

| | |
|---|---|
| $a$, $\mathbf{a}$ | acceleration, m/s$^2$ |
| $c_p$ | specific heat at constant pressure, J/(kg K) |
| $c_v$ | specific heat at constant volume, J/(kg K) |
| $Ec$ | Eckert number $= U_\infty/c_p\lvert T_\infty - T_w\rvert$ |
| $F$, $\mathbf{F}$ | force, N |
| $f$, $\mathbf{f}$ | body force per unit mass, N/kg |
| $g$ | gravitational acceleration, m/s$^2$ |
| $h$ | heat-transfer coefficient, W/(m$^2$ K) |
| $i$ | enthalpy per unit mass, J/kg |
| $k$ | thermal conductivity, W/(m K) |
| $L$ | characteristic length, m |
| $\mathbf{M}$ | momentum, kg m/s |
| $m$ | mass, kg |
| $Nu$ | Nusselt number $= hL/k$ |
| $P$ | pressure, N/m$^2$ |
| $Pe$ | Péclêt number $= RePr = VL/\alpha$ |
| $Pr$ | Prandtl number $= \mu C_p/k = v/\alpha$ |
| $q$ | heat-transfer rate, W |
| $q''$, $\mathbf{q}''$ | heat flux, W/m$^2$ |
| $\dot{q}$ | volumetric heat generation rate, W/m$^3$ |
| $r$ | radial coordinate, m |
| $Re$ | Reynolds number $= \rho VL/\mu$ |
| $T$ | temperature, °C, K |
| $t$ | time, s |
| $U$ | free-stream velocity, m/s |
| $u$ | velocity component in $x$-direction, m/s |
| $V$, $\mathbf{V}$ | velocity, m/s |
| $v$ | velocity component in $y$-direction, m/s |
| $w$ | velocity component in $z$-direction, m/s |
| $x$ | rectangular coordinate; distance parallel to surface, m |
| $y$ | rectangular coordinate; distance perpendicular to surface, m |
| $z$ | rectangular coordinate |

**Greek Symbols**

| | |
|---|---|
| $\alpha$ | thermal diffusivity $= k/\rho c_p$, m$^2$/s |
| $\Delta$ | finite increment |
| $\theta$ | latitude angle in cylindrical coordinates, rad, deg |
| $\mu$ | dynamic viscosity, Pa s |
| $\mathcal{U}$ | internal energy per unit mass, J/kg |
| $\nu$ | kinematic viscosity, m$^2$/s |
| $\rho$ | density, kg/m$^3$ |
| $\sigma$ | normal stress acting on an element of fluid, N/m$^2$ |
| $\tau$ | shear stress between fluid layers, N/m$^2$ |
| $\Phi$ | viscous dissipation function, W/m$^3$ |
| $\phi$ | azimuth angle in spherical coordinates, rad, deg |

**Subscripts**

| | |
|---|---|
| $r$ | radial direction |
| $w$ | wall condition |
| $x$ | $x$-direction |
| $y$ | $y$-direction |
| $z$ | $z$-direction |
| $\theta$ | $\theta$-direction |
| $\phi$ | $\phi$-direction |
| $\infty$ | free-stream conditions |

## 2.1 Introduction

The energy transfer process that is observed to occur in fluids and between a solid surface and a fluid by means of the combined action of heat conduction (and radiation) and fluid motion has already been defined as convection in Chapter 1. In the special case when the fluid is everywhere at rest, the process is equivalent to heat transfer in solid bodies, that is, conduction.

Experimental observations show that the motion of fluid particles in contact with a solid surface is almost identical with that of the solid surface. Hence, the physical mechanisms by which heat is transferred between a solid surface and a fluid, whether the fluid is in a state of motion or not, can only be conduction (and radiation). In most cases, especially at moderate temperatures, however, the heat transferred by radiation is usually negligible compared to the heat transferred by conduction.

When the state of the fluid motion is independent of heat transfer, we speak of forced convection. In that case, the fluid motion is caused by some external means in the form of a pressure difference. If the fluid motion is caused by means of any body force within the fluid, then we speak of free (or natural) convection. This can be, for example, as a result of density gradients near a solid surface due to the heat transfer.

We shall be concerned in this text with fluids that behave as a continuum. Ordinary fluids, such as air, water, and oils, behave as a continuum, and they also exhibit a linear relationship between the applied shear stress and the rate of strain. Such fluids are called *Newtonian* fluids.

The expression relating the shear stress $\tau$ to the rate of strain (velocity gradient $du/dy$) for a Newtonian fluid in *simple shear flow*, where only one velocity component is different from zero, is given by

$$\tau = \mu \frac{du}{dy}, \tag{2.1}$$

where the proportionality constant $\mu$ is called the *absolute viscosity, dynamic viscosity, coefficient of viscosity*, or, more simply, *viscosity* of the fluid. The viscosity is constant for each Newtonian fluid at a given temperature and pressure. Non-Newtonian fluids are the ones in which the viscosity at a given pressure and temperature is a function of the velocity gradient. Colloidal suspensions and emulsions are examples of non-Newtonian fluids. The previous relation (2.1) is a particular law in fluid mechanics and is also known as *Newton's law of viscosity*. For 2D and 3D flows, the expressions relating the shear stresses to the strain rates are more complicated and will be discussed later in this chapter.

An *ideal fluid* is defined as the one which is incompressible and has zero viscosity. Thus, in an ideal fluid, shear stresses are absent despite shear deformations in the fluid. No real fluids are, in fact, inviscid, that is, having zero viscosity, but the concept of an ideal fluid is useful in as much as it provides a simple model, which, at the same time, approximates real fluids in many situations. An ideal fluid also glides past solid boundaries, which constitutes the essential difference between an ideal and a real fluid.

The main objective of the theory of the convective heat transfer is to permit the determination of the temperature distribution in a fluid and, thus, to establish the basis for the calculation of the heat flux, that is, the time rate of heat flow per unit area, transferred between the fluid and a solid surface in contact with it. It is desirable that such a calculation be possible for any specified surface under any boundary, initial, and inlet conditions. For a given flow, using the law of conservation of linear momentum together with the law of conservation of mass and the first law of thermodynamics, one can set up five simultaneous partial differential equations for the three velocity components (e.g., for $u$, $v$, and $w$ in the $x$-, $y$-, and $z$-directions, respectively), the pressure $p$, and the temperature $T$. If the density of the fluid changes with pressure and temperature, that is, if the fluid is compressible, then a sixth equation has to be introduced relating density to the temperature and pressure, such as the equation of state for a perfect gas. Finally, if the viscosity changes with temperature, then an empirical viscosity law $\mu(T)$ will be the seventh equation of the system.

Given the boundary and initial conditions for the flow and the temperature field and given the properties of the fluid, it is only necessary to find mathematical methods, exact or approximate, that lead to the determination of the local heat flux at the wall, which is given by

$$q''_w = -k \left( \frac{\partial T}{\partial y} \right)_w, \tag{2.2}$$

where
  $k$ is the thermal conductivity of the fluid
  $y$ represents the direction perpendicular to the wall

Difficulties in finding the temperature distribution in a fluid are of mathematical nature. Especially, in the study of transient turbulent convection as compared with laminar convection, a solvable mathematical formulation is usually based on a number of assumptions, some of which cannot be justified by experiments.

Convective heat transfer is affected by the mechanics of fluid flow occurring adjacent to the solid surface. Therefore, in the following sections, the momentum and energy transfer will be considered together. The momentum and energy transfer within a flow field subjected to specified boundary and initial conditions are governed by the principle of conservation of mass, Newton's second law of motion, and the laws of thermodynamics. For our purposes here, the first law of thermodynamics together with the conservation of mass and Newton's second law of motion will be sufficient to determine the temperature distribution in a continuous fluid uniquely; particular laws will be used to relate fluxes to gradients and the equation of state will be assumed known.

In the following sections, the fundamental equations of convective heat transfer are discussed.

## 2.2 Continuity Equation

The mathematical expression of the principle of conservation of mass applied to an elemental control volume within a fluid in motion is called the continuity equation and is given by Equation 1.10. It is instructive to derive this equation in another way illustrating a direct application of Equation 1.7a or Equation 1.7b.

Consider the flow of a single-phase single-component fluid (homogeneous and invariant in composition) with the velocity $V$ having the components $u$, $v$, and $w$ in the $x$-, $y$-, and $z$-directions, and define an elemental control volume with dimensions $\Delta x$, $\Delta y$, and $\Delta z$ in this fluid as shown in Figure 2.1. In this figure, the mass flow rates entering and leaving

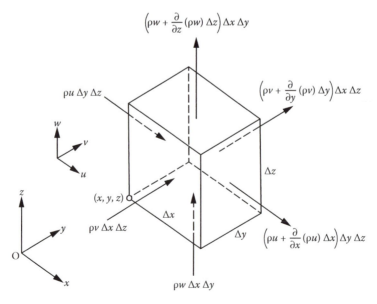

**FIGURE 2.1**
Elemental control volume in a flow field for the derivation of the continuity equation indicated.

the control volume in the *x*-, *y*-, and *z*-directions are also indicated. Hence, the net rate of mass leaving the control volume in the *x*-direction is given by

$$\frac{\partial(\rho u)}{\partial X} \Delta x \Delta y \Delta z. \tag{2.3a}$$

Similarly, the net rates of mass leaving the control volume in the *y*- and *z*-directions may be written as

$$\frac{\partial(\rho v)}{\partial y} \Delta y \Delta z \Delta x \tag{2.3b}$$

and

$$\frac{\partial(\rho w)}{\partial z} \Delta z \Delta x \Delta y. \tag{2.3c}$$

Hence, the net rate of mass entering the control volume becomes

$$-\left[\frac{\partial(\rho u)}{\partial x} + \frac{\partial(\rho v)}{\partial y} + \frac{\partial(\rho w)}{\partial z}\right]\Delta x \Delta y \Delta z. \tag{2.4}$$

The rate of increase of mass within the control volume may be written as

$$\frac{\partial \rho}{\partial t} \Delta x \Delta y \Delta z. \tag{2.5}$$

The law of conservation of mass (1.7b) leads, therefore, to

$$\frac{\partial \rho}{\partial t} \Delta x \Delta y \Delta z = -\left[\frac{\partial(\rho u)}{\partial x} + \frac{\partial(\rho v)}{\partial y} + \frac{\partial(\rho w)}{\partial z}\right]\Delta x \Delta y \Delta z \tag{2.6}$$

or

$$\frac{\partial \rho}{\partial t} + \frac{\partial(\rho u)}{\partial x} + \frac{\partial(\rho v)}{\partial y} + \frac{\partial(\rho w)}{\partial z} = 0, \tag{2.7}$$

which is called the *continuity equation* and is the mathematical expression of the law of conservation of mass for an elemental control volume within a continuous flow field. This result is general in the sense that it is valid for *unsteady* flows of *compressible* fluids. Equation 2.7 can also be written as

$$\frac{\partial \rho}{\partial t} + \nabla \cdot (\rho \mathbf{V}) = 0, \tag{2.8}$$

which is the same result as Equation 1.10. Equation 2.7 or 2.8 may be rearranged as

$$\frac{D\rho}{Dt} + \rho\left(\frac{\partial u}{\partial x} + \frac{\partial v}{\partial y} + \frac{\partial w}{\partial z}\right) = 0 \tag{2.9}$$

or

$$\frac{D\rho}{Dt} + \rho\nabla \cdot \mathbf{V} = 0, \tag{2.10}$$

where we have introduced the following differential operator (also see Problem 1.6):

$$\frac{D}{Dt} \equiv \frac{\partial}{\partial t} + u\frac{\partial}{\partial x} + v\frac{\partial}{\partial y} + w\frac{\partial}{\partial z}, \tag{2.11}$$

which is often called the *substantial* derivative or the derivative following the motion of fluid.

For *steady* flows, $\partial p/\partial t = 0$, and therefore, the continuity equation (2.8) reduces to

$$\nabla \cdot (\rho\mathbf{V}) = 0. \tag{2.12}$$

For *incompressible* flows, the continuity equation (2.8) becomes

$$\nabla \cdot \mathbf{V} = 0, \tag{2.13}$$

which is valid for steady as well as unsteady flows. Equation 2.13 can be written in rectangular coordinates as

$$\frac{\partial u}{\partial x} + \frac{\partial v}{\partial y} + \frac{\partial w}{\partial z} = 0. \tag{2.14}$$

In many problems, the continuity equation may be required in coordinate systems other than the rectangular coordinates. Of course, the continuity equation in any other coordinate system can be derived just as we did in the rectangular coordinates, or they can also be obtained from the previous result for the rectangular coordinates by coordinate transformation. In cylindrical coordinates, the continuity equation takes the following form:

$$\frac{\partial \rho}{\partial t} + \frac{1}{r}\frac{\partial}{\partial r}(\rho r v_r) + \frac{1}{r}\frac{\partial}{\partial \theta}(\rho v_\theta) + \frac{\partial}{\partial z}(\rho w) = 0, \tag{2.15}$$

where $v_r$, $v_\theta$, and $w$ are the velocity components in the $r$-, $\theta$-, and $z$-directions.

In Table 2.1, the continuity equation is tabulated in the rectangular, cylindrical, and spherical coordinates. In addition, the velocity components in the cylindrical and spherical coordinates are shown in Figure 2.2.

**TABLE 2.1**

Continuity Equation in Several Coordinate Systems

| General | Compressible | $\dfrac{\partial \rho}{\partial t} + \nabla \cdot (\rho \cdot \mathbf{V}) = 0$ |
|---|---|---|
| | Incompressible | $\nabla \cdot \mathbf{V} = 0$ |
| Rectangular coordinates $(x, y, z)$ | Compressible | $\dfrac{\partial \rho}{\partial t} + \dfrac{\partial}{\partial x}(\rho u) + \dfrac{\partial}{\partial y}(\rho v) + \dfrac{\partial}{\partial z}(\rho w) = 0$ |
| | Incompressible | $\dfrac{\partial u}{\partial x} + \dfrac{\partial v}{\partial y} + \dfrac{\partial w}{\partial z} = 0$ |
| Cylindrical coordinates $(r, \theta, z)$ | Compressible | $\dfrac{\partial \rho}{\partial t} + \dfrac{1}{r}\dfrac{\partial}{\partial r}(\rho r v_r) + \dfrac{1}{r}\dfrac{\partial}{\partial \theta}(\rho v_\theta) + \dfrac{\partial}{\partial z}(\rho w) = 0$ |
| | Incompressible | $\dfrac{\partial v_r}{\partial r} + \dfrac{v_r}{r} + \dfrac{1}{r}\dfrac{\partial v_\theta}{\partial \theta} + \dfrac{\partial W}{\partial Z} = 0$ |
| Spherical coordinates $(r, \theta, \phi)$ | Compressible | $\dfrac{\partial \rho}{\partial t} + \dfrac{1}{r^2}\dfrac{\partial}{\partial r}(\rho r^2 v_r) + \dfrac{1}{r \sin \theta}\dfrac{\partial}{\partial \theta}(\rho v_\theta \sin \theta)$ |
| | | $+ \dfrac{1}{r \sin \theta}\dfrac{\partial}{\partial \phi}(\rho v \phi) = 0$ |
| | Incompressible | $\dfrac{1}{r}\dfrac{\partial}{\partial r}(r^2 v_r) + \dfrac{1}{\sin \theta}\dfrac{\partial}{\partial \theta}(v_\theta \sin \theta) + \dfrac{1}{\sin \theta}\dfrac{\partial v_\phi}{\partial \phi} = 0$ |

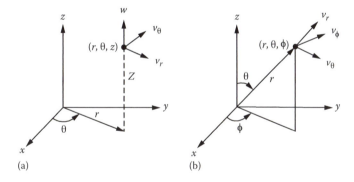

**FIGURE 2.2**
Velocity components in (a) cylindrical and (b) spherical coordinate systems.

## 2.3 Momentum Equations

The dynamic behavior of fluid motion is governed by a set of equations called the *momentum equations* or the *equations of motion*. These equations are obtained by applying either Newton's second law of motion, Equation 1.11, to an infinitesimal mass of fluid or the law of conservation of linear momentum, Equation 1.17a or Equation 1.17b, to an infinitesimal control volume in the fluid. In distinction to the approach followed in the derivation of the continuity equation where we have considered an elemental volume element, we shall consider here an elementary fluid particle of definite mass and follow its motion. Hence, Newton's second law of motion, Equation 1.11, for the fluid particle of definite mass may be rewritten as

$$\mathbf{F} = \frac{d\mathbf{M}}{dt} = \frac{d(m\mathbf{V})}{dt} = m\mathbf{a}, \tag{2.16}$$

where
  **F** is the net force acting on the fluid particle
  $\mathbf{a} = d\mathbf{V}/dt$ is the acceleration of the fluid particle

A fluid particle situated at the point $x$, $y$, and $z$ at time $t$ will be situated at a neighboring point $x + \Delta x$, $y + \Delta y$, and $z + \Delta z$ at the instant $t + \Delta t$. The total change in the $x$-component of velocity of the particle can be written as

$$\Delta u = \frac{\partial u}{\partial x} \Delta x + \frac{\partial u}{\partial y} \Delta y + \frac{\partial u}{\partial z} \Delta z + \frac{\partial u}{\partial t} \Delta t. \tag{2.17}$$

The $x$-component of acceleration of the particle situated at the point $x$, $y$, and $z$ at the instant $t$, therefore, becomes

$$a_x = \lim_{\Delta t \to 0} \frac{\Delta u}{\Delta t} = u \frac{\partial u}{\partial x} + v \frac{\partial v}{\partial y} + w \frac{\partial u}{\partial z} + \frac{\partial u}{\partial t} \equiv \frac{Du}{Dt}. \tag{2.18}$$

Similarly, the $y$- and $z$-components of acceleration at time $t$ are given by

$$a_y = u \frac{\partial v}{\partial x} + v \frac{\partial v}{\partial y} + w \frac{\partial v}{\partial z} + \frac{\partial u}{\partial t} \equiv \frac{Dv}{Dt}, \tag{2.19}$$

$$a_z = u \frac{\partial w}{\partial x} + v \frac{\partial w}{\partial y} + w \frac{\partial w}{\partial z} + \frac{\partial w}{\partial t} \equiv \frac{Dw}{Dt}. \tag{2.20}$$

As we have already discussed in Chapter 1, the forces acting on a particle of fluid can be of two types, namely, body forces such as forces of gravitational, electrical, or magnetic origin and surface forces (contact forces). We assume that a body force $\mathbf{f} = \hat{\mathbf{i}} f_x + \hat{\mathbf{J}} f_y + \hat{\mathbf{K}} f_z$ per unit mass acts on the fluid element at the point $x$, $y$, $z$. Also, let us denote the surface stresses (force per unit surface area) that lie in the plane of the surface by the symbol $\tau$ (shear stress) and the surface stresses normal to the plane of the surface by the symbol $\sigma$ (normal stress). Further, two subscripts are attached to each of the stress symbols: the first indicates the direction of the normal of the surface on which the stress acts and the second indicates the direction in which the stress acts. The normal and shear stresses on a surface are reported in terms of a right-handed coordinate system in which the outwardly directed surface normal indicates the positive direction as illustrated in Figure 2.3.

The state of stress in a fluid at a point, which is similar to that occurring in an elastic solid, is determined when each element of the following stress tensor is known:

$$\begin{bmatrix} \sigma_{xx} & \tau_{xy} & \tau_{xz} \\ \tau_{yx} & \sigma_{yy} & \tau_{yz} \\ \tau_{zx} & \tau_{zy} & \sigma_{zz} \end{bmatrix}$$

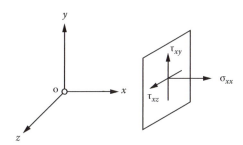

**FIGURE 2.3**
Normal and shear stresses acting on a surface element whose normal is in the *x*-direction.

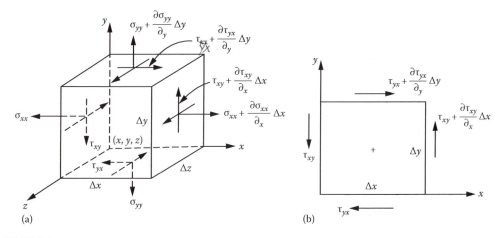

**FIGURE 2.4**
Normal and shear stresses acting on an element of fluid. (a) All stresses acting on an element of fluids and (b) only the shear stresses in the *x*- and *y*-directions (both positive and negative).

Consider now an element of fluid as shown in Figure 2.4a. Let us concentrate our attention on the shear stresses that contribute to a torque about an axis through the center of the element and parallel to the *z*-axis as indicated in Figure 2.4b. Neglecting the higher-order terms, the torque produced by these forces about this axis can be directly computed and is given by

$$T = (\tau_{xy} - \tau_{yx})\Delta x \Delta y \Delta z. \tag{2.21}$$

According to the laws of motion, this torque will be equal to the product of the moment of inertia and angular acceleration $\ddot{\alpha}_z$ of the fluid element, both to be taken with respect to the previously specified axis of rotation. We then obtain

$$(\tau_{xy} - \tau_{yx})\Delta x \Delta y \Delta z = \frac{\rho}{12}\left[(\Delta x)^2 + (\Delta y)^2\right]\Delta x \Delta y \Delta z \ddot{\alpha}_z \tag{2.22a}$$

or

$$(\tau_{xy} - \tau_{yx}) = \frac{\rho}{12}\left[(\Delta x)^2 + (\Delta y)^2\right]\ddot{\alpha}_z. \tag{2.22b}$$

Since $\Delta x$ and $\Delta y$ are infinitesimal quantities, it is seen from the previous equation that the angular acceleration of every infinitesimal element would tend toward infinity as $\Delta x$ and $\Delta y$ approach zero unless $T_{xy} = T_{yx}$. It is therefore concluded that $T_{xy}$ must be equal to $T_{yx}$. The argument for the shear stresses in the other two directions is, of course, identical, that is, $T_{xz} = T_{zx}$ and $T_{yz} = T_{zy}$. Hence, these relationships reduce the stress tensor so as to consist of six rather than nine elements.

Consider now a fluid particle of mass $(p\Delta x\Delta y\Delta z)$ situated at the point $x$, $y$, and $z$ at time $t$ in a flow field. This particle, as we have seen, has the following acceleration in the $x$-direction:

$$\frac{Du}{Dt} = \frac{\partial u}{\partial t} + u\frac{\partial u}{\partial x} + v\frac{\partial u}{\partial y} + w\frac{\partial u}{\partial z}. \tag{2.23}$$

Referring to Figure 2.4a, the net force acting on this fluid particle in the $x$-direction is

$$F_x = \left( pf_x + \frac{\partial\sigma_{xx}}{\partial X} + v\frac{\partial\tau_{yx}}{\partial y} + \frac{\partial\tau_{zx}}{\partial z} \right)\Delta x\Delta y\Delta z. \tag{2.24}$$

Equation 2.16 is a vectorial relation and the $x$-component of this equation is

$$F_x = m\frac{Du}{Dt}. \tag{2.25}$$

Hence, equating the net force in the $x$-direction, Equation 2.24, to the mass times the acceleration in the $x$-direction, Equation 2.23, the $x$-component of Newton's second law of motion (2.16) gives

$$\rho\frac{Du}{Dt} = pf_x + \frac{\partial\sigma_{xx}}{\partial X} + \frac{\partial\tau_{yx}}{\partial y} + \frac{\partial\tau_{zx}}{\partial z}. \tag{2.26}$$

Similar considerations in the $y$- and $z$-directions result in

$$\rho\frac{Dv}{Dt} = pf_y + \frac{\partial\tau_{xy}}{\partial x} + \frac{\partial\sigma_{yy}}{\partial y} + \frac{\partial\tau_{zy}}{\partial z} \tag{2.27}$$

and

$$\rho\frac{Dw}{Dt} = pf_z + \frac{\partial\tau_{xz}}{\partial x} + \frac{\partial\tau_{yz}}{\partial y} + \frac{\partial\sigma_{zz}}{\partial z}. \tag{2.28}$$

These Equations 2.26 through 2.28, which are called *momentum equations* or *equations of motion*, have a very general validity since they are based on the fundamental law of mechanics. In order to make use of them, the relations between the stresses and the deformation of the fluid particle must be introduced. It has been found experimentally that to a high degree of accuracy, the stresses in many fluids are related linearly to the rates of strains (derivatives of the velocities). It can be shown [1,2] that the most general relations between stresses and rates of strains in a Newtonian fluid are given by

$$\sigma_{xx} = -p + \lambda \left( \frac{\partial u}{\partial x} + \frac{\partial v}{\partial y} + \frac{\partial w}{\partial z} \right) + 2\mu \frac{\partial u}{\partial x}, \tag{2.29a}$$

$$\sigma_{yy} = -p + \lambda \left( \frac{\partial u}{\partial x} + \frac{\partial v}{\partial y} + \frac{\partial w}{\partial z} \right) + 2\mu \frac{\partial v}{\partial y}, \tag{2.29b}$$

$$\sigma_{zz} = -p + \lambda \left( \frac{\partial u}{\partial x} + \frac{\partial v}{\partial y} + \frac{\partial w}{\partial z} \right) + 2\mu \frac{\partial w}{\partial z}, \tag{2.29c}$$

$$\tau_{xy} = \tau_{yx} = \mu \left( \frac{\partial v}{\partial x} + \frac{\partial u}{\partial y} \right), \tag{2.29d}$$

$$\tau_{xz} = \tau_{zx} = \mu \left( \frac{\partial w}{\partial x} + \frac{\partial u}{\partial z} \right), \tag{2.29e}$$

$$\tau_{yz} = \tau_{zy} = \mu \left( \frac{\partial w}{\partial y} + \frac{\partial v}{\partial z} \right). \tag{2.29f}$$

In these equations, $\mu$ is the absolute viscosity, and $\lambda$ is the so-called second coefficient of viscosity. In general, $3\lambda + 2\mu \geq 0$. Stokes [3] in 1845 proposed that $3\lambda + 2\mu = 0$, and the kinetic theory of gases proves that this assumption is valid for monatomic gases. It appears that the second coefficient of viscosity is still a controversial quantity [4]. However, we use Stokes' hypothesis as an approximation and take $\lambda = -2/3\mu$. Further, in the previous relations $p$ represents the *average pressure* defined as

$$p = -\frac{1}{3}(\sigma_{xx} + \sigma_{yy} + \sigma_{zz}), \tag{2.30}$$

which is related to the thermodynamic pressure $p_{th}$ by

$$p = p_{th} + \frac{2}{3}\mu \left( \frac{\partial u}{\partial X} + \frac{\partial v}{\partial y} + \frac{\partial w}{\partial z} \right). \tag{2.31}$$

The thermodynamic pressure is the one related to the density and temperature by the equation of state. As can be seen, for incompressible fluids, the average and the thermodynamic pressures are identical. For compressible fluids, the difference is negligibly small in most of the cases.

Combining these equations, relating stresses to the derivatives of the velocity components, with the momentum equations (2.26 through 2.28), we obtain the following:

$$\rho \frac{Du}{Dt} = \rho f_x + \frac{\partial p}{\partial x} + \frac{\partial}{\partial x}\left[\mu\left(2\frac{\partial u}{\partial x} - \frac{2}{3}\nabla \cdot \mathbf{V}\right)\right] + \frac{\partial}{\partial y}\left[\mu\left(\frac{\partial v}{\partial x} + \frac{\partial u}{\partial y}\right)\right]$$
$$+ \frac{\partial}{\partial z}\left[\mu\left(\frac{\partial w}{\partial x} + \frac{\partial u}{\partial z}\right)\right], \tag{2.32a}$$

$$\rho \frac{Dv}{Dt} = \rho f_y + \frac{\partial p}{\partial y} + \frac{\partial}{\partial x}\left[\mu\left(\frac{\partial v}{\partial x} + \frac{\partial u}{\partial y}\right)\right] + \frac{\partial}{\partial y}\left[\mu\left(2\frac{\partial v}{\partial x} + \frac{2}{3}\nabla \cdot \mathbf{V}\right)\right]$$
$$+ \frac{\partial}{\partial z}\left[\mu\left(\frac{\partial w}{\partial y} + \frac{\partial v}{\partial z}\right)\right], \tag{2.32b}$$

$$\rho \frac{Dw}{Dt} = \rho f_z + \frac{\partial p}{\partial z} + \frac{\partial}{\partial x}\left[\mu\left(\frac{\partial w}{\partial x} + \frac{\partial u}{\partial z}\right)\right] + \frac{\partial}{\partial y}\left[\mu\left(\frac{\partial w}{\partial y} + \frac{\partial v}{\partial z}\right)\right]$$
$$+ \frac{\partial}{\partial z}\left[\mu\left(2\frac{\partial w}{\partial z} + \frac{2}{3}\nabla \cdot \mathbf{V}\right)\right], \tag{2.32c}$$

where

$$\nabla \cdot \mathbf{V} = \frac{\partial u}{\partial X} + \frac{\partial v}{\partial y} + \frac{\partial w}{\partial z}.$$

The previous equations are the famous Navier–Stokes equations in rectangular coordinates, and nearly all analytical work involving a viscous fluid is based on them. These equations are general in the sense that they are valid for any viscous, compressible Newtonian fluid with varying viscosity.

When the density and viscosity are constant, that is, when the fluid is incompressible and the temperature variations are small, the Navier–Stokes equations simplify to

$$\rho \frac{Du}{Dt} = Pf_x - \frac{\partial p}{\partial x} + \mu\left(\frac{\partial^2 u}{\partial x^2} + \frac{\partial^2 u}{\partial y^2} + \frac{\partial^2 u}{\partial z^2}\right), \tag{2.33a}$$

$$\rho \frac{Dv}{Dt} = Pf_y - \frac{\partial p}{\partial y} + \mu\left(\frac{\partial^2 v}{\partial x^2} + \frac{\partial^2 v}{\partial y^2} + \frac{\partial^2 v}{\partial z^2}\right), \tag{2.33b}$$

$$\rho \frac{Dw}{Dt} = Pf_z - \frac{\partial p}{\partial y} + \mu\left(\frac{\partial^2 w}{\partial x^2} + \frac{\partial^2 w}{\partial y^2} + \frac{\partial^2 w}{\partial z^2}\right). \tag{2.33c}$$

These equations may conveniently be summarized in vector notation as

$$\frac{D\mathbf{V}}{Dt} = \mathbf{f} - \frac{1}{\rho}\nabla p + \nu\nabla^2\mathbf{V},$$ (2.34)

where

$\nu = \mu/\rho$ is the *kinematic viscosity* of the fluid
$\mathbf{f}$ is the body force vector per unit mass
$\nabla^2$ is the *Laplacian operator*, that is,

$$\nabla^2 = \nabla\cdot\nabla = \frac{\partial^2}{\partial x^2} + \frac{\partial^2}{\partial y^2} + \frac{\partial^2}{\partial z^2}.$$ (2.35)

In many problems, it may be more convenient to use coordinate systems other than the rectangular coordinates; for example, for flows through circular tubes, the cylindrical coordinates are most convenient, and for problems involving flows past a sphere, the spherical coordinates may be more convenient. Of course, the equations of motion for any other coordinate system may be derived just as we did for the rectangular coordinates, or they can also be obtained from the previous results by coordinate transformations, which is a straightforward but tedious procedure. In cylindrical coordinates, the Navier–Stokes equations for incompressible Newtonian fluids with constant viscosity take the following form:

In *r*-direction,

$$\frac{\partial v_r}{\partial t} + v_r\frac{\partial v_r}{\partial r} + \frac{v_\theta}{r}\frac{\partial v_r}{\partial\theta} - \frac{v_\theta^2}{r} + w\frac{\partial v_r}{\partial z} = f_r - \frac{1}{p}\frac{\partial p}{\partial r}$$

$$+ V\left\{\frac{\partial}{\partial r}\left[\frac{1}{r}\frac{\partial}{\partial r}(rv_r)\right] + \frac{1}{r^2}\frac{\partial^2 v_r}{\partial\theta^2} - \frac{2}{r^2}\frac{\partial v_\theta}{\partial\theta} + \frac{\partial^2 v_r}{\partial z^2}\right\},$$ (2.36a)

in θ-direction,

$$\frac{\partial v_\theta}{\partial t} + v_r\frac{\partial v_\theta}{\partial r} + \frac{v_\theta}{r}\frac{\partial v_\theta}{\partial\theta} - \frac{v_r v_\theta}{r} + w\frac{\partial v_\theta}{\partial z} = f_\theta - \frac{1}{rp}\frac{\partial p}{\partial\theta}$$

$$+ V\left\{\frac{\partial}{\partial r}\left[\frac{1}{r}\frac{\partial}{\partial r}(rv_\theta)\right] + \frac{1}{r^2}\frac{\partial^2 v_\theta}{\partial\theta^2} - \frac{2}{r^2}\frac{\partial v_r}{\partial\theta} + \frac{\partial^2 v_\theta}{\partial z^2}\right\},$$ (2.36b)

in *z*-direction,

$$\frac{\partial w}{\partial t} + v_r\frac{\partial w}{\partial r} + \frac{v_\theta}{r}\frac{\partial w}{\partial\theta} + w\frac{\partial w}{\partial z} = f_z - \frac{1}{p}\frac{\partial p}{\partial z}$$

$$+ V\left\{\frac{1}{r}\frac{\partial}{\partial r}\left[r\frac{\partial w}{\partial r}\right] + \frac{1}{r^2}\frac{\partial^2 w}{\partial\theta^2} + \frac{\partial^2 w}{\partial z^2}\right\},$$ (2.36c)

where

$v_r$, $v_\theta$, and $w$ are the velocity components in the *r*-, θ-, and *z*-directions
$f_r$, $f_\theta$, and $f_z$ are the components of the body force vector per unit mass

In Table 2.2, the Navier–Stokes equations for an incompressible fluid with constant viscosity are tabulated in the rectangular as well as in the cylindrical and spherical coordinates [5,6].

**TABLE 2.2**

Equations of Motion for an Incompressible Newtonian Fluid with Constant Viscosity in Several Coordinate Systems

---

### Rectangular coordinates (x, y, z)

x-component

$$\frac{\partial u}{\partial t} + u\frac{\partial u}{\partial x} + v\frac{\partial u}{\partial y} + w\frac{\partial u}{\partial z} = f_x - \frac{1}{\rho}\frac{\partial p}{\partial x} + \nu\left(\frac{\partial^2 u}{\partial x^2} + \frac{\partial^2 u}{\partial y^2} + \frac{\partial^2 u}{\partial z^2}\right)$$

y-component

$$\frac{\partial v}{\partial t} + u\frac{\partial v}{\partial x} + v\frac{\partial v}{\partial y} + w\frac{\partial v}{\partial z} = f_y - \frac{1}{\rho}\frac{\partial p}{\partial y} + \nu\left(\frac{\partial^2 v}{\partial x^2} + \frac{\partial^2 v}{\partial y^2} + \frac{\partial^2 v}{\partial z^2}\right)$$

z-component

$$\frac{\partial w}{\partial t} + u\frac{\partial w}{\partial x} + v\frac{\partial w}{\partial y} + w\frac{\partial w}{\partial z} = f_z - \frac{1}{\rho}\frac{\partial p}{\partial y} + \nu\left(\frac{\partial^2 w}{\partial x^2} + \frac{\partial^2 w}{\partial y^2} + \frac{\partial^2 w}{\partial z^2}\right)$$

### Cylindrical coordinates (r, θ, z)

r-component

$$\frac{\partial v_r}{\partial t} + v_r\frac{\partial v_r}{\partial r} + \frac{v_\theta}{r}\frac{\partial v_r}{\partial \theta} - \frac{v_\theta^2}{r} + w\frac{\partial v_r}{\partial z}$$

$$= f_r - \frac{1}{\rho}\frac{\partial p}{\partial r} + \nu\left\{\frac{\partial}{\partial r}\left[\frac{1}{r}\frac{\partial}{\partial r}(rv_r)\right] + \frac{1}{r^2}\frac{\partial^2 v_r}{\partial \theta^2} - \frac{2}{r^2}\frac{\partial v_\theta}{\partial \theta} + \frac{\partial^2 v}{\partial z^2}\right\}$$

θ-component

$$\frac{\partial v_\theta}{\partial t} + v_r\frac{\partial v_\theta}{\partial r} + \frac{v_\theta}{r}\frac{\partial v_\theta}{\partial \theta} + \frac{v_r v_\theta}{r} + w\frac{\partial v_\theta}{\partial z}$$

$$= f_\theta - \frac{1}{\rho r}\frac{\partial p}{\partial \theta} + \nu\left\{\frac{\partial}{\partial r}\left[\frac{1}{r}\frac{\partial}{\partial r}(rv_\theta)\right] + \frac{1}{r^2}\frac{\partial^2 v_\theta}{\partial \theta^2} - \frac{2}{r^2}\frac{\partial v_r}{\partial \theta} + \frac{\partial^2 v_\theta}{\partial z^2}\right\}$$

z-component

$$\frac{\partial w}{\partial t} + v_r\frac{\partial w}{\partial r} + \frac{v_\theta}{r}\frac{\partial w}{\partial \theta} + w\frac{\partial w}{\partial z} = f_z - \frac{1}{\rho}\frac{\partial p}{\partial z} + \nu\left\{\frac{1}{r}\frac{\partial}{\partial r}\left(r\frac{\partial w}{\partial r}\right) + \frac{1}{r^2}\frac{\partial^2 w}{\partial \theta^2} + \frac{\partial^2 w}{\partial z^2}\right\}$$

### Spherical coordinates (r, θ, φ)

r-component[a]

$$\rho\left(\frac{\partial v_r}{\partial t} + v_r\frac{\partial v_r}{\partial r} + \frac{v_\theta}{r}\frac{\partial v_r}{\partial \theta} + \frac{v_\phi}{r\sin\theta}\frac{\partial v_r}{\partial \phi} - \frac{v_\theta^2 + v_\phi^2}{r}\right)$$

$$= \rho f_r - \frac{\partial p}{\partial r} + \mu\left(\nabla^2 v_r - \frac{2v_r}{r^2} - \frac{2}{r^2}\frac{\partial v_\theta}{\partial \theta} - \frac{2}{r^2}v_\theta\cot\theta - \frac{2}{r^2\sin\theta}\frac{\partial v_\phi}{\partial \phi}\right)$$

Θ-component[a]

$$\rho\left(\frac{\partial v_\theta}{\partial t} + v_r\frac{\partial v_\theta}{\partial r} + \frac{v_\theta}{r}\frac{\partial v_\theta}{\partial \theta} + \frac{v_\phi}{r\sin\theta}\frac{\partial v_\theta}{\partial \phi} + \frac{v_r + v_\theta}{r} - \frac{v_\theta^2\cot\theta}{r}\right)$$

$$= \rho f_\theta - \frac{1}{r}\frac{\partial p}{\partial \theta} + \mu\left(\nabla^2 v_\theta + \frac{2}{r^2}\frac{\partial v_r}{\partial \theta} - \frac{v_\theta}{r^2\sin\theta} - \frac{2\cos\theta}{r^2\sin\theta}\frac{\partial v_\phi}{\partial \phi}\right)$$

Φ-component[a]

$$\rho\left(\frac{\partial v_\phi}{\partial t} + v_r\frac{\partial v_\phi}{\partial r} + \frac{v_\theta}{r}\frac{\partial v_\phi}{\partial \theta} + \frac{v_\phi}{r\sin\theta}\frac{\partial v_\phi}{\partial \phi} + \frac{v_\phi + v_r}{r} - \frac{v_\theta v_\phi}{r}\cot\theta\right)$$

$$= \rho f_\phi - \frac{1}{r\sin\theta}\frac{\partial p}{\partial \phi} + \mu\left(\nabla^2 v_\phi - \frac{v_\phi}{r^2\sin^2\theta} + \frac{2}{r^2\sin\theta}\frac{\partial v_r}{\partial \phi} + \frac{2\cos\theta}{r^2\sin^2\theta}\frac{\partial v_\phi}{\partial \phi}\right)$$

---

[a] In these relations, $\nabla^2 \equiv \frac{1}{r^2}\frac{\partial}{\partial r}\left(r^2\frac{\partial}{\partial r}\right) + \frac{1}{r^2\sin\theta}\frac{\partial}{\partial \theta}\left(\sin\theta\frac{\partial}{\partial \theta}\right) + \frac{1}{r^2\sin^2\theta}\frac{\partial^2}{\partial \phi^2}$.

The model of a fluid with constant density and viscosity is a reasonably good approximation as long as the temperature differences are moderate and the pressure variations are small.

Any problem that involves the determination of fluid velocity as a function of space, time, and the pressure distribution requires the simultaneous solution of the continuity and the Navier–Stokes equations under suitable boundary and initial conditions. Although these equations are, in general, too complicated to be solved analytically in most cases, they may be solved by numerical methods. Fortunately, in many cases the nature of the flow is such that these equations can be simplified considerably and an analytical solution can be obtained [4,6,7].

The Navier–Stokes equations cannot be used directly to solve turbulent flow problems. This will be discussed in Chapter 7.

## 2.4 Energy Equation

The energy equation may be obtained by applying the first law of thermodynamics to an element of fluid of mass $(\rho\Delta x\Delta y\Delta z)$ situated at the point $x$, $y$, and $z$ at time $t$ as shown in Figure 2.5. The first law of thermodynamics, Equation 1.21, states that the rate of heat transfer to the element minus the rate of work done by the element is equal to the rate of increase of energy of the element. The net rate of heat transfer to the element (ignoring radiation heat transfer within the fluid) is

$$\left[\frac{\partial}{\partial x}\left(k\frac{\partial T}{\partial x}\right)+\frac{\partial}{\partial y}\left(k\frac{\partial T}{\partial y}\right)+\frac{\partial}{\partial z}\left(k\frac{\partial T}{\partial z}\right)\right]\Delta x\Delta y\Delta z. \tag{2.37}$$

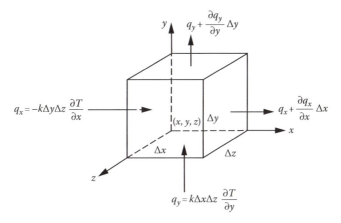

**FIGURE 2.5**
Heat transfer by conduction to a fluid element. Only the $x$- and $y$-components are shown.

Referring to Figure 2.4, the net rate of work done by the fluid element against the surface and body forces is given by

$$
-\left[ \frac{\partial}{\partial x}(u\sigma_{xx} + v\tau_{xy} + w\tau_{xz}) + \frac{\partial}{\partial y}(u\tau_{yx} + v\sigma_{yy} + w\tau_{yz}) \right.
$$

$$
\left. + \frac{\partial}{\partial z}(u\tau_{zx} + v\tau_{zy} + w\sigma_{zz}) + \rho(uf_x + vf_y + wf_z) \right] \Delta x \Delta y \Delta z. \tag{2.38}
$$

The rate of increase of internal and kinetic energies of the element can be written as

$$
\rho \Delta x \Delta y \Delta z \frac{D}{Dt}\left[ \mathcal{U} + \frac{u^2 + v^2 + w^2}{2} \right], \tag{2.39}
$$

where $\mathcal{U}$ is the internal energy per unit mass of the fluid.

Noting that we have already included the change in potential energy in the work term (2.38), the first law of thermodynamics for the fluid particle under consideration becomes

$$
\rho \frac{D}{Dt}\left[ \mathcal{U} + \frac{u^2 + v^2 + w^2}{2} \right] = \frac{\partial}{\partial x}\left( k \frac{\partial T}{\partial x} \right) + \frac{\partial}{\partial y}\left( k \frac{\partial T}{\partial y} \right) + \frac{\partial}{\partial z}\left( k \frac{\partial T}{\partial z} \right)
$$

$$
+ \left[ \frac{\partial}{\partial x}(u\sigma_{xx} + v\tau_{xy} + w\tau_{xz}) + \frac{\partial}{\partial y}(u\tau_{yx} + v\sigma_{yy} + w\tau_{yz}) \right.
$$

$$
\left. + \frac{\partial}{\partial z}(u\tau_{zx} + v\tau_{zy} + w\sigma_{zz}) + \rho(uf_x + vf_y + wf_z) \right], \tag{2.40}
$$

which is also known as the *total energy equation* since it comprises both thermal and mechanical energies. ·

Multiplying both sides of the momentum equations (2.26 through 2.28) by $u$, $v$, and $w$, respectively, we obtain

$$
\rho u \frac{Du}{Dt} = u\left( \rho f_x + \frac{\partial \sigma_{xx}}{\partial x} + \frac{\partial \tau_{yx}}{\partial y} + \frac{\partial \tau_{zx}}{\partial z} \right), \tag{2.41a}
$$

$$
\rho v \frac{Dv}{Dt} = v\left( \rho f_y + \frac{\partial \tau_{xy}}{\partial x} + \frac{\partial \sigma_{yy}}{\partial y} + \frac{\partial \tau_{zy}}{\partial z} \right), \tag{2.41b}
$$

$$
\rho w \frac{Dw}{Dt} = w\left( \rho f_z + \frac{\partial \tau_{xz}}{\partial x} + \frac{\partial \tau_{yz}}{\partial y} + \frac{\partial \sigma_{zz}}{\partial z} \right). \tag{2.41c}
$$

Summing these equations, we get

$$\rho \frac{D}{Dt}\left[\frac{u^2 + v^2 + w^2}{2}\right] = u\left(\frac{\partial \sigma_{xx}}{\partial x} + \frac{\partial \tau_{yx}}{\partial y} + \frac{\partial \tau_{zx}}{\partial z}\right)$$

$$+ v\left(\frac{\partial \tau_{xy}}{\partial x} + \frac{\partial \sigma_{yy}}{\partial y} + \frac{\partial \tau_{zy}}{\partial z}\right) + w\left(\frac{\partial \tau_{xz}}{\partial x} + \frac{\partial \tau_{yz}}{\partial y} + \frac{\partial \sigma_{zz}}{\partial z}\right)$$

$$+ p\left(uf_x + vf_y + wf_z\right)\Big], \tag{2.42}$$

which is an energy equation obtained directly from the laws of mechanics and is appropriately called the *mechanical energy equation*.

Subtracting the mechanical energy equation (2.42) from the total energy equation (2.40), we obtain

$$\rho \frac{D\mathcal{U}}{Dt} = \frac{\partial}{\partial x}\left(k\frac{\partial T}{\partial x}\right) + \frac{\partial}{\partial y}\left(k\frac{\partial T}{\partial y}\right) + \frac{\partial}{\partial z}\left(k\frac{\partial T}{\partial z}\right) + \sigma_{xx}\frac{\partial u}{\partial x} + \sigma_{yy}\frac{\partial v}{\partial y} + \sigma_{zz}\frac{\partial w}{\partial z}$$

$$+ \tau_{xy}\left(\frac{\partial v}{\partial x} + \frac{\partial u}{\partial y}\right) + \tau_{yz}\left(\frac{\partial w}{\partial y} + \frac{\partial v}{\partial z}\right) + \tau_{zx}\left(\frac{\partial u}{\partial z} + \frac{\partial w}{\partial x}\right), \tag{2.43}$$

which is referred to as the *thermal energy equation* or, in short, the *energy equation*.

Utilizing the relations between the stresses and the rates of strains, that is, Equation 2.29, for a viscous, Newtonian fluid, we get

$$\sigma_{xx}\frac{\partial u}{\partial x} + \sigma_{yy}\frac{\partial v}{\partial y} + \sigma_{zz}\frac{\partial w}{\partial z} = -p\nabla \cdot \mathbf{V} - \frac{2}{3}\mu(\nabla \cdot \mathbf{V})^2 + 2\mu\left[\left(\frac{\partial u}{\partial x}\right)^2 + \left(\frac{\partial v}{\partial y}\right)^2 + \left(\frac{\partial w}{\partial z}\right)^2\right], \tag{2.44}$$

and making use of the continuity equation

$$\frac{Dp}{Dt} + p\nabla \cdot \mathbf{V} = 0, \tag{2.45}$$

we can rewrite Equation 2.44 as

$$\sigma_{xx}\frac{\partial u}{\partial x} + \sigma_{yy}\frac{\partial v}{\partial y} + \sigma_{zz}\frac{\partial w}{\partial z} = \frac{p}{p}\frac{Dp}{Dt} - \frac{2}{3}\mu(\nabla \cdot \mathbf{V})^2 + 2\mu\left[\left(\frac{\partial u}{\partial x}\right)^2 + \left(\frac{\partial v}{\partial y}\right)^2 + \left(\frac{\partial w}{\partial z}\right)^2\right]. \tag{2.46}$$

Also from Equations 2.29d through f, we get

$$\tau_{xy}\left(\frac{\partial v}{\partial x} + \frac{\partial u}{\partial y}\right) = \mu\left(\frac{\partial v}{\partial x} + \frac{\partial u}{\partial y}\right)^2, \tag{2.47a}$$

$$\tau_{yz}\left(\frac{\partial w}{\partial y} + \frac{\partial v}{\partial z}\right) = \mu\left(\frac{\partial w}{\partial y} + \frac{\partial v}{\partial z}\right)^2, \tag{2.47b}$$

$$\tau_{zx}\left(\frac{\partial u}{\partial z} + \frac{\partial w}{\partial x}\right) = \mu\left(\frac{\partial u}{\partial z} + \frac{\partial w}{\partial x}\right)^2. \tag{2.47c}$$

When these relations are substituted into the energy equation (2.43), it reduces to

$$\rho\frac{D\mathcal{U}}{Dt} = \frac{\partial}{\partial x}\left(k\frac{\partial T}{\partial x}\right) + \frac{\partial}{\partial y}\left(k\frac{\partial T}{\partial y}\right) + \frac{\partial}{\partial z}\left(k\frac{\partial T}{\partial z}\right) + \frac{p}{\rho}\frac{D\rho}{Dt} + \Phi, \qquad (2.48)$$

where we have introduced

$$\Phi = 2\mu\left[\left(\frac{\partial u}{\partial x}\right)^2 + \left(\frac{\partial v}{\partial y}\right)^2 + \left(\frac{\partial w}{\partial z}\right)^2 + \frac{1}{2}\left(\frac{\partial v}{\partial x} + \frac{\partial u}{\partial y}\right)^2 \right.$$
$$\left. + \frac{1}{2}\left(\frac{\partial w}{\partial y} + \frac{\partial v}{\partial z}\right)^2 + \frac{1}{2}\left(\frac{\partial u}{\partial z} + \frac{\partial w}{\partial x}\right)^2 - \frac{1}{3}(\nabla\cdot\mathbf{V})^2 \right], \qquad (2.49)$$

which is called the *dissipation junction* and is the rate at which the viscous forces do irreversible work on the fluid particles per unit volume.

Finally, the energy equation (2.48) may also be written in terms of the fluid enthalpy defined by $i = \mu + p/p$ as follows:

$$\rho\frac{Di}{Dt} = \frac{\partial}{\partial x}\left(k\frac{\partial T}{\partial x}\right) + \frac{\partial}{\partial y}\left(k\frac{\partial T}{\partial y}\right) + \frac{\partial}{\partial z}\left(k\frac{\partial T}{\partial z}\right) + \frac{Dp}{Dt} + \Phi \qquad (2.50a)$$

or

$$\rho\frac{Di}{Dt} = \nabla\cdot(k\nabla T) + \frac{Dp}{Dt} + \Phi. \qquad (2.50b)$$

For a perfect gas,

$$di = C_p dT, \qquad (2.51)$$

where $C_p$ is the specific heat at constant pressure. Hence, for a perfect gas, the energy equation (2.50b) takes the form

$$\rho c_p\frac{DT}{Dt} = \nabla\cdot(k\nabla T) + \frac{Dp}{Dt} + \Phi. \qquad (2.52)$$

In Equation 2.52, the left-hand side represents the convective terms, and the terms on the right-hand side are, respectively, the rate of heat diffusion to the fluid particles, the rate of reversible work done on the fluid particles by compression, and the rate of viscous dissipation per unit volume. The work of compression, $Dp/Dt$, is usually negligible except previous sonic velocities. For low-speed flows with constant thermal conductivity, the energy equation (2.52) for perfect gases becomes

$$\frac{DT}{Dt} = \alpha\nabla^2 T + \frac{1}{\rho c_p} + \Phi, \qquad (2.53)$$

where

$$\alpha = \frac{k}{\rho c_p} \tag{2.54}$$

is the *thermal diffusivity* of the fluid and

$$\nabla^2 T = \nabla \cdot \nabla T = \frac{\partial^2 T}{\partial x^2} + \frac{\partial^2 T}{\partial y^2} + \frac{\partial^2 T}{\partial z^2}. \tag{2.55}$$

For an incompressible fluid,

$$d\mathcal{U} = cdT, \tag{2.56}$$

where $c = c_v \cong C_p$. Hence, for an incompressible fluid, the energy equation (2.48) takes the form

$$\rho c \frac{DT}{Dt} = \frac{\partial}{\partial x}\left(k\frac{\partial T}{\partial x}\right) + \frac{\partial}{\partial y}\left(k\frac{\partial T}{\partial y}\right) + \frac{\partial}{\partial z}\left(k\frac{\partial T}{\partial z}\right) + \Phi, \tag{2.57}$$

with

$$\Phi = 2\mu\left[\left(\frac{\partial u}{\partial x}\right)^2 + \left(\frac{\partial v}{\partial y}\right)^2 + \left(\frac{\partial w}{\partial z}\right)^2 + \frac{1}{2}\left(\frac{\partial v}{\partial x}+\frac{\partial u}{\partial y}\right)^2 + \frac{1}{2}\left(\frac{\partial w}{\partial y}+\frac{\partial v}{\partial z}\right)^2 + \frac{1}{2}\left(\frac{\partial u}{\partial z}+\frac{\partial w}{\partial x}\right)^2\right]. \tag{2.58}$$

When the thermal conductivity is constant, the energy equation (2.57) for incompressible fluids reduces to

$$\frac{DT}{Dt} = \alpha\nabla^2 T + \frac{1}{\rho c} + \Phi. \tag{2.59}$$

For a *steady* flow of an incompressible fluid with constant thermal conductivity, the energy equation becomes

$$u\frac{\partial T}{\partial x} + v\frac{\partial T}{\partial y} + w\frac{\partial T}{\partial z} = \alpha\left(\frac{\partial^2 T}{\partial x^2} + \frac{\partial^2 T}{\partial y^2} + \frac{\partial^2 T}{\partial z^2}\right) + \frac{1}{\rho c}\Phi. \tag{2.60}$$

For a 2D flow, Equation 2.60 reduces to

$$u\frac{\partial T}{\partial x} + v\frac{\partial T}{\partial y} = \alpha\left(\frac{\partial^2 T}{\partial x^2} + \frac{\partial^2 T}{\partial y^2}\right) + \frac{1}{\rho c}\Phi, \tag{2.61a}$$

with

$$\Phi = 2\mu \left[ \left( \frac{\partial u}{\partial x} \right)^2 + \left( \frac{\partial v}{\partial y} \right)^2 + + \frac{1}{2} \left( \frac{\partial v}{\partial x} + \frac{\partial u}{\partial y} \right)^2 \right]. \tag{2.61b}$$

For a steady and fully developed incompressible laminar flow between two parallel walls (see Chapter 6), Equation 2.60 becomes

$$u \frac{\partial T}{\partial x} = \alpha \left( \frac{\partial^2 T}{\partial x^2} + \frac{\partial^2 T}{\partial y^2} \right) + \frac{\mu}{\rho c} \left( \frac{du}{dy} \right)^2, \tag{2.62}$$

An energy equation in any other coordinate system such as cylindrical coordinates can be derived by following the same approach, or it can be obtained from the previous results for rectangular coordinates through coordinate transformations. In cylindrical coordinates, the energy equation for an incompressible fluid with constant thermal conductivity takes the following form:

$$\rho c \left( \frac{\partial T}{\partial t} + v_r \frac{\partial T}{\partial r} + \frac{v_\theta}{r} \frac{\partial T}{\partial \theta} + w \frac{\partial T}{\partial z} \right) = k \nabla^2 \mathbf{T} + \Phi, \tag{2.63a}$$

where

$$\nabla^2 \mathbf{T} = \frac{1}{r} \frac{\partial}{\partial r} \left( r \frac{\partial T}{\partial r} \right) + \frac{1}{r^2} \frac{\partial^2 T}{\partial \theta^2} + \frac{\partial^2 T}{\partial z^2}, \tag{2.63b}$$

and

$$\Phi = 2\mu \left\{ \left( \frac{\partial v r}{\partial r} \right) + \left[ \frac{1}{r} \left( \frac{\partial v_\theta}{\partial \theta} + v_r \right) \right]^2 + \left( \frac{\partial w}{\partial z} \right)^2 \right\}$$

$$+ \mu \left\{ \left( \frac{\partial v_\theta}{\partial z} + \frac{1}{r} \frac{\partial w}{\partial \theta} \right)^2 + \left( \frac{\partial w}{\partial r} + \frac{\partial v_r}{\partial z} \right)^2 + \left[ \frac{1}{r} \frac{\partial v_r}{\partial \theta} + r \frac{\partial}{\partial r} \left( \frac{v_\theta}{r} \right) \right]^2 \right\}. \tag{2.63c}$$

In Equation 2.63, $v_r$, $v_\theta$, and $w$ represent the velocity components in $r$-, $\theta$-, and $z$-directions, respectively (Figure 2.2).

The energy equation for Newtonian fluids in rectangular, cylindrical, and spherical coordinates is listed in Table 2.3. In this table, an additional term $\dot{q}$ representing the rate of internal energy generation per unit volume within the fluid due to chemical, nuclear, electrical, etc., sources has also been included for completeness [5,6].

**TABLE 2.3**

Energy Equation for Newtonian Fluids in Several Coordinate Systems

**Rectangular coordinates**

$$\boxed{\rho \frac{Di}{Dt} = \nabla \cdot (k \nabla T) + \frac{Dp}{Dt} + \dot{q} + \Phi}$$

$$\frac{D}{Dt} \equiv \frac{\partial}{\partial t} + u \frac{\partial}{\partial x} + v \frac{\partial}{\partial y} + w \frac{\partial}{\partial z}$$

$$\nabla \cdot (k \nabla T) = \frac{\partial}{\partial x} \left( k \frac{\partial T}{\partial x} \right) + \frac{\partial}{\partial y} \left( k \frac{\partial T}{\partial y} \right) + \frac{\partial}{\partial z} \left( k \frac{\partial T}{\partial z} \right)$$

$$\Phi = 2\mu \left[ \left( \frac{\partial u}{\partial x} \right)^2 + \left( \frac{\partial v}{\partial y} \right)^2 + \left( \frac{\partial w}{\partial z} \right)^2 + \frac{1}{2} \left( \frac{\partial v}{\partial x} + \frac{\partial u}{\partial y} \right)^2 + \frac{1}{2} \left( \frac{\partial w}{\partial y} + \frac{\partial v}{\partial z} \right)^2 + \frac{1}{2} \left( \frac{\partial u}{\partial z} + \frac{\partial w}{\partial x} \right)^2 - \frac{1}{3} \left( \frac{\partial u}{\partial x} + \frac{\partial v}{\partial y} + \frac{\partial w}{\partial z} \right)^2 \right]$$

**Rectangular coordinates (perfect gas)**

$$\boxed{\rho c_p \frac{DT}{Dt} = \nabla \cdot (k \nabla T) + \frac{Dp}{Dt} + \dot{q} + \Phi}$$

$$\frac{D}{Dt} \equiv \frac{\partial}{\partial t} + u \frac{\partial}{\partial x} + v \frac{\partial}{\partial y} + w \frac{\partial}{\partial z}$$

$$\nabla \cdot (k \nabla T) = \frac{\partial}{\partial x} \left( k \frac{\partial T}{\partial x} \right) + \frac{\partial}{\partial y} \left( k \frac{\partial T}{\partial y} \right) + \frac{\partial}{\partial z} \left( k \frac{\partial T}{\partial z} \right)$$

$$\Phi = 2\mu \left[ \left( \frac{\partial u}{\partial x} \right)^2 + \left( \frac{\partial v}{\partial y} \right)^2 + \left( \frac{\partial w}{\partial z} \right)^2 + \frac{1}{2} \left( \frac{\partial v}{\partial x} + \frac{\partial u}{\partial y} \right)^2 + \frac{1}{2} \left( \frac{\partial w}{\partial y} + \frac{\partial v}{\partial z} \right)^2 + \frac{1}{2} \left( \frac{\partial u}{\partial z} + \frac{\partial w}{\partial x} \right)^2 - \frac{1}{3} \left( \frac{\partial u}{\partial x} + \frac{\partial v}{\partial y} + \frac{\partial w}{\partial z} \right)^2 \right]$$

**Rectangular coordinates (incompressible fluid)**

$$\boxed{\rho c \frac{DT}{Dt} = \nabla \cdot (k \nabla T) + \dot{q} + \Phi}$$

$$\frac{D}{Dt} \equiv \frac{\partial}{\partial t} + u \frac{\partial}{\partial x} + v \frac{\partial}{\partial y} + w \frac{\partial}{\partial z}$$

$$\nabla \cdot (k \nabla T) = \frac{\partial}{\partial x} \left( k \frac{\partial T}{\partial x} \right) + \frac{\partial}{\partial y} \left( k \frac{\partial T}{\partial y} \right) + \frac{\partial}{\partial z} \left( k \frac{\partial T}{\partial z} \right)$$

$$\Phi = 2\mu \left[ \left( \frac{\partial u}{\partial x} \right)^2 + \left( \frac{\partial v}{\partial y} \right)^2 + \left( \frac{\partial w}{\partial z} \right)^2 + \frac{1}{2} \left( \frac{\partial v}{\partial x} + \frac{\partial u}{\partial y} \right)^2 + \frac{1}{2} \left( \frac{\partial w}{\partial y} + \frac{\partial v}{\partial z} \right)^2 + \frac{1}{2} \left( \frac{\partial u}{\partial z} + \frac{\partial w}{\partial x} \right)^2 \right]$$

**Cylindrical coordinates (incompressible fluid)**

$$\boxed{\rho c \frac{DT}{Dt} = \nabla \cdot (k \nabla T) + \dot{q} + \Phi}$$

$$\frac{D}{Dt} = \frac{\partial}{\partial t} + v_r \frac{\partial}{\partial r} + \frac{v_\theta}{r} \frac{\partial}{\partial \theta} + w \frac{\partial}{\partial z}$$

*(continued)*

**TABLE 2.3 (continued)**

Energy Equation for Newtonian Fluids in Several Coordinate Systems

**Cylindrical coordinates (incompressible fluid)**

$$\nabla \cdot (k\nabla T) = \frac{1}{r}\frac{\partial}{\partial r}\left(rk\frac{\partial T}{\partial r}\right) + \frac{1}{r^2}\frac{\partial}{\partial \theta}\left(k\frac{\partial T}{\partial \theta}\right) + \frac{\partial}{\partial z}\left(k\frac{\partial T}{\partial z}\right)$$

$$\Phi = 2\mu\left\{\left(\frac{\partial v_r}{\partial r}\right)^2 + \left[\frac{1}{r}\left(\frac{\partial v_\theta}{\partial \theta}+v_r\right)\right]^2 + \left(\frac{\partial w}{\partial z}\right)^2 + \frac{1}{2}\left(\frac{\partial v_\theta}{\partial z}+\frac{1}{r}\frac{\partial w}{\partial \theta}\right)^2 + \frac{1}{2}\left(\frac{\partial w}{\partial r}+\frac{\partial v_r}{\partial z}\right)^2 + \frac{1}{2}\left[\frac{1}{r}\frac{\partial v_r}{\partial \theta}+r\frac{\partial}{\partial r}\left(\frac{v_\theta}{r}\right)\right]^2\right\}$$

**Spherical coordinates (incompressible fluid)**

$$\boxed{\rho c\frac{DT}{Dt} = \nabla \cdot (k\nabla T) + \dot q + \Phi}$$

$$\frac{D}{Dt} \equiv \frac{\partial}{\partial t} + v_r\frac{\partial}{\partial r} + \frac{v_\theta}{r}\frac{\partial}{\partial \theta} + \frac{v_\phi}{r\sin\theta}\frac{\partial}{\partial \phi}$$

$$\nabla \cdot (k\nabla T) = \frac{1}{r^2}\frac{\partial}{\partial r}\left(r^2 k\frac{\partial T}{\partial r}\right) + \frac{1}{r^2\sin\theta}\frac{\partial}{\partial \theta}\left(k\sin\theta\frac{\partial T}{\partial \theta}\right) + \frac{1}{r^2\sin^2\theta}\frac{\partial}{\partial \phi}\left(k\frac{\partial T}{\partial \phi}\right)$$

$$\Phi = 2\mu\left\{\left(\frac{\partial v_r}{\partial r}\right)^2 + \frac{1}{r^2}\left(\frac{\partial v_\theta}{\partial \theta}+v_r\right)^2 + \frac{1}{r^2}\left(\frac{1}{\sin\theta}\frac{\partial v_\phi}{\partial \phi}+v_r+v_\theta\cot\theta\right)^2 + \frac{1}{2}\left[r\frac{\partial}{\partial r}\left(\frac{v_\theta}{r}\right)+\frac{1}{r}\frac{\partial v_r}{\partial \theta}\right]^2\right.$$

$$\left. + \frac{1}{2}\left[\frac{1}{r\sin\theta}\frac{\partial v_r}{\partial \phi}+r\frac{\partial}{\partial r}\left(\frac{v_\phi}{r}\right)\right]^2 + \frac{1}{2}\left[\frac{\sin\theta}{r}\frac{\partial}{\partial \theta}\left(\frac{v_\phi}{\sin\theta}\right)+\frac{1}{r\sin\theta}\frac{\partial v_\theta}{\partial \phi}\right]^2\right\}$$

## 2.5 Discussion of the Fundamental Equations

The continuity, the Navier–Stokes, and the energy equations derived in the preceding sections provide a comprehensive description of the energy transfer in a moving fluid. These equations, however, are so involved that they present insurmountable mathematical difficulties because of the number of equations to be simultaneously satisfied and the nonlinear terms in these equations. This nonlinearity is exhibited by terms such as $u(\partial u/\partial x)$ in the expressions of the substantial derivative. Because of the nonlinearities, the superposition principle is not applicable and complex flows may not be compounded from simple flows. Some exact solutions to these equations have been found that represent simple flows [2]. In these cases, the nonlinear terms are either extremely small or identically zero.

If the effect of viscosity is neglected and the flow pattern is assumed irrotational, the so-called potential flows may be solved for. When viscous effects are neglected, the order of the Navier–Stokes equations is thereby reduced, and it becomes impossible to satisfy all physical boundary conditions of the fluid. In particular, the condition fixing the velocity along the boundary must be relaxed. When heat-transfer rates and shearing stresses at the boundaries are of interest, the potential flow approximation becomes unsuitable.

If the troublesome nonlinear terms of the Navier–Stokes equations are assumed to be small compared to the other terms in these equations, then solutions may be possible. The flows represented by such solutions are referred to as *slow motions* or *creeping flows*. These solutions are important in the theory of lubrication and in the settling of small particles in fluids. In most cases in practice, flows of such fluids as air or water result in behavior quite different from creeping flows, and the nonlinear terms are most often of greater magnitude than the other terms in the Navier–Stokes equations.

A dimensionless number that measures the relative magnitude of the inertia effects in a fluid compared to viscous effects is the Reynolds number. This quantity is defined as

$$Re = \rho \frac{VL}{\mu},$$

where
$\rho$ is the fluid density
$V$ is the fluid velocity
$L$ represents a characteristic dimension in the fluid flow field
$\mu$ is the fluid viscosity

Creeping flows are characterized by small Reynolds numbers, whereas most practical flows are often characterized by Reynolds numbers that are large compared to unity. For example, experiments have indicated that the theory of slow motions is able to predict the drag force exerted on a sphere moving at constant speed relative to a fluid when the Reynolds number (utilizing the sphere diameter as the characteristic dimension) is less than about 1–2.

Prandtl in 1904 [8] made a significant advance in fluid mechanics (and, therefore, in heat transfer) when he introduced the boundary-layer approximations that allowed flows at high Reynolds numbers to be studied mathematically. The use of these approximations in studying various fluid flows results in the so-called boundary-layer theory that will be discussed in Chapter 3.

For the reasons mentioned previously, the full equations cannot be considered for further analysis in all their generality, owing to their complexity, and the impossibility of postulating realistic boundary, initial, and inlet conditions for them. However, two important observations are worth mentioning. First, the flow field depends upon the variation of viscosity and density with temperature, more generally with position too. Therefore, the two fields, that is, the velocity and temperature fields, are coupled. Secondly, it is possible that the temperature field under certain conditions can become similar to the velocity field. As can be seen from Equations 2.34 and 2.59, the terms that arise from the pressure gradient $\Delta p$, $\Phi$, and $f$ prevent the similarity between these two equations. Further, the viscosity $\mu$ and the thermal conductivity $k$ may be different functions of temperature. If the pressure gradient $\nabla p$, $\Phi$, and $f$ are zero and if the Prandtl number $Pr = v/\alpha = 1$, the solutions for the velocity and temperature fields will be similar if their corresponding boundary conditions are also similar.

## 2.6 Similarities in Fluid Flow and Heat Transfer

For liquids within which the temperature differences are not too large and for gases within which the temperature differences and the differences in flow speed are not too great, an enormously simplifying approximation of constant density is applicable. In the discussions to follow, we shall further assume that the other properties are also constant and neglect the effects of body forces and viscous dissipation. Under these conditions, we shall determine the conditions for similarity in fluid flow and heat transfer.

With these assumptions, for a steady and 2D flow of a viscous fluid, the continuity equation, the equations of motion, and the energy equation in rectangular coordinates can be written as

$$\frac{\partial u}{\partial x} + \frac{\partial v}{\partial y} = 0, \tag{2.64}$$

$$\rho \left( u \frac{\partial u}{\partial x} + v \frac{\partial u}{\partial y} \right) = -\frac{\partial p}{\partial x} + \mu \nabla^2 u, \tag{2.65}$$

$$\rho \left( u \frac{\partial v}{\partial x} + v \frac{\partial v}{\partial y} \right) = -\frac{\partial p}{\partial y} + \mu \nabla^2 v, \tag{2.66}$$

$$\rho c_p \left( u \frac{\partial T}{\partial x} + v \frac{\partial T}{\partial y} \right) = k \nabla^2 T. \tag{2.67}$$

Consider now a 2D body, with the characteristic dimensions $L$ and constant surface temperature $T_w$, to be immersed in an essentially infinite extent of fluid that moves toward the body with the uniform and steady velocity $U_\infty$, pressure $p_\infty$, and temperature $T_\infty$ far from the body as illustrated in Figure 2.6. Introduce the following dimensionless quantities:

$$\bar{u} = \frac{u}{U_\infty}, \quad \bar{v} = \frac{v}{U_\infty}, \quad \bar{p} = \frac{p - p_\infty}{\rho U_\infty^2}, \quad \bar{T} = \frac{T - T_w}{T_\infty - T_w},$$

$$\bar{x} = \frac{x}{L}, \quad \bar{y} = \frac{y}{L}, \quad Re_L = \frac{\rho U_\infty L}{\mu}, \quad Pr = \frac{c_p \mu}{k},$$

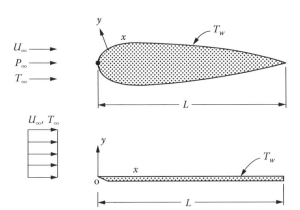

**FIGURE 2.6**
Flow over external surfaces.

where $Pr$ is the so-called *Prandtl number*. In terms of these dimensionless quantities, the continuity equation (2.64), the equations of motion (2.65) and (2.66), and the energy equation (2.67) may be rewritten as

$$\frac{\partial \bar{u}}{\partial \bar{x}} + \frac{\partial \bar{v}}{\partial \bar{y}} = 0, \tag{2.68}$$

$$\bar{u}\frac{\partial \bar{v}}{\partial \bar{x}} + \bar{v}\frac{\partial \bar{u}}{\partial \bar{y}} = -\frac{\partial \bar{p}}{\partial \bar{x}} + \frac{1}{Re_L}\nabla^2\bar{u}, \tag{2.69}$$

$$\bar{u}\frac{\partial \bar{v}}{\partial \bar{x}} + \bar{v}\frac{\partial \bar{u}}{\partial \bar{y}} = -\frac{\partial \bar{p}}{\partial \bar{y}} + \frac{1}{Re_L}\nabla^2\bar{v}, \tag{2.70}$$

$$\bar{u}\frac{\partial \bar{T}}{\partial \bar{x}} + \bar{v}\frac{\partial \bar{T}}{\partial \bar{y}} = \frac{1}{Re_L Pr}\nabla^2\bar{T}, \tag{2.71}$$

where $\bar{\nabla}^2$ is the 2D Laplacian operator given by

$$\bar{\nabla}^2 = \frac{\partial^2}{\partial \bar{x}^2} + \frac{\partial^2}{\partial \bar{y}^2}. \tag{2.72}$$

Boundary conditions for the previous equations can be written as

$$\text{at } \bar{y} = 0 \,(\text{on the surface of the body}), \quad \bar{u} = \bar{v} = \bar{T} = 0,$$

$$\text{as } \bar{y} \to \infty \,(\text{far from the body}), \quad \bar{u} = \bar{T} = 1, \bar{p} = 0.$$

For the given boundary conditions, the solutions of the previous equations (for dimensionless velocity, pressure, and temperature distributions) will depend on the dimensionless independent variables $\bar{x}$ and $\bar{y}$ and the dimensionless parameters $Re_L$ and $Re_L Pr$. From Equations 2.68 through 2.70, we conclude that

$$\bar{u} = \Psi_1(\bar{x}, \bar{y}, Re_L), \tag{2.73}$$

$$\bar{v} = \Psi_2(\bar{x}, \bar{y}, Re_L), \tag{2.74}$$

$$\bar{p} = \Psi_3(\bar{x}, \bar{y}, Re_L). \tag{2.75}$$

We see that geometrically similar bodies at the same corresponding points will have the same dimensionless pressure and velocity and, therefore, the same shear stress distribution when the Reynolds number of the flows is the same.

We may also note that in the case of a constant-density, constant-property flow, the velocity field is independent of the temperature distribution and can be determined once and for all, regardless of the heat-transfer conditions imposed on the flow.

The results for the velocity distribution may now be substituted into the energy equation and the temperature distribution may be determined. In view of Equations 2.73 and 2.74, from the energy equation (2.71) we conclude that the solution for the dimensionless temperature distribution depends upon the independent variables $\bar{x}$ and $\bar{y}$, the Reynolds number, and the product of the Reynolds number and the Prandtl number, that is, the Péclêt number ($pe = Re_L Pr = U_\infty L / \alpha$), which is the measure of the relative magnitude of heat transfer by convection to heat transfer by conduction. Since the Reynolds number is required independently for dynamical similarity, it is customary to work with the Reynolds and Prandtl numbers separately rather than the Péclêt number, and hence,

$$\bar{T} = \Psi_4(\bar{x}, \bar{y}, Re_L, Pr). \tag{2.76}$$

We can therefore deduce from the previous equations that the condition for complete similarity in fluid flow and heat transfer between two different cases of forced convection with geometrically similar boundaries is that the Reynolds and the Prandtl numbers should each have the same values in the two systems (in moderate velocities).

We have already defined the heat-transfer coefficient by Equation 1.60 as

$$h = \frac{-k(\partial T / \partial y)_{y=0}}{T_w - T_\infty}. \tag{2.77}$$

It becomes convenient to nondimensionalize the heat-transfer coefficient as

$$Nu = \frac{hL}{k}, \tag{2.78}$$

which is called the *Nusselt number*, and in terms of the nondimensional parameters, it becomes

$$Nu = \left(\frac{\partial \bar{T}}{\partial \bar{y}}\right)_{\bar{y}=0}. \tag{2.79}$$

Hence, we conclude that

$$Nu = \Psi_5(\bar{x}, Re_L, Pr). \tag{2.80}$$

Alternately, the heat-transfer coefficient can also be put into a nondimensional form as follows:

$$St = \frac{h}{\rho c_p U_\infty} = \frac{Nu}{Re_L Pr} = \Psi_6(\bar{x}, Re_L, Pr), \tag{2.81}$$

which is known as the *Stanton number*.

If the energy equation 2.53 or 2.59 for a fluid with constant thermal conductivity is expressed in dimensionless form, we get

$$Nu = \psi_7(\bar{x}, Re_L, Pr, Ec),$$ (2.82)

where the dimensionless parameter $Ec$ is defined as

$$Ec = \frac{U_\infty^2}{c_p |T_\infty - T_w|},$$ (2.83)

which is called the *Eckert number*. If the flow velocity is not very large, the Eckert number is usually small and, therefore, the effect of viscous dissipation becomes negligible.

## Problems

**2.1** Consider a steady (laminar) flow of a viscous fluid with constant thermophysical properties through a pipe of radius $r_o$ under fully developed conditions (far from the pipe entrance). By solving the Navier–Stokes equations, obtain an expression of the velocity distribution within the pipe.

**2.2** Solve Problem 2.1 for the flow between two parallel plates. What can you deduce about the pressure gradients?

**2.3** Write down the Navier–Stokes equation for the $y$-direction (taken to be the vertical direction) by assuming that the only body force that acts is that due to gravity. Put the equation into dimensionless form and obtain the conditions for dynamic similarity.

**2.4** Derive the continuity equation (2.15) in cylindrical coordinates by applying the law of conservation of mass to an elemental control volume of dimensions $\Delta r$, $r\Delta\theta$, and $\Delta z$ in $r$-, $\theta$-, and $z$-directions, respectively.

**2.5** From the Navier–Stokes equations, obtain the equation of motion in the $x$-direction for the parallel flow of a constant-viscosity fluid (i.e., $v = w = 0$). Simplify this equation for the case of steady incompressible flow.

**2.6** Calculate the mean kinetic energy in terms of the density and the mean velocity for both of the profiles obtained in Problems 2.1 and 2.2. Also, calculate the mean kinetic energy for the case of turbulent flow in a pipe of circular cross section assuming that $u/U_c = (y/r_o)^{1/7}$, where $U_c$ = velocity on the axis of the pipe, $y$ = distance from the pipe wall, and $r_o$ = pipe radius.

**2.7** A liquid film of density $p$ and viscosity $\mu$ flows steadily under the influence of gravity down an inclined surface as illustrated in Figure 2.7. After a certain distance, the velocity component $u$ in the $x$-direction and thickness $b$ of the film become independent of $x$. Assuming constant density and viscosity, obtain the velocity $u$ as a function of $y$, and determine the film thickness $b$ in terms of a specified liquid volume flow rate $\dot{Q}$ per unit width of the surface.

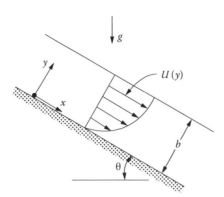

**FIGURE 2.7**
Problem 2.7.

**2.8** Derive the continuity equation given in Table 2.1 for spherical coordinates by applying the law of conservation of mass to an elemental control volume of dimensions $\Delta r$, $r \sin \phi \, \Delta\phi$, and $r \, \Delta\theta$ in $r$-, $\phi$-, and $\theta$-directions, respectively.

**2.9** Show that the energy equation (2.50a) can be expressed for a *compressible* viscous fluid in terms of temperature in the following form:

$$\rho c_p \frac{DT}{Dt} = \nabla \cdot (k\nabla T) + \beta T \frac{Dp}{Dt} + \Phi,$$

where $\beta$ is the coefficient of thermal expansion.

**2.10** Obtain Equation 2.50a from Equation 2.48 by the use of the definition of enthalpy.

**2.11** Starting with the energy equation (2.48), show by a succession of steps how and why it reduces to the general heat conduction equation for a solid and finally to Laplace's equation.

**2.12** Consider the flow of a viscous fluid between two parallel flat plates, one of which is stationary and the other is moving with velocity $U$ parallel to the stationary plate. Starting with the Navier–Stokes equations, obtain the velocity distribution for this flow.

**2.13** Solve Problem 2.12 assuming the pressure gradient is zero (Couette flow).

**2.14** Show that

$$p(\text{div } V) = \rho \frac{D}{Dt}\left(\frac{p}{\rho}\right) - \frac{Dp}{Dt}.$$

**2.15** A lubricating oil of viscosity $\mu$ and thermal conductivity $k$ fills the clearance $L$ between a journal and its bearing, which can be regarded as two parallel plates for the purpose of analysis. The velocity of the rotating surface is $U$, and the journal and the bearing are both maintained at the same temperature $T_o$. Develop an expression for the maximum temperature rise in the oil.

# References

1. Lamb, H., *Hydrodynamics*, Dover Publications, New York, 1945.
2. Schlichting, H., *Boundary-Layer Theory*, Translated into English by J. Kestin, 7th edn., McGraw-Hill, New York, 1979.
3. Stokes, G. G., On the theories of internal friction of fluids in motion, *Trans. Camb. Phil. Soc.*, 8, 287, 1845.
4. Rosenhead, L., The second coefficient of viscosity: A brief review of fundamentals, *Proc. R. Soc. Lond.*, A226, 1–6, 1954.
5. Kakaç, S., Shah, R. K., and Aung, W. (Eds.), *Handbook of Single Phase Convective Heat Transfer*, John Wiley, New York, 1987.
6. Bird, R. B., Steward, W. E., and Lightfoot, E. N., *Transport Phenomena*, John Wiley, New York, 1960.
7. Bejan, A., *Convection Heat Transfer*, 2nd edn., John Wiley & Sons, New York, 1995.
8. Prandtl, L., Über Flüssigkeitsbewegung bei sehr kleiner Reibung, *Proceedings of the Third International Mathematical Congress*, Heidelberg, Germany, pp. 484–491, Teuber, Leipzig, 1904; also in English as motion of fluids with very little viscosity, *NACA TM 452*, 1928.

# 3

## Boundary-Layer Approximations for Laminar Flow

### Nomenclature

$c_p$  specific heat at constant pressure, J/(kg K)
$Ec$  Eckert number $= U_o/c_p(T_\infty - T_w)$
$k$  thermal conductivity, W/(m·K)
$L$  length of the plate, m
$p$  pressure, N/m²
$Pr$  Prandtl number $= c_p\mu/k$
$q_w^n$  wall heat flux, W/m²
$Re$  Reynolds number $= U_o L/\nu$
$Re_{cr}$  critical Reynolds number $= U_o x_{cr}/\nu$
$T$  temperature, °C, K
$T_w$  wall temperature, °C, K
$T_\infty$  free-stream temperature, °C, K
$U_\infty$  free-stream velocity, m/s
$U_0$  free-stream velocity at $x = 0$, m/s
$u$  velocity component parallel to the plate, m/s
$v$  velocity component normal to the plate, m/s
$x$  rectangular coordinate; distance parallel to surface, m
$x_{cr}$  critical length, m
$y$  rectangular coordinate; distance normal to surface, m

### Greek Symbols

$\delta$  velocity boundary-layer thickness, m
$\delta_T$  thermal boundary-layer thickness, m
$\mu$  dynamic viscosity, Pa s
$\nu$  kinematic viscosity, m²/s
$\rho$  density, kg/m³

### Other Symbols

$O$  order of magnitude
over-bar dimensionless quantity

## 3.1 Introduction

In this chapter we discuss the simplification of the Navier–Stokes and the energy equations for laminar boundary layers.

Consider the flow of a viscous fluid over a stationary smooth plate surface as illustrated in Figure 3.1. As we have already discussed, at the surface, the fluid particles adhere to it, and the frictional forces between the fluid layers retard the motion of the fluid within a thin layer near the surface. In this thin layer, which is termed the *boundary layer*, the velocity of the fluid decreases from its free-stream value $U_\infty$ to a value of zero at the surface (no-slip condition). Figure 3.2 shows diagrammatically the variation of the velocity component parallel to the surface in the boundary layer at a given location on the surface.

Ludwig Prandtl (1875–1953), who is often spoken as the founder of modern aerodynamics, introduced the boundary-layer concept for the first time in 1904 [1]. According to the Prandtl boundary-layer concept, under certain conditions viscous forces are of importance

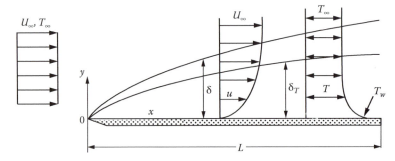

**FIGURE 3.1**
Flow of a viscous fluid parallel to a flat surface.

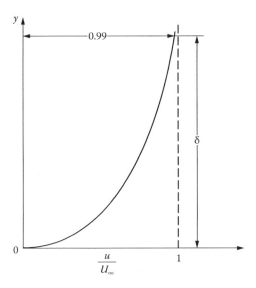

**FIGURE 3.2**
Velocity boundary-layer thickness.

only in the immediate vicinity of a solid surface where velocity gradients are large. This region near the surface is referred to as the boundary layer. In regions removed from the solid surface where there exists no large gradients in fluid velocity, the fluid motion may be considered frictionless, that is, potential flow. Prandtl, with his introduction of the boundary-layer concept, made enormous contributions to the advancement of the science of fluid mechanics and, therefore, of convective heat transfer [2,3].

There is, in fact, no precise dividing line between the potential flow region, where friction is negligible, and the boundary-layer region, because the velocity component parallel to the surface, $u$, approaches the free-stream value $U_\infty$ asymptotically. However, it is customary to define the boundary layer as that region where the velocity component parallel to the surface is less than 99% of the free-stream velocity.

Since the velocity component parallel to the surface, $u$, is varying, the continuity requires that there should also be a velocity component $v$ perpendicular to the surface. So the originally parallel flow becomes, at least, a 2D flow in the boundary layer.

Initially the boundary-layer development is *laminar*, but at some critical distance from the leading edge, small disturbances in the flow begin to be amplified and a transition process takes place until the flow in the boundary layer becomes *turbulent* as illustrated in Figure 3.3. Experiments have shown that the transition will normally occur at a distance $x_{cr}$ measured from the leading edge, where the value of a *local Reynolds number* based on $x_{cr}$, that is, $Re_{cr} = U_\infty x_{cr}/\nu$, has a value approximately equal to $5 \times 10^5$. Depending on the flow conditions, fluid properties, and surface conditions, the transition to turbulent flow may start at a Reynolds number as small as $10^5$ or the laminar boundary layer may exist up to a Reynolds number of $3 \times 10^6$.

If one measures the velocity component $u$ in turbulent boundary-layer region at a fixed position as a function of time, the result will be a plot like the one presented in Figure 3.4. We see that while the flow in the laminar boundary-layer region may be steady and 2D, the flow in the transition and turbulent regions is inherently unsteady and 3D. The structure of turbulence is a complex phenomenon that is still not fully understood, and the theory that has been developed is so limited that semiempirical methods are invariably used for the solution of engineering problems. In this chapter we shall simplify the Navier–Stokes and the energy equations for laminar boundary layers and obtain the so-called laminar boundary-layer equations. In the following three chapters, we shall deal with the solutions of these simplified equations. We shall discuss turbulent boundary layers later in Chapter 7.

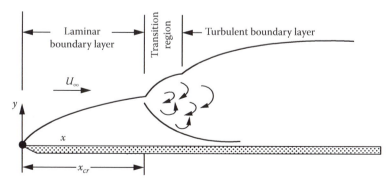

**FIGURE 3.3**
Laminar and turbulent boundary layers on a flat plate.

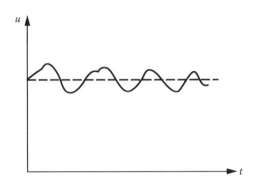

**FIGURE 3.4**
Timewise variation of $u$ in turbulent region.

## 3.2 Momentum Equations of the Boundary Layer

Consider now the case of a steady 2D incompressible laminar boundary layer over a surface with a free-stream velocity $U_\infty(x)$, which may vary with $x$ in an arbitrary manner. Assume that the body forces are negligible and the viscosity of the fluid is constant.

The continuity Equation 2.13 and the Navier–Stokes equations (2.33) for this case reduce to

$$\frac{\partial u}{\partial x} + \frac{\partial v}{\partial y} = 0, \tag{3.1}$$

$$u\frac{\partial u}{\partial x} + v\frac{\partial u}{\partial y} = -\frac{1}{\rho}\frac{\partial p}{\partial x} + \nu\left(\frac{\partial^2 u}{\partial x^2} + \frac{\partial^2 u}{\partial y^2}\right), \tag{3.2}$$

and

$$u\frac{\partial v}{\partial x} + v\frac{\partial v}{\partial y} = -\frac{1}{\rho}\frac{\partial p}{\partial y} + \nu\left(\frac{\partial^2 v}{\partial x^2} + \frac{\partial^2 v}{\partial y^2}\right). \tag{3.3}$$

These equations may now be written in the following dimensionless forms:

$$\frac{\partial \bar{u}}{\partial \bar{x}} + \frac{\partial \bar{v}}{\partial \bar{y}} = 0, \tag{3.4}$$

$$\bar{u}\frac{\partial \bar{u}}{\partial \bar{x}} + \bar{v}\frac{\partial \bar{u}}{\partial \bar{y}} = -\frac{\partial \bar{p}}{\partial \bar{x}} + \frac{1}{Re}\left(\frac{\partial^2 \bar{u}}{\partial \bar{x}^2} + \frac{\partial^2 \bar{u}}{\partial \bar{y}^2}\right), \tag{3.5}$$

and

$$\bar{u}\frac{\partial \bar{v}}{\partial \bar{x}} + \bar{v}\frac{\partial \bar{v}}{\partial \bar{y}} = -\frac{\partial \bar{p}}{\partial \bar{y}} + \frac{1}{Re}\left(\frac{\partial^2 \bar{v}}{\partial \bar{x}^2} + \frac{\partial^2 \bar{v}}{\partial \bar{y}^2}\right), \tag{3.6}$$

where

$$\bar{x} = \frac{x}{L}, \quad \bar{y} = \frac{y}{L}, \quad Re = \frac{U_o L}{\nu \nu},$$

$$\bar{u} = \frac{u}{U_o}, \quad \bar{v} = \frac{v}{U_o}, \quad \bar{p} = \frac{p}{\rho U_o^2}.$$

Here, $U_0 = U_\infty(0)$ is taken to be the reference velocity and $L$ is the length of the surface in the $x$-direction.

We now wish to simplify Equations 3.4 through 3.6 by utilizing the experimentally observed fact that $\delta = \delta/L \ll 1$. We do this by subjecting these equations to an *order-of-magnitude* analysis with the purpose of dropping those terms that we feel are negligibly small in the computations as compared to the other terms in the same equations.

In most of the boundary layer, $\bar{u} = O(1)$ and $\bar{y} = O(\bar{\delta})$, and in locations removed from the leading edge, $\bar{x} = O(1)$, where the symbol $O(\ )$ stands for the "order of." Considering the order of magnitudes of $\bar{u}$ and $\bar{x}$, we observe that

$$\frac{\partial \bar{u}}{\partial \bar{x}} \sim \frac{O(1)}{O(1)} = O(1),$$

and therefore, it follows that

$$\frac{\partial^2 \bar{u}}{\partial \bar{x}^2} = O(1).$$

Similarly, since $\bar{y} = O(\bar{\delta})$, we also have

$$\frac{\partial \bar{u}}{\partial \bar{y}} = O\left(\frac{1}{\bar{\delta}}\right) \quad \text{and} \quad \frac{\partial^2 \bar{u}}{\partial \bar{y}^2} = O\left(\frac{1}{\bar{\delta}^2}\right).$$

On the other hand, from the continuity Equation 3.4, we know that $\partial \bar{u}/\partial \bar{x}$ and $\partial \bar{v}/\partial \bar{y}$ are to be of the same order of magnitude. Thus,

$$\frac{\partial \bar{v}}{\partial \bar{y}} = O(1),$$

and therefore, $\bar{v} = O(\bar{\delta})$. Hence, we also conclude that

$$\frac{\partial \bar{v}}{\partial \bar{x}} = O(\bar{\delta}), \quad \frac{\partial^2 \bar{v}}{\partial \bar{y}^2} = O\left(\frac{1}{\bar{\delta}}\right), \quad \text{and} \quad \frac{\partial^2 \bar{v}}{\partial \bar{x}^2} = O(\bar{\delta}).$$

We now rewrite Equations 3.4 through 3.6 and indicate under each term its order of magnitude:

$$\frac{\partial \bar{u}}{\partial \bar{x}} + \frac{\partial \bar{v}}{\partial \bar{y}^2} = 0,$$

$$\frac{1}{1} \qquad \frac{\delta}{\delta}$$

(3.7)

$$\bar{u}\frac{\partial \bar{u}}{\partial \bar{x}}+\bar{v}\frac{\partial \bar{u}}{\partial \bar{y}}\ ,=-\frac{\partial \bar{p}}{\partial \bar{x}}+\frac{1}{Re}\left(\frac{\partial^2 \bar{u}}{\partial \bar{x}^2}+\frac{\partial^2 \bar{u}}{\partial \bar{y}^2}\right),$$

$$\frac{1}{1}+\bar{\delta}\frac{1}{\bar{\delta}}\qquad\qquad\qquad \left(\frac{1}{1}+\frac{1}{\bar{\delta}^2}\right)$$

(3.8)

$$\bar{u}\frac{\partial \bar{v}}{\partial \bar{x}}+\bar{v}\frac{\partial \bar{v}}{\partial \bar{y}}=-\frac{\partial \bar{p}}{\partial \bar{y}}+\frac{1}{Re}\left(\frac{\partial^2 \bar{v}}{\partial \bar{x}^2}+\frac{\partial^2 \bar{v}}{\partial \bar{y}^2}\right).$$

$$1\frac{\bar{\delta}}{1}+\bar{\delta}\frac{\bar{\delta}}{\bar{\delta}}\qquad\qquad\qquad \left(\frac{\bar{\delta}}{1}+\frac{\bar{\delta}}{\bar{\delta}^2}\right)$$

(3.9)

Since $\bar{\delta} \ll 1$, from Equation 3.8 we conclude that

$$\frac{\partial^2 \bar{u}}{\partial \bar{x}^2}\ll\frac{\partial^2 \bar{u}}{\partial \bar{y}^2},$$

and therefore, the first term in the parenthesis on the right-hand side can be neglected compared to the second term. Note that the order of magnitude of the left-hand side of Equation 3.8 is unity. In order for the viscous effects to be of the same order of magnitude as the inertia effects on the left-hand side of this equation, the Reynolds number should be of order $1/\bar{\delta}^2$:

$$Re=O\left(\frac{1}{\bar{\delta}^2}\right),$$

which means that the Reynolds number must be very large for the boundary-layer approximation to be applicable. Inserting this order of magnitude into Equation 3.9, we see that all terms are of order of magnitude $\bar{\delta}$ and consequently conclude that

$$\frac{\partial \bar{p}}{\partial \bar{y}}=O(\bar{\delta}),$$

and therefore, the change in pressure across the boundary layer is of the order of $\bar{\delta}^2$, which is very small. This implies that at any $x$, the pressure is practically constant in the $y$-direction within the boundary layer but can vary in the $x$-direction. Thus, the pressure within the boundary layer may be considered to be determined by the outer potential flow.

As a result of the foregoing order-of-magnitude analysis, the momentum Equations 3.2 and 3.3 reduce to

$$u\frac{\partial u}{\partial x}+v\frac{\partial u}{\partial y}=-\frac{1}{\rho}\frac{\partial p}{\partial x}+v\left(\frac{\partial^2 u}{\partial y^2}\right),$$

(3.10)

$$\frac{\partial p}{\partial y}=0,$$

(3.11)

which shows that pressure can be assumed as constant across the boundary layer. Then Equations 3.10 and 3.11 can be combined into one as

$$u \frac{\partial u}{\partial x} + v \frac{\partial u}{\partial y} = -\frac{1}{\rho} \frac{\partial p}{\partial x} + v \left( \frac{\partial^2 u}{\partial y^2} \right). \tag{3.12}$$

Equation 3.12 in the limit as $y \to \infty$ (outside the edge of the velocity boundary layer) reduces to

$$U_\infty \frac{dU_\infty}{dx} = -\frac{1}{\rho} \frac{dp}{dx} \tag{3.13}$$

in the outer flow region. In the boundary-layer analysis, the free-stream velocity $U_\infty(x)$ is assumed to be available from the solution of the potential flow outside the boundary layer, and thus, the pressure gradient term $dp/dx$ is considered to be known from Equation 3.13.

Hence, Equations 3.1 through 3.3 may now be replaced by the following *Prandtl boundary-layer equations:*

$$\frac{\partial u}{\partial x} + \frac{\partial v}{\partial y} = 0, \tag{3.14}$$

$$u \frac{\partial u}{\partial x} + v \frac{\partial u}{\partial y} = -\frac{1}{\rho} \frac{dp}{dx} + v \left( \frac{\partial^2 u}{\partial y^2} \right), \tag{3.15}$$

with the following boundary conditions:

$$\text{at } y = 0: \quad u = v = 0, \tag{3.16a,b}$$

$$\text{as } y \to \infty: \quad u = U_\infty(x). \tag{3.16c}$$

In addition, the velocity distribution at $x = 0$ must also be specified.

It may be noted that although one of the viscous terms in Equation 3.8 has been dropped, the order of this equation has not been reduced. Also, one of the equations of motion has been dropped completely. As a result, the number of unknowns has been reduced by one.

A similar analysis has been carried out for the boundary-layer flow along a curved wall, and it has been concluded that the previous equations may be applied to a curved wall as long as no large variations in curvature occur [3].

It should be noted that the boundary-layer approximations are valid for large values of the Reynolds number and the no-slip condition is assumed on the solid surface, that is, the fluid layer at $y = 0$ sticks to the solid surface.

## 3.3 Boundary-Layer Energy Equation

If the plate temperature is different from the fluid temperature, a thermal boundary layer will also develop over the plate as illustrated in Figure 3.1, indicating a significant temperature variation over a narrow zone in the immediate vicinity of the plate. The thermal

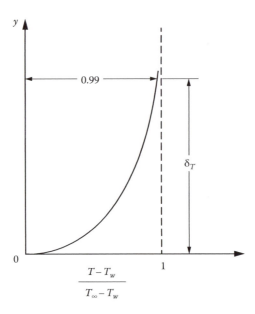

**FIGURE 3.5**
Thermal boundary-layer thickness.

boundary-layer thickness $\delta_T$ may be defined as that distance from the surface where $(T - T_w) = 0.99(T_\infty - T_w)$ as shown in Figure 3.5.

For a steady, 2D, and incompressible viscous flow with constant thermophysical properties, the energy Equation 2.61 reduces to

$$\rho c_p \left( u \frac{\partial T}{\partial x} + v \frac{\partial T}{\partial y} \right) = k \left( \frac{\partial^2 T}{\partial x^2} + \frac{\partial^2 T}{\partial y^2} \right) + 2\mu \left[ \left( \frac{\partial u}{\partial x} \right)^2 + \left( \frac{\partial v}{\partial y} \right)^2 + \frac{1}{2} \left( \frac{\partial u}{\partial y} + \frac{\partial v}{\partial x} \right)^2 \right]. \qquad (3.17)$$

Introducing the following dimensionless variables

$$\bar{x} = \frac{x}{L}; \quad \bar{y} = \frac{y}{L}; \quad \bar{p} = \frac{p}{\rho U_o^2}; \quad \bar{u} = \frac{u}{U_o}; \quad \bar{v} = \frac{v}{U_o}; \quad \bar{T} = \frac{T - T_w}{T_\infty - T_w},$$

where $U_o = U_\infty(0)$, the energy Equation 3.17 may be written in the following dimensionless form:

$$\bar{u} \frac{\partial \bar{T}}{\partial \bar{x}} + \bar{v} \frac{\partial \bar{T}}{\partial \bar{y}} = \frac{1}{RePr} \left( \frac{\partial^2 \bar{T}}{\partial \bar{x}^2} + \frac{\partial^2 \bar{T}}{\partial \bar{y}^2} \right) + 2 \frac{Ec}{Re} \left[ \left( \frac{\partial \bar{u}}{\partial \bar{x}} \right)^2 + \left( \frac{\partial \bar{v}}{\partial \bar{y}} \right)^2 + \frac{1}{2} \left( \frac{\partial \bar{u}}{\partial \bar{y}} + \frac{\partial \bar{v}}{\partial \bar{x}} \right)^2 \right], \qquad (3.18)$$

where we have introduced

$$Re = \frac{U_o L}{\nu}, \quad Pr = \frac{c_p \mu}{k}, \quad Ec = \frac{U_o^2}{c_p (T_\infty - T_w)}.$$

Following the discussions of the previous section, we now again conclude that in the energy equation (3.18),

$$\bar{u} = O(1), \quad \frac{\partial \bar{u}}{\partial \bar{x}} = O(1), \quad \frac{\partial \bar{u}}{\partial \bar{y}} = O\left(\frac{1}{\bar{\delta}}\right)$$

$$\bar{v} = O(\bar{\delta}), \quad \frac{\partial \bar{v}}{\partial \bar{x}} = O(\bar{\delta}), \quad \frac{\partial \bar{v}}{\partial \bar{y}} = O(1)$$

and $Re = O(1/\bar{\delta}^2)$, where, as before, $\bar{x} = O(1), \bar{y} = O(\bar{\delta}),$ and $\bar{\delta} = \delta/L.$

Comparing the order of magnitudes of the dissipation terms, the energy equation (3.18) can be rewritten as

$$\bar{u}\frac{\partial \bar{T}}{\partial \bar{x}} + \bar{v}\frac{\partial \bar{T}}{\partial \bar{y}} = \frac{1}{RePr}\left(\frac{\partial^2 \bar{T}}{\partial \bar{x}^2} + \frac{\partial^2 \bar{T}}{\partial \bar{y}^2}\right) + \frac{Ec}{Re}\left(\frac{\partial \bar{u}}{\partial \bar{y}}\right)^2. \tag{3.19}$$
$$\quad 1 \quad 1 \quad \bar{\delta}1/\bar{\delta}_T \quad\quad\quad 1 \quad 1/\bar{\delta}_T^2 \quad\quad\quad 1/\bar{\delta}^2$$

Furthermore, we note that in the thermal boundary layer, $\bar{T} = O(1)$ because $\bar{T}$ varies from zero at the surface of the plate to almost unity at $y = \delta_T,$ and also $\bar{y} = O(\bar{\delta}_T),$ where $\bar{\delta}_T = \delta_T/L.$ Hence, we observe that

$$\frac{\partial \bar{T}}{\partial \bar{x}} = O(1) \quad \text{and} \quad \frac{\partial^2 \bar{T}}{\partial \bar{x}^2} = O(1),$$

$$\frac{\partial \bar{T}}{\partial \bar{y}} = O\left(\frac{1}{\bar{\delta}_T}\right) \quad \text{and} \quad \frac{\partial^2 \bar{T}}{\partial \bar{y}^2} = O\left(\frac{1}{\bar{\delta}_T^2}\right).$$

The order of magnitude of various terms in Equation 3.19 is indicated under each term. Comparing the orders of magnitude of the conduction terms, we conclude that

$$\frac{\partial^2 \bar{T}}{\partial \bar{x}^2} \ll \frac{\partial^2 \bar{T}}{\partial \bar{y}^2}.$$

It can also be seen from Equation 3.19 that the conduction and convection terms are to be of the same order of magnitude if the product $RePr$ is of order $1/\bar{\delta}_T^2$; that is,

$$RePr = O\left(\frac{1}{\bar{\delta}_T^2}\right).$$

Combining the previously obtained estimation for the order of magnitude of $Re$ number with the previous result gives

$$\frac{\delta_T}{\delta} \sim \frac{1}{\sqrt{Pr}}. \tag{3.20}$$

In Equation 3.19, the dissipation term will be important if

$$\frac{Ec}{Re} = O(\overline{\delta}^2),$$

or since

$$Re = O\left(\frac{1}{\overline{\delta}^2}\right),$$

$$Ec = \frac{U_o^2}{c_p(T_\infty - T_w)} = O(1).$$

In view of the foregoing order-of-magnitude analysis, the boundary-layer energy equation for a steady, 2D, and incompressible laminar flow with constant thermophysical properties becomes

$$u\frac{\partial T}{\partial x} + v\frac{\partial T}{\partial y} = \frac{k}{\rho c_p}\frac{\partial^2 T}{\partial y^2} + \frac{\mu}{\rho c_p}\left(\frac{\partial u}{\partial y}\right)^2, \tag{3.21}$$

with the boundary conditions

$$\text{at } y = 0: \quad T = T_w \quad \text{or} \quad -k\frac{\partial T}{\partial y} = q_w'' \tag{3.22a}$$

$$\text{as } y \to \infty: \quad T = T_\infty. \tag{3.22b}$$

As stated in Equations 3.22a,b, at the surface of the plate, either the wall temperature $T_w$ or the heat flow rate per unit area from the surface to the fluid $q_w''$ is specified.

In conclusion, for 2D, steady, and incompressible laminar boundary-layer flows with constant thermophysical properties, we have only three equations, that is, Equations 3.14, 3.15, and 3.21, for the three unknowns $u$, $v$, and $T$. Solutions to these equations will be discussed in the following chapters. It should also be noted that for constant property flows, the velocity field becomes independent of the temperature distribution. More analyses can be found in the literature [4–8].

---

## Problems

**3.1**   State the conditions under which Equation 3.15 is applicable.

**3.2**   Under what conditions would Equations 3.15 and 3.21 have similar solutions?

**3.3**   The influence of pressure variations along a solid surface is smaller in heat transfer than on flow parameters like the drag. Why?

**3.4**   State the conditions under which Equation 3.21 is applicable.

**3.5** Consider a stationary volume element of dimensions $\Delta x$ and $\Delta y$ in the $x$ and $y$ directions, respectively, and located at any point $(x,y)$ within a 2D laminar boundary layer. Introducing the appropriate assumptions, obtain Equations 3.15 and 3.21 by applying the principles of the conservation of momentum and the conservation of thermal energy to the volume element.

**3.6** The term $\mu/\rho c_p (\partial u/\partial y)^2$ in the energy Equation 3.21 is of importance only at high-velocity flows. Making an order-of-magnitude analysis shows that this term is negligible compared to the other terms in the energy equation for flow of air at $T_\infty = 20°C$ and atmospheric pressure with $U_\infty = 5$ m/s over a solid surface maintained at $T_w = 100°C$.

# References

1. Prandtl, L., Über Flüssigkeitsbewegung bei sehr kleiner Reibung, *Proceedings of the Third International Mathematical Congress*, Heidelberg, Germany, pp. 484–491, Teuber, Leipzig, 1904; also in English as Motion of Fluids with Very Little Viscosity, *NACA TM 452*, 1928.
2. Lamb, H., *Hydrodynamics*, Dover Publications, New York, 1945.
3. Schlichting, H., *Boundary-Layer Theory*, Translated into English by J. Kestin, 7th edn., McGraw-Hill, New York, 1979.
4. Kays, W. M. and Crawford, M. E., *Convective Heat and Mass Transfer*, 3rd edn., McGraw-Hill, New York, 1993.
5. Goldstein, S., *Modern Developments in Fluid Dynamics*, Vol. 1, Dover Publications, New York, 1965.
6. Carrier, G. F. and Lin. C. C., On the nature of the boundary layer near the leading edge of a flat plate, *Quart. Appl. Math.*, VI, 63–68, 1948.
7. Cebeci, T., *Convective Heat Transfer*, Horizons Publ., Long Beach, CA, 2002.
8. Bejan, A., *Convection Heat Transfer*, 2nd edn., John Wiley & Sons, New York, 1995.

# 4

# Heat Transfer in Incompressible Laminar External Boundary Layers: Similarity Solutions

## Nomenclature

$A$     surface area, m²; constant defined by Equation 4.14a

$a$     speed of sound, m/s

$B$     constant defined by Equation 4.14b

$b$     reference width, m

$C$     constant defined by Equation 4.14c

$C_{fx}$     local skin friction coefficient $= 2\tau_w(x)/\rho_\infty U^2_\infty$

$\bar{C}_f$     average skin friction coefficient $= 2\tau_w/\rho_\infty U^2_\infty$

$F_D$     drag force, N

$h$     heat-transfer coefficient, W/(m² K)

$h_x$     local heat-transfer coefficient, W/(m² K)

$k$     thermal conductivity, W/(m K)

$L$     reference length, m

$M_\infty$     Mach number $= U_\infty/a$

$m$     exponent defined by Equation 4.77

$N_u$     average Nusselt number $= hL/k$

$Nu_x$     local Nusselt number $= h_x x/k$

$P$     pressure, N/m²

$Pr$     Prandtl number $= c_p\mu/k$

$q$     heat-transfer rate, W

$q^n_w$     wall heat flux, W/m²

$Re_L$     Reynolds number based on $L$, $= \rho U_\infty L/\mu$

$Re_x$     local Reynolds number based on $x$, $= \rho U_\infty x/\mu$

$St$     Stanton number $= Nu/(RePr)$

$T$     temperature, K

$T_f$     film temperature, K

$T_w$     wall temperature, K

$T_\infty$     free-stream temperature, K

$U_\infty$     free-stream velocity, m/s

$u$     velocity component in $x$ direction, m/s

$v$      velocity component in $y$ direction, m/s
$x$      rectangular coordinate parallel to surface, m
$y$      rectangular coordinate normal to surface, m

## Greek Symbols

$\alpha$      thermal diffusivity = $k/pc$, m²/s; constant given by Equation 4.33
$\beta$      wedge angle parameter defined by Equation 4.77
$\delta$      velocity boundary-layer thickness, m
$\delta_T$      thermal boundary-layer thickness
$\eta$      similarity variable $y\sqrt{U_\infty/(vx)}$
$\theta$      dimensionless temperature profile = $(T - T_\infty/)(T_W - T_\infty)$
$\mu$      dynamic viscosity, Pa s
$v$      kinematic viscosity, m²/s
$\rho$      density, kg/m³
$\tau_w$      average wall shear stress = $F_D/A$
$\tau_w(x)$      local wall shear stress, Pa
$\psi$      stream function

## 4.1 Introduction

Even though the boundary-layer approximations result in great simplifications of the complete Navier–Stokes and energy equations, exact analytical solutions have been obtained only for a few simple cases [1]. One such case, for example, involves the determination of the velocity and temperature distributions exactly in the vicinity of a heated flat plate aligned parallel with a stream of infinite extent, which will be discussed in this chapter. Today, numerical solutions to the boundary-layer equations can be obtained for almost any prescribed boundary conditions and pressure distribution with the use of modern electronic computers, but they still require a considerable effort because the equations to be solved involve nonlinear terms.

Prandtl [2] was the first to point out in 1904 that the boundary-layer momentum equation can be transformed into an ordinary differential equation for flows over a flat plate with a uniform free-stream velocity. In 1908, Blasius [3] obtained in this way the first solution to the boundary-layer momentum equation. In 1921, Pohlhausen [4], following the works of Prandtl and Blasius, gave the first solution to the boundary-layer energy equation. Falkner and Skan [5] in 1930 demonstrated the possibility of the same transformation for a family of problems. Later, in 1939, Goldstein [6] investigated in detail the conditions under which such a transformation can be carried out.

In order to illustrate the method of solution introduced by Prandtl, Blasius, and Pohlhausen, we shall consider in this chapter the flow of a viscous fluid over a semi-infinite flat plate as shown in Figure 4.1. It is assumed that the fluid, which is at temperature $T_\infty$, is infinite in extent and flows with a steady uniform velocity $U_\infty$ parallel to the plate maintained at a constant temperature $T_w$ ($\neq T_\infty$). If the plate moves through an infinite medium, which is uniform in properties and at rest, the solution will still be the same since it is only the relative motion that matters; it is, however, convenient to attach the coordinate reference frame to the body and consider the fluid to be moving past the body at rest with uniform velocity.

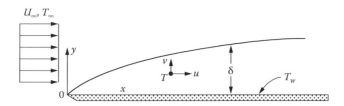

**FIGURE 4.1**
Boundary-layer along a flat plate.

Let the thermophysical properties of the fluid ($\rho$, $\mu$, $k$, and $c_p$) be constant. As we have discussed in the previous chapter, the physical requirement of the no-slip condition at the surface of the plate results in the development of a velocity boundary layer starting at the leading edge of the plate, and the flow in the boundary layer is initially laminar. Since the free-stream velocity $U_\infty$ is constant, the pressure gradient is zero. Assuming that the flow in the laminar boundary layer is 2D, the boundary-layer equations, from Equations 3.14, 3.15, and 3.21, become

$$\frac{\partial u}{\partial x} + \frac{\partial v}{\partial y} = 0,$$ (4.1)

$$u\frac{\partial u}{\partial x} + v\frac{\partial u}{\partial y} = \nu\frac{\partial^2 u}{\partial y^2},$$ (4.2)

$$u\frac{\partial T}{\partial x} + v\frac{\partial T}{\partial y} = \alpha\frac{\partial^2 T}{\partial y^2} + \frac{\mu}{\rho c_p}\left(\frac{\partial u}{\partial y}\right)^2,$$ (4.3)

and the boundary conditions are given by

$$\text{at } y = 0 : u = v = 0 \quad \text{and} \quad T = T_w,$$ (4.4a,b)

$$\text{as } y \to \infty : u \to U_\infty \quad \text{and} \quad T = T_\infty,$$ (4.4c,d)

$$\text{at } x = 0 : u = U_\infty \quad \text{and} \quad T = T_\infty.$$ (4.4e,f)

The following sections will be concerned with the solution of the previous problem for the velocity components $u(x,y)$ and $v(x,y)$ and the temperature distribution $T(x,y)$ in the vicinity of the plate.

## 4.2 Laminar Velocity Boundary Layer

Since the fluid properties have been assumed to be constant, the velocity field will be independent of the temperature field. Hence, Equations 4.1 and 4.2 can be solved simultaneously for the velocity components $u(x,y)$ and $(x,y)$ independently of the energy Equation 4.3.

The system under consideration has no characteristic length in any direction. Therefore, it is reasonable to assume that the $x$-component of velocity has similar profiles at all locations along the plate in the sense that

$$\frac{u}{U_\infty} = g\left(\frac{y}{\delta}\right),$$

(4.5)

where $\delta = \delta(x)$ is the local boundary-layer thickness.

We now proceed to estimate the boundary-layer thickness $\delta(x)$: Integration of the boundary-layer momentum Equation 4.2 with respect to $y$ from $y = 0$ to $y = \delta(x)$ yields

$$\int_0^\delta u \frac{\partial u}{\partial x} dy + \int_0^\delta v \frac{\partial u}{\partial y} dy = \nu \frac{\partial u}{\partial y} \Big|_{y=0}^{y=\delta}.$$

(4.6)

The second term on the left-hand side of this result can be integrated by parts as follows:

$$\int_0^\delta v \frac{\partial u}{\partial y} dy = vu \Big|_0^\delta - \int_0^\delta u \frac{\partial v}{\partial y} dy.$$

(4.7)

Integrating the continuity Equation 4.1 over $y$ from $y = 0$ to $y$ gives

$$v(x, y) = -\int_0^y \frac{\partial u}{\partial x} dy.$$

(4.8)

Since $U(x,\delta) = 0.99\, U_\infty$ and

$$\frac{\partial v}{\partial y} = -\frac{\partial u}{\partial x}.$$

(4.9)

Equation 4.7 can be rewritten as

$$\int_0^\delta v \frac{\partial u}{\partial y} dy = -0.99 U_\infty \int_0^\delta \frac{\partial u}{\partial x} dy + \int_0^\delta u \frac{\partial u}{\partial x} dy.$$

(4.10)

Therefore, Equation 4.6 becomes

$$2\int_0^\delta u \frac{\partial u}{\partial y} dy = -0.99 U_\infty \int_0^\delta \frac{\partial u}{\partial x} dy = v \frac{\partial u}{\partial y} \Big|_{y=0}^{y=\delta}.$$

(4.11)

Making use of Equation 4.5, we can write

$$\left( \frac{\partial u}{\partial x} = U_\infty g' \frac{\partial}{\partial x}\left(\frac{y}{\delta}\right) = -U_\infty \frac{y}{\delta^2} g' \frac{d\delta}{dx} \right)$$

(4.12a)

and

$$\frac{\partial u}{\partial y} = U_\infty g' \frac{\partial}{\partial y}\left(\frac{y}{\delta}\right) = \frac{U_\infty}{\delta} g', \tag{4.12b}$$

where primes denote differentiation with respect to $y/\delta$. Equation 4.11 can now be rewritten as

$$(A-B)\delta\frac{d\delta}{dx} = \frac{v}{U_\infty}C, \tag{4.13}$$

where we have introduced the following constants:

$$A = 0.99\int_0^1 g'\frac{y}{\delta}d\left(\frac{y}{\delta}\right), \tag{4.14a}$$

$$B = 2\int_0^1 gg'\frac{y}{\delta}d\left(\frac{y}{\delta}\right), \tag{4.14b}$$

$$C = g'(1) - g'(0). \tag{4.14c}$$

Integrating Equation 4.13 yields

$$\delta = \sqrt{\frac{2C}{A-B}}\sqrt{\frac{vx}{U_\infty}}, \tag{4.15a}$$

where we have used the fact that $\delta = 0$ at $x = 0$. Since, from Equation 4.15a,

$$\delta \sim \sqrt{\frac{vx}{U_\infty}}, \tag{4.15b}$$

the similarity assumption (4.5) can also be expressed as

$$\frac{u}{U_\infty} = \bar{g}(\eta), \tag{4.16}$$

where

$$\eta = y\sqrt{\frac{U_\infty}{vx}}, \tag{4.17}$$

which is named as the similarity *variable*.

Introduce a stream function $\Psi(x,y)$ such that

$$u = \frac{\partial \Psi}{\partial y} \quad \text{and} \quad v = -\frac{\partial \Psi}{\partial x}, \tag{4.18a,b}$$

with the condition $\Psi(x,0) = 0$, so that the continuity equation is identically satisfied and the momentum equation reduces to

$$\frac{\partial \Psi}{\partial y}\frac{\partial^2 \Psi}{\partial x \partial y} - \frac{\partial \Psi}{\partial x}\frac{\partial^2 \Psi}{\partial y^2} = v\frac{\partial^3 \Psi}{\partial y^3}. \tag{4.19}$$

Integrating Equation 4.18a yields

$$\Psi = \int_0^y u\,dy = \sqrt{U_\infty vx}\int_0^\eta \overline{g}(\eta)d\eta = \sqrt{U_\infty vx}f(\eta), \tag{4.20}$$

where we have introduced

$$f(\eta) = \int_0^\eta \overline{g}(\eta)d\eta. \tag{4.21}$$

Thus, Equation 4.16 becomes

$$\frac{u}{U_\infty} = \frac{df}{d\eta}, \tag{4.22}$$

and from Equation 4.18b, we obtain

$$\frac{v}{U_\infty}\frac{1}{2}\sqrt{\frac{v}{U_\infty x}}\left[\eta\frac{df}{d\eta} - f\right]. \tag{4.23}$$

After substituting the derivatives

$$\frac{\partial^2 \Psi}{\partial x \partial y} = -\frac{1}{2}U_\infty\frac{\eta}{x}\frac{d^2 f}{d\eta^2}, \tag{4.24a}$$

$$\frac{\partial^2 \Psi}{\partial y^2} = U_\infty\sqrt{\frac{U_\infty}{vx}}\frac{d^2 f}{d\eta^2}, \tag{4.24b}$$

$$\frac{\partial^3 \Psi}{\partial y^3} = \frac{U_\infty^2}{vx}\frac{d^3 f}{d\eta^3} \tag{4.24c}$$

into Equation 4.19 and rearranging the terms, we get

$$\frac{d^3 f}{d\eta^3} + \frac{1}{2} f \frac{d^2 f}{d\eta^2} = 0. \tag{4.25}$$

As seen from Equations 4.4, 4.22 and 4.23, the boundary conditions on $f(\eta)$ are given by

$$\text{at } \eta = 0 : f = \frac{df}{d\eta} = 0, \tag{4.26a,b}$$

$$\text{as } \eta \to \infty : \frac{df}{d\eta} = 1, \tag{4.26c}$$

which satisfy the conditions that the plate is impermeable, the plate has no slip on the surface, and the velocity far from the plate is the free-stream velocity.

Thus, the partial differential Equations 4.1 and 4.2 representing the velocity boundary-layer problem have been transformed into a third-order nonlinear ordinary differential equation. The three boundary conditions (4.26) are, therefore, sufficient to determine the solution for $f(\eta)$.

A closed-form solution to Equation 4.25 has not been obtained yet for the boundary conditions (4.26). The solution can, however, be obtained either by numerical methods or by a series expansion technique. Blasius in 1908 [3] obtained the solution in the form of power series expansion as

$$f(\eta) = A_0 + A_1\eta + \frac{A_2}{2!}\eta^2 + \frac{A_3}{3!}\eta^3 + \frac{A_4}{4!}\eta^4 + \cdots \tag{4.27}$$

The boundary conditions at $\eta = 0$ give $A_0 = A_1 = 0$. Substituting Equation 4.27 into the differential Equation 4.25, we find

$$2A_3 + 2A_4\eta + (\alpha^2 + 2A_5)\frac{\eta^2}{2!} + (4\alpha A_3 + 2A_6)\frac{\eta^3}{3!} + \cdots = 0, \tag{4.28}$$

where we have substituted $\alpha = A_2$. In order for this equation to be true, the coefficients of the various powers of $\eta$ must be identically zero, yielding

$$A_3 = A_4 = 0$$

$$A_5 = -\frac{1}{2}\alpha^2$$

$$A_6 = A_7 = 0$$

$$A_8 = \frac{11}{4}\alpha^3$$

·········

Thus, it is necessary to determine only α. Substituting these coefficients into the series (4.27), we get

$$f(\eta) = \frac{\alpha}{2!}\eta^2 - \frac{1}{2}\frac{\alpha^2}{5!}\eta^5 + \frac{1}{4}\frac{11\alpha^3}{8!}\eta^8 - \frac{1}{8}\frac{375\alpha^4}{11!}\eta^{11} + \cdots \tag{4.29}$$

Equation 4.29 satisfies the two boundary conditions at $\eta = 0$. The constant α can be determined by the use of the boundary condition (4.26c), which still remains to be satisfied. Now, we note that Equation 4.29 can be written in the form

$$f(\eta) = \alpha^{1/3} F\left(\alpha^{1/3}\eta\right), \tag{4.30}$$

where we have introduced

$$F(\zeta) = \frac{\zeta^2}{2!} - \frac{1}{2}\frac{\zeta^5}{5!} + \frac{11}{4}\frac{\zeta^8}{8!} - \frac{375}{8}\frac{\zeta^{11}}{11!} + \cdots \tag{4.31}$$

Applying the boundary condition (4.26c), we have

$$\lim_{\eta \to \infty} f'(\eta) = \alpha^{2/3} \lim_{\eta \to \infty} F'\left(\alpha^{1/3}\eta\right) = \alpha^{2/3} \lim_{\eta \to \infty} F'(\eta) = 1, \tag{4.32}$$

or

$$\alpha = \left[ \frac{1}{\lim_{\eta \to \infty} F'(\eta)} \right]^{3/2}. \tag{4.33}$$

The value of α can be determined numerically from this equation to any desired approximation (see Problem 4.1). The numerical evaluation of α from Equation 4.33 was first given by Blasius [3] in 1908. Additional solutions have been obtained by others, the most accurate one being the work of Howarth [7] who gave $\alpha = 0.33206$. With this value of α, the numerical values of $f, f',$ and $f''$ have been tabulated in Table 4.1 [1,7]. The variation of the velocity distribution parallel to the surface, that is, $u/U_\infty = f'(\eta)$, is also seen plotted in Figure 4.2.

From this velocity distribution, the thickness of the boundary layer as well as the local shear stress at the surface of the plate can be determined. If the boundary-layer thickness is taken to be the distance from the plate at which the velocity reaches 99% of the free-stream value, then from Table 4.1 or Figure 4.2, it is found that

$$\delta = 5.0 \sqrt{\frac{\upsilon x}{U_\infty}}, \tag{4.34a}$$

**TABLE 4.1**

Velocity Distribution in the Laminar Boundary
Layer over a Flat Plate

| $\eta = y\sqrt{\dfrac{U_\infty}{vx}}$ | $f$ | $f' = \dfrac{u}{U_\infty}$ | $F''$ |
|---|---|---|---|
| 0 | 0 | 0 | 0.33206 |
| 0.2 | 0.00664 | 0.06641 | 0.33199 |
| 0.4 | 0.02656 | 0.13277 | 0.33147 |
| 0.6 | 0.05974 | 0.19894 | 0.33008 |
| 0.8 | 0.10611 | 0.26471 | 0.32739 |
| 1.0 | 0.16557 | 0.32979 | 0.32301 |
| 1.2 | 0.23795 | 0.39378 | 0.31659 |
| 1.4 | 0.32298 | 0.45627 | 0.30787 |
| 1.6 | 0.42032 | 0.51676 | 0.29667 |
| 1.8 | 0.52952 | 0.57477 | 0.28293 |
| 2.0 | 0.65003 | 0.62977 | 0.26675 |
| 2.2 | 0.78120 | 0.68132 | 0.24835 |
| 2.4 | 0.92230 | 0.72899 | 0.22809 |
| 2.6 | 1.07252 | 0.77246 | 0.20646 |
| 2.8 | 1.23099 | 0.81152 | 0.18401 |
| 3.0 | 1.39682 | 0.84605 | 0.16136 |
| 3.2 | 1.56911 | 0.87609 | 0.13913 |
| 3.4 | 1.74696 | 0.90177 | 0.11788 |
| 3.6 | 1.92954 | 0.92333 | 0.09809 |
| 3.8 | 2.11605 | 0.94112 | 0.08013 |
| 4.0 | 2.30576 | 0.95552 | 0.06424 |
| 4.2 | 4.49806 | 0.96696 | 0.05052 |
| 4.4 | 2.69238 | 0.97587 | 0.03897 |
| 4.6 | 2.88826 | 0.98269 | 0.02948 |
| 4.8 | 3.08534 | 0.98779 | 0.02187 |
| 5.0 | 3.28329 | 0.99155 | 0.01591 |
| 5.2 | 3.48189 | 0.99425 | 0.01134 |
| 5.4 | 3.68094 | 0.99616 | 0.00793 |
| 5.6 | 3.88031 | 0.99748 | 0.00543 |
| 5.8 | 4.07990 | 0.99838 | 0.00365 |
| 6.0 | 4.27964 | 0.99898 | 0.00240 |
| 6.2 | 4.47948 | 0.99937 | 0.00155 |
| 6.4 | 4.67938 | 0.99961 | 0.00098 |
| 6.6 | 4.87931 | 0.99977 | 0.00061 |
| 6.8 | 5.07928 | 0.99987 | 0.00037 |
| 7.0 | 5.27926 | 0.99992 | 0.00022 |
| 7.2 | 5.47925 | 0.99996 | 0.00013 |
| 7.4 | 5.67924 | 0.99998 | 0.00007 |
| 7.6 | 5.87924 | 0.99999 | 0.00004 |
| 7.8 | 6.07923 | 1.00000 | 0.00002 |

*(continued)*

**TABLE 4.1 (continued)**

Velocity Distribution in the Laminar Boundary
Layer over a Flat Plate

| $\eta = y\sqrt{\dfrac{U_\infty}{vx}}$ | $f$ | $f' = \dfrac{u}{U_\infty}$ | $F''$ |
|---|---|---|---|
| 8.0 | 6.27923 | 1.00000 | 0.00001 |
| 8.2 | 6.47923 | 1.00000 | 0.00001 |
| 8.4 | 6.67923 | 1.00000 | 0.00000 |
| 8.6 | 6.87923 | 1.00000 | 0.00000 |
| 8.8 | 7.07923 | 1.00000 | 0.00000 |

*Source:* Howarth, L., *Proc. Roy. Soc.*, 164A, 547, 1938.

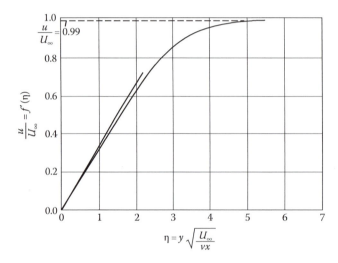

**FIGURE 4.2**
Velocity distribution in laminar boundary layer over a flat plate.

which can also be expressed as

$$\frac{\delta}{x} = \frac{5.0}{\sqrt{Re_x}},$$

(4.34b)

where $Re_x$ is the local Reynolds number defined as

$$Re_x = \frac{U_\infty x}{v}.$$

(4.35)

**Example 4.1**

Air at 20°C and $0.5 \times 10^5$ N/m² (0.5 bar) flows over a flat plate with a velocity of 60 m/s.
The length of the plate is 25 cm. Calculate the boundary-layer thickness at various locations along the plate.

**Solution**

From Appendix B, the kinematic viscosity of air at 20° C and atmospheric pressure is

$$v_0 = 15.11 \times 10^{-6} \text{ m}^2/\text{s}.$$

Since the variation of the dynamic viscosity with pressure is negligible, the kinematic viscosity of air at $0.5 \times 10^5$ N/m$^2$ and 20° C can be found from the equation of state, that is,

$$v = v_0 \frac{p_0}{p} = 2v_0 = 2 \times 15.11 \times 10^{-6} \text{ m}^2/\text{s} = 30.22 \times 10^{-6} \text{ m}^2/\text{s}.$$

The thickness of the boundary layer can now be calculated from Equation 4.34a:

$$\delta = 5.0 \sqrt{\frac{(30.22 \times 10^{-6})x}{60}} = 3.55 \times 10^{-3} \sqrt{x} \text{ m}.$$

The following table gives the boundary-layer thickness at various points along the plate:

| X (cm) | δ (mm) |
|--------|--------|
| 0 | 0 |
| 2 | 0.502 |
| 4 | 0.710 |
| 10 | 1.122 |
| 15 | 1.374 |
| 25 | 1.775 |

The local shear stress at the surface of the plate is given by

$$\tau_w(x) = \mu \left( \frac{\partial u}{\partial y} \right)_{y=0} = \mu U_\infty \sqrt{\frac{U_\infty}{vx}} \frac{d^2 f(0)}{d\eta^2}. \tag{4.36a}$$

From Table 4.1, it is seen that $f''(0) = 0.332$. Hence, Equation 4.36a can be rewritten as

$$\tau_w(x) = 0.332 \mu U_\infty \sqrt{\frac{U_\infty}{vx}}. \tag{4.36b}$$

A dimensionless wall shear stress, which is usually referred to as *the drag coefficient* or, since drag is due only to wall shear, *the friction coefficient*, may be defined as

$$C_{fx} = \frac{\tau_w(x)}{(1/2)\rho U_\infty^2} = \frac{0.664}{\sqrt{Re_x}}, \tag{4.37}$$

where $Re_x$ is the local Reynolds number defined by Equation 4.35.

This solution results in an infinite shear stress at the leading edge of the plate. This is due to the fact that the boundary-layer assumptions are not valid in the vicinity of the leading edge where the derivative $\partial^2 u/\partial x^2$ is not small compared to $\partial^2 u/\partial x^2$. Because the region in which this is true is usually very small compared to the size of the entire plate, the effect of the infinite shear stress usually becomes negligible when average values over the entire plate are considered. It has been found that the earlier results are in good agreement with measurements made at sufficiently high Reynolds numbers [8]. Some results for low Reynolds numbers are given in the reference [1].

The total drag force exerted on a plate of length L and width b, wetted only on one side, may be obtained by integrating the local shear stress over the plate length, that is,

$$F_D = \int_0^L \tau_w(x)b\,dx = 0.332b\mu U_\infty \sqrt{\frac{U_\infty}{v}} \int_0^L \frac{dx}{\sqrt{x}} = 0.664bU_\infty \sqrt{\rho\mu L U_\infty}. \tag{4.38}$$

An average drag or friction coefficient may be defined as

$$\bar{C}_f = \frac{F_D/(bL)}{(1/2)\rho U_\infty} = \frac{1.32824}{\sqrt{Re_L}}, \tag{4.39}$$

where $Re_L$ is the Reynolds number based on the length $L$, that is,

$$Re_L = \frac{U_\infty L}{v}. \tag{4.40}$$

The previous results appear to be valid for length Reynolds numbers greater than about $10^3$. At higher Reynolds numbers, the flow in the boundary layer becomes turbulent and the earlier theory for laminar flows does not apply. As we have already discussed, the critical Reynolds number at which turbulence appears in the boundary layer on a flat plate varies with the amount of disturbances present in the external flow. In practice, the critical Reynolds number is usually found to be in the range from $2 \times 10^5$ to $5 \times 10^5$. Somewhat higher values have also been obtained in specially designed wind tunnels.

Exact solutions similar to the one described in this section can also be obtained for a number of other constant-property laminar boundary-layer situations, including the cases where the free-stream velocity $U_\infty$ varies as a power of distance from the leading edge, that is,

$$U_\infty(x) = Cx^m. \tag{4.41}$$

This corresponds to the case of flow past a wedge (see Section 4.5). For the special case of the flat plate, $m = 0$, and therefore, $U_\infty$ is constant for all $x$.

---

## 4.3 Thermal Boundary Layer

The temperature field in the thermal boundary layer over the flat plate of Figure 4.3 can now be determined by solving Equation 4.3 under the boundary conditions (4.4b, d, and f) after inserting the velocity components found in the preceding section. Since the properties were assumed to be constant, the energy equation (4.3) is a linear differential equation. Therefore, it is possible to express the temperature distribution as a superposition of two parts. One part would represent the contribution due to viscous dissipation in the velocity

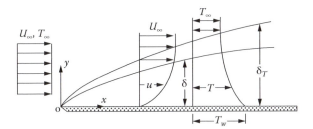

**FIGURE 4.3**
Velocity and thermal boundary layers on a flat plate.

boundary layer and the remaining part due to the heating or cooling of the plate. For low-speed flows, the temperature changes occurring due to viscous dissipation are small and will be neglected here. We shall also assume that the temperature difference between the free stream and the plate is small enough to neglect the variation of properties with temperature and that the flow is steady. Under these conditions, the boundary-layer energy Equation 4.3 reduces to

$$u\frac{\partial T}{\partial x}+v\frac{\partial T}{\partial y}=\alpha\frac{\partial^2 T}{\partial y^2},\qquad(4.42)$$

with the boundary conditions

$$\text{at } y=0: T=T_w,\qquad(4.43a)$$

$$\text{as } y\to\infty: T=T_\infty,\qquad(4.43b)$$

$$\text{at } x=0: T=T_\infty.\qquad(4.43c)$$

In terms of a dimensionless temperature defined as

$$\theta=\frac{T-T_w}{T_\infty-T_w},\qquad(4.44)$$

the energy equation (4.42) may also be written as

$$u\frac{\partial\theta}{\partial x}+v\frac{\partial\theta}{\partial y}=\alpha\frac{\partial^2\theta}{\partial y^2},\qquad(4.45)$$

and the boundary conditions become

$$\text{at } y=0: \theta=0,\qquad(4.46a)$$

$$\text{as } y\to\infty: \theta=1,\qquad(4.46b)$$

$$\text{at } x=0: \theta=1.\qquad(4.46c)$$

It may be noted that when $\theta$ is replaced by $u/U_\infty$, Equation 4.45 and the conditions (4.46) become identical to those of the velocity boundary-layer problem if $\alpha = v$, that is, if $Pr = 1$. Thus,

$$\frac{T - T_w}{T_\infty - T_w} = \frac{u}{U_\infty}, (Pr = 1).\tag{4.47}$$

The fact that the dimensionless temperature profile is identical to the dimensionless velocity profile when $Pr = 1$ suggests that similarity solutions may also exist for this problem when $Pr \neq 1$. We may then try a procedure similar to the one we have used in the previous section. Accordingly, let us assume that

$$\theta = \theta(\eta),\tag{4.48}$$

where

$$\eta = y\sqrt{\frac{U_\infty}{vx}}.\tag{4.49}$$

The validity of this assumption will depend upon whether we are successful in reducing the partial differential Equation 4.45 to an ordinary differential equation with sufficient number of boundary conditions.

Now, we can write

$$\frac{\partial \theta}{\partial x} = -\frac{1}{2}\frac{\eta}{x}\frac{d\theta}{d\eta},\tag{4.50a}$$

$$\frac{\partial \theta}{\partial y} = \sqrt{\frac{U_\infty}{vx}}\frac{d\theta}{d\eta},\tag{4.50b}$$

$$\frac{\partial^2 \theta}{\partial y^2} = \frac{U_\infty}{vx}\frac{d^2\theta}{d\eta^2}.\tag{4.50c}$$

After substituting these derivatives and the velocity components from the Blasius solution into Equation 4.45 and a rearrangement of the terms, the following ordinary differential equation is obtained:

$$\frac{d^2\theta}{d\eta^2} + \frac{1}{2}Prf\frac{d\theta}{d\eta} = 0,\tag{4.51}$$

with the boundary conditions

$$\text{at } \eta = 0 : \theta = 0,\tag{4.52a}$$

$$\text{at } \eta \to \infty : \theta = 1.\tag{4.52b}$$

The fact that we were successful in reducing the partial differential Equation 4.45 to an ordinary differential Equation 4.51, together with the boundary conditions (4.52), indicates that similarity solutions do indeed exist.

Equation 4.51 may be rewritten as

$$\frac{\theta''}{\theta'} = -\frac{1}{2} Pr f. \tag{4.53}$$

From the momentum Equation 4.25, we have

$$\frac{1}{2} f = -\frac{f'''}{f''}, \tag{4.54}$$

where primes denote differentiation with respect to $\eta$. Hence, combining these two equations, we get

$$\frac{\theta''}{\theta'} = Pr \frac{f'''}{f''}. \tag{4.55}$$

Integrating this equation twice yields

$$\theta(\eta, Pr) = C_1 \int_0^\eta \left[ f'' \right]^{Pr} d\eta + C_2. \tag{4.56}$$

From the boundary condition (4.52a), we have $C_2 = 0$, and the second boundary condition (4.52b) gives

$$C_1 = \frac{1}{\int_0^\infty \left[ f'' \right]^{Pr} d\eta}. \tag{4.57}$$

Therefore, the dimensionless temperature distribution becomes

$$\theta(\eta, Pr) = \frac{\int_0^\eta \left[ f'' \right]^{Pr} d\eta}{\int_0^\infty \left[ f'' \right]^{Pr} d\eta}. \tag{4.58}$$

With $f(\eta)$ known (at least in numerical form) from the solution of the Blasius problem, Equation 4.58 can now readily be solved numerically for any particular Prandtl number. The temperature distributions for a range of Prandtl numbers were first obtained from Equation 4.58 by Pohlhausen [4], and they are shown here in Figure 4.4.

For the special case in which $Pr = 1$, Equation 4.58 yields

$$\theta(\eta) = f'(\eta) = \frac{u}{U_\infty}. \tag{4.59}$$

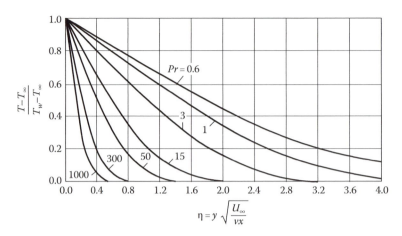

**FIGURE 4.4**
Temperature profiles in the laminar boundary on a flat plate.

Hence, this result shows, as predicted, that the dimensionless temperature and velocity distributions are identical for $Pr = 1$. Also, the thermal boundary-layer thickness and the velocity boundary-layer thickness are therefore equal.

We now can proceed to evaluate the local heat-transfer rate and the local heat-transfer coefficient. The local heat-transfer rate from the plate per unit surface area, that is, the local heat flux, is given by

$$q_w''(x) = -k\left(\frac{\partial T}{\partial y}\right)_{y=0} = -k(T_\infty - T_w)\left(\frac{d\theta}{d\eta}\right)_{\eta=0}\frac{\partial \eta}{\partial y} = k(T_w - T_\infty)\sqrt{\frac{U_\infty}{vx}}\left(\frac{d\theta}{d\eta}\right)_{\eta=0}. \qquad (4.60)$$

From Equation 4.56, we have

$$\left(\frac{d\theta}{d\eta}\right)_{\eta=0} = \frac{\left[f''(0)\right]^{Pr}}{\displaystyle\int_0^\infty \left[f''\right]^{Pr} d\eta}. \qquad (4.61a)$$

This has been evaluated numerically for a range of Prandtl numbers by Pohlhausen [4], and his results can be approximated by

$$\left(\frac{d\theta}{d\eta}\right)_{\eta=0} = \begin{cases} 0.564(Pr)^{1/2} & Pr \to 0 \\ 0.332(Pr)^{1/3} & 0.6 < Pr < 15 \\ 0.339(Pr)^{1/3} & Pr \to \infty \end{cases} \qquad (4.61b)$$

Therefore, for example, in the range $0.6 < Pr < 15$, we have

$$q_w''(x) = 0.332k(T_w - T_\infty)Pr^{1/3}\sqrt{\frac{U_\infty}{vx}}. \qquad (4.62)$$

Note that an infinitely large heat flux is predicted near the leading edge where the boundary-layer approximations are not valid.

The local heat-transfer coefficient by the use of Newton's law of cooling is defined as

$$h_x = \frac{q_w''(x)}{T_w - T_\infty}.$$ 

(4.63)

After substituting $q_w''(x)$ from Equation 4.62, we get

$$h_x = 0.332 k Pr^{1/3} \sqrt{\frac{U_\infty}{vx}}.$$ 

(4.64)

It becomes convenient, and also is a general practice, to nondimensionalize the local heat-transfer coefficients as

$$Nu_x = \frac{h_x x}{k},$$ 

(4.65)

which is called the *local Nusselt number*. Substituting Equation 4.64 into Equation 4.65 gives

$$Nu_x = 0.332 Pr^{1/3} Re_x^{1/2},$$ 

(4.66)

where $Re_x$ is the local Reynolds number.

The average heat-transfer coefficient over the entire plate of length $L$ can be obtained as

$$h = \frac{1}{L}\int_0^L h_x dx = \frac{1}{L}\int_0^L 0.332 k Pr^{1/3}\sqrt{\frac{U_\infty}{vx}}dx = 0.664 k Pr^{1/3}\sqrt{\frac{U_\infty}{vL}} = 2h_{x=L}.$$ 

(4.67)

Thus, the average heat-transfer coefficient for a plate of length $L$ is twice the local value at the end of the plate.

Defining an average Nusselt number in terms of the average heat-transfer coefficient and the plate length, we obtain

$$Nu = \frac{hL}{k} = 0.664 Pr^{1/3} Re_L^{1/2}$$ 

(4.68)

where

$$Re_L = \frac{U_\infty L}{v}.$$ 

Prandtl number is a thermophysical property and is equal to the ratio of the kinematic viscosity (i.e., momentum diffusivity) of the fluid to its thermal diffusivity:

$$Pr = \frac{\text{Momentum diffusivity}}{\text{Thermal diffusivity}} = \frac{v}{\alpha} = \frac{\mu c_p}{k}.$$ 

As seen from Figure 4.4, the ratio of the thermal boundary-layer thickness to the velocity boundary-layer thickness changes with Prandtl number. The thermal and velocity boundary layers are of the same thickness for $Pr = 1$. Denoting the thermal boundary-layer thickness by $\delta_T$ and the velocity boundary-layer thickness by $\delta$, we observe the following:

$$\text{if } Pr < 1 \quad \text{then } \delta_T > \delta,$$

$$\text{if } Pr = 1 \quad \text{then } \delta_T = \delta,$$

$$\text{if } Pr > 1 \quad \text{then } \delta > \delta_T.$$

From Figure 4.4, the ratio $\delta/\delta_T$ can also be estimated to be

$$\frac{\delta}{\delta_T} = (Pr)^{1/3}. \tag{4.69}$$

Gases such as air have Prandtl numbers close to unity ($\sim 0.7$), and the velocity and thermal boundary layers are therefore of the same thickness. Oils have large Prandtl numbers ($\sim 1000$) and exhibit much thinner thermal boundary layers as compared to the velocity boundary layer. Liquid metals have small Prandtl numbers ($\sim 0.01$) and exhibit the opposite behavior.

Since the fluid temperature varies from $T_w$ to $T_\infty$, there will be some variations in the fluid properties that were assumed, in this analysis, to be constants. For small temperature differences (say <5°C for liquids and <50°C for gases between the free-stream and the wall temperatures), these variations will not be large and can be taken into account by evaluating the properties at an average temperature, called the *film temperature* defined as

$$T_f = \frac{T_w - T_\infty}{2}. \tag{4.70}$$

The use of this reference temperature for property evaluation in gases is only recommended for low-speed ($M_\infty < 0.3$) flows.

Equation 4.63 gives an infinite local heat-transfer coefficient at the leading edge of the plate. This is again because of the fact that the boundary-layer assumptions are not valid in the vicinity of the leading edge. The previous results also do not apply when the boundary-layer flow is turbulent, which we will discuss in Chapters 7 and 11.

Other exact solutions to the boundary-layer energy equation may be found in literature [9]. These solutions, besides being important in heat-transfer applications, are even more useful in serving as a means of indicating the accuracy of various approximate methods, which are used in calculating the properties of boundary layers.

The flow and heat transfer in the entrance region of tubes and channels are also qualitatively the same as for the flat plate.

**Example 4.2**

Air at 15°C and atmospheric pressure flows over a flat plate at a velocity of 6 m/s. The plate is maintained isothermal over its entire length at a temperature of 105°C.

  a. Derive an expression for the local heat-transfer coefficient in the laminar region along the plate.
  b. Calculate the total rate of heat transfer in the first 1 m of the plate for unit depth.
  c. What is the boundary-layer thickness at a distance of 1 m from the leading edge of the plate?

**Solution**

The properties of air should be evaluated at the film temperature that is

$$T_f = \frac{15 + 105}{2} = 60°C.$$

Then, from Appendix B, the properties are

$$v = 18.90 \times 10^{-6} \, \text{m}^2/\text{s}, \quad k = 0.0285 \, \text{W}/(\text{m K}), \quad Pr = 0.709.$$

a. We shall assume that the critical Reynolds number is

$$Re_{cr} = \frac{U_\infty L_{cr}}{v} = 5 \times 10^5.$$

Hence, the critical length $L_{cr}$ over which the flow in the boundary layer is laminar is

$$L_{cr} = 5 \times 10^5 \frac{18.90 \times 10^{-6}}{6} = 1.575 \, \text{m}.$$

From Equation 4.64, the local heat-transfer coefficient is

$$h_x = 0.332 k Pr^{1/3} \sqrt{\frac{U_\infty}{vx}}$$

$$= 0.332 \times (0.0285) \times (0.709)^{1/3} \sqrt{\frac{6}{18.90 \times 10^{-6} x}}$$

$$= 4.75(x)^{-1/2} \, \text{W}/(\text{m}^2 \, \text{K}).$$

where $x$ is in m.

b. At 1 m from the leading edge, the local heat-transfer coefficient is

$$h_{x=1m} = 4.75 \, \text{W}/(\text{m}^2 \, \text{K}).$$

Hence, the average heat-transfer coefficient in the first 1 m of the plate from Equation 4.67 is given by

$$h = 2 \times 4.75 = 9.5 \, \text{W}/(\text{m}^2 \, \text{K}).$$

We now can calculate the total rate of heat transfer in the first 1 m of the plate for unit depth as

$$q = hA(T_w - T_\infty)$$

$$= (9.5) \times (1) \times (105 - 15)$$

$$= 855 \, \text{W}.$$

c. The boundary-layer thickness at a distance 1 m from the leading edge is

$$\delta = 5.0\sqrt{\frac{vx}{U_\infty}}$$

$$= 5.0\sqrt{\frac{18.90\times10^{-6}\times1}{6}}$$

$$= 8.87\times10^{-3}\,\text{m}.$$

## 4.4 Fluid Friction and Heat Transfer

In the laminar boundary layer on an isothermal flat plate, assuming constant thermophysical properties and no viscous dissipation, with uniform free-stream velocity and temperature, the local Nusselt number is given by

$$Nu_x = 0.332Pr^{1/3}Re_x^{1/2}, \tag{4.71}$$

which may be rewritten as

$$\frac{Nu_x}{PrRe_x} = 0.332Pr^{-2./3}Re_x^{-1/2}. \tag{4.72}$$

The dimensionless group on the left-hand side is called the *local Stanton number*,

$$St_x = \frac{Nu_x}{PrRe_x} = \frac{h_x}{\rho c_p U_\infty}. \tag{4.73}$$

Thus, Equation 4.72 can also be written as

$$St_x Pr^{2/3} = 0.332Re_x^{-1/2}. \tag{4.74}$$

Upon comparing Equations 4.74 and 4.37, we observe that

$$St_x Pr^{2/3} = \frac{1}{2}C_{fx}, \tag{4.75}$$

which is known as the *Colburn analogy* between the fluid friction and heat transfer for laminar flows on a flat plate. Equation 4.75 permits the calculation of the local heat-transfer coefficient when the local friction coefficient is known on a flat plate under the conditions in which no heat transfer is involved.

When $Pr = 1$, the previous analogy is referred to as the *Reynolds analogy* in literature, because Reynolds [10] developed the analogy for $Pr = 1$ before Colburn [11].

The real usefulness of analogies between fluid friction and heat transfer is not in laminar flows but in turbulent flows, where theoretical analysis is usually absent and where experimental data on friction coefficients are widely available. This will be discussed in more detail in Chapter 7.

All of the heat-transfer results presented in this chapter were obtained under the assumption that the fluid properties are not temperature dependent. The properties of most fluids, in fact, do depend on temperature and thus will vary throughout the thermal boundary layer. The effects of the property variations in external flow will be discussed in Chapter 11.

## 4.5 Flows with Pressure Gradients

In the preceding sections, similarity solutions for laminar boundary layers on a flat plate were given under the assumption that the pressure gradient is negligible. Falkner and Skan [5] discovered a similarity transformation appropriate for 2D wedge flows for which the potential flow theory (inviscid solution) gives the solution for the free-stream velocity as

$$U_\infty(x) = Cx^m,$$ (4.76)

where $C$ is a constant, the exponent m is related to the wedge angle $\beta\pi$ by

$$m = \frac{\beta}{2-\beta},$$ (4.77)

and $x$ is measured from the tip of the wedge (see Figure 4.5).

Outside the edge of the velocity boundary layer, the momentum Equation 3.13 reduces to

$$U_\infty \frac{dU_\infty}{dx} = -\frac{1}{\rho}\frac{dp}{dx}.$$ (4.78)

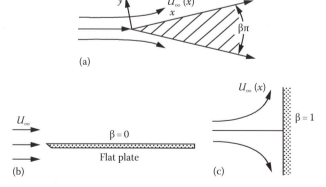

**FIGURE 4.5**
Wedge flow: (a) general configuration, (b) flat plate, and (c) 2D stagnation flow.

Then the Prandtl boundary-layer continuity and momentum equations can be written as

$$\frac{\partial u}{\partial x} + \frac{\partial v}{\partial y} = 0, \tag{4.79}$$

$$u\frac{\partial u}{\partial x} + v\frac{\partial u}{\partial y} = \frac{m}{x}U_\infty^2 + v\frac{\partial^2 u}{\partial y^2}, \tag{4.80}$$

with the boundary conditions

$$\text{at } y = 0: u = v = 0, \tag{4.81a}$$

$$\text{as } y \rightarrow \infty : u = U_\infty(x) = Cx^m. \tag{4.81b}$$

It can be shown that with the similarity variable

$$\eta = y\sqrt{\frac{U_\infty}{vx}} = y\sqrt{\frac{C}{v}}x^{(m-1)/2} \tag{4.82}$$

and the stream function defined as $\psi = \sqrt{U_\infty vx}\,f(\eta)$, so that

$$u = \frac{\partial \Psi}{\partial y} \quad \text{and} \quad v = -\frac{\partial \Psi}{\partial x}, \tag{4.83a,b}$$

the continuity Equation 4.79 is automatically satisfied, and the momentum Equation 4.80 reduces to the following nonlinear ordinary differential equation:

$$\frac{d^3f}{d\eta^3} + \frac{m+1}{2}f\frac{d^2f}{d\eta^2} + m\left[1 - \left(\frac{df}{d\eta}\right)^2\right] = 0, \tag{4.84}$$

with the boundary conditions

$$\text{at } \eta = 0: f = \frac{df}{d\eta} = 0, \tag{4.85a,b}$$

$$\text{as } \eta \rightarrow \infty : \frac{df}{d\eta} = 1. \tag{4.85c}$$

Equation 4.84 is known as the *Falkner–Skan equation* after Falkner and Skan [5] who were the first to develop and solve this equation numerically. Hartee [9,12] carried out a more detailed study of this problem and gave some accurate results. The *skin friction parameter*, $1/2C_{fx}Re_x^{1/2}$, from the calculations of Hartee [12,13] for some typical flows is given in

**TABLE 4.2**

Laminar Wedge Flow Results

| $\beta$ | $m$ | $C_{fx}Re_x^{1/2}/2$ | Case |
|---------|-----|----------------------|------|
| 1.6 | 5 | 2.6344 | |
| 1.0 | 1 | 1.2326 | 2D stagnation |
| 0.5 | 1/3 | 0.75746 | |
| 0 | 0 | 0.33206 | Flat plate |
| −0.14 | −0.06542 | 0.16372 | |
| −0.1988 | −0.09041 | 0 | Separation |

Table 4.2. Using the information given in Table 4.2, the local shear stress can be calculated from Equation 4.36a. The total drag force exerted on a wedge of length $L$ and width $b$, as well as an average drag coefficient or average coefficient of friction, can be determined for various types of wedge flows. The 2D stagnation flow ($m = 1$) and the flat plate ($m = 0$) are the two special cases of the wedge flow that commonly occur in practice.

As it is seen from Table 4.2, at $\beta = -0.1988$, flow separation occurs (velocity gradient is zero). When $\beta > 0$, the free-stream velocity increases along the wedge surface; for $\beta < 0$, the flow decelerates.

The temperature distribution as a function of $\eta = y\sqrt{U_\infty/vx}$ can be obtained from the laminar boundary-layer equation by assuming constant thermophysical properties and neglecting viscous dissipation. The boundary-layer energy Equation 4.45 under constant wall temperature boundary condition reduces to the following linear ordinary differential equation:

$$\frac{d^2\theta}{d\eta^2} + \frac{m+1}{2}Pr\,f\,\frac{d\theta}{d\eta} = 0, \tag{4.86}$$

with the boundary conditions

$$\text{at } \eta = 0 : \theta = 0, \tag{4.87a}$$

$$\text{as } \eta \to \infty : \theta = 1. \tag{4.87b}$$

The solution of Equation 4.86 satisfying the boundary conditions (4.87a,b) was first given by Eckert [14] and later by Evans [15]. Some of these heat-transfer results are presented in Table 4.3 for various values of $Pr$ and $m$.

The heat-transfer coefficient at the stagnation point in any symmetric 2D flow about a circular cylinder is given by the similarity solution for $m = 1$.

Similarity solutions also exist for flows around a body of revolution (axisymmetric flows). See Figure 4.6.

The relationship between similarity solutions for 2D boundary-layer equations and rotationally symmetric flows can be developed by the use of the Mangler coordinate transformation [16]. Details of solutions for rotationally symmetric flows can be found in [17,18]. The Mangler transformation indicates that the stagnation-point solution for the flow past a body of revolution with a blunt nose (Figure 4.6, or a sphere) can be obtained

**TABLE 4.3**

$Nu_x/Re_x^{1/2}$ for Laminar Wedge Flow

| | $Nu_x/Re_x^{1/2}$ | | | | |
|---|---|---|---|---|---|
| $m$ | $Pr = 0.7$ | 0.8 | 1.0 | 5.0 | 10.0 |
| −0.0753 | 0.242 | 2.53 | 0.272 | 0.457 | 0.570 |
| 0 | 0.292 | 0.307 | 0.332 | 0.585 | 0.730 |
| 0.111 | 0.331 | 0.348 | 0.378 | 0.669 | 0.851 |
| 0.333 | 0.384 | 0.403 | 0.440 | 0.792 | 1.013 |
| 1.0 | 0.496 | 0.523 | 0.570 | 1.043 | 1.344 |
| 4.0 | 0.813 | 0.858 | 0.938 | 1.736 | 2.236 |

*Note:* $T_w$ = const., $T_\infty$ = const., and negligible viscous dissipation.

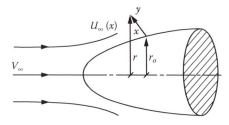

**FIGURE 4.6**
Coordinate system for boundary-layer flow over an axisymmetric body.

from the 2D Falkner–Skan solution with $m = 1/3$. The heat-transfer results for a body of revolution can be expressed as [13]

$$\frac{Nu_x}{Pr^{0.4}Re_x^{1/2}} = 0.76. \tag{4.88}$$

Near the stagnation point, $U_\infty$ used in $Re_x$ is proportional to $x$ so that $x$ cancels out of Equation 4.88. For a sphere, the limiting expression for $U_\infty(x)$ becomes $U_\infty(x) = 3/2$ $V_\infty x/R$ as $x \to 0$, where $V_\infty$ is the incoming undisturbed flow velocity and $R$ is the radius of the sphere.

The heat-transfer results at the stagnation line on a uniform-temperature circular cylinder in cross flow were obtained by a similarity solution of the boundary-layer equations for compressible flow by Cohen [19].

The heat-transfer literature in incompressible laminar external boundary-layer flows is very extensive. Interested readers are referred to the information given in References 13,20,21.

## Problems

**4.1**  Determine the value of $\alpha$ given by Equation 4.33 by a numerical procedure.

**4.2**  Plot the distribution of the velocity $u$ in the boundary layer of Example 4.1 at a distance 15 cm from the leading edge of the plate.

**4.3** In the 2D steady laminar boundary layer on a flat plate with constant free-stream velocity $U_\infty$, the boundary-layer equations are given, assuming constant thermophysical properties and negligible viscous dissipation, by

$$\frac{\partial u}{\partial x} + \frac{\partial v}{\partial y} = 0,$$

$$u\frac{\partial u}{\partial x} + v\frac{\partial u}{\partial y} = \nu\frac{\partial^2 u}{\partial y^2},$$

$$u\frac{\partial T}{\partial x} + v\frac{\partial T}{dy} = \alpha\frac{\partial^2 T}{\partial y^2},$$

where
$\nu$ is the kinematic viscosity
$\alpha$ the thermal diffusivity of the fluid

Let the free stream and the plate be isothermal at constant temperatures $T_\infty$ and $T_w$, respectively. Show that, without solving these equations, if $Pr = 1$, then

$$Nu_x = \frac{1}{2}\,C_{fx}Re_x,$$

where $Nu_x$, $C_{fx}$, and $Re_x$ are the local Nusselt number, friction coefficient, and Reynolds number, respectively.

**4.4** Consider a 2D steady laminar boundary-layer flow over a flat plate with a constant free-stream velocity $U_\infty$, constant thermophysical properties, and negligible viscous dissipation. Let the free stream and the plate be isothermal at constant temperatures $T_\infty$ and $T_w$, respectively. Show that when $Pr \gg 1$,

$$Nu_x = 0.339Pr^{1/3}Re_x^{1/2},$$

where $Nu_x$ and $Re_x$ are the local Nusselt and Reynolds numbers, respectively, at any distance $x$ from the leading edge of the plate.

**4.5** Consider the 2D laminar boundary-layer flow in Problem 4.4. Show that if $Pr \ll 1$, then

$$Nu_x = 0.564Pr^{1/2}Re_x^{1/2}.$$

**4.6** A liquid film of constant thickness $b$ flows under the influence of gravity in a steady laminar motion down an inclined solid surface as illustrated in Figure 4.7. The downcoming liquid film and the solid surface are both isothermal at temperature $T_0$, but from $x = 0$ downward, the surface is maintained at a different constant temperature $T_w$. Obtain, by a similarity solution, an expression for the local heat flux from the surface to the liquid film for those $x$ locations where the temperature of the liquid changes appreciably only in the immediate vicinity of the surface (i.e., where $\delta \ll b$). Assume constant thermophysical properties and neglect viscous dissipation. Also develop an expression for the local heat-transfer coefficient $h_x$.

**4.7** The behavior of a laminar boundary layer can be substantially influenced by either adding or removing fluid at the solid surface by injection or suction processes.

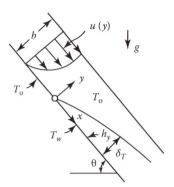

**FIGURE 4.7**
Problem 4.6.

Consider the steady flow of a constant-property fluid parallel to a porous flat plate with constant and uniform free-stream velocity $U_\infty$ and temperature $T_\infty$, such that the fluid is sucked uniformly at a velocity $v_0$ by the plate. Assume that the quantity of fluid removed from the stream is so small that only those fluid particles in the immediate neighborhood of the plate are sucked away and the no-slip condition as well as the expression $\tau_w = \mu(\partial u/\partial y)_w$ for the shear stress at the plate's surface is retained with suction present. Also let the plate be isothermal at temperature $T_w$.

   a. Assuming that the flow in the boundary layer is laminar and 2D, show that the corresponding boundary-layer equations yield solutions for the velocity components $u$ and $v$ in the $x$- and $y$-directions, respectively, and the temperature distribution $T$ that is independent of $x$ for large $x$.

   b. Using the exact solution for $u = u(y)$, develop an expression for the local friction coefficient for large $x$.

   c. Using the exact solution for $T = T(y)$, develop an expression for local heat-transfer coefficient for large $x$.

**4.8**  Obtain the similarity forms of the boundary-layer momentum and energy equations given by Equations 4.84 and 4.86.

**4.9**  Consider the steady laminar flow of a very low Prandtl number fluid (e.g., a liquid metal) over a flat plate with a constant and uniform free-stream velocity $U_\infty$ and temperature $T_\infty$. Let the heat flux from the plate to the fluid be prescribed as a given constant $q_w''$. Assuming constant thermophysical properties, obtain an exact solution for the variation of wall temperature along the surface of the plate. Also, find an expression for the local Nusselt number.

---

## References

1. Schlichting, H., *Boundary-Layer Theory*, Translated into English by J. Kestin, 7th edn., McGraw-Hill, New York, 1979.
2. Prandtl, L., Über Flüssigkeitsbewegung bei sehr kleiner Reibung, *Proceedings of the 3rd International Mathematical Congress*, Heidelberg, Germany, pp. 484–491, Teuber, Leipzig, 1904; also in English as Motion of fluids with very little viscosity, *NACA TM 452*, 1928.

3. Blasius, H., Grenzschichten in Flussigkeiten mit kleiner Reibung, *Z. Math. Phys.*, 56, 1–37, 1908; also in English as The boundary layers in fluids with little friction, *NACA TM 1256*.

4. Pohlhausen, K., Der Warmeaustausch zwichen festen Korpern und Flussigkeiten mit kleiner Reibung und kleiner Warmeleitung, *Z. Angew. Math. Mech. (ZAMM)*, 1, 115–121, 1921.

5. Falkner, W. M. and Skan, S. W., Solutions of the boundary–layer equations, *Phil. Mag.*, 12, 865–896, 1931.

6. Goldstein, S., A note on boundary layer equations, *Proc. Cambridge Phil. Soc.*, 35, 338–340, 1939.

7. Howarth, L., On the solution of the laminar boundary layer equations, *Proc. R. Soc.*, 164A, 547–579, 1938.

8. Lamb, H., *Hydrodynamics*, Dover Publications, New York, 1945.

9. Eckert, E. R. G. and Drake, R. M., Jr., *Analysis of Heat and Mass Transfer*, McGraw Hill, New York, 1972.

10. Reynolds, O., On the extent and action of the heating surface for steam boilers, *Proc. Manchester Lit. Phil. Soc.*, 14, 7, 1874.

11. Colburn, A. P., A method of correlating forced convection heat transfer data and a comparison with fluid friction, *Trans. AlChE*, 29, 174–210, 1933.

12. Hartree, D. R., On an equation occurring in Falkner and Skan's approximate treatment of the equations of the boundary layer, *Proc. Cambridge Phil. Soc.*, 33, 223–239, 1937.

13. Kakaç, S., Shah, R. K., and Aung, W. (Eds.), *Handbook of Single-Phase Convective Heat Transfer*, John Wiley, New York, Chapter 2, 1987.

14. Eckert, E., Die Berechnung des Wärmeiiberganges in der Laminaren Grenzschicht um strömter Körper, *VDl-Forschungsheft*, No. 416, Berlin, Germany, 1942.

15. Evans, H. L., Mass transfer through laminar boundary layers. Further similar solutions to the B-equation for the case B = 0, *Int. J. Heat Mass Transfer*, 5, 35–57, 1962.

16. Mangler, W., Zusammenhang zwischen ebenen und Rotationssymmetrichen Grenzchichten in Kompressiblen Flüssigkeiten, *Z. Angew. Math. Mech.*, 28, 97–103, 1948.

17. Burmeister, L. C., *Convective Heat Transfer*, 2nd edn., Wiley-Interscience Publication, New York, 1993.

18. Cebeci, T. and Bradshaw, P., *Physical and Computational Aspect of Convective Heat Transfer*, Springer-Verlag, New York, 1984.

19. Cohen, N. B., Boundary-layer similar solutions and correlation equations for laminar heat transfer distribution in equilibrium air at velocities up to 41,000ft/s, NASA Tech. Rep. R-l 18, 1961.

20. Cebeci, T., *Convective Heat Transfer*, Horizons Publ., Long Beach, CA, 2002.

21. Bejan, A., *Convection Heat Transfer*, 2nd edn., John Wiley & Sons, New York, 1995.

# 5

# *Integral Method*

## Nomenclature

| | |
|---|---|
| $b$ | liquid film thickness, m |
| $C_f$ | coefficient of friction |
| $c_p$ | specific heat at constant pressure, J/(kg K) |
| $F$ | defined by Equation 124a |
| $f_1, f_2$ | defined by Equations 5.130b and c |
| $G$ | defined by Equation 5.124b |
| $H$ | distance $> \delta, \delta_T, m$; defined by Equation 5.133 |
| $h$ | heat-transfer coefficient, W/(m$^2$ K) |
| $K$ | defined by Equation 5.129a |
| $k$ | thermal conductivity, W/(m K) |
| $m$ | exponent defined in Problem 5.24 |
| $Nu$ | Nusselt number |
| $Pr$ | Prandtl number |
| $p$ | pressure, N/m$^2$ |
| $q$ | rate of heat transfer, W |
| $q''$ | heat flux, W/m$^2$ |
| $Re$ | Reynolds number |
| $r_o$ | radius, m |
| $T$ | temperature, °C, K |
| $U$ | velocity, m/s |
| $u$ | velocity component in $x$-direction, m/s |
| $V$ | velocity, m/s |
| $v$ | velocity component in $y$-direction, m/s |
| $x$ | rectangular coordinate; distance parallel to surface, m |
| $x_{cr}$ | critical length, m |
| $x_o$ | unheated starting length, m |
| $y$ | rectangular coordinate; distance normal to surface, m |
| $z$ | defined by Equation 5.129a |

## Greek Symbols

| | |
|---|---|
| $\alpha$ | thermal diffusivity, m$^2$/s |
| $\beta$ | beta function; wedge angle |
| $\delta_1, \delta_2$ | displacement and momentum thicknesses, m |

| | |
|---|---|
| $\delta, \delta_T$ | velocity and thermal boundary-layer thicknesses, m |
| $\zeta$ | $\delta_T/\delta = \eta/\eta_T$ |
| $\eta$ | $\eta_T\, y/\delta,\, y/\delta_T$ |
| $\Gamma$ | gamma function |
| $\theta$ | dimensionless temperature; angle measured from stagnation point |
| $\lambda$ | defined by Equation 5.124c |
| $\mu$ | dynamic viscosity, Pa s |
| $\nu$ | kinematic viscosity, m$^2$/s |
| $\xi$ | dummy variable |
| $\rho$ | density, kg/m$^3$ |
| $\tau$ | shear stress, N/m$^2$ |
| $\phi$ | dimensionless temperature |

## Subscripts

| | |
|---|---|
| $L$ | local value at $L$ |
| $w$ | wall value |
| $x$ | local value at $x$ |
| $\infty$ | free-stream value |
| | over-bar average value; dimensionless quantity |

## 5.1 Introduction

The problems discussed in the previous chapter belong to a very special group where the boundary-layer equations can be solved analytically to yield exact solutions within the limitations of the underlying assumptions. Only a few other such exact solutions are available in the literature [1]. In general, solving the boundary-layer equations analytically to obtain exact solutions is very cumbersome and time consuming. Although numerical solutions to the boundary-layer equations may be obtained for almost any prescribed pressure variation and boundary conditions with the use of today's modern computers, they still require a considerable effort because the equations contain nonlinear terms. For engineering applications, it is often more convenient if approximate analytical solutions can be obtained when the boundary-layer equations cannot be solved exactly by analytical or numerical means. One of the most useful approximate methods is the *integral method*. This method was introduced by von Karman [2] and developed by Pohlhausen [3].

The boundary-layer integral equations provide the basis for the integral method. These equations can be obtained by integrating the boundary-layer equations across the boundary layer. They can also be derived directly by applying the basic principles to a control volume, which extends through the whole boundary layer. The integral method relaxes the condition that the continuity, momentum, and energy equations are to be satisfied at each point within the boundary layer and instead requires that these equations be satisfied on the average across the boundary layer.

## 5.2 Momentum Integral Equation

Consider the steady flow of an incompressible fluid with constant viscosity over a solid surface as illustrated in Figure 5.1. Let the flow in the boundary layer be laminar and 2D. The velocity components $u(x,y)$ and $v(x,y)$ will satisfy the following continuity and momentum equations:

$$\frac{\partial u}{\partial x} + \frac{\partial v}{\partial y} = 0, \tag{5.1a}$$

$$u\frac{\partial u}{\partial x} + v\frac{\partial u}{\partial y} = U_\infty \frac{dU_\infty}{dx} + v\frac{\partial^2 u}{\partial y^2}, \tag{5.1b}$$

with the boundary conditions:

$$\text{at } y = 0 : u = v = 0, \tag{5.1c}$$

$$\text{as } y \to \infty : u = U_\infty(x). \tag{5.1d}$$

As an approximation, the boundary condition (5.1d) can be replaced by

$$\text{at } y = \delta(x) : u = U_\infty(x). \tag{5.2a}$$

With this approximation, $\partial u/\partial y$ becomes zero for $y > \delta(x)$. It is also reasonable to require that the derivative of $u$ with respect to $y$ be continuous in the neighborhood of $y = \delta(x)$. Thus,

$$\text{at } y = \delta(x) : \frac{\partial u}{\partial y} = 0. \tag{5.2b}$$

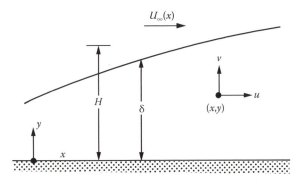

**FIGURE 5.1**
Velocity boundary layer over a solid surface.

Under these approximations, integrating the momentum Equation 5.1b at any $x$ along the surface with respect to $y$ from $y = 0$ to a constant height $y = H > \delta(x)$, we obtain

$$\int_0^H u \frac{\partial u}{\partial x} dy + \int_0^H v \frac{\partial u}{\partial y} dy = U_\infty \frac{dU_\infty}{dx} H - v \left( \frac{\partial u}{\partial y} \right)_{y=0}. \qquad (5.3)$$

First integrating it by parts, and then using the boundary conditions on $u$ and $v$ and the continuity Equation 5.1a, the second term on the left-hand side of Equation 5.3 can be written as

$$\int_0^H v \frac{\partial u}{\partial y} dy = v(x,H)U_\infty + \int_0^H u \frac{\partial u}{\partial x} dy. \qquad (5.4)$$

In addition, from Equation 4.8 we have

$$v(x,H) = -\int_0^H \frac{\partial u}{\partial x} dy. \qquad (5.5)$$

Hence, Equation 5.4 becomes

$$\int_0^H v \frac{\partial u}{\partial y} dy = -U_\infty \int_0^H \frac{\partial u}{\partial x} dy + \int_0^h u \frac{\partial u}{\partial x} dy. \qquad (5.6)$$

Substituting Equation 5.6 into Equation 5.3, we get

$$2 \int_0^H u \frac{\partial u}{\partial x} dy - U_\infty \int_0^H \frac{\partial u}{\partial x} dy = U_\infty \frac{dU_\infty}{dx} H - v \left( \frac{\partial u}{\partial y} \right)_{y=0}. \qquad (5.7)$$

Since

$$2 \int_0^H u \frac{\partial u}{\partial x} dy = \int_0^H \frac{\partial (u^2)}{\partial x} dy = \frac{d}{dx} \int_0^H u^2 dy \qquad (5.8a)$$

and

$$U_\infty \int_0^H \frac{\partial u}{\partial x} dy = \frac{d}{dx} \int_0^H U_\infty u \, dy - \frac{dU_\infty}{dx} \int_0^H u \, dy, \qquad (5.8b)$$

Equation 5.7 can be rewritten as

$$\frac{d}{dx} \int_0^H (U_\infty - u)u \, dy + \frac{dU_\infty}{dx} \int_0^H (U_\infty - u) dy = v \left( \frac{\partial u}{\partial y} \right)_{y=0}. \qquad (5.9)$$

Furthermore, because $u = U_\infty$ for $y \geq \delta$, Equation 5.9 reduces to

$$\frac{d}{dx} \int_0^\delta (U_\infty - u) u \, dy + \frac{dU_\infty}{dx} \int_0^\delta (U_\infty - u) dy = v \left( \frac{\partial u}{\partial y} \right)_{y=0}. \tag{5.10}$$

This is the *momentum integral equation* for steady 2D incompressible boundary layers. For a prescribed $U_\infty(x)$, once a velocity distribution $u$ is assumed in the boundary layer, this equation reduces to a first-order ordinary differential equation for the boundary-layer thickness $\delta$.

Before going ahead with the application of this result, it is worthwhile to derive it in another way by applying the principles of conservation of mass and momentum to a volume element in the boundary layer. For this, consider a control volume of length $\Delta x$, height $H$, and unit depth as shown in Figure 5.2. The rate of mass flow into the control volume across the control surface at $x$ is given by

$$\int_0^H \rho u \, dy. \tag{5.11}$$

The rate of mass flow out of the control volume across the control surface at $x + \Delta x$ is

$$\int_0^H \rho u \, dy + \frac{d}{dx} \left[ \int_0^H \rho u \, dy \right] \Delta x. \tag{5.12}$$

Since no mass enters or leaves the control volume at $y = 0$, the difference between the mass flow rates across the surfaces at $x + \Delta x$ and $x$ must be equal to the rate of mass flow entering the control volume across the control surface at $y = H$. Therefore, the rate of mass flow into the control volume across the surface at $y = H$ is given by

$$\frac{d}{dx} \left[ \int_0^H \rho u \, dy \right] \Delta x. \tag{5.13}$$

The law of conservation of linear momentum states that under steady-state conditions the sum of all forces acting on a control volume in any direction is equal to the net rate of

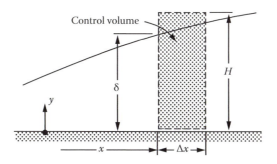

**FIGURE 5.2**
Control volume for the derivation of the momentum integral equation.

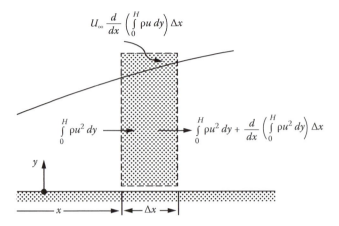

**FIGURE 5.3**
The rates of momentum flow in the $x$-direction into and out of the control volume.

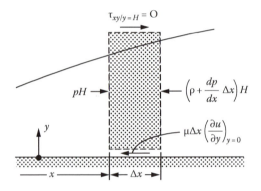

**FIGURE 5.4**
Forces acting on the surfaces of the control volume.

momentum leaving the control volume in the same direction. The net rate of momentum leaving the control volume in the $x$-direction, referring to Figure 5.3, is

$$\frac{d}{dx}\left[\int_0^H \rho u^2 \, dy\right]\Delta x - U_\infty \frac{d}{dx}\left[\int_0^H \rho u \, dy\right]\Delta x. \tag{5.14}$$

Neglecting the body forces, the forces acting on the control volume are the pressure forces and the shear force along the wall as illustrated in Figure 5.4. There is no shear force at $y = H$ (why?). Therefore, the net force acting on the control volume in the $x$-direction is

$$\frac{dp}{dx}H\Delta x - \mu\left(\frac{\partial u}{\partial y}\right)_{y=0}\Delta x. \tag{5.15}$$

Equating these forces to the net rate of momentum flow out of the control volume (5.14), we obtain

$$\frac{d}{dx}\left[\int_0^H \rho u^2 \, dy\right] - U_\infty \frac{d}{dx}\left[\int_0^H \rho u \, dy\right] = -\frac{dp}{dx}H - \mu\left(\frac{\partial u}{\partial y}\right)_{y=0}. \tag{5.16}$$

Since the flow is incompressible and

$$\frac{dp}{dx}H = -\rho U_\infty \frac{dU_\infty}{dx}H = -\rho U_\infty \frac{dU_\infty}{dx}\int_0^H dy,$$ (5.17)

Equation 5.16 may be rearranged as

$$\frac{d}{dx}\int_0^H (U_\infty - u)u\,dy + \frac{dU_\infty}{dx}\int_0^H (U_\infty - u)dy = v\left(\frac{\partial u}{\partial y}\right)_{y=0}.$$ (5.18)

Again, since $U_\infty - u = 0$ for $y \geq \delta$, Equation 5.18 reduces to Equation 5.10.

## 5.3 Energy Integral Equation

Consider the same 2D incompressible laminar boundary layer discussed in the previous section. Let the free-stream be at isothermal temperature $T_\infty$, and assume that the fluid has constant thermophysical properties. In order to derive the energy integral equation, let us consider a control volume of length $\Delta x$, height $H$ that exceeds the thermal boundary-layer thickness, and unit depth as shown in Figure 5.5. The rate of thermal energy carried into the control volume by the fluid motion across the control surface at $x$ is given by

$$\int_0^H \rho c_p Tu\,dy.$$ (5.19)

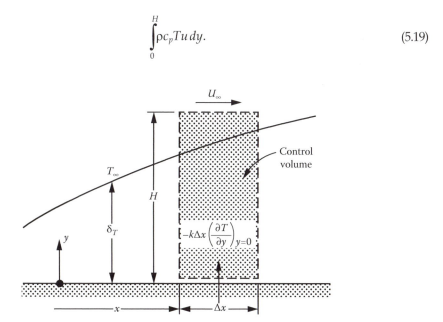

**FIGURE 5.5**
Control volume for the derivation of the energy integral equation.

The rate of thermal energy flow out of the control volume across the surface at $x + \Delta x$ is

$$\int_0^H \rho c_p T u \, dy + \frac{d}{dx}\left[\int_0^H \rho c_p T u \, dy\right]\Delta x. \tag{5.20}$$

The rate of mass entering the control volume across the upper surface of the control volume is given by Equation 5.13. Therefore, the rate of thermal energy entering the control volume with this mass is given by

$$c_p T_\infty \frac{d}{dx}\left[\int_0^H \rho u \, dy\right]\Delta x. \tag{5.21}$$

In addition, the rate of heat conducted into the control volume at the solid boundary is

$$-k\Delta x \left(\frac{\partial T}{\partial y}\right)_{y=0}. \tag{5.22}$$

Hence, neglecting both the internal energy generation due to viscous dissipation and the heat conducted into and out of the control volume across the surfaces at $x$ and $x + \Delta x$, the principle of conservation of thermal energy gives

$$-k\left(\frac{\partial T}{\partial y}\right)_{y=0} + c_p T_\infty \frac{d}{dx}\left[\int_0^H \rho u \, dy\right] = \frac{d}{dx}\left[\int_0^H \rho c_p T u \, dy\right], \tag{5.23}$$

which, since the properties are constant, can be rearranged as

$$\rho c_p \frac{d}{dx}\int_0^H (T - T_\infty)u \, dy = -k\left(\frac{\partial T}{\partial y}\right)_{y=0}. \tag{5.24}$$

Again, as an approximation, if we let $T = T_\infty$ for $y \geq \delta_T$, then Equation 5.24 reduces to

$$\rho c_p \frac{d}{dx}\int_0^{\delta_T} (T - T_\infty)u \, dy = -k\left(\frac{\partial T}{\partial y}\right)_{y=0}, \tag{5.25}$$

or

$$\frac{d}{dx}\int_0^{\delta_T} (T_\infty - T)u \, dy = \alpha\left(\frac{\partial T}{\partial y}\right)_{y=0}. \tag{5.26}$$

This is the *energy integral equation* for steady 2D laminar boundary-layer flows with constant thermophysical properties and isothermal free stream. For assumed velocity and temperature distributions in the velocity and thermal boundary layers, Equation 5.26 reduces to a first-order ordinary differential equation for the thermal boundary-layer thickness $\delta_T$.

The energy integral Equation 5.26 can also be obtained by integrating the boundary-layer energy equation with respect to $y$ from $y = 0$ to $y = H$ (see Problem 5.2).

## 5.4 Laminar Forced Flow over a Flat Plate

Consider a steady incompressible and 2D laminar boundary layer along a flat plate. Let the free-stream velocity, $U_\infty$, be constant. The momentum integral Equation 5.10 therefore simplifies to

$$\frac{d}{dx}\int_0^\delta (U_\infty - u)u\,dy = v\left(\frac{\partial u}{\partial y}\right)_{y=0}. \tag{5.27}$$

The integral method is based on representing the velocity distribution $u$ in the boundary layer by a suitable assumed profile. Once a velocity profile is assumed, the momentum integral equation reduces to a first-order ordinary differential equation for $\delta$.

Before going ahead with the application of this method, it is convenient to rewrite the momentum integral Equation 5.27 as follows:

$$\frac{d}{dx}\left[\delta\int_0^1 (1-\bar{u})\bar{u}\,d\eta\right] = \frac{v}{U_\infty\delta}\left(\frac{\partial \bar{u}}{\partial \eta}\right)_{\eta=0}, \tag{5.28}$$

where we have introduced that $\bar{u} = u/U_\infty$ and $\eta = y/\delta$. Suppose that, at any $x$ location, the velocity $u$ varies linearly across the boundary layer, that is,

$$u = a + by. \tag{5.29}$$

This profile should satisfy the following physical requirements:

$$\text{at } y = 0 : u = 0, \tag{5.30a}$$

$$\text{at } y = \delta : u = U_\infty. \tag{5.30b}$$

Hence, Equation 5.29 reduces to

$$\frac{u}{U_\infty} = \frac{y}{\delta}, \tag{5.31a}$$

or

$$\bar{u} = \eta. \tag{5.31b}$$

Substituting the linear profile (5.31b) into Equation 5.28 gives

$$\frac{d}{dx}\left[\delta\int_0^1(1-\eta)\eta\,d\eta\right]=\frac{v}{U_\infty\delta}. \tag{5.32}$$

The integral in this equation may readily be evaluated to yield

$$\int_0^1(1-\eta)\eta\,d\eta=\frac{1}{6}. \tag{5.33}$$

Therefore, Equation 5.32 reduces to the following ordinary differential equation for $\delta$:

$$\delta\frac{d\delta}{dx}=\frac{6v}{U_\infty}. \tag{5.34}$$

Integrating this equation, we obtain

$$\delta^2=\frac{12v}{U_\infty}x+C, \tag{5.35}$$

where $C$ is a constant of integration. Let the leading edge of the plate be at $x = 0$. Therefore, $C = 0$ because $\delta(0) = 0$. Thus, the boundary-layer thickness is found to be

$$\delta=\sqrt{\frac{12vx}{U_\infty}}=3.47\sqrt{\frac{vx}{U_\infty}}, \tag{5.36}$$

which can also be written in a dimensionless form as

$$\frac{\delta}{x}=\frac{3.47}{\sqrt{Re_x}}, \tag{5.37}$$

where $Re_x$ is the local Reynolds number defined as

$$Re_x=\frac{U_\infty x}{v}. \tag{5.38}$$

The local wall shear stress as computed from Equation 4.36a is given by

$$\tau_w(x)=\frac{\mu U_\infty}{\delta}=\frac{\mu U_\infty}{3.47}\sqrt{\frac{U_\infty}{vx}}=0.288\mu U_\infty\sqrt{\frac{U_\infty}{vx}}. \tag{5.39}$$

In addition, the local friction coefficient is given, from the definition (4.37), by

$$C_{f_x}=\frac{0.576}{\sqrt{Re_x}}, \tag{5.40}$$

and the average friction coefficient over a length $L$ from the leading edge, from the definition (4.39), by

$$\bar{C}_f = \frac{1.152}{\sqrt{Re_L}}, \tag{5.41}$$

where $Re_L$ is the local Reynolds number at $x = L$.

This extremely simple calculation has provided estimates of the boundary-layer thickness, wall shear stress, and friction coefficient. They have the same functional forms as the exact solutions, Equations 4.34a and b, 4.37, and 4.39, respectively, but differ from them only by a constant factor. The estimate of the boundary-layer thickness is about 31% less than the exact result, whereas the estimates of the local wall shear stress and friction coefficient (both the local and the average values) are about 13% less than the exact values.

Improvements in the earlier estimates can be made by choosing a velocity profile that more closely fits the physical conditions. In fact, the velocity profile approaches the free-stream value with a decreasing slope and is uniform outside the boundary layer. That is, the velocity profile should satisfy the following:

$$\text{at } y = 0 : u = 0, \tag{5.42a}$$

$$\text{at } y = \delta : u = U_\infty \quad \text{and} \quad \frac{\partial u}{\partial y} = 0. \tag{5.42b,c}$$

Further conditions on the velocity profile can be deduced from the boundary-layer momentum equation. Consider the boundary-layer momentum equation for zero pressure gradient:

$$u \frac{\partial u}{\partial x} + v \frac{\partial u}{\partial y} = v \frac{\partial^2 u}{\partial y^2}. \tag{5.43}$$

Since $u = v = 0$ at $y = 0$, this equation, for example, yields

$$\text{at } y = 0 : \frac{\partial^2 u}{\partial y^2} = 0, \tag{5.44}$$

which means that the curvature of the velocity profile at the wall has to be zero. Although additional conditions can be deduced from the momentum Equation 5.43, the four conditions (5.42a through c) and (5.44) are sufficient to assume a third-degree polynomial for the velocity $u$ in the boundary layer. With these four conditions, the assumed velocity profile becomes

$$\frac{u}{U_\infty} = \frac{3}{2} \frac{y}{\delta} - \frac{1}{2} \left( \frac{y}{\delta} \right)^3 \tag{5.45a}$$

or

$$\bar{u} = \frac{3}{2} \eta - \frac{1}{2} \eta^3. \tag{5.45b}$$

Substituting this result into Equation 5.28, we get

$$\frac{d}{dx}\left\{\delta\int_0^1\left(1-\frac{3}{2}\eta+\frac{1}{2}\eta^3\right)\left(\frac{3}{2}\eta-\frac{1}{2}\eta^3\right)d\eta\right\}=\frac{3}{2}\frac{v}{U_\infty\delta}. \tag{5.46}$$

After the integral in Equation 5.46 has been evaluated, it reduces to the following differential equation for $\delta$:

$$\delta\frac{d\delta}{dx}=\frac{140}{13}\frac{v}{U_\infty}. \tag{5.47}$$

Integrating this equation and utilizing the condition that $\delta = 0$ at $x = 0$, we obtain

$$\delta^2=\frac{280}{13}\frac{vx}{U_\infty}. \tag{5.48a}$$

Thus,

$$\delta=4.64\sqrt{\frac{vx}{U_\infty}}, \tag{5.48b}$$

or, in dimensionless form,

$$\frac{\delta}{x}=\frac{4.64}{\sqrt{Re_x}}. \tag{5.49}$$

Now, we can determine the local wall shear stress by employing Equation 4.36a:

$$\tau_w(x)=\frac{3\mu U_\infty}{2\delta}=0.323\mu U_\infty\sqrt{\frac{U_\infty}{vx}}. \tag{5.50}$$

The local and the average friction coefficients then become

$$C_{f_x}=\frac{0.646}{\sqrt{Re_x}}, \tag{5.51}$$

$$\bar{C}_f=\frac{1.292}{\sqrt{Re_L}}. \tag{5.52}$$

The use of a third-degree polynomial approximation for the velocity profile has resulted in an estimate of the boundary-layer thickness that is about 7% less than the exact result. The estimates of the wall shear stress and friction coefficients are about 3% less than the exact values.

## 5.5 Thermal Boundary Layer on an Isothermal Flat Plate

Consider the same 2D laminar boundary-layer flow discussed in the previous section. Let the free-stream temperature $T_\infty$ be constant, and assume that the fluid has constant thermophysical properties. Since the velocity distribution becomes independent of the temperature distribution when the properties are constant, the results of the previous section may readily be applied to the determination of the thermal boundary layer. We also consider that the flat plate has an unheated starting length that extends a distance $x_o$ from the leading edge as shown in Figure 5.6. The plate is kept isothermal at a temperature $T_w$ for $x > x_0$. We further assume that thermal energy generation due to viscous dissipation is negligible.

As in the use of the momentum integral equation, the application of the energy integral equation for an approximate solution is based on representing the temperature distribution within the thermal boundary layer by a suitable profile. Any assumed temperature profile should satisfy the following three conditions:

$$\text{at } y = 0 : T = T_w, \tag{5.53a}$$

$$\text{at } y = \delta_T : T = T_\infty \quad \text{and} \quad \frac{\partial T}{\partial y} = 0. \tag{5.53b,c}$$

Additional conditions on the temperature profile can be deduced from the energy equation. For example, the energy Equation 3.21 yields

$$\text{at } y = 0 : \frac{\partial^2 T}{\partial y^2} = 0, \tag{5.53d}$$

which means that the curvature of the temperature profile at the surface of the plate (in the absence of viscous dissipation) has to be zero. Although more conditions can be deduced from the energy Equation 3.21, we are now in a position to assume a third-degree polynomial for the temperature distribution. Under the application of the four conditions (5.53a through d), the assumed third-degree temperature profile becomes

$$\frac{T - T_w}{T_\infty - T_w} = \frac{3}{2}\frac{y}{\delta_T} - \frac{1}{2}\left(\frac{y}{\delta_T}\right)^3. \tag{5.54}$$

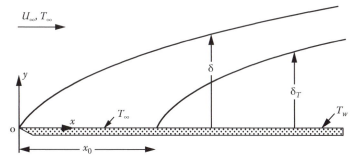

**FIGURE 5.6**
A flat plate with an unheated starting section.

The energy integral Equation 5.26 can also be written as

$$\frac{d}{dx}\left[\delta_T \int_0^1 (1-\theta)\bar{u}\,d\eta_T\right] = \frac{\alpha}{U_\infty \delta_T}\left(\frac{\partial\theta}{\partial\eta_T}\right)_{\eta_T=0}, \tag{5.55}$$

where, in addition to $\bar{u} = u/U_\infty$, we have introduced

$$\theta = \frac{T-T_w}{T_\infty - T_w} \quad \text{and} \quad \eta_T = \frac{y}{\delta_T}. \tag{5.56a,b}$$

Substituting the assumed temperature profile (5.54) and the cubic velocity distribution (5.45) into Equation 5.55, we obtain

$$\frac{d}{dx}\left[\delta_T \int_0^1 \left(1-\frac{3}{2}\eta_T + \frac{1}{2}\eta_T^3\right)\left(\frac{3}{2}\eta - \frac{1}{2}\eta^3\right)d\eta_T\right] = \frac{3}{2}\frac{\alpha}{U_\infty \delta_T}, \tag{5.57}$$

where we have assumed $\delta_T \le \delta$ and used $\eta = y/\delta$. Introducing

$$\zeta = \frac{\delta_T}{\delta} = \frac{\eta}{\eta_T} \tag{5.58}$$

and carrying out the integration in Equation 5.57, we obtain

$$\frac{d}{dx}\left[\zeta\delta_T\left(\frac{3}{20} - \frac{3}{280}\zeta^2\right)\right] = \frac{3}{2}\frac{\alpha}{U_\infty \delta_T}. \tag{5.59}$$

Since $\zeta \le 1$ (i.e., $\delta_T \le \delta$), the second term in the parenthesis in Equation 5.59 may be neglected compared to the first term. After substituting $\zeta\delta$ for $\delta_T$, Equation 5.59 reduces to

$$\frac{d}{dx}\left[\zeta^2\delta\right] = \frac{10\alpha}{U_\infty \zeta\delta}. \tag{5.60}$$

Recalling from Equation 5.49 that $\delta = 4.64\sqrt{vx/U_\infty}$, Equation 5.60 can be rewritten as

$$\zeta^3 + \frac{4}{3}x\frac{d\zeta^3}{dx} = \frac{0.929}{Pr}. \tag{5.61}$$

The solution of this equation is given by

$$\zeta^3 = Cx^{-3/4} + \frac{0.929}{Pr}, \tag{5.62}$$

where $C$ is an arbitrary constant of integration. Since $\zeta = 0$ at $x = x_0$ as illustrated in Figure 5.6, we have

$$C = -\frac{0.929}{Pr}x_0^{3/4}. \tag{5.63}$$

Hence, the solution for $\zeta$ becomes

$$\zeta = \frac{0.976}{\sqrt[3]{Pr}}\left[1-\left(\frac{x}{x_0}\right)^{-3/4}\right]^{1/3}.$$  (5.64a)

When the plate has no unheated starting section, that is, when $x_0 = 0$,

$$\zeta = \frac{0.976}{\sqrt[3]{Pr}}.$$  (5.64b)

Here we note that when $x_0 = 0$, the ratio of the boundary-layer thicknesses $\zeta = \delta_T/\delta$ becomes a constant.

The local heat flux at the surface of the plate can be determined as

$$q_w'' = -k\left(\frac{\partial T}{\partial y}\right)_{y=0} = \frac{3}{2}\frac{k}{\delta_T}(T_w - T_\infty) = \frac{3}{2}\frac{k(T_w - T_\infty)}{\zeta\delta}.$$  (5.65a)

Substituting $\delta$ and $\zeta$ from Equations 5.49 and 5.64a into Equation 5.65a, we obtain

$$q_w'' = 0.331\frac{k(T_w - T_\infty)}{x}Pr^{1/3}Re_x\left[1-\left(\frac{x}{x_0}\right)^{-3/4}\right]^{-1/3}.$$  (5.65b)

The local heat-transfer coefficient can now be obtained from

$$h_x = \frac{q_w''}{T_w - T_\infty} = 0.331\frac{k}{x}Pr^{1/3}Re_x^{1/2}\left[1-\left(\frac{x}{x_0}\right)^{-3/4}\right]^{-1/3}.$$  (5.66)

The local Nusselt number then becomes

$$Nu_x = \frac{h_x x}{k} = 0.331Pr^{1/3}Re_x^{1/3}\left[1-\left(\frac{x}{x_0}\right)^{-3/4}\right]^{-1/3}.$$  (5.67)

When the plate has no unheated starting section (i.e., when $x_0 = 0$), the earlier results reduce to

$$q_w'' = 0.331\frac{k(T_w - T_\infty)}{x}Pr^{1/3}Re_x^{1/2},$$  (5.68)

$$h_x = 0.331\frac{k}{x}Pr^{1/3}Re_x^{1/2},$$  (5.69)

$$Nu_x = 0.331Pr^{1/3}Re_x^{1/2}.$$  (5.70)

Moreover, when $x_0 = 0$, the average heat-transfer coefficient over a length $L$ from the leading edge is given, from the definition (4.63), by

$$\bar{h} = 0.662 \frac{k}{L} Pr^{1/3} Re_x^{1/2};$$  (5.71)

and the average Nusselt number becomes

$$\overline{Nu} = \frac{hL}{k} = 0.662 Pr^{1/3} Re_L^{1/2}.$$  (5.72)

The agreement between the earlier results for a plate without unheated starting section and the corresponding exact results given by Equations 4.34, 4.36, 4.37, 4.64, and 4.68 is completely satisfactory.

It should further be noted that whereas no exact solution exists for a plate with an unheated starting section, a solution by the integral method was obtained quite easily. Later on, it will be seen that this solution can be used in building up solutions for more complicated wall temperature variation problems by utilizing superposition techniques.

The earlier results apply to those cases where the thermal boundary layer is smaller than the velocity boundary layer. If the entire plate is at a uniform temperature $T_w$ so that the thermal and velocity boundary layers start to develop at the leading edge, then this implies $Pr \geq 1$. For gases, such as air, with Prandtl numbers less than but close to unity, these results are still applicable as very good approximations.

When the thermal boundary layer is thicker than the velocity boundary layer, a complication arises due to the fact that the integral in Equation 5.26 must be split up into two parts: one from $y = 0$ to $y = \delta$ and the other from $y = \delta$ to $y = \delta_T$, that is,

$$\frac{d}{dx}\left[ \int_0^\delta (T_\infty - T)u\,dy + U_\infty \int_0^{\delta_T} (T_\infty - T)\,dy \right] = \alpha \left( \frac{\partial T}{\partial y} \right)_{y=0}.$$  (5.73)

Using the cubic expressions given by Equations 5.45 and 5.54 for the velocity and temperature profiles, we can obtain the following equation for the local Nusselt number for a flat plate kept at a constant temperature over its entire length as shown in Figure 5.7:

$$Nu_x = \frac{\sqrt{Re_x Pr}}{1.547\sqrt{Pr} + 1.885\sqrt{1 - 0.404Pr}}.$$  (5.74)

If $Pr$ is very small compared to unity, then the earlier result reduces to

$$Nu_x = 0.530\sqrt{Re_x Pr}.$$  (5.75)

This result is particularly important because it applies to liquid metals that have Prandtl numbers between 0.005 and 0.05. In addition, Equation 5.75 yields an estimate of $Nu_x$ that is 6% less than the exact value.

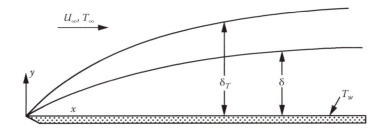

**FIGURE 5.7**
Flat plate kept at a constant temperature $T_w$ over its entire length. The thermal boundary-layer thickness $\delta_T$ is thicker than the velocity boundary-layer thickness $\delta$.

### Example 5.1

Figure 5.8 shows a flat plate, 100 cm in length and 30 cm in width, which is used as a heating element. Air at 20°C and atmospheric pressure flows over the plate with a velocity of 6 m/s. The plate has unheated sections at both ends, each 25 cm in length, and the heated section is maintained at 140°C. Calculate the total heat-transfer rate from the plate to the air.

### Solution

The properties of air should be evaluated at the film temperature, which is

$$T_f = \frac{20 + 140}{2} = 80°C.$$

Hence, from Appendix B, the properties are

$$k = 0.0299 \text{ W}/(\text{m K})$$

$$v = 20.94 \times 10^{-6} \text{ m}^2/\text{s}$$

$$Pr = 0.708.$$

At $x = 0.75$ m,

$$Re_x = \frac{U_\infty x}{v} = \frac{6 \times 0.75}{20.94 \times 10^{-6}} = 2.15 \times 10^5 < 5 \times 10^5,$$

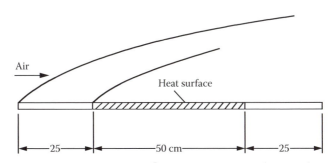

**FIGURE 5.8**
Heating element of Example 5.1.

and

$$\zeta = \frac{\delta_T}{\delta} = \frac{0.976}{\sqrt[3]{Pr}}\left[1-\left(\frac{x}{x_0}\right)^{-3/4}\right]^{1/3}$$

$$= \frac{0.976}{\sqrt[3]{0.708}}\left[1-\left(\frac{0.75}{0.25}\right)^{-3/4}\right]^{1/3} = 0.903 < 1.$$

Thus, the flow is laminar and $\delta_T < \delta$ at $x = 0.75$ m. The local heat-transfer coefficient in 0.25 m $< x <$ 0.75 m is then given, from Equation 5.66, by

$$h_x = 0.331\frac{k}{x}Pr^{1/3}Re^{1/2}\left[1-\left(\frac{x}{x_0}\right)^{-3/4}\right]^{-1/3}$$

$$= (0.331)\frac{(0.0299)}{x}(0.708)^{1/3}\sqrt{\frac{6x}{20.94\times10^{-6}}}\left[1-\left(\frac{x}{x_0}\right)^{-3/4}\right]^{-1/3}$$

$$= 4.72(x)^{-1/2}\left[1-\left(\frac{x}{x_0}\right)^{-3/4}\right]^{-1/3},$$

where $x_0 = 0.25$ m. The average heat-transfer coefficient can now be calculated as

$$\bar{h} = \frac{1}{0.5}\int_{0.25}^{0.75}h_x dx = \frac{4.72}{0.5}\int_{0.25}^{0.75}(x)^{-1/2}\left[1-\left(\frac{x}{x_0}\right)^{-3/4}\right]^{-1/3}dx.$$

Let $z = x^{3/4} - x_0^{3/4}$, then $dz = 3/4\ x^{-1/4}dx$. Therefore, the average heat-transfer coefficient becomes

$$\bar{h} = 12.59\int_0^{0.452}z^{-1/3}dz = 18.89\times(0.452)^{2/3} = 11.13\,\mathrm{W/(m^2\,K)}.$$

Hence, the total rate of heat transfer is

$$q = \bar{h}A(T_w - T_\infty)$$

$$= (11.13)(0.5\times0.3)(140-20) = 200.34\,\mathrm{W}.$$

## 5.6 Thermal Boundary Layer on a Flat Plate with Constant Surface Heat Flux

We now consider the flat plate of the previous section with a constant heat flux $q_w''$ applied at its surface for $x > x_0$. The starting point is again the energy integral Equation 5.26, which can also be written as

$$\frac{d}{dx}\int_0^{\delta_T}(T - T_\infty)u\,dy = \frac{1}{\rho c_p}q_w''. \tag{5.76}$$

A third-degree polynomial approximation for the temperature profile yields

$$T - T_\infty = \frac{q_w''\delta_T}{3k}\left[2 - 3\frac{y}{\delta_T} + \left(\frac{y}{\delta_T}\right)^3\right], \tag{5.77}$$

which satisfies the following conditions:

$$\text{at } y = 0: \frac{\partial T}{\partial y} = -\frac{q_w''}{k} \quad \text{and} \quad \frac{\partial^2 T}{\partial y^2} = 0, \tag{5.78a,b}$$

$$\text{at } y = \delta_T: T = T_\infty \quad \text{and} \quad \frac{\partial T}{\partial y} = 0. \tag{5.78c,d}$$

Here, it should be noted that the condition (5.78b) is obtained by evaluating the energy Equation 3.21 at $y = 0$. For the velocity distribution, we again assume a cubic polynomial, that is,

$$\frac{u}{U_\infty} = \frac{3}{2}\frac{y}{\delta} - \frac{1}{2}\left(\frac{y}{\delta}\right)^3. \tag{5.79}$$

Substituting the temperature profile (5.77) and the velocity profile (5.79) into the energy integral Equation 5.76, we obtain

$$\frac{d}{dx}\left[\delta_T^2\int_0^1\left(2 - 3\eta_T + \eta_T^3\right)\left(3\zeta\eta_T - \zeta^3\eta_T^3\right)d\eta_T\right] = \frac{6\alpha}{U_\infty}, \tag{5.80}$$

where we have assumed that $\delta_T \le \delta$ and introduced

$$\eta_T = \frac{y}{\delta_T} \quad \text{and} \quad \zeta = \frac{\delta_T}{\delta}. \tag{5.81a,b}$$

Carrying out the integration in Equation 5.80, we get

$$\frac{d}{dx}\left[\delta_T^2\left(\zeta - \frac{1}{14}\zeta^3\right)\right] = \frac{10\alpha}{U_\infty}. \tag{5.82}$$

Since $\zeta \leq 1$, we neglect $\zeta^3/14$ as compared to $\zeta$. Hence, Equation 5.82 reduces to

$$\frac{d}{dx}\left[\delta_T^2\zeta\right] = \frac{10\alpha}{U_\infty}, \tag{5.83a}$$

which can be rewritten as

$$\frac{d}{dx}\left[\delta^2\zeta^3\right] = \frac{10\alpha}{U_\infty}, \tag{5.83b}$$

Integrating this equation, we obtain

$$\delta^2\zeta^3 = \frac{10\alpha}{U_\infty}x + C, \tag{5.84}$$

where $C$ is a constant of integration. For a plate with an unheated starting length $x_0$, this constant is given by

$$C = -\frac{10\alpha}{U_\infty}x_0. \tag{5.85}$$

Thus,

$$\delta^2\zeta^3 = \frac{10\alpha}{U_\infty}x\left[1 - \frac{x_0}{x}\right]. \tag{5.86}$$

Since $\delta = 4.64\sqrt{vx/U_\infty}$, Equation 5.86 gives

$$\zeta = \frac{0.774}{\sqrt[3]{Pr}}\left[1 - \frac{x_0}{x}\right]^{1/3}, \tag{5.87}$$

which, when $x_0 = 0$, reduces to

$$\zeta = \frac{0.774}{\sqrt[3]{Pr}}. \tag{5.88}$$

Again, we note that when $x_0 = 0$ the ratio $\zeta = \delta_T/\delta$ becomes independent of $x$. The local surface temperature for $x > x_0$ is obtained from Equation 5.77 as

$$T_w - T_\infty = \frac{2}{3}\frac{q_w''\delta_T}{k}. \tag{5.89}$$

Since

$$\delta_T = \zeta\delta = 3.591 x Pr^{-1/3} Re_x^{-1/2}\left[1 - \frac{x_0}{x}\right]^{1/3}, \tag{5.90}$$

Equation 5.89 reduces to

$$T_w - T_\infty = 2.394\frac{q_w'' x}{k}Pr^{-1/3}Re_x^{-1/2}\left[1 - \frac{x_0}{x}\right]^{1/3}. \tag{5.91}$$

The local heat coefficient can now be obtained from its definition as

$$h_x = \frac{q_w''}{T_w - T_\infty} = 0.418\frac{k}{x}Pr^{1/3}Re_x^{1/2}\left[1 - \frac{x_0}{x}\right]^{-1/3}. \tag{5.92}$$

The local Nusselt number then becomes

$$Nu_x = \frac{h_x x}{k} = 0.418 Pr^{1/3}Re_x^{1/2}\left[1 - \frac{x_0}{x}\right]^{-1/3}. \tag{5.93}$$

When the plate has no unheated starting section (i.e., when $x_0 = 0$), the earlier results reduce to

$$h_x = 0.418\frac{k}{x}Pr^{1/3}Re_x^{1/2}, \tag{5.94}$$

$$Nu_x = 0.418 Pr^{1/3}Re_x^{1/2}, \tag{5.95}$$

which are valid when $Pr \geq 1$.

The earlier result for the local Nusselt number is about 10% less than the exact result obtained by a numerical integration of the boundary-layer energy equation [4]. It is also interesting to note that the Nusselt number for a plate with uniform heat flux exceeds the Nusselt number for an isothermal plate by about 26%.

## 5.7 Flat Plate with Varying Surface Temperature

The solutions we have considered so far are either for a constant wall temperature or for a constant wall heat flux boundary condition with and without an unheated starting length. In constant-property flows, the velocity field is independent of the temperature field, and therefore, the velocity field can be determined once and for all without regard to the heat-transfer process. Moreover, for such flows, the boundary-layer energy equation is linear in temperature. Thus, the superposition principle may be used to construct solutions to problems with arbitrarily varying wall temperatures and heat fluxes from the simple solutions we have obtained in the previous sections.

### 5.7.1 Simple Application of Superposition

Consider a steady flow of a viscous fluid with constant free-stream velocity $U_\infty$ and temperature $T_\infty$ along a flat plate with the surface temperature $T_w(x)$ as given in Figure 5.9. Let the flow in the boundary layer be laminar and 2D. Assuming constant fluid properties and negligible viscous dissipation, the boundary-layer energy equation can be written as

$$u\frac{\partial T}{\partial x}+v\frac{\partial T}{\partial y}=\alpha\frac{\partial^2 T}{\partial y^2}, \tag{5.96}$$

with the following conditions:

$$\text{at } y=0: T=\begin{cases} T_{w_1}, x<x_0, \\ T_{w_2}, x>x_0, \end{cases} \tag{5.97a}$$

$$\text{as } y\to\infty: T=T_\infty, \tag{5.97b}$$

$$\text{at } x=0: T=T_\infty. \tag{5.97c}$$

The energy Equation 5.96 and the conditions (5.97) are linear. Hence, if the solution $T(x,y)$ of the earlier problem is assumed to be of the form

$$T(x,y)-T_\infty=T_1(x,y)+T_2(x,y), \tag{5.98}$$

then the problem can be written as a superposition of the following two simple problems:

$$u\frac{\partial T_1}{\partial x}+v\frac{\partial T_1}{\partial y}=\alpha\frac{\partial^2 T_1}{\partial y^2}, \tag{5.99}$$

$$\text{at } y=0: T_1=T_{w_1}-T_\infty, \quad x>0, \tag{5.100a}$$

$$\text{as } y\to\infty: T_1=0, \tag{5.100b}$$

$$\text{at } x=0: T_1=0, \tag{5.100c}$$

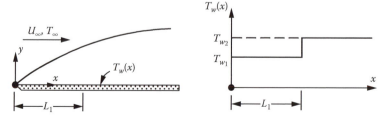

**FIGURE 5.9**
A flat plate with stepwise variation in wall temperature.

and

$$u\frac{\partial T_2}{\partial x} + v\frac{\partial T_2}{\partial y} = \alpha\frac{\partial^2 T_2}{\partial y^2},$$ (5.101)

$$\text{at } y = 0 : T_2 = \begin{cases} 0 & 0 < x < x_0, \\ T_{w_2} - T_{w_1} & x > x_0, \end{cases}$$ (5.102a)

$$\text{as } y \to \infty : T_2 = 0,$$ (5.102b)

$$\text{at } x = x_0 : T_2 = 0.$$ (5.102c)

We have already obtained the solutions to these simple problems by the integral method in this chapter. Indeed, we have also obtained an exact solution to the first simple problem in Chapter 4. Assuming third-degree polynomials for both $T_1(x,y)$ and $T_2(x,y)$, the integral method gives

$$T_1(x,y) = (T_{w_1} - T_\infty)\left[1 - \frac{3}{2}\frac{y}{\delta_{T_1}} + \frac{1}{2}\left(\frac{y}{\delta_{T_1}}\right)^3\right], \quad x > 0,$$ (5.103a)

$$T_2(x,y) = (T_{w_2} - T_{w_1})\left[1 - \frac{3}{2}\frac{y}{\delta_{T_2}} + \frac{1}{2}\left(\frac{y}{\delta_{T_2}}\right)^3\right], \quad x > x_0.$$ (5.103b)

Hence, the temperature distribution $T(x,y)$ is given by

$$T(x,y) - T_\infty = \begin{cases} T_1(x,y), & 0 < x < x_0, \\ T_1(x,y) + T_2(x,y) & x > x_0. \end{cases}$$ (5.104)

In these relations, $\delta_{T1}$ and $\delta_{T2}$ can be calculated from

$$\delta_{T_1} = \delta\zeta_1, \quad \zeta_1 = \frac{0.976}{\sqrt[3]{Pr}},$$

$$\delta_{T_2} = \delta\zeta_2, \quad \zeta_2 = \frac{0.976}{\sqrt[3]{Pr}}\left[1 - \left(\frac{x}{x_0}\right)^{-3/4}\right]^{1/3},$$

where we have assumed $\delta_{T_1} < \delta$, and the velocity boundary-layer thickness $\delta$ is given by Equation 5.49.

The local heat flux from the plate is now given, from Equation 4.60, by

$$q_w'' = 0.331\frac{k}{x}(T_{w_1} - T_\infty)Pr^{1/3}Re_x^{1/2} \quad 0 < x < x_0,$$ (5.105a)

$$q_w'' = 0.331\frac{k}{x}Pr^{1/3}Re_x^{1/2}\left\{(T_{w_1} - T_\infty) + (T_{w2} - T_{w_1})\left[1 - \left(\frac{x}{x_0}\right)^{-3/4}\right]^{-1/3}\right\}, \quad x > x_0.$$ (5.105b)

Any finite number of stepwise jumps in wall temperature can be accounted for similarly by superposing simple solutions. The same technique also applies to the calculation of wall temperatures when stepwise jumps in wall heat flux occur.

**Example 5.2**

Figure 5.10 shows the wall temperature variation over a flat plate.

a. For a uniform free-stream velocity $U_\infty$ and temperature $T_\infty$, calculate the heat flux from the plate as a function of $x$ and the properties of the flow.
b. If the fluid is air at atmospheric pressure with $U_\infty = 5$ m/s and $T_\infty = 10°C$, evaluate the heat flux at $x = 20$ cm.

**Solution**

a. The heat flux from the plate in $0 < x < 6$ cm is given by

$$q_w'' = 0.331\frac{k}{x}Pr^{1/3}Re_x^{1/2}(30 - T_\infty).$$

In $6$ cm $< x < 12$ cm,

$$q_w'' = 0.331\frac{k}{x}Pr^{1/3}Re_x^{1/2}\left\{(30 - T_\infty) + 40\left[1 - \left(\frac{\xi_1}{x}\right)^{3/4}\right]^{-1/3}\right\}.$$

In $12$ cm $< x < 18$ cm,

$$q_w'' = 0.331\frac{k}{x}Pr^{1/3}Re_x^{1/2}\left\{(30 - T_\infty) + 40\sum_{i-1}^{3}\left[1 - \left(\frac{\xi_i}{x}\right)^{3/4}\right]^{-1/3}\right\}.$$

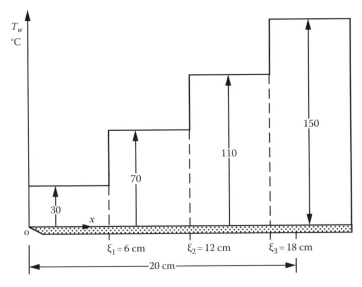

**FIGURE 5.10**
Wall temperature variation in Example 5.2. $T_w(x)$.

For $x > 18$ cm,

$$q_w'' = 0.331 \frac{k}{x} Pr^{1/3} Re_x^{1/2} \left\{ (30 - T_\infty) + 40 \sum_{i-1}^{3} \left[ 1 - \left( \frac{\xi_i}{x} \right)^{3/4} \right]^{-1/3} \right\}.$$

b. At $x = 20$ cm, the film temperature is

$$T_f = \frac{150 + 10}{2} = 80°C.$$

Then, from Appendix B, the properties of air are

$$k = 0.0299\,W/(m\,K)$$

$$v = 20.94 \times 10^{-6}\,m^2/s$$

$$Pr = 0.708.$$

At $x = 20$ cm,

$$Re_x = \frac{U_\infty x}{v} = \frac{5 \times 0.2}{20.94 \times 10^{-6}} = 4.78 \times 10^4,$$

and therefore, the flow is laminar. Substituting the numerical values into the heat flux equation for $x = 20$ cm, we get

$$q_w'' = 0.331 \times \frac{0.0299}{0.20} \times (0.708)^{1/3} \times (4.78 \times 10^4)^{1/2}$$

$$\times 40 \left\{ \frac{1}{2} + \left[ 1 - \left( \frac{6}{20} \right)^{3/4} \right]^{-1/3} + \left[ 1 - \left( \frac{12}{20} \right)^{3/4} \right]^{-1/3} + \left[ 1 - \left( \frac{18}{20} \right)^{3/4} \right]^{-1/3} \right\}$$

$$= 2126\,W/m^2.$$

## 5.7.2  Duhamel's Method

Consider the same 2D laminar boundary-layer problem we have discussed in the previous section. Let the surface temperature of the plate vary arbitrarily as illustrated in Figure 5.11. The boundary-layer energy equation and the related conditions can be written as

$$u \frac{\partial T}{\partial x} + v \frac{\partial T}{\partial y} = \alpha \frac{\partial^2 T}{\partial y^2}, \tag{5.106}$$

$$\text{at } y = 0 : T = T_w(x), \tag{5.107a}$$

$$\text{as } y \to \infty : T = T_\infty, \tag{5.107b}$$

$$\text{at } x = 0 : T = T_\infty. \tag{5.107c}$$

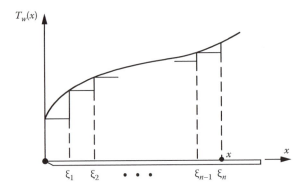

**FIGURE 5.11**
A flat plate with arbitrarily varying wall temperature.

Following the method of the previous section, it may be possible to construct a solution for $T(x,y)$ by merely breaking the surface temperature, $T_w(x)$, up into a number of constant temperature steps and then superposing the constant surface temperature solutions for each step. Hence, noting the linearity of the problem, we may write the temperature distribution $T(x,y)$ at a distance $x$ from the leading edge in the form

$$T(x,y) - T_\infty = [T_w(0) - T_\infty]\phi(0,x,y) + [T_w(\xi_1) - T_w(0)]\phi(\xi_1,x,y)$$

$$+ [T_w(\xi_2) - T_w(\xi_1)]\phi(\xi_2,x,y)$$

$$+ \cdot \quad \cdot \quad \cdot$$

$$+ [T_w(\xi_n) - T_w(\xi_{n-1})]\phi(\xi_n,x,y), \tag{5.108}$$

where $\phi(\xi,x,y)$ is the solution of the following auxiliary problem:

$$u\frac{\partial \phi}{\partial x} + v\frac{\partial \phi}{\partial y} = \alpha\frac{\partial^2 \phi}{\partial y^2}, \tag{5.109}$$

$$\text{at } y = 0 : \phi = \begin{cases} 0, & 0 < x < \xi, \\ 1 & x > \xi, \end{cases} \tag{5.110a}$$

$$\text{as } y \to \infty : \phi = 0, \tag{5.110b}$$

$$\text{at } x = \xi : \phi = 0. \tag{5.110c}$$

Introducing $\Delta T_{w,m} = T_w(\xi_m) - T_w(\xi_{m-1})$, Equation 5.108 can be written as

$$T(x,y) - T_\infty = [T_w(0) - T_\infty]\phi(0,x,y) + \sum_{m=1}^{n}\phi(\xi_m,x,y)\Delta T_{w,m}, \tag{5.111}$$

which can also be rearranged as

$$T(x,y) - T_\infty = [T_w(0) - T_\infty]\phi(0,x,y) + \sum_{m=1}^{n} \phi(\xi_m,x,y)\frac{\Delta T_{w,m}}{\Delta\xi_m}\Delta\xi_m, \tag{5.112}$$

where we have introduced $\Delta\xi_m = \xi_m - \xi_{m-1}$. Thus, in the limit as $n \to \infty$, Equation 5.112 becomes

$$T(x,y) - T_\infty = [T_w(0) - T_\infty]\phi(0,x,y) + \int_0^x \phi(\xi,x,y)\frac{dT_w}{d\xi}d\xi. \tag{5.113}$$

This result is also known as *Duhamel's superposition integral.*

The local heat flux from the plate can now be calculated to be

$$q_w'' = [T_w(0) - T_\infty]h(0,x) + \int_0^x h(\xi,x)\frac{dT_w(\xi)}{d\xi}d\xi, \tag{5.114}$$

where $h(\xi,x)$ is the local heat-transfer coefficient given by

$$h(\xi,x) = -k\frac{\partial\phi(\xi,x,0)}{\partial y}. \tag{5.115}$$

As we have already seen, with a third-degree polynomial assumption for $\phi(\xi,x,y)$, the integral method gives

$$\phi(\xi,x,y) = 1 - \frac{3}{2}\frac{y}{\delta_T} + \frac{1}{2}\left(\frac{y}{\delta_T}\right)^3, \quad x > \xi, \tag{5.116}$$

where

$$\delta_T = \delta\zeta, \quad \zeta = \frac{0.976}{\sqrt[3]{Pr}}\left[1 - \left(\frac{\xi}{x}\right)^{3/4}\right]^{1/3}, \tag{5.117a,b}$$

and $\delta$ is given by Equation 5.49. Hence, the local heat-transfer coefficient as defined by Equation 5.15 reduces to

$$h(\xi,x) = 0.331\frac{k}{x}Pr^{1/3}Re_x^{1/2}\left[1 - \left(\frac{\xi}{x}\right)^{3/4}\right]^{-1/3}. \tag{5.118}$$

Substituting Equation 5.118 into Equation 5.114 then yields

$$q_w'' = 0.331\frac{k}{x}Pr^{1/3}Re_x^{1/2}\left\{[T_w(0) - T_\infty] + \int_0^x \left[1 - \left(\frac{\xi}{x}\right)^{3/4}\right]^{-1/3}\frac{dT_w}{d\xi}d\xi\right\}. \tag{5.119}$$

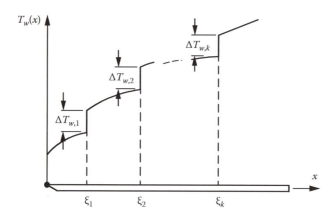

**FIGURE 5.12**
A flat plate with finite jumps in its arbitrarily varying surface temperature.

If finite jumps occur in the wall temperature simultaneously with continuous variation as shown in Figure 5.12, then it can be shown that the local heat flux is given by

$$q_w'' = [T_w(0) - T_\infty]h(0, x) + \sum_{i=1}^{k} \Delta T_{w,i} h(\xi_i, x) + \int_0^x h(\xi, x) \frac{dT_w}{d\xi} d\xi. \tag{5.120}$$

The same technique also applies to the calculation of surface temperature for the case of arbitrarily prescribed surface heat flux distribution. This is discussed in the reference [4], which also contains tabulations of some special functions useful in this work.

### Example 5.3

A flat plate has a linearly varying surface temperature described by

$$T_w(x) = a + bx,$$

where
 $a$ and $b$ are two given constants
 $x$ is the distance measured parallel to the surface from the leading edge

For a given uniform free-stream velocity $U_\infty$ and a constant free-stream temperature $T_\infty$, obtain expressions for the local heat flux from the plate and the local heat-transfer coefficient in terms of $x$ and the properties of the fluid in the laminar region.

### Solution
Since $dT_w/d\xi = b$, from Equation 5.119, we have

$$q_w''(x) = 0.331 \frac{k}{x} Pr^{1/3} Re_x^{1/2} \left\{ (a - T_\infty) + \int_0^x \left[ 1 - \left( \frac{\xi}{x} \right)^{3/4} \right]^{-1/3} b d\xi \right\}.$$

If we let

$$z = \left( \frac{\xi}{x} \right)^{3/4} \quad \text{and} \quad dz = \frac{3}{2} x^{-3/4} \xi^{-1/4} d\xi,$$

then the earlier relation for the local heat flux can be rewritten as

$$q_w''(x) = 0.331 \frac{k}{x} Pr^{1/3} Re_x^{1/2} \left\{ (a - T_\infty) + \frac{4}{3} bx \int_0^1 (1 - z)^{-1/3} z^{1/3} dz \right\},$$

or

$$q_w''(x) = 0.331 \frac{k}{x} Pr^{1/3} Re_x^{1/2} \left\{ (a - T_\infty) + \frac{4}{3} bx \beta \left( \frac{4}{3}, \frac{2}{3} \right) \right\},$$

where $\beta(p,q)$ is the *beta function* defined as

$$\beta(p,q) = \int_0^1 z^{p-1} (1 - z)^{q-1} dz \quad \text{for } p > 0, \quad q < \infty.$$

The beta function can also be written as

$$\beta(p,q) = \frac{\Gamma(p)\Gamma(q)}{\Gamma(p+q)},$$

where $\Gamma$ is the gamma junction defined as

$$\Gamma(s) = \int_0^\infty e^{-x} x^{s-1} dx.$$

The gamma function is tabulated in most of the calculus books [5], and we can write

$$\beta \left( \frac{4}{3}, \frac{2}{3} \right) = \frac{\Gamma(4/3)\Gamma(2/3)}{\Gamma(2)} = 1.2087.$$

Hence, the local heat flux becomes

$$q_w''(x) = 0.331 \frac{k}{x} Pr^{1/3} Re_x^{1/2} [(a - T_\infty) + 1.612 bx].$$

A local heat-transfer coefficient can now be obtained as

$$h_x = \frac{q_w''(x)}{T_w - T_\infty}$$

$$= \frac{0.331(k/x) Pr^{1/3} Re_x^{1/2} [(a - T_\infty) + 1.612 bx]}{a - T_\infty + bx}.$$

It is noted that for $b = 0$, the earlier result reduces to

$$h_x = 0.331 \frac{k}{x} Pr^{1/3} Re_x^{1/2},$$

which is the local heat-transfer coefficient for the constant wall temperature case.

For a more complicated wall temperature distribution, a solution can be obtained by the same method if the wall temperature can be expressed as a power series, that is,

$$T_w(x) = a + \sum_{n=1}^{N} b_n x^n.$$

Then, the local heat flux is given by

$$q_w''(x) = 0.331 \frac{k}{x} Pr^{1/3} Re_x^{1/2} \left[ (a - T_\infty) + \frac{4}{3} \sum_{n=1}^{N} n b_n x^n \beta_n \right],$$

where

$$\beta_n = \frac{\Gamma((4/3)n)\Gamma(2/3)}{\Gamma((4/3)n + (2/3))}.$$

If the wall temperature cannot be expressed as a power series, the integral in Equation 5.119 or in Equation 5.120 has to be evaluated numerically [6].

## 5.8 Flows with Pressure Gradient

The boundary-layer integral equations may also be used to treat boundary-layer flows with pressure gradient. The method to be discussed here was first introduced by Pohlhausen [3] at the suggestion of von Karman [2] and, therefore, is known as the *von Karman–Pohlhausen method*. It was later simplified by Holstein and Bohlen [7] and Walz [8].

### 5.8.1 von Karman–Pohlhausen Method

Following Pohlhausen, we assume a fourth-degree polynomial for the velocity distribution u as follows:

$$\frac{u}{U_\infty} = a + b\eta + c\eta^2 + d\eta^3 + e\eta^4, \quad 0 < \eta < 1, \tag{5.121}$$

where
$\eta = y/\delta(x)$
$u/U_\infty = 1$ for $\eta > 1$

In order to determine the five coefficients in Equation 5.121, we utilize the following five conditions:

$$\text{at } y = 0 : u = 0, \quad v\frac{\partial^2 u}{\partial y^2} = \frac{1}{\rho}\frac{dp}{dx} = -U_\infty \frac{dU_\infty}{dx}, \tag{5.122a,b}$$

$$\text{at } y = \delta : u = U_\infty(x), \quad \frac{\partial u}{\partial y} = 0, \quad \frac{\partial^2 u}{\partial y^2} = 0. \tag{5.122c,d,e}$$

The first condition (5.122a) is the no-slip condition at the wall, whereas the second condition (5.122b) is obtained from the momentum Equation 5.1b when this equation is evaluated at $y = 0$. The conditions (5.122c and d) are introduced following the approximations introduced in Section 5.2, and the condition (5.122e) is obtained from the momentum Equation 5.1b when it is written at $y = \delta$. With these five conditions, the assumed velocity profile (5.121) becomes

$$\frac{u}{U_\infty} = F(\eta) + \lambda G(\eta), \tag{5.123}$$

where

$$F(\eta) = 2\eta - 2\eta^3 + \eta^4, \tag{5.124a}$$

$$G(\eta) = \frac{1}{6}\eta(1-\eta)^3, \tag{5.124b}$$

and

$$\lambda = \frac{\delta^2}{v}\frac{dU_\infty}{dx} = -\frac{\delta^2}{\mu U_\infty}\frac{dp}{dx}. \tag{5.124c}$$

Equation 5.123 constitutes a one-parameter family of curves for the velocity profile with a parameter $\lambda$ that depends on the pressure gradient. The parameter $\lambda$ may be interpreted physically as the ratio of pressure forces to viscous forces.

The velocity profiles for various values of $\lambda$ are shown in Figure 5.13. For the values of $\lambda$ greater than 12, $u$ is seen to exceed $U_\infty$, in the boundary layer. In steady flows, this

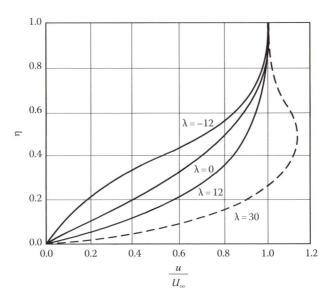

**FIGURE 5.13**
Velocity profiles for various values of $\lambda$.

is not possible physically. From Equation 5.123, the slope of the velocity profiles in the boundary layer is given by

$$\frac{\partial u}{\partial y} = \frac{U_\infty}{\delta}\left[(2-6\eta^2+4\eta^3)+\frac{\lambda}{6}(1-6\eta+9\eta^2-4\eta^3)\right],$$

which reduces at $y = 0$ to

$$\left(\frac{\partial u}{\partial y}\right)_{y=0} = \frac{U_\infty}{\delta}\left(2+\frac{\lambda}{6}\right).$$

Hence, $\lambda = -12$ corresponds to the velocity profile at separation, that is, $(\partial u/\partial y)_{y=0} = 0$, and for the values of $\lambda < -12$, the boundary-layer concept loses its significance. Therefore, the parameter $\lambda$ is restricted to the range

$$-12 < \lambda < 12.$$

For $\lambda = 0$, the profile corresponds to either a flat plate with $U_\infty$ = constant or to the case where the velocity of the potential flow passes through a minimum or a maximum. In the case of a flat plate, the profile represents a fourth-degree polynomial approximation for the Blasius solution.

We are now in a position to calculate the boundary-layer thickness from Equation 5.10 with the aid of the assumed velocity profile (5.123). Before we proceed to calculate the boundary-layer thickness from the momentum integral Equation 5.10, let us rewrite it in the following form:

$$\frac{d}{dx}\left[U_\infty^2\delta_2\right]+\delta_1 U_\infty\frac{dU_\infty}{dx}=v\left(\frac{\partial u}{\partial y}\right)_{y=0}, \tag{5.125}$$

where $\delta_1$ and $\delta_2$ are the *displacement and momentum* thicknesses, respectively, defined as

$$\delta_1 = \frac{1}{U_\infty}\int_0^\delta [U_\infty - u]\,dy \text{ (displacement thickness)}, \tag{5.126a}$$

$$\delta_2 = \frac{1}{U_\infty^2}\int_0^\delta [U_\infty - u]\,dy \text{ (momentum thickness)}. \tag{5.126b}$$

Substituting the assumed profile (5.123) into Equations 5.126a and b, we get

$$\frac{\delta_1}{\delta} = \int_0^1 [1-F(\eta)-\lambda G(\eta)]\,d\eta = \frac{1}{10}\left(3-\frac{\lambda}{12}\right), \tag{5.127a}$$

$$\frac{\delta_1}{\delta} = \int_0^1 [F(\eta)-\lambda G(\eta)][1-F(\eta)-\lambda G(\eta)]\,d\eta = \frac{1}{63}\left(\frac{37}{5}-\frac{\lambda}{15}-\frac{\lambda^2}{144}\right). \tag{5.127b}$$

Multiplying by $U_\infty \delta_2/v$ and simplifying, we can also write the momentum integral Equation 5.125 in the following dimensionless form:

$$\frac{U_\infty \delta_2}{v}\frac{d\delta_2}{dx} + \left(2 + \frac{\delta_1}{\delta_2}\right)\frac{\delta_2^2}{v}\frac{dU_\infty}{dx} = \frac{\delta_2}{U_\infty}\left(\frac{\partial u}{\partial y}\right)_{y=0}, \qquad (5.128)$$

in which the boundary-layer thickness does not appear explicitly. We now introduce the following parameters:

$$z = \frac{\delta_2^2}{v} \quad \text{and} \quad K = z\frac{dU_\infty}{dx}. \qquad (5.129a,b)$$

With the aid of Equation 5.124c and Equation 5.127, we get

$$K = \left(\frac{\delta_2}{\delta}\right)^2 \lambda = \frac{1}{3969}\left(\frac{37}{5} - \frac{\lambda}{15} - \frac{\lambda^2}{144}\right)^2 \lambda, \qquad (5.130a)$$

$$\frac{\delta_1}{\delta_2} = f_1(K) = \frac{63(3 - \lambda/12)}{10(37/5 - \lambda/15 - \lambda^2/144)}, \qquad (5.130b)$$

$$\frac{\delta_2}{U_\infty}\left(\frac{\partial u}{\partial y}\right)_{y=0} = f_2(K) = \frac{1}{63}\left(2 + \frac{\lambda}{6}\right)\left(\frac{37}{5} - \frac{\lambda}{15} - \frac{\lambda^2}{144}\right). \qquad (5.130c)$$

Substituting $z$, $K$, $f_1(K)$, and $f_2(K)$ from earlier into the momentum integral Equation 5.128, we obtain

$$\frac{1}{2}U_\infty\frac{dz}{dx} + [2 + f_1(K)]K = f_2(K), \qquad (5.131)$$

which can also be written as

$$\frac{dz}{dx} = \frac{H(K)}{U_\infty}, \qquad (5.132)$$

where we have introduced

$$H(K) = 2f_2(K) - 2[2 + f_1(K)]K. \qquad (5.133)$$

Equation 5.132 is a nonlinear differential equation of the first order for $z$. It is further to be noted that the function $H(K)$ is a universal function, that is, it can be calculated once and for all. The functions $K(\lambda), f_1(K), f_2(K)$, and $H(K)$ have been tabulated by Holstein and Bohlen [7] and can also be found in the reference [1]. An application of the earlier results to the determination of the heat-transfer coefficient at the stagnation point of a circular cylinder will be discussed in the next section.

### 5.8.2  Example: Heat Transfer at the Stagnation Point of an Isothermal Cylinder

The method discussed earlier will now be used to evaluate the heat-transfer coefficient at the stagnation point of a circular cylinder, which is a good approximation for the stagnation point of any rounded-nose 2D body. Referring to Figure 5.14, the potential-flow solution for the velocity along the surface of a circular cylinder with flow normal at a velocity $V_\infty$ is given by

$$U_\infty = 2V_\infty \sin\theta = 2V_\infty \sin\left(\frac{x}{r_0}\right), \tag{5.134}$$

where $\theta$ and $x$ are measured from the stagnation point [1]. We now assume that the boundary layer around the stagnation point is thin and the free-stream velocity in the $x$-direction just outside the boundary layer may be approximated by the velocity at the surface as computed from Equation 5.134. For small values of $x$, that is, for $x/r_0 \le 1$, Equation 5.134 may, however, be written as

$$U_\infty = 2V_\infty \frac{x}{r_0} \quad \text{for } \frac{x}{r_0} \le 1. \tag{5.135}$$

The velocity distribution in the boundary layer, on the other hand, can be approximated by Equation 5.123.

At the stagnation point $x = 0$, $U_\infty = 0$. Since $dz/dx$ cannot be infinite, $H(K)$ must be zero at the stagnation point. Hence,

$$2f_2(K) - 2[2 + f_1(K)]K = 0, \tag{5.136}$$

which reduces to

$$\lambda^3 + 147.4\lambda^2 - 1670.4\lambda + 9072 = 0. \tag{5.137}$$

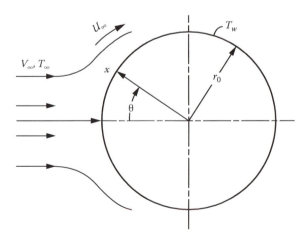

**FIGURE 5.14**
Flow around an isothermal cylinder.

There are three roots to Equation 5.137. These are

$$\lambda_1 = 7.052, \quad \lambda_2 = 17.75, \quad \text{and} \quad \lambda_3 = -70.$$

Since $\lambda$ is to be restricted to the range $-12 < \lambda < 12$, $\lambda = 7.052$ is the only acceptable root of Equation 5.137.

From Equation 5.124c, for $x \leq r_0$, we obtain

$$\delta^2 = \frac{v\lambda}{dU_\infty/dx} = \frac{v\lambda r_0}{2V_\infty}. \tag{5.138}$$

Hence, at the stagnation point $x = 0$ with $\lambda = 7.052$, we get

$$\delta^2 = 3.526 \frac{v r_0}{V_\infty}. \tag{5.139}$$

In the thermal boundary layer, we assume a third-degree polynomial to approximate the temperature profile as

$$\theta = \frac{T - T_w}{T_\infty - T_w} = A + B\eta_T + C\eta_T^2 + D\eta_T^3, \tag{5.140}$$

where $\eta_T = y/\delta_T$. The boundary conditions to be satisfied are

$$\text{at } \eta_T = 0: \theta = 0, \quad \frac{\partial^2 \theta}{\partial \eta_T^2} = 0, \tag{5.141a,b}$$

$$\text{at } \eta_T = 1: \theta = 1, \quad \frac{\partial \theta}{\partial \eta_T} = 0. \tag{5.141c,d}$$

With the application of the earlier conditions, the approximate profile for the temperature distribution becomes

$$\theta = \frac{3}{2}\eta_T - \frac{1}{2}\eta_T^3. \tag{5.142}$$

Substituting this approximation for the temperature distribution and the velocity distribution from Equation 5.123 into Equation 5.55, we obtain

$$\frac{d}{dx}\left\{ U_\infty \delta_T \int_0^1 \left[ 1 - \frac{3}{2}\eta_T^3 \right][F(\eta) + \lambda G(\eta)]d\eta_T \right\} = \frac{3}{2}\frac{\alpha}{\lambda}, \tag{5.143}$$

where we have assumed $\delta_T < \delta$. After performing the integration, Equation 5.143 becomes

$$\zeta\delta \frac{d}{dx}[U_\infty \zeta \delta(M + \lambda N)] = \frac{3}{2}\alpha, \tag{5.144}$$

where we have introduced $\zeta = \delta_T / \delta$ and

$$M = \frac{1}{5}\zeta - \frac{30}{70}\zeta^3 + \frac{1}{80}\zeta^4, \tag{5.145a}$$

$$N = \frac{1}{6}\left[ \frac{1}{10}\zeta - \frac{1}{8}\zeta^2 + \frac{9}{140}\zeta^3 - \frac{1}{80}\zeta^4 \right]. \tag{5.145b}$$

Since $\zeta < 1$, we neglect the higher-order terms in the earlier relations. Thus, for $x \le r_0$, Equation 5.144 can be rewritten as

$$\zeta \frac{d}{dx}[x\zeta^2 \sqrt{\lambda}(12 + \lambda)] = \frac{90}{Pr\sqrt{\lambda}}, \tag{5.146}$$

where we have substituted the values of $U_\infty$ and $\delta$ from Equations 5.135 and 5.138, respectively. At the stagnation point with $\lambda = 7.052$, Equation 5.146 reduces to

$$x\frac{d\zeta^3}{dx} + \frac{3}{2}\zeta^3 = \frac{1.0048}{Pr}. \tag{5.147}$$

The solution of this equation can be written as

$$\zeta^3 = Cx^{-3/2} + \frac{0.6699}{Pr}. \tag{5.148}$$

Since $\zeta$ must be finite at $x = 0$, the constant of integration $C = 0$. Thus,

$$\zeta = \frac{0.875}{\sqrt[3]{Pr}}. \tag{5.149}$$

The heat-transfer coefficient can now be determined from its definition as follows:

$$h = \frac{-k(\partial T/\partial y)_{y=0}}{T_w - T_\infty} = \frac{3}{2}\frac{k}{\zeta\delta} \tag{5.150a}$$

$$= 0.645\frac{k}{r_0}Pr^{1.3}Re^{1/2}, \tag{5.150b}$$

where $Re = 2V_\infty r_0/\nu$. The Nusselt number then becomes

$$Nu = \frac{2hr_0}{k} = 1.291\,Pr^{1/3}Re^{1/2}. \tag{5.151}$$

In the following section, an approximate relation for the heat-transfer coefficient in the stagnation region will be obtained.

### 5.8.3 Walz Approximation

Walz [8] pointed out that the function $H(K)$ in Equation 5.133 can be approximated closely by a straight line

$$H(K) = 0.47 - 6K. \tag{5.152}$$

Substituting Equation 5.152 into Equation 5.132, we get

$$U_\infty \frac{dz}{dx} = 0.47 - 6K \tag{5.153}$$

or

$$U_\infty \frac{dz}{dx} = 0.47 - 6z \frac{dU_\infty}{dx}, \tag{5.154}$$

which can be rearranged as

$$\frac{d}{dx}\left(\frac{u_\infty^6 \delta_2^2}{v}\right) = 0.47 U_\infty^5, \tag{5.155}$$

where we have used the definitions of $z$ and $K$ from Equations 5.129a,b. The solution of Equation 5.155, with the condition that $u_\infty^6 \delta_2^2 / v = 0$ at $x = 0$, gives

$$\delta_2^2 = \frac{0.47 v}{U_\infty^6(x)} \int_0^x U_\infty^5(\xi)\, d\xi. \tag{5.156}$$

Thus, the solution of Equation 5.132 reduces to a simple quadrature from which the momentum thickness can easily be found, provided that $U_\infty(x)$ is available. Once $\delta_2$ is determined, Equation 5.127b readily gives $\delta$.

As an application, let us reconsider the isothermal cylinder problem of the previous section. Substituting $U_\infty$ from Equation 5.135 into (5.156), we get

$$\delta_2^2 = \frac{0.235 v r_0}{V_\infty x^6} \int_0^x \xi^5 d\xi = 0.0391 \frac{v r_0}{V_\infty}. \tag{5.157}$$

Thus, it is found that the momentum thickness $\delta_2$ is constant in the stagnation region. Further, Equation 5.138 in the stagnation region gives

$$\delta^2 = \frac{\lambda}{2} \frac{v r_0}{V_\infty}. \tag{5.158}$$

Thus, from Equations 5.157 and 5.158, we obtain

$$\left(\frac{\delta_2}{\delta}\right)^2 = \frac{0.0782}{\lambda}. \tag{5.159}$$

Substituting the expression for $\delta_2/\delta$ from Equation 5.127b into the earlier equation, we get

$$\frac{0.2796}{\sqrt{\lambda}} = \frac{1}{63}\left(\frac{37}{5} - \frac{\lambda}{15} - \frac{\lambda^2}{144}\right). \tag{5.160}$$

The acceptable root of this equation is $\lambda = 7.25$. With this value of $\lambda$, Equation 5.146 yields

$$\zeta = \frac{0.864}{\sqrt[3]{Pr}}. \tag{5.161}$$

In addition, Equation 5.158 gives

$$\delta = \frac{2.693 r_0}{\sqrt{Re}}. \tag{5.162}$$

Thus, using Equation 5.150a, we obtain

$$h = 0.645 \frac{k}{r_0} Pr^{1/3} Re^{1/2}. \tag{5.163}$$

Thus, it is found that the Walz approximation yields a constant heat-transfer coefficient in the stagnation region, and it is equal to the previously found value at the stagnation point, that is, Equation 5.150b. More information can be found in the literatures [9–18].

## Problems

**5.1** The momentum integral Equation 5.10 was obtained in the text for a constant viscosity fluid. However, it is also applicable when the viscosity is variable. In that case, $v$ in Equation 5.10 represents the kinematic viscosity at $y = 0$, that is, at the wall temperature $T_w$. Explain.

**5.2** Obtain the energy integral Equation 5.26 by integrating the boundary-layer Equation 3.21 at any $x$ over the thermal boundary-layer thickness.

**5.3** Consider the steady laminar flow of a constant-property fluid parallel to a flat plate with a constant and uniform free-stream velocity $U_\infty$ and temperature $T_\infty$. Let the plate be isothermal at $T_w$. Assume that in the thermal boundary layer $u = U_\infty$ and the temperature distribution is given by a third-degree polynomial in $y$, where $y$ is the distance measured normal to the surface of the plate. Develop an expression for the thermal boundary-layer thickness $\delta_T$ as a function of the distance $x$ along the surface of the plate. Also, obtain an expression for the local Nusselt number $Nu_x$. This represents an approximate expression for $Nu_x$ in the laminar flow of a liquid metal over a flat surface. Why?

**5.4** Consider the steady and 2D laminar boundary-layer flow of a very low Prandtl number fluid (e.g., a liquid metal) of constant thermophysical properties with a constant and uniform free-stream velocity $U_\infty$ and temperature $T_\infty$ over a flat plate. Let the plate be isothermal at $T_w$. Assume that the temperature profile in the thermal boundary layer at any $x$ from the leading edge can be approximated as a function of the distance $y$ normal to the plate by the relation

$$T(x,y) = A + B\sin(Cy).$$

Develop (a) an expression for the thermal boundary-layer thickness $\delta_T(x)$ and (b) an expression for the local Nusselt number $Nu_x$ and compare the result with Equation 5.75.

**5.5** Water at 15°C flows over a flat plate with a velocity of 3 m/s. Estimate the mass flow rate across the boundary layer at a distance of 5 cm from the leading edge of the plate.

**5.6** Air at 20°C and atmospheric pressure flows over a flat plate with a velocity of 2 m/s. Estimate the $y$-component of velocity, v, at the outer edge of the boundary layer at 15 and 25 cm from the leading edge.

**5.7** Consider the steady and 2D laminar boundary-layer flow of a constant-property fluid over a flat plate with a constant and uniform free-stream velocity $U_\infty$ and temperature $T_\infty$. The plate is maintained isothermal at $T_w$. Assume that the velocity and temperature profiles in the respective boundary layers at any $x$ from the leading edge of the plate can be approximated as a function of the distance $y$ normal to the plate by the relations:

$$\frac{u}{U_\infty} = \frac{y}{\delta} \quad \text{and} \quad \frac{T - T_w}{T_\infty - T_w} = \frac{y}{\delta_T},$$

where $\delta$ and $\delta_T$ are the velocity and thermal boundary-layer thicknesses, respectively.

a. Develop an expression for $\delta$.

b. It is estimated that $\delta_T/\delta = 10$. Develop an expression for the local Nusselt number $Nu_x$.

c. Estimate the Prandtl number of the fluid.

**5.8** Consider a steady 2D laminar boundary-layer flow of a constant-property fluid over a flat plate with a constant and uniform free-stream velocity $U_\infty$ and temperature $T_\infty$. Let there be an unheated starting region that extends a distance $x_0$ from the leading edge (see Figure 5.6) and the plate be maintained at a constant temperature $T_w$ for $x > x_0$. Assuming linear velocity and temperature profiles in the respective boundary layers and neglecting viscous dissipation, develop an expression for the local Nusselt number $Nu_x$ applicable to such $x > x_0$ where $\delta_T < \delta$.

**5.9** When the velocity boundary layer is thinner than the thermal boundary layer, the local Nusselt number for the case of a 2D constant-property laminar boundary-layer flow over an isothermal flat plate with no unheated starting length is given by Equation 5.74. Obtain this result.

**5.10** Figure 5.15 shows a thin flat plate used as a heating element. Air at 10°C and atmospheric pressure flows over the plate with a velocity of 4 m/s. Steam temperature is 160°C and the thermal conductivity of the plate is very high. Calculate the total rate of heat transfer to the air per unit depth.

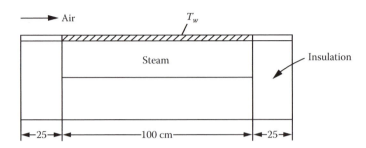

**FIGURE 5.15**
Problem 5.10.

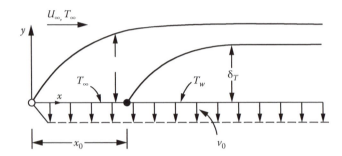

**FIGURE 5.16**
Problem 5.11.

**5.11** Consider the steady flow of a constant-property fluid parallel to a porous flat plate with a constant and uniform free stream $U_\infty$ and temperature $c$, such that the fluid is sucked uniformly at $f$ velocity $v_0$ by the plate as shown in Figure 5.16. Assume that the quantity of fluid removed from the stream is so small that only fluid particles in the immediate neighborhood of the plate are sucked away and the no-slip condition is retained with suction present.

a. Assuming that the flow in the boundary layer is laminar and 2D, develop an expression for the momentum integral equation.

b. Assuming a linear velocity profile in the boundary layer, obtain an expression for the boundary-layer thickness $\delta(x)$. Also, show that the boundary-layer thickness approaches a constant value for large $x$.

c. Develop an expression for the energy integral equation for a 2D laminar boundary-layer flow with suction.

d. Suppose that at some large $x = x_0$, where $\delta$ has become constant, a step change in wall temperature from $\delta$ to $T_w$ occurs. Assuming a linear temperature profile in the thermal boundary layer, derive the differential equation that relates the thermal boundary-layer thickness $\delta_T$ to $x$. Show that $\delta_T$ also approaches a constant value as $x$ becomes much larger than $x_0$ and also show what happens to the local heat-transfer coefficient for $x \gg x_0$.

**5.12** A liquid film of constant thickness b and temperature $T_0$ flows in a steady laminar motion down an inclined solid surface under the influence of gravity as illustrated in Figure 5.17. At $x = 0$, a step change in surface temperature from $T_0$ to $T_w$ occurs. Obtain, by the integral method, an expression for the local heat flux from

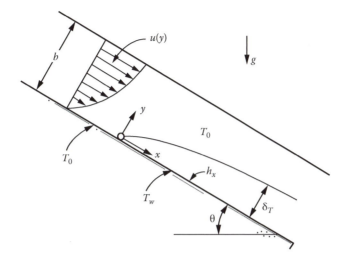

**FIGURE 5.17**
Problem 5.12.

the surface to the liquid film at those $x$ locations where $\delta_T \ll b$. Use a third-degree polynomial in $y$ for the temperature distribution in the thermal boundary layer. Assume constant thermophysical properties and neglect viscous dissipation. Also, develop an expression for the local heat-transfer coefficient $h_x$ based on the temperature difference $T_w - T_0$.

**5.13** Re-solve Problem 5.12 if the liquid film is subjected to a constant heat flux $q''_w$ for $x \geq 0$.

**5.14** Consider a 2D steady and laminar boundary-layer flow of a constant-property fluid over a flat plate with a constant and uniform free-stream velocity $U_\infty$ and temperature $T_\infty$. Let the heat flux from the plate to the fluid be prescribed as

$$q''_w = \begin{cases} 0 & \text{when } 0 < x < x_0, \\ mx & \text{when } x > x_0, \end{cases}$$

where
   $m$ is a given constant
   $x$ denotes the distance from the leading edge of the plate

Assuming linear velocity and temperature profiles in the respective boundary layers and neglecting viscous dissipation, develop an expression for the local heat-transfer coefficient applicable to those $x > x_0$ where $\delta_T < \delta$.

**5.15** Re-solve Problem 5.14 by assuming third-degree polynomials for the velocity and temperature profiles in the respective boundary layers.

**5.16** Re-solve Problem 5.14 if the wall heat flux is prescribed as

$$q''_w = \begin{cases} 0 & \text{when } 0 < x < x_0 \\ mx^a & \text{when } x > x_0 \end{cases}$$

where $m$ and $a(> 0)$ are two given constants.

**5.17** Re-solve Problem 5.9 for the case $q''_w$ = constant.

**5.18** Consider the steady laminar flow of a constant-property fluid parallel to a flat plate with a constant and uniform free-stream velocity $U_\infty$ and temperature $T_\infty$. The surface temperature of the plate varies linearly as described by

$$T_w(x) = T_\infty + Cx,$$

where
  C is a given constant
  x is the distance measured along the surface of the plate from the leading edge

Assume that in the thermal boundary layer $u = U_\infty$ and the temperature distribution is given by a third-degree polynomial in $y$, where $y$ is the distance measured normal to the surface of the plate. Develop an expression for the local Nusselt number as a function of $x$.

**5.19** Consider the steady flow of a constant-property fluid with a constant and uniform free-stream velocity $U_\infty$ over a flat plate of length $L$. The free stream is isothermal at $T_\infty$, whereas the plate temperature $T_w$ varies with the distance $x$ from the leading edge according to

$$T_w(x) = T_\infty \left( 1 + \frac{x}{L} \right).$$

Assume that the flow is laminar over the whole length of the plate, that is, $L < x_{cr}$, and the fluid Prandtl number $Pr \geq 1$. Derive an expression for the local heat flux from the plate to the fluid (a) by using Duhamel's superposition integral and (b) by decomposing the linearly increasing plate temperature into three steps as shown in Figure 5.18. Compare and comment on the results obtained by the two methods of approach.

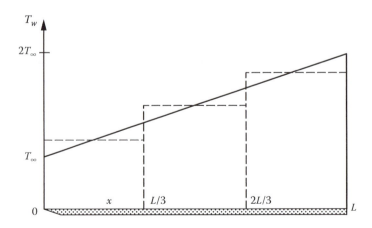

**FIGURE 5.18**
Problem 5.19.

**5.20** Consider the steady laminar flow of a constant-property fluid parallel to a flat plate of length $L$. The free-stream velocity and temperature are $U_\infty$ = constant and 0°C, respectively, while the plate temperature $T_w$ varies according to

$$T_w = T_0\left(1 - m\frac{x}{L}\right), \quad L < x_{cr},$$

where
  $T_0$ is a known temperature in °C
  $m$ is a given constant
  $x$ is the distance measured from the leading edge

Determine the value of m for which the heat transfer will be in the direction from the plate to the fluid for $0 < x < 3L/4$ and from the fluid to the plate for $3L/4 < x < L$. Plot the variations of the wall heat flux $q_w''$ and the local heat-transfer coefficient along the plate.

**5.21** Consider a flat plate with the surface heat flux specified as

$$q_w'' = \begin{cases} 20 \text{ W/m}^2 & \text{when } 0 < x \leq 15 \text{ cm}, \\ 10 \text{ W/m}^2 & \text{when } x > 15 \text{ cm}. \end{cases}$$

Air at atmospheric pressure and 20°C flows over this plate at 3 m/s. Estimate the variation of wall temperature with the distance $x$ from the leading edge in the laminar boundary-layer region, and evaluate the wall temperature numerically at 30 cm.

**5.22** Air at –20°C and atmospheric pressure flows at 10 m/s along a flat surface, which is 30 cm long in flow direction. A 5 cm wide strip of the surface, located between 5 and 10 cm from the leading edge, is uniformly heated, and the rest of the surface is maintained adiabatic. What must be the minimum heat flux from this strip so that the temperature of the surface at the trailing edge remains above 0°C? Plot the temperature distribution along the entire surface.

**5.23** Air at atmospheric pressure flows, with different velocities and at different temperatures, on both sides of a very thin flat plate as illustrated in Figure 5.19.

   a. Determine the variation of the plate's temperature as a function of $x$.

   b. Find the rate of heat transfer from the hot air to the cold air across the plate per unit depth.

**5.24** The potential-flow solution for the velocity distribution along the surface of a wedge-shaped wall is given by (see Figure 4.5)

$$U_\infty(x) = Cx^m,$$

where $C$ is a constant and the exponent $m$ is related to the wedge angle $\beta$ by

$$m = \frac{\beta}{2 - \beta}.$$

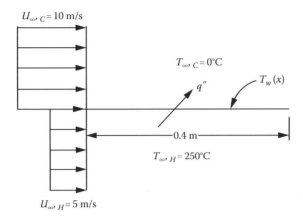

**FIGURE 5.19**
Problem 5.23.

Let the free stream and the wedge surface be isothermal at temperatures $T_\infty$ and $T_w$, respectively. Obtain, by the use of the Walz approximation, expressions for the local heat-transfer coefficient in the laminar boundary-layer region along the surfaces of the wedge for $\beta = 0$, $\beta = 1/2$, and $\beta = 1$, and discuss the results. Assume constant thermophysical properties and neglect viscous dissipation.

## References

1. Schlichting, H., *Boundary-Layer Theory*, Translated into English by J. Kestin, 7th edn., McGraw-Hill, New York, 1979.
2. von Karman, T., Über Laminare und Turbulente Reibung, *Z. Angew. Math. Mech.*, 1, 233–252, 1921.
3. Pohlhausen, K., Zur Nähemngsweise Integration der Differential Gleichung der Laminanen Reibungschict, *Z. Angew. Math. Mech.*, 1, 252–268, 1921.
4. Eckert, E. R. G., Hartnett, J. P., and Birkebak, R., The calculation of the wall temperature along surfaces which are exposed to a fluid stream when the local heat flow through the surfaces is prescribed, WADC TER 57-315, ASTIA No. 118333.
5. Hildebrand, F. B., *Advanced Calculus for Applications*, 2nd edn., Prentice Hall, Upper Saddle River, NJ, 1976.
6. Haitnett, J. P., Eckert, E. R. G., Birkebak. R., and Sampson, R. L., Simplified procedures for the calculation of heat transfer to surfaces with non-uniform temperatures, WADC Technical Report 50-373, 1957.
7. Holstein, H. and Bohlen, T., Ein einfaches Verfahren zur Berechnung Laminaren Reibungsschichten, die dem Naherung-sansatz von K. Pohlhausen genügen, *Lilienthal-Bericht*, 10, 5, 1940.
8. Walz, A., Ein neuer Ansatz für das Greschwindigkeitsprofil der laminaren Reibungsschicht, *Lilienthal-Bericht*, 141, 8, 1941.
9. Spalding, D. B., Heat transfer from surfaces of non-uniform temperature, *J. Fluid Mech.*, 4, 22, 1958.
10. Eckert, E. R. G. and Drake, R. M., Jr., *Analysis of Heat and Mass Transfer*, McGraw Hill, New York, 1972.
11. Goldstein, S., *Modern Developments in Fluid Dynamics*, Vol. 1, Dover Publications, New York, 1965.

12. Rohsenow, W. M. and Choi, H. Y., *Heat, Mass and Momentum Transfer*, Prentice Hall, Upper Saddle River, NJ, 1961.

13. Pai, S. I., *Viscous Flow Theory—Laminar Flow*, D. Van Nostrand, New York, 1956.

14. Kay, J. M., *An Introduction to Fluid Mechanics and Heat Transfer*, Cambridge University Press, New York, 1963.

15. Rohsenow, W. M. (Ed.), *Developments in Heat Transfer*, Edward Arnold, London, U.K., 1964.

16. Arpaci, V. S. and Larsen, P. S., *Convection Heat Transfer*, Prentice Hall, Upper Saddle River, NJ, 1984.

17. Burmeister, L. C., *Convective Heat Transfer*, Wiley-Interscience, New York, 1983.

18. Cebeci, T. and Bradshaw, P., *Physical and Computational Aspects of Convective Heat Transfer*, Springer, New York, 1984.

# 6

## Laminar Forced Convection in Pipes and Ducts

## Nomenclature

| | |
|---|---|
| $A$ | flow area or surface area, m² |
| $a$ | dimension of a rectangular duct, m |
| $b$ | dimension of a rectangular duct, m |
| $c_p$ | specific heat of the fluid at constant pressure, J/(kg K) |
| $d$ | diameter of the circular duct or half distance between parallel plates, m |
| $d_H$ | hydraulic diameter of the duct $= 4A/P$, m |
| $f$ | Fanning friction factor $= \tau_w/\left(\rho U_m^2/2\right)$ |
| $f_{app}$ | apparent Fanning friction factor |
| $h$ | convective heat-transfer coefficient, W/(m² K) |
| $J_i(\ )$ | Bessel function of the first kind and orders 0 or 1 corresponding to $i = 0$ or 1 |
| $K(x)$ | incremental pressure drop number defined by Equation 6.40 |
| $k$ | thermal conductivity, W/(m K) |
| $L$ | length of the duct, m |
| $L_h$ | hydrodynamic entrance length, m |
| $L_t$ | thermal entrance length, m |
| $l_o$ | modified Bessel function of the first kind and zero order |
| $m$ | an exponent |
| $\dot{m}$ | fluid mass flow rate through the duct, kg/s |
| $Nu$ | mean Nusselt number $= hd/k$ |
| $Nu_x$ | local Nusselt number $= h_x d/k$ |
| $n$ | an exponent |
| $P$ | wetted perimeter |
| $Pe$ | Péclet number $= U_m d_H/\alpha = RePr$ |
| $Pr$ | Prandtl number, $v/\alpha$ |
| $p$ | fluid static pressure, Pa |
| $\Delta p$ | fluid static pressure drop, Pa |
| $q''$ | wall heat flux, heat-transfer rate per unit area, W/m² |
| $Re$ | Reynolds number $= U_m d/v$ or $U_m d_H/v$ |
| $r$ | radial coordinate in the cylindrical coordinate system, m |
| $r_0$ | radius of the circular duct, m |
| $T$ | fluid temperature, K |
| $T_m$ | fluid bulk mean temperature, defined by Equation 6.48, K |
| $T_w$ | wall temperature at the inside duct periphery, K |
| $U_m$ | fluid mean axial velocity across the duct cross section, m/s |
| $u$ | fluid axial velocity, fluid velocity component in $x$-direction, m/s |

| | |
|---|---|
| $v$ | fluid velocity component in $y$- or $r$-direction, m/s |
| $x$ | axial (streamwise) coordinate in Cartesian or cylindrical coordinate system, m |

## Greek Symbols

| | |
|---|---|
| $\alpha$ | fluid thermal diffusivity = $k/\rho c$, m²/s |
| $\delta$ | hydrodynamic boundary-layer thickness, m |
| $\delta_T$ | thermal boundary-layer thickness, m |
| $\varepsilon$ | roughness |
| $\eta$ | dimensionless distance, $r/r_0$ or $y/\delta_T$ |
| $\theta$ | angular coordinate in the cylindrical coordinate system, rad, deg; also dimensionless fluid temperature = $(T - T_w)/(T_m - T_w)$ or $(T - T_w)/(T_i - T_w)$ |
| $\lambda_n$ | eigenvalues |
| $\mu$ | fluid dynamic viscosity, Pa s |
| $\nu$ | fluid kinematic viscosity, $\mu/\rho$, m²/s |
| $\xi$ | dimensionless distance = $(x/r_0)/Pe$ or $(4x/d)/RePr$ |
| $\rho$ | fluid density, kg/m³ |
| $\tau_w$ | wall shear stress due to skin friction, Pa |

## Subscripts

| | |
|---|---|
| $c$ | center |
| $cr$ | critical |
| $f$ | fluid |
| $fd$ | fully developed |
| $i$ | inner surface, inlet |
| $m$ | mean value |
| $max$ | maximum |
| $min$ | minimum |
| $o$ | outer surface, outer |
| $T$ | constant wall temperature boundary condition, temperature |
| $x$ | local value |
| $w$ | wall or value at the wall |
| $\infty$ | value at $x \to \infty$ |

## 6.1 Introduction

Heat and mass transfer in almost every operation of fluid transport involves flow of a viscous fluid in some form of a closed conduit. In such internal flows, a complete knowledge of the mechanism of the flow of fluids in pipes or ducts is basic to the understanding of the heat-transfer processes. In internal convection problems, where the flow is confined within a finite-size passage, all the boundary conditions on velocity, temperature, or fluxes are imposed at the solid surfaces of the passage, in contrast to external convection problems where the fluid velocity and temperature approach prescribed free-stream values at large distances from the solid surface.

When a viscous fluid flows in a duct, a velocity boundary layer develops along the inside surfaces of the duct as illustrated in Figure 6.1. Gradually the boundary layer fills

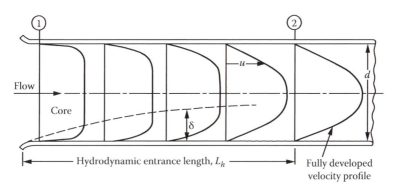

**FIGURE 6.1**
Velocity boundary-layer growth along a duct.

the entire duct. Referring to Figure 6.1, at section 1 the velocity profile is uniform across the cross section of the duct. At section 2, the velocity profile is completely developed, and from this point downward, the boundary-layer thickness δ has a constant thickness ($\delta = d/2$). The region where the velocity profile is developing is designated as the *velocity* or *hydrodynamic entrance region*. The distance $L_h$ is called the *hydrodynamic entrance length*. The region beyond the entrance region (section 2) is referred to as the *hydrodynamically fully developed region*. In this region, the boundary layer completely fills the duct and the velocity profile becomes invariant with the axial coordinate along the duct. Theoretically, however, the approach to the fully developed velocity profile is asymptotic, and it is, therefore, impossible to describe a definite location where the boundary layer fills the entire duct.

It is possible that transition to turbulence may occur before the boundary layer fills the duct. If such a transition occurs, then the flow in the fully developed region remains turbulent. Otherwise it is laminar. This chapter deals only with laminar forced convection heat transfer in ducts.

The flow is referred to as laminar when the velocities are free of macroscopic fluctuations at any point in the flow field. Laminar flow is also referred to as streamlined flow. In a fully developed laminar flow in a constant cross-sectional duct, the fluid particles move along definite paths called streamlines parallel to the surfaces, and therefore, there are no transverse components of the fluid velocity across the duct.

An approximate estimate of the magnitude of the hydrodynamic entrance length in laminar pipe flows is given by the following relation [1,2]:

$$\frac{L_h}{d} = 0.056\,Re. \tag{6.1}$$

In Equation 6.1, *Re* is the Reynolds number defined in terms of the pipe diameter *d* as

$$Re = \frac{U_m d}{\nu}, \tag{6.2}$$

where
$U_m$ is the mean flow velocity in the pipe
$\nu$ is the kinematic viscosity of the fluid

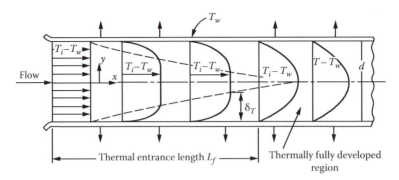

**FIGURE 6.2**
Thermal boundary-layer growth along a duct.

For turbulent flows in smooth pipes, on the other hand, the hydrodynamic entrance length may be estimated from [3]

$$\frac{L_h}{d} = 1.359\, Re^{1/4}. \tag{6.3}$$

If the walls of the duct are heated (or cooled), then a thermal boundary layer will also develop along the inner surfaces of the duct as illustrated in Figure 6.2. At a certain location downward from the inlet, we can talk about fully developed thermal conditions. The thermal entrance length $L_t$ is defined, somewhat arbitrarily, as the duct length required to achieve a value of local Nusselt number equal to 1.05 times the Nusselt number for fully developed flow, when the entering fluid temperature profile is uniform. The thermal entrance length for laminar flow in circular pipes varies with the Reynolds and Prandtl numbers as well as with the type of the boundary condition imposed on the pipe wall and is given approximately by the following relation [1]:

$$\frac{L_t}{d} \cong 0.050\, Re\, Pr. \tag{6.4}$$

For a more accurate discussion on thermal entrance length in ducts under various laminar flow conditions (i.e., hydrodynamically fully developed or developing) and boundary conditions (i.e., constant wall temperature or heat flux), the reader is referred to Reference 2. The thermal entrance length for turbulent flows in smooth pipes in general would be

$$\frac{L_t}{d} < 30. \tag{6.5}$$

Both hydrodynamically and thermally fully developed flow in a duct is defined as *fully developed flow*. The region where the velocity profile is developed and the temperature profile is developing is referred to as the *hydrodynamically developed and thermally developing region* or, simply, *thermally developing region*.

If heating (or cooling) starts from the inlet of the duct, then both velocity and temperature profiles develop simultaneously. The associated heat-transfer problem is then referred to as the *simultaneously developing flow* (i.e., combined hydrodynamic and thermal entrance region) problem. In this case, both heat and momentum diffuse simultaneously from the duct wall, starting at $x = 0$. Depending on the Prandtl number, *Pr*, the two effects diffuse at different rates as explained in Section 4.3. When $Pr = 1$, the viscous and thermal effects diffuse through the fluid at the same rate in external flows. It should be noted, however, that the applicable momentum and energy equations in internal flows do not become analogous when $Pr = 1$, and therefore, the diffusion of heat and momentum will not result in $\delta = \delta_T$, even though the boundary conditions for the momentum and heat-transfer problems could be similar.

## 6.2 Laminar and Turbulent Flows in Ducts

It had been observed that under some conditions the flow in a pipe contains, in the words of Osborne Reynolds [4], "… a mass of eddies, that is, in motion." In this type of flow, containing seemingly random fluctuations, the pressure drop was observed to be nearly proportional to the square of the mean velocity. It was his investigation of this pressure drop–velocity relation that led Reynolds to the investigation of the conditions for transition from laminar to turbulent flow.

Laminar flow occurs in practice only as long as the Reynolds number, $Re$, is less than a critical value, the so-called critical Reynolds number $Re_{cr}$. By injecting a dye at the centerline of a glass tube with a smooth trumpet-shaped entrance submerged in a calming tank of water as illustrated in Figure 6.3, Reynolds obtained experimentally in 1883 the value of the critical Reynolds number. At low velocities, when the water flow is laminar, the dye moves along a straight line. But at higher velocities, the dye begins to mix with the surrounding water and the entire volume of the tube soon becomes colored with dye. Such experiments with different velocities, diameters, temperatures (hence different viscosities), and fluids (hence different densities and viscosities) led to the establishment of the critical Reynolds number as

$$Re_{cr} = \frac{U_m d}{v} \cong 2300. \tag{6.6}$$

The value of the critical Reynolds number strongly depends on the duct inlet conditions and surface roughness, as well as on the noise and vibration imposed on the exterior of the

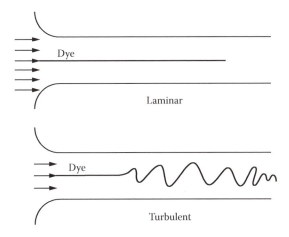

**FIGURE 6.3**
Reynolds dye experiment.

duct wall. In some cases, critical Reynolds numbers as high as 10,000 have been observed, but values around 2,300 are the more common.

The magnitude of $Re_{cr}$ is not 2300 for noncircular ducts but is somewhat dependent upon the geometry of the duct cross section. The values of $Re_{cr}$ can be found in Reference 5 for various duct geometries. But, for practical applications, the relation (6.6) can be used to obtain $Re_{cr}$ for noncircular ducts provided that the diameter $d$ is replaced by the hydraulic diameter $d_H$ as defined later in this chapter.

At Reynolds numbers beyond the critical value, the laminar flow becomes unstable in the presence of small disturbances. The fluid particles do not travel in a well-ordered manner, and the streamline structure disappears. In turbulent flow, the fluid particles have velocities with macroscopic fluctuations at any point in the flow field, and the transport of momentum and heat to the main flow direction is greatly enhanced.

## 6.3 Some Exact Solutions of Navier–Stokes Equations

The continuity and the Navier–Stokes equations derived in Chapter 2 provide a comprehensive description of the flow of viscous fluids. These equations, however, are of limited use to the engineer because of insurmountable mathematical difficulties encountered in finding solutions. This is primarily a consequence of their being nonlinear partial differential equations. Although there exist some exact solutions to these equations in the literature, the number is limited to situations of mathematical interest rather than of engineering importance [6]. In this section, we discuss two exact solutions to illustrate the applications of these equations.

### 6.3.1 Flow between Two Parallel Walls

The simplest case of engineering interest is that of a steady and fully developed laminar flow of a viscous fluid with constant thermophysical properties between two parallel walls. The problem is illustrated diagrammatically in Figure 6.4. The continuity and the Navier–Stokes equations from Equations 2.14 and 2.33 in 2D flow, in the absence of body forces, reduce to (see Problem 6.1)

$$\frac{\partial u}{\partial x} + \frac{\partial v}{\partial y} = 0, \tag{6.7}$$

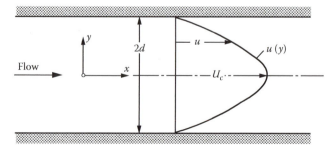

**FIGURE 6.4**
Steady and fully developed laminar flow between two parallel plates.

$$u \frac{\partial u}{\partial x} + v \frac{\partial u}{\partial y} = -\frac{1}{\rho} \frac{\partial p}{\partial x} + v \left( \frac{\partial^2 u}{\partial x^2} + \frac{\partial^2 u}{\partial y^2} \right), \tag{6.8a}$$

$$u \frac{\partial v}{\partial x} + v \frac{\partial v}{\partial y} = -\frac{1}{\rho} \frac{\partial p}{\partial y} + v \left( \frac{\partial^2 v}{\partial x^2} + \frac{\partial^2 v}{\partial y^2} \right). \tag{6.8b}$$

For steady fully developed flow, $\partial u / \partial x = 0$ or $u = u(y)$ only, and the transverse velocity component must be zero everywhere. Therefore, the continuity equation (6.7) is identically satisfied and the Navier–Stokes equations (6.8a,b) reduce to

$$-\frac{1}{\rho} \frac{\partial p}{\partial x} + v \frac{d^2 u}{dy^2} = 0, \tag{6.9}$$

$$-\frac{1}{\rho} \frac{\partial p}{\partial x} = 0. \tag{6.10}$$

From Equation 6.10, we conclude that pressure $p$ must be constant across any cross section perpendicular to the flow, that is, $p = p(x)$ only. Hence, Equation 6.9 can be rewritten as

$$\frac{d^2 u}{dy^2} = \frac{1}{\mu} \frac{dp}{dx}, \tag{6.11}$$

and therefore, since $u$ is a function of $y$ only and the viscosity $\mu$ is constant, the pressure gradient in the flow direction must be constant, that is,

$$\frac{dp}{dx} = \text{constant} = -\frac{\Delta p}{L}, \tag{6.12}$$

where $\Delta p$ represents the pressure drop over a length $L$ of the channel. Hence, Equation 6.11 becomes

$$\frac{d^2 u}{dy^2} = \frac{\Delta p}{\mu L}, \tag{6.13}$$

with the boundary conditions: $u = 0$ at $y = \mp d$. Integrating Equation 6.13 over $y$ twice and making use of the boundary conditions at $y = \mp d$, we obtain

$$u = \frac{\Delta p}{2\mu L} (d^2 - y^2). \tag{6.14}$$

The resulting velocity profile is thus parabolic as illustrated in Figure 6.4. Taking a unit width in the $z$-direction, the total volume flow rate between the walls is given by

$$\dot{Q} = \int_{-d}^{d} u \, dy = \frac{2}{3} = \frac{\Delta p}{\mu L} d^3. \tag{6.15}$$

The mean flow velocity in the channel can be obtained as

$$U_m = \frac{\dot{Q}}{2d} = \frac{\Delta p}{3\mu L} d^2 = \frac{2}{3} U_c, \tag{6.16}$$

where $U_c = (\Delta p d^2)/(2\mu L)$ is the maximum velocity in the channel, that is, the velocity at $y = 0$. Then, the velocity distribution (6.14) becomes

$$u = U_c \left[ 1 - \left( \frac{y}{d} \right)^2 \right], \tag{6.17a}$$

$$u = \frac{3}{2} U_m \left[ 1 - \left( \frac{y}{d} \right)^2 \right]. \tag{6.17b}$$

### 6.3.2 Flow in a Circular Pipe

The equation of motion for a steady and fully developed laminar flow of a viscous fluid in a circular pipe, from the Navier–Stokes equations in cylindrical coordinates, Equation 2.36, can be written as

$$-\frac{1}{\rho}\frac{dp}{dx} + v\left[ \frac{1}{r}\frac{d}{dr}\left( r\frac{du}{dr} \right) \right] = 0. \tag{6.18}$$

The problem is illustrated diagrammatically in Figure 6.5. Following the same procedure as in the previous case, one can obtain the desired velocity distribution in the pipe by integrating Equation 6.18. But it will be more instructive to obtain the velocity distribution in this case by the application of the momentum theorem (1.17) in the x-direction to the stationary elemental control volume shown in Figure 6.5. Since the flow is steady and fully developed, the momentum will not be stored in the control volume and there will be no net change in the rate of the momentum flow in the x-direction across the surfaces of the control volume. Hence, the net force acting on the control volume must be zero.

In a circular pipe, far enough downstream from its entrance, the velocity distribution does not change in the flow direction, and the pressure at any cross section is uniform

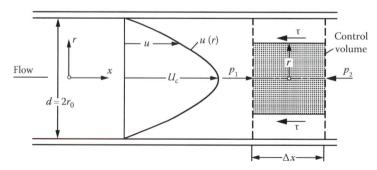

**FIGURE 6.5**
Steady and hydrodynamically fully developed flow in a circular duct.

and decreases only in the flow direction. A force balance on the volume element equates the pressure drop force $(p_1 - p_2)\pi r^2$ to the flow-retarding shear force $2\pi r \Delta x \tau$, where $\tau$ is the shear stress at radius $r$. Thus, we have

$$\Delta p \pi r^2 = 2\pi r \Delta x \tau, \tag{6.19a}$$

where

$$\Delta p = p_1 - p_2. \tag{6.19b}$$

From Equation 6.19a, we get

$$\tau = \frac{r}{2} \frac{\Delta p}{\Delta x}. \tag{6.20}$$

The shear stress is also related to the velocity gradient through Newton's law of viscosity (2.1) by

$$\tau = -\mu \frac{du}{dr}, \tag{6.21}$$

where $\mu$ is the coefficient of viscosity (dynamic viscosity). In Equation 6.21, we have introduced the minus sign because the velocity gradient at $r$ is negative. Combining Equations 6.20 and 6.21 and then integrating the resulting expression over $r$, we get

$$u = \frac{1}{4\mu} \frac{\Delta p}{\Delta x} \left( r_0^2 - r^2 \right), \tag{6.22}$$

where we have used the condition that $u = 0$ at $r = r_0$. Equation 6.22 gives a parabolic velocity distribution that may also be written as

$$u = U_c \left[ 1 - \left( \frac{r}{r_0} \right)^2 \right], \tag{6.23}$$

where

$$U_c = \frac{r_0^2}{4\mu} \frac{\Delta p}{\Delta x}, \tag{6.24}$$

is the maximum velocity (i.e., centerline velocity) in the pipe. The total volume flow rate is then given by or the mass flow rate is

$$\dot{Q} = \int_0^{r_0} u 2\pi r \, dr = \frac{\pi r_0^4}{8\mu} \frac{\Delta p}{\Delta x}, \tag{6.25}$$

or the mass flow rate is

$$\dot{m} = \rho \dot{Q} = \frac{\pi d^4 \rho}{128\mu} \frac{\Delta p}{\Delta x}, \qquad (6.26)$$

where $d = 2r_0$ is the diameter of the pipe. The mean flow velocity is then obtained as

$$U_m = \frac{\dot{Q}}{\pi r_0^2} = \frac{r_0^2}{8\mu} \frac{\Delta p}{\Delta x} = \frac{1}{2} U_c. \qquad (6.27)$$

The parabolic velocity profile (6.23) can now be rewritten in terms of the mean velocity as

$$\frac{u}{U_m} = 2 \left[ 1 - \left( \frac{r}{r_0} \right)^2 \right], \qquad (6.28)$$

and the pressure drop per unit length is

$$\frac{\Delta p}{\Delta x} = \frac{32\mu U_m}{d^2}, \qquad (6.29)$$

which is the Hagen–Poiseuille law for pressure drop in a long circular pipe with laminar flow. Hence, for laminar flow the pressure drop per unit length $\Delta p/\Delta x$ is proportional to the average velocity.

Equation 6.20 shows that shear stress $\tau$ is linear in $r$ since $\Delta p$ over $\Delta x$ is the same at all radii. We can therefore write

$$\tau_w = \frac{r_0}{2} \frac{\Delta p}{\Delta x}, \qquad (6.30)$$

where $\tau_w$ is the shear stress at the pipe wall. From Equations 6.20 and 6.30, we get

$$\frac{\tau}{\tau_w} = \frac{r}{r_0}.$$

Therefore, in a fully developed laminar pipe flow, shear stress varies linearly from a maximum value at the pipe wall to zero at the centerline.

Although we have not yet described a shear mechanism for turbulent flows, we can state here that Equation 6.30 is equally applicable to a fully developed turbulent flow as long at $\tau$ refers to apparent turbulent shear stress rather than a viscous shear stress.

## 6.4 Friction Factor

In fully developed pipe flow, either laminar or turbulent, we assume that the pressure drop $\Delta p$ is proportional to the length $L$ and the following functional relationship is valid

$$\frac{\Delta p}{L} = \pi(U_m, d, \rho, \mu, \varepsilon), \qquad (6.31)$$

where the quantity $\varepsilon$ is a statistical measure of surface roughness of the pipe and has dimension of length. With $F$ (force), $M$ (mass), $L$ (length), and $t$ (time) as fundamental dimensions, $U_m$, $d$, and $\rho$ form the set of maximum number of quantities that in themselves cannot form a dimensionless group. Hence, the $\pi$-theorem leads to (for further details, see Chapter 9)

$$\frac{\Delta p}{4(L/d)\left(\rho U_m^2/2\right)} = \psi\left(\frac{U_m d\rho}{\mu}, \frac{\varepsilon}{d}\right), \tag{6.32}$$

where the numerical constants 4 and 2 are added for convenience.

The previous dimensionless group involving $\Delta p$ has been defined as Fanning friction factor, $f$, that is,

$$f = \frac{\Delta p}{4(L/d)\left(\rho U_m^2/2\right)}. \tag{6.33}$$

Then, Equation 6.32 becomes

$$f = \psi\left(Re, \frac{\varepsilon}{d}\right), \tag{6.34}$$

which means that $f$ depends on the Reynolds number and the roughness of the pipe surface. Since from Equation 6.30 we can write

$$\frac{\Delta p}{L} = \frac{2}{r_0}\tau_w = \frac{4}{d}\tau_w, \tag{6.35}$$

the friction factor given by Equation 6.33 may also be written as

$$f = \frac{\tau_w}{(1/2)\rho U_m^2}. \tag{6.36}$$

Thus, the friction factor represents the ratio of the wall shear stress, $\tau_w$, to the flow kinetic energy per unit volume (the dynamic velocity head). For a fully developed laminar pipe flow, Equation 6.36, in view of Equations 6.29 and 6.30, results in (see Problem 6.2)

$$f = \frac{16}{Re}. \tag{6.37}$$

Hence, in fully developed laminar flow, $f$ is independent of $\varepsilon/d$. Figure 6.6 shows the relationship (6.34) as deducted by Moody [7] from experimental data for fully developed flow.

Equation 6.33 can also be written in terms of total pressure drop as

$$\Delta p_t = \frac{\Delta p}{\rho U_m^2/2} = 4f\left(\frac{L}{d}\right). \tag{6.38}$$

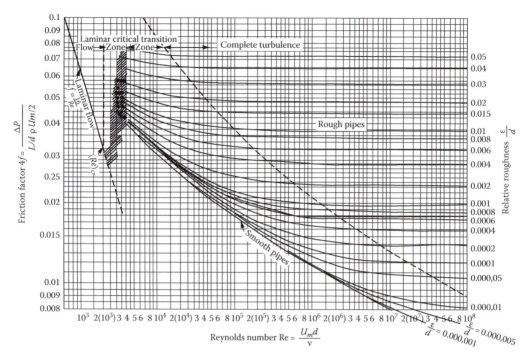

**FIGURE 6.6**
Moody diagram.

In the hydrodynamic entrance region, since the velocity profile varies, the momentum rate changes along the flow direction. In this region, based on the total pressure drop from the inlet of the duct to the point of interest, an apparent Fanning friction factor $f_{app}$ is defined as

$$\Delta p_t = \frac{\Delta p}{\rho U_m^2/2} = 4 f_{app}\left(\frac{L}{d}\right), \tag{6.39}$$

which incorporates the combined effects of wall shear and the change in momentum flow rate due to the developing velocity profile. The incremental pressure drop number $K(x)$ in the hydrodynamic entrance region is defined as [8]

$$K(x) = 4(f_{app} - f_{fd})\left(\frac{L}{d}\right), \tag{6.40}$$

where $f_{fd}$ is the Fanning friction factor in the hydrodynamically fully developed region.

Values of the friction factor $f$ for commercial pipes are difficult to determine precisely, because, even for the so-called smooth tubes, $f$ may vary as much as 5%. In addition, the change in roughness with age is difficult to predict.

In curved pipes, the friction factor may rise considerably above the values shown in Figure 6.6, and the critical Reynolds number may be considerably different than 2300. In turbulent flow, the effect of pipe curvature has not been studied extensively; however, the effect is less significant.

In nonisothermal flow, the effect of temperature on viscosity produces a distortion of the velocity profile. In liquids, $\mu$ decreases with the temperature, and in gases, $\mu$ increases with

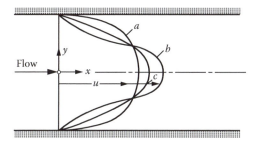

**FIGURE 6.7**
Velocity distribution in nonisothermal flow.

the temperature. Figure 6.7 shows the effect of temperature on the velocity distribution, where curve *a* represents heating of a liquid or cooling of a gas since μ near the wall would be decreased. Curve *b*, on the other hand, represents cooling a liquid or heating a gas. The effect of temperature-dependent properties on heat transfer is discussed in Chapter 9.

## 6.5 Noncircular Cross-Sectional Ducts

A duct of noncircular cross section is not geometrically similar to a circular pipe. Hence, dimensional analysis does not relate the performance of these two geometries. However, in turbulent flow, the friction factor for noncircular cross-sectional ducts (annular spaces, rectangular and triangular ducts, etc.) may be evaluated from the data for circular pipes if *d* is replaced by a *hydraulic* diameter, $d_H$, defined by

$$d_H = \frac{4A}{P} = \frac{4 \times (\text{flow area})}{\text{Wetted perimeter}}. \tag{6.41}$$

For example, the hydraulic diameter of an annulus of inner and outer diameters $d_i$ and $d_o$, respectively, is

$$d_H = \frac{4(\pi/4)\left(d_o^2 - d_i^2\right)}{\pi(d_o + d_i)} = (d_o - d_i). \tag{6.42}$$

For a circular duct of diameter $d$, Equation 6.42 reduces to $d_H = d$.

The critical Reynolds number $Re_{cr} = \rho U_m d_H / \mu$ for noncircular ducts, as discussed in Section 6.2, is also found to be approximately 2300 as for circular pipes.

For laminar flow, however, the results for noncircular cross-sectional ducts are not universally correlated. In a thin annulus, for example, in which spacing $Z$ is very much less than the mean diameter of the annulus, the flow has a parabolic velocity distribution and has the same distribution at every circumferential position. Treating this as a flow between two parallel flat plates with no shear stress at the side edges, one can arrive at the relation

$$\frac{\Delta p}{\Delta x} = \frac{12\mu U_m}{Z^2}, \tag{6.43}$$

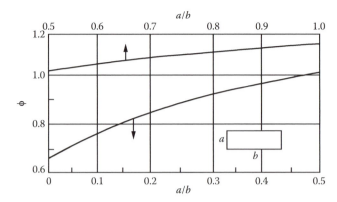

**FIGURE 6.8**
Values of $\phi$ for rectangular ducts. (From McAdams, W.H., *Heat Transmission*, 3rd edn., McGraw-Hill, New York, 1954.)

following a reasoning similar to that which led to Equation 6.29. Here, since $d_H = 2Z$, Equation 6.43 can be rewritten in the form

$$f = \frac{24}{Re} \tag{6.44}$$

with $d_H$ replacing $d$ in the definitions of $f$ and $Re$. This equation is obviously a variation of Equation 6.37 that applies to laminar flow in circular pipes.

Flow in a rectangular duct (of dimensions $a$ by $b$) in which $b \gg a$ is similar to this annular flow. For rectangular ducts of other aspect ratios, the friction factor would be

$$f = \frac{16}{\phi Re}, \tag{6.45}$$

where the hydraulic diameter in the definitions of $f$ and $Re$ is

$$d_H = \frac{2ab}{a+b},$$

and the values of $\phi$ are given in graphical form in Figure 6.8.

## 6.6 Laminar Forced Convection in Ducts

Now we proceed to discuss laminar convective heat transfer in ducts. We shall assume that all body forces are negligible and that the fluid is forced through the duct by some external means unrelated to the temperature field in the fluid. We will further assume that thermophysical properties of the fluid are constant. The duct walls in all cases will be considered to be smooth, nonporous, and rigid. They will also be assumed to be uniformly thin so that the temperature distribution within the solid duct wall has negligible influence on the heat transfer within the fluid flow.

According to Newton's law of cooling, heat flow from a surface to a fluid will increase as the temperature difference between the surface and the fluid increases. The constant of proportionality, that is, the heat flux divided by the temperature difference between the surface and the fluid, is called, as it was already defined in the previous chapters, the heat-transfer coefficient and is denoted by $h$. The heat-transfer coefficient (or the film coefficient) is an important parameter in the analysis and design of heat-exchange equipment and energy conversion devices. In this chapter, we will define $h$ in a manner more appropriate for flows inside pipes and ducts.

We will start with a case where both hydrodynamically and thermally fully developed conditions exist (i.e., fully developed conditions). We will then consider problems in which the velocity profile has already fully developed and the temperature profile is developing (i.e., thermal entrance region problems). We will also present some results for simultaneously developing conditions (i.e., combined hydrodynamic and thermal entrance region problems).

## 6.7 Thermal Boundary Conditions

Various thermal boundary conditions can be imposed at the inside surface of ducts. A systematic exposition of the thermal boundary conditions has been provided by Shah and London [8]. The heat-transfer coefficient in laminar flow is strongly dependent on the thermal boundary conditions. We give here only three important thermal boundary conditions, namely, the $T$, $H_1$, and $H_2$ types. The $T$ boundary condition refers to constant wall temperature both axially and peripherally throughout the channel (or passage) length. This boundary condition is approximated in condensers, evaporators, and liquid-to-gas heat exchangers with high-velocity liquid flows. The heat-transfer coefficient and Nusselt number for this boundary condition are designated with a subscript $T$.

The $H_1$ boundary condition refers to constant wall heat-transfer rate along the axial direction, but having constant wall temperature in the peripheral direction at any cross section, as shown in Figure 6.9. In contrast, the $H_2$ boundary condition refers to constant wall heat-transfer rate in the axial as well as in the peripheral direction at every cross section, as shown in Figure 6.9. The $H_1$ boundary condition is realized for ducts constructed from highly conductive materials for which the temperature gradients in the peripheral direction are minimum. The $H_2$ boundary condition, on the other hand, is realized for ducts of very low conductive materials in which temperature gradients exist in the peripheral direction. For intermediate values of thermal conductivity, the boundary condition will be somewhat between $H_1$ and $H_2$. These two boundary conditions become the same and are referred to as $H$ for symmetrically heated straight flow passages having constant peripheral curvature and no corners, for example, circular tubes and concentric annular pipes. These boundary conditions are different for noncircular flow passages, such as rectangular, triangular, elliptical, and polygonal passages. The Nusselt numbers for these boundary conditions are designated with subscripts $H_1$ and $H_2$, respectively. For noncircular passages, $Nu_{H_2}$ is always lower than $Nu_{H_1}$ and $Nu_T$ is always lower than $Nu_{H_1}$; however, $Nu_{H_2}$ can be higher or lower than $Nu_T$ depending upon the duct geometry. The $T$, $H_1$, and $H_2$ boundary conditions are defined in Figure 6.9.

Unless otherwise specified, in this text the $T$ and $H_1$ boundary conditions are referred to simply as constant wall temperature (isothermal wall) and constant heat-flux boundary conditions, respectively.

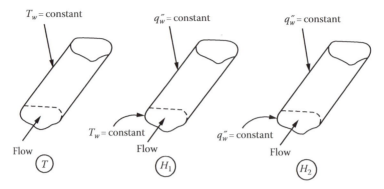

**FIGURE 6.9**
Three important thermal boundary conditions for duct flows.

## 6.8 Laminar Forced Convection in Circular Pipes with Fully Developed Conditions

As we have already obtained in Section 6.3.2, in a steady fully developed laminar flow in a circular pipe, the velocity profile will have the shape of a parabola. This type of flow is usually referred to as Poiseuille flow. Under these conditions, the energy equation, Equation 2.63a, reduces, in the absence of viscous dissipation, to

$$\frac{u}{\alpha}\frac{\partial T}{\partial x} = \frac{1}{r}\frac{\partial}{\partial r}\left(r\frac{\partial T}{\partial r}\right) + \frac{\partial^2 T}{\partial x^2}. \tag{6.46}$$

If axial conduction is neglected relative to conduction in the radial direction, the energy equation for steady, laminar, constant-property, viscous, incompressible, and fully developed flow in a circular pipe becomes

$$\frac{u}{\alpha}\frac{\partial T}{\partial x} = \frac{1}{r}\frac{\partial}{\partial r}\left(r\frac{\partial T}{\partial r}\right). \tag{6.47}$$

In order to define a heat-transfer coefficient $h$, as we have already mentioned before, three quantities are required, namely, the heat flux across the surface, the temperature of the surface, and the temperature of the fluid. The temperature of the fluid at any axial location will vary across the pipe cross section and we face, therefore, the problem of defining an appropriate mean temperature. The most convenient mean value for the temperature is that which, when multiplied by the mass flow rate of the fluid and its specific heat, gives the rate of thermal energy (i.e., enthalpy) transport along the pipe. With this definition, changes in the mean temperature of the fluid along the pipe may be calculated from the heat input to the fluid. Thus,

$$T_m C_{pm}\dot{m} = \int_A T c_p \rho u \, dA \tag{6.48}$$

or

$$T_m = \frac{\int_A T c_p \rho u \, dA}{c_{pm}\dot{m}} = \frac{\int_A T c_p \rho u \, dA}{c_{pm}\int_A \rho u \, dA} \tag{6.49}$$

which, for an incompressible fluid with constant specific heat, reduces to

$$T_m = \frac{\int_A T u \, dA}{\int_A u \, dA}. \tag{6.50}$$

The definition of $T_m$ by Equation 6.49 or Equation 6.50 is general in the sense that it applies to pipes as well as to ducts of any cross-sectional shape, and in these definitions, $A$ represents the cross-sectional area of the duct. The mean temperature $T_m$ is also referred to as the *bulk temperature* or *mixing-cup temperature,* and it is the temperature that characterizes the average thermal energy state of the fluid.

The term fully developed temperature profile implies that there exists, under certain conditions, a generalized nondimensional temperature profile defined as

$$\Theta = \frac{T - T_w}{T_m - T_w}, \tag{6.51}$$

which is invariant with the axial coordinate along the duct; thus, as illustrated in Figure 6.10,

$$\frac{\partial \Theta}{\partial x} = \frac{\partial}{\partial x}\left(\frac{T - T_w}{T_m - T_w}\right) = 0 \tag{6.52a}$$

or

$$\Theta = \Theta(r). \tag{6.52b}$$

Hence, while the local fluid temperature $T$ varies with two space coordinates $r$ and $x$ and the fluid bulk temperature $T_m$ is a function only of the axial coordinate $x$ (and, if not

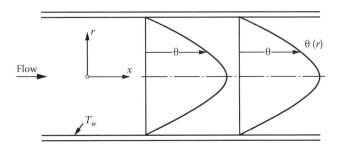

**FIGURE 6.10**
Fully developed temperature profile.

constant, the wall temperature $T_w$ is also a function of $x$), the dimensionless temperature distribution $\Theta$ as defined by Equation 6.51 varies only with the radial coordinate $r$.

Alternatively, a fully developed temperature profile could also be defined by a nondimensional equation similar to Equation 6.51 with $(T_m - T_w)$ replaced by $(T_c - T_w)$. These two definitions would, in fact, be identical since one follows from the other (why?).

We now define the local heat-transfer coefficient at any axial location $x$ in terms of the mean temperature $T_m$ at the same location as

$$h_x = \frac{q_w''}{T_w - T_m}, \tag{6.53a}$$

where $q_w''$ represents the local heat flux to the fluid.

In the case of a circular pipe of radius $r_0$, the local heat-transfer coefficient from Equation 6.53a can be written as

$$h_x = \frac{k(\partial T/\partial r)_{r=r_0}}{T_w - T_m} = -k\frac{\partial}{\partial r}\left[\frac{T - T_w}{T_m - T_w}\right]_{r=r_0} = -k\left(\frac{\partial \Theta}{\partial r}\right)_{r=r_0}. \tag{6.53b}$$

Hence, when the temperature profile is fully developed, $(\partial \Theta/\partial r)_{r=r_0} = $ constant, and therefore, the heat-transfer coefficient becomes constant along the pipe.

Differentiating Equation 6.51 with respect to $x$ and solving for $\partial T/\partial x$, we now get

$$\frac{\partial T}{\partial x} = \frac{dT_w}{d_x} - \left(\frac{T - T_w}{T_m - T_w}\right)\frac{dT_w}{dx} + \left(\frac{T - T_w}{T_m - T_w}\right)\frac{dT_m}{dx}. \tag{6.54}$$

The following two cases are of interest.

### 6.8.1 Uniform Heat-Flux Boundary Condition

If the wall heat flux is constant, then from the definition of heat-transfer coefficient (6.53a), we have

$$q_w'' = h\left(T_w - T_m\right) = \text{constant}. \tag{6.55}$$

Since $h = $ constant when the temperature profile is fully developed, then $T_w - T_m = $ constant. Hence,

$$\frac{dT_w}{dx} = \frac{dT_m}{dx}. \tag{6.56}$$

An energy balance applied to a control volume as shown in Figure 6.11 gives

$$q_w'' = \frac{r_0 U_m \rho c_p}{2}\frac{dT_m}{dx}. \tag{6.57}$$

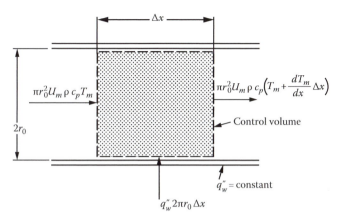

**FIGURE 6.11**
A control volume in the fully developed region in a circular duct.

Therefore, it follows from Equations 6.56 and 6.57 that

$$\frac{dT_w}{dx} = \frac{dT_m}{dx} = \text{constant.} \tag{6.58}$$

Substituting this result into Equation 6.54, we obtain

$$\frac{\partial T}{\partial x} = \frac{dT_w}{dx} = \frac{dT_m}{dx} = \text{constant.} \tag{6.59}$$

Thus, when the wall heat flux is constant, the bulk temperature of a constant-property fluid varies linearly in the flow direction as illustrated in Figure 6.12, with the slope being proportional to $q_w''$. For thermally fully developed flow, this must also hold true for the temperature at any distance $r$ as indicated by Equation 6.59.

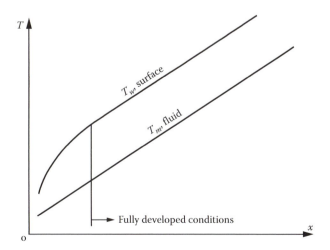

**FIGURE 6.12**
Variation of bulk temperature with $q_w'' = \text{constant}$.

The boundary conditions for this case may be written as

$$\frac{\partial T(x,0)}{\partial r} = 0, \tag{6.60a}$$

$$k\frac{\partial T(x,r_0)}{\partial r} = q''_w = \text{constant.} \tag{6.60b}$$

The energy equation to be solved under these boundary conditions is Equation 6.47. Inserting the velocity distribution from Equation 6.23 into Equation 6.47 and then integrating it at any $x$ twice with respect to $r$, we get

$$T(r,x) = \frac{1}{\alpha}\frac{\partial T}{\partial x}U_c\left(\frac{r^2}{4} - \frac{r^4}{16r_0^2}\right) + C_1\ln r + c_2. \tag{6.61}$$

From the boundary condition at $r = 0$, we obtain $C_1 = 0$. Therefore, the temperature distribution in terms of $T(0,x) = T_c$ becomes

$$T - T_c = \frac{1}{\alpha} = \frac{\partial T}{\partial x}\frac{U_c r_0^2}{4}\left[\left(\frac{r}{r_0}\right)^2 - \frac{1}{4}\left(\frac{r}{r_0}\right)^4\right]. \tag{6.62}$$

By the use of Equations 6.23 and 6.62, calculating the bulk temperature from Equation 6.50, we obtain

$$T_m = T_c + \frac{7}{96}\frac{U_c r_0^2}{\alpha}\frac{\partial T}{\partial x}. \tag{6.63}$$

The wall temperature and the gradient of the fluid temperature at the wall from Equation 6.62 are

$$T_w = T_c + \frac{3}{16}\frac{U_c r_0^2}{\alpha}\frac{\partial T}{\partial x} \tag{6.64}$$

and

$$\left(\frac{\partial T}{\partial r}\right)_{r=r_0} = \frac{U_c r_0}{4\alpha}\frac{\partial T}{\partial x}. \tag{6.65}$$

From Equations 6.63 and 6.64, we also get

$$T_w - T_m = \frac{11}{96}\frac{U_c r_0^2}{\alpha}\frac{\partial T}{\partial x}. \tag{6.66}$$

Substituting Equations 6.65 and 6.66 into Equation 6.53, we obtain the heat-transfer coefficient in fully developed flow under uniform heat-flux condition as

$$h = \frac{24}{11}\frac{k}{r_0} = \frac{48}{11}\frac{k}{d} \tag{6.67}$$

or expressing the result in terms of a Nusselt number, we get

$$Nu = \frac{hd}{k} = 4.364. \tag{6.68}$$

## 6.8.2 Constant Wall Temperature Boundary Condition

This is also a common boundary condition. It is encountered in heat exchangers such as evaporators and condensers. For constant wall temperature,

$$\frac{dT_w}{dx} = 0 \tag{6.69}$$

and therefore, Equation 6.54 becomes

$$\frac{\partial T}{\partial x} = \left( \frac{T - T_w}{T_m - T_w} \right) \frac{dT_m}{dx}. \tag{6.70}$$

If Equation 6.70 is substituted into the energy equation (6.47), we obtain

$$\frac{1}{r}\frac{\partial}{\partial r}\left( r\frac{\partial T}{\partial r} \right) = \frac{u}{\alpha}\left( \frac{T - T_w}{T_m - T_w} \right)\frac{dT_m}{dx}. \tag{6.71}$$

The boundary conditions are as follows:

$$T(x, r_0) = T_w = \text{constant}, \tag{6.72a}$$

$$\frac{\partial T(x, 0)}{\partial r} = 0. \tag{6.72b}$$

If the parabolic velocity profile for $u$ given by Equation 6.28 is substituted into Equation 6.71, we get

$$\frac{1}{r}\frac{\partial}{\partial r}\left( r\frac{\partial T}{\partial r} \right) = \frac{2U_m}{\alpha}\left( 1 - \frac{r^2}{r_0^2} \right)\left( \frac{T - T_w}{T_m - T_w} \right)\frac{dT_m}{dx}. \tag{6.73}$$

One way to solve this equation is by the use of a successive approximation method (see Problem 6.5): As a first approximation, the temperature distribution obtained for the constant heat-flux case can be substituted for the dimensionless temperature $(T - T_w)/(T_m - T_w)$

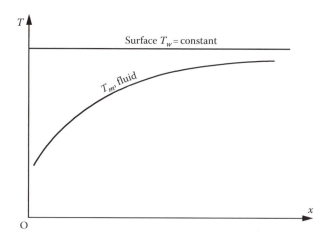

**FIGURE 6.13**
Variation of bulk temperature with $T_w$ = constant.

on the right-hand side, and then a new temperature profile is obtained by integrating this equation in the $r$-direction. This temperature profile is next used as a second approximation on the right-hand side of Equation 6.73, and another integration gives another new profile. For each new temperature profile obtained, the mean temperature $T_m$ and the Nusselt number are calculated, and this procedure is repeated until the Nusselt number converges to a limit, which for this case is

$$Nu = 3.658.$$

It is again seen that the heat-transfer coefficient depends on the type of the boundary condition, in other words, on the type of surface temperature variation; the constant wall temperature result for the Nusselt number is 16% less than the result for the constant wall heat-flux boundary condition. Variation of the fluid bulk temperature for the constant wall temperature case is shown in Figure 6.13.

A considerable amount of information is available on analytical solutions for various channel shapes [8]. The results include information on solutions for developed and developing velocity and temperature profiles, as well as for the entry lengths required for fully developed flows. The Nusselt numbers, values for $fRe$, and the entry lengths for some noncircular ducts are summarized in Table 6.1. In this table, $K(\infty)$ is the value of the incremental pressure drop number $K(x)$ for fully developed conditions.

It is common practice to consider laminar flows to be fully developed if gas flow prevails in a duct having $L/d_H > 100$, as also noted in Table 6.1. This is a good guideline for low Reynolds number flows, $Re < 1000$. However, for $Re > 1000$, one should compare the duct length $L$ with $L_h$ in order to ensure that the flow is fully developed. The entrance length effect will be negligible only when the actual duct length $L$ is much greater than the minimum length required for the flow to be fully developed [9,10].

Table 6.1 also indicates that there is a strong influence of thermal boundary conditions on the heat-transfer coefficient that is independent of flow velocity under fully developed conditions. Therefore, in a thermally and hydrodynamically fully developed laminar flow, an increase in $h$ can be achieved by reducing the hydraulic diameter $d_H$ or by changing the duct geometry.

**TABLE 6.1**

Nusselt Numbers, Friction Coefficient, and Hydrodynamic Entry Lengths for Fully Developed Flow Conditions in Laminar Flow through Ducts of Different Cross Sections

| Geometry($Ld_H > 100$) | $Nu_{H1}$ | $Nu_{H2}$ | $Nu_T$ | $fRe$ | $K(\infty)$ | $L_h/d_H Re$ |
|---|---|---|---|---|---|---|
| $2b$ ☐ $\frac{2b}{2a} = 1$  $2a$ | 3.608 | 3.091 | 2.976 | 14.227 | 0.286 | 0.090 |
| ⬡ | 4.002 | 3.862 | 3.34 | 15.054 | 0.288 | 0.086 |
| $2b$ ◯ $\frac{2b}{2a} = \frac{1}{2}$  $2a$ | 4.364 | 4.364 | 3.657 | 16.000 | 0.307 | 0.056 |
| ▭ | 4.123 | 3.017 | 3.391 | 15.548 | 0.299 | 0.085 |
| $2b$ ▭ $\frac{2b}{2a} = \frac{1}{4}$  $2a$ | 5.331 | 2.94 | 4.439 | 18.233 | 0.329 | 0.078 |
| $2b$ ▭ $\frac{2b}{2a} = \frac{1}{6}$  $2a$ | 6.049 | 2.93 | 5.137 | 19.702 | 0.346 | 0.070 |
| $2b$ ▭ $\frac{2b}{2a} = \frac{1}{8}$  $2a$ | 6.490 | 2.94 | 5.597 | 20.585 | 0.355 | 0.063 |
| ▭ $\frac{2b}{2a} = 0$ | 8.235 | 8.235 | 7.541 | 24.000 | 0.386 | 0.011 |

*Sources:* Shah, R.K. and Bhatti, M.S., Laminar convective heat transfer in ducts, in *Handbook of Single-Phase Convective Heat Transfer*, S. Kakaç, R.K. Shah, and W. Aung, Eds., John Wiley & Sons, New York, 1987, Chapter 3; Shah, R.K. and London, A.L., *Laminar Flow Forced Convection in Ducts*, Supplement 1, *Advances in Heat Transfer*, Academic Press, New York, 1978; Shah, R.K., Fully developed laminar flow forced convection in channels, in *Low Reynolds Number Flow Heat Exchanger*, S. Kakaç, R.K. Shah, and A.E. Bergles, Eds., Hemisphere Publishing Co., New York, 1983.

## 6.9 Laminar Forced Convection in the Thermal Entrance Region of a Circular Duct

The first solutions to this case were given by Graetz [11] in 1885. Nusselt [12] in 1910 solved the same problem independently for the Poiseuille flow case. This problem is now widely known as the *Graetz problem*. Graetz in his first solution made the following assumptions:

1. The radial velocity component is everywhere zero.
2. $k/c_p \rho_u$ is uniform and constant throughout the fluid.
3. $T_i$ is constant.
4. $T_w$ is constant.
5. $T = T_i$ if $x < 0, T = T_w$ at $r = r_0$ if $x > 0$.
6. The thermal conductivity in the flow direction is zero.

It is apparent that Graetz treated the fluid as a solid rod moving through the circular pipe. Using the energy equation in cylindrical coordinates, he obtained the temperature distribution in terms of the Bessel function $J_0$.

In his second attempt, Graetz solved the same problem under the same assumptions for Poiseuille flow. He further assumed that $K$, $\rho$, and $c_p$ were constant. This solution, however, required considerable numerical work to be complete.

Later Jakob [13], partly in cooperation with Eckert, recalculated and corrected the constants of the Graetz problem. With the advancement of computer usage, various investigators were able to obtain the solution with an accuracy sufficient for practical use. Brown [14] gave a good description of a computer solution. He solved the eigenvalue problems associated with the solution of the energy equation in a parallel-plate channel and a circular tube with uniform wall temperature to 10-figure accuracy. Clark and Kays [15] presented solutions for rectangular and triangular ducts.

The method described in Reference 14 was applied to problems in parallel-plate channels and pipes with prescribed symmetrical and unsymmetrical heat transfer at the walls [16–18].

The complete set of eigenvalues and eigenfunctions for the classical Graetz problem has been presented by Sellars et al. [19], and the solution has been extended to include wall temperature or heat-flux variations.

### 6.9.1 Graetz Solution for Uniform Velocity

The assumptions have already been given in Section 6.9. When the first assumption is taken together with the second, it is apparent that the fluid is being treated as a solid rod moving through the pipe (rodlike flow or slug flow). Graetz originally did not state the sixth assumption, but he tacitly introduced it in the course of his work. The equation to be solved for the temperature distribution is

$$U_m \frac{\partial \Theta}{\partial x} = \frac{k}{\rho c_p} \left( \frac{\partial^2 \Theta}{\partial r^2} + \frac{1}{r} \frac{\partial \Theta}{\partial r} \right), \tag{6.74}$$

where
$\Theta = (T - T_w)/(T_i - T_w)$
$U_m =$ constant and the inlet and the boundary conditions are

$$\Theta(0,r) = 1, \tag{6.75a}$$

$$\Theta(x, r_0) = 0, \tag{6.75b}$$

$$\frac{\partial \Theta(x,0)}{\partial r} = 0. \tag{6.75c}$$

One of the methods of solving a problem of the previous type is to try an assumption of the following form [20]:

$$\Theta(x,r) = R(r)X(x), \tag{6.76}$$

where
$R$ is a function of $r$ only
$X$ is a function of $x$ only

Substituting Equation 6.76 into the energy equation (6.74) and noting that $\Theta(\infty,r) \to 0$, we get

$$\frac{U_m}{\alpha}\frac{1}{X}\frac{dX}{dx} = \frac{1}{R}\left[\frac{d^2R}{dr^2} + \frac{1}{r}\frac{dR}{dr}\right] = -\lambda^2 \qquad (6.77)$$

where $\lambda$ is a constant. From Equation 6.77, we obtain the following two ordinary differential equations:

$$\frac{dX}{dx} + \frac{\alpha}{U_m}\lambda^2 X = 0, \qquad (6.78)$$

$$\frac{d^2R}{dr^2} + \frac{1}{r}\frac{dR}{dr} + \lambda^2 R = 0. \qquad (6.79)$$

Equation 6.79 is Bessel's differential equation of order zero (see Appendix C). Solutions of the Equations 6.78 and 6.79 lead to the following general solution

$$\Theta(x,r) = e^{(\alpha/U_m)\lambda^2 x}[AJ_0(\lambda r) + BY_0(\lambda r)] \qquad (6.80)$$

for the partial differential equation (6.74). The boundary condition (6.75c) gives $B = 0$, and from the boundary condition (6.75b), we get

$$J_0(\lambda r_0) = 0, \qquad (6.81)$$

which yields an infinite number of values for $\lambda$, that is, $\lambda_1, \lambda_2, \ldots, \lambda_n \ldots$ Hence, for any value of $\lambda$, from this set, say $\lambda_n$, the following solution

$$\Theta_n(x,r) = A_n J_0(\lambda_n r)e^{(\alpha/U_m)\lambda_n^2 x} \qquad (6.82)$$

satisfies the partial differential equation (6.74) and the boundary conditions (6.75b,c) and so is true for any linear combination of such solutions. Therefore, the most general solution of the partial differential equation (6.74) that satisfies the boundary conditions (6.75b and c) may be written as

$$\Theta(x,r) = \sum_{n=1}^{\infty} A_n J_0(\lambda_n r)e^{(\alpha/U_m)\lambda_n^2 x}. \qquad (6.83)$$

For this equation to be the solution of the problem under consideration, it should also satisfy the inlet condition (6.75a). Applying the inlet condition (6.75a) gives

$$1 = \sum_{n=1}^{\infty} A_n J_0(\lambda_n r) \qquad (6.84)$$

which is a Fourier–Bessel expansion [20] of 1 with the values of $\lambda_n$ determined from Equation 6.81. Thus, the initial condition (6.75a) is also satisfied by Equation 6.83.

The constants $A_n$ may be found by multiplying both sides of Equation 6.84 by $rJ(\lambda_n r)$ and then integrating the resulting equation over $r$ from 0 to $r_0$, which yields

$$\int_0^{r_0} rJ_0(\lambda_n r)dr = A_n \int_0^{r_0} rJ_0^2(\lambda_n r)dr, \tag{6.85}$$

where we have used the following orthogonality relation of the Bessel functions of the first kind of zero order

$$\int_0^{r_0} rJ_0(\lambda_n r)J_0(\lambda_m r)dr = 0, \quad \text{if } m \neq n. \tag{6.86}$$

Performing the integrations in Equation 6.85, we obtain [20]

$$A_n = \frac{2}{\lambda_n r_0} \frac{1}{J_1(\lambda_n r_0)}. \tag{6.87}$$

Hence, the temperature distribution is given by

$$\Theta(x,r) = 2\sum_{n=1}^{\infty} \frac{1}{\lambda_n r_0} \frac{J_0(\lambda_n r)}{J_1(\lambda_n r_0)} e^{(\alpha/U_m)\lambda_n^2 x}. \tag{6.88}$$

For the average or bulk temperature $T_m$ of the stream at any distance $x$, that is, the temperature averaged over the pipe cross section with respect to the radius $r$, Equation 6.50 yields

$$\Theta_m \frac{T_m - T_m}{T_i - T_w} = 4\sum_{n=1}^{\infty} \frac{1}{(\lambda_n r_0)} e^{(\alpha/U_m)\lambda_n^2 x} \tag{6.89a}$$

or

$$\Theta_m \frac{T_m - T_i}{T_w - T_i} = 4\sum_{n=1}^{\infty} \frac{1}{(\lambda_n r_0)^2} e^{(\alpha/U_m)\lambda_n^2 x}, \tag{6.89b}$$

which may also be written as

$$\frac{T_m - T_i}{T_w - T_i} = 1 - 4\sum_{n=1}^{\infty} \frac{1}{\beta_n^2} e^{-4\beta_n^2 \frac{(x/d)}{Pe}}, \tag{6.89c}$$

where
  $Pe = U_m id/\alpha$ is the *Péclet number*
  $\beta_n = \lambda_n r_0$ are the zeros of

$$J_0(\beta_n) = 0 \tag{6.90}$$

Bessel function for Equation 6.90 is given in Appendix C. The first six zeros are as follows:

$$\beta_1 = 2.4048 \quad \beta_4 = 11.7915$$
$$\beta_2 = 5.5207 \quad \beta_5 = 14.9309$$
$$\beta_3 = 8.6537 \quad \beta_6 = 18.071$$

Equation 6.53b, together with the use of Equations 6.88 and 6.89a, can be used to determine the local heat-transfer coefficient as a function of $x$. The local Nusselt number in the thermal entrance region of a circular pipe for laminar uniform velocity flow then becomes

$$Nu_x = \frac{h_x d}{k} = \frac{\sum_{n=1}^{\infty} e^{-4\beta_n^2 (x/d)/Pe}}{\sum_{n=1}^{\infty} \left(1/\beta_n^2\right) e^{-4\beta_n^2 (x/d)/Pe}}. \tag{6.91}$$

Since both series in this result are increasingly more convergent, for large values of $(x/d)/Pe$ only the first term is of significance. Thus,

$$Nu_\infty - Nu_{x \to \infty} = \beta_1^2 = (2.4048)^2 = 5.783,$$

which is the Nusselt number for a fully developed temperature profile under constant wall temperature boundary condition.

### 6.9.2 Graetz Solution for Parabolic Velocity Profile

The applicable differential energy equation is again Equation 6.47. It will be convenient to work with dimensionless quantities. Let us introduce the following dimensionless variables:

$$\Theta = \frac{T - T_w}{T_i - T_w}, \quad \eta = \frac{r}{r_0}, \quad \xi = \frac{x/r_0}{RePr} = \frac{x/r_0}{Pe}, \tag{6.92a,b,c}$$

where
$T_w$ is the constant surface temperature
$T_i$ is the uniform entering fluid temperature

The inverse of $\xi$ is called the Graetz modulus and is denoted by $G_z = 1/\xi = Pe/(x/r_0)$. The problem is illustrated in Figure 6.14.

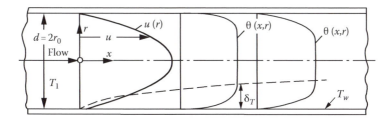

**FIGURE 6.14**
Hydrodynamically fully developed and thermally developing circular duct flow.

For hydrodynamically fully developed laminar flow in a circular pipe, the velocity profile, as indicated by Equation 6.28, is parabolic and given by

$$\frac{u}{U_m} = 2\left[1 - \left(\frac{r}{r_0}\right)^2\right] = 2(1 - \eta^2).$$

(6.93)

Substituting Equation 6.93 into Equation 6.47, the final form of the energy equation in terms of the dimensionless variables is obtained as

$$(1 - \eta^2)\frac{\partial\Theta}{\partial\xi} = \frac{1}{\eta}\frac{2}{\partial\eta}\left(\eta\frac{\partial\Theta}{\partial\eta}\right),$$

(6.94)

which implies that solutions are similar for given values of $RePr$ (Péclet number). The inlet and the boundary conditions may also be written in dimensionless form as

$$\Theta(0,\eta) = 1,$$

(6.95a)

$$\frac{\partial\Theta(\xi,0)}{\partial\eta} = 0,$$

(6.95b)

$$\Theta(\xi,1) = 0.$$

(6.95c)

This problem can also be solved by the use of the method of separation of variables by assuming

$$\Theta(\xi,\eta) = X(\xi)R(\eta),$$

(6.96)

which, when substituted into Equation 6.94 and noting that $\Theta(\infty,\eta) \to 0$, gives

$$\frac{1}{X}\frac{dX}{d\xi} = \frac{1}{R(1 - \eta^2)}\frac{1}{\eta}\frac{d}{d\eta}\left(\eta\frac{dR}{d\eta}\right) = -\lambda^2,$$

(6.97)

where $\lambda$ is constant. From Equation 6.97 for the $X(\xi)$ function, we get

$$\frac{dX}{d\xi} + \lambda 2X = 0,$$

(6.98)

which has the solution in the form

$$X(\xi) = Ae^{-\lambda^2\xi},$$

(6.99)

where $A$ is any arbitrary constant. For the $R(\eta)$ function, we have

$$\frac{d^2R}{d\eta^2} + \frac{1}{\eta}\frac{dR}{d\eta} + \lambda^2(1 - \eta^2)R = 0.$$

(6.100)

The boundary conditions (6.95b and c) give

$$\frac{dR(0)}{d\eta} = 0 \tag{6.101a}$$

and

$$R(1) = 0. \tag{6.101b}$$

The second-order ordinary differential equation (6.100) and the boundary conditions (6.101a and b) constitute a Sturm–Liouville system [20]. Therefore, the eigenvalues $\lambda_n$ are real numbers and the eigenfunctions $R_n(\eta)$ form a complete orthogonal set with respect to the weight function $\eta(1 - \eta^2)$ over the interval (0,1), that is (see Problems 6.26 and 6.27),

$$\int_0^1 \eta(1-\eta^1)R_m(\eta)d\eta = 0, \quad \text{if } m \neq n \tag{6.102a}$$

with

$$\int_0^1 \eta(1-\eta^2)R_n^2(\eta)d\eta = \frac{1}{2\lambda_n}\left(\frac{\partial R_n}{\partial \lambda_n}\frac{dR_n}{d\eta}\right)_{\eta=1}. \tag{6.102b}$$

A general solution of Equation 6.100 in terms of tabulated functions is not available. Graetz [12] used a series method of solution; he derived a recurrence relation for the series and was able to calculate the first two eigenfunctions. Nusselt's [12] calculations extended the number of eigenfunctions to three. Brown [14] did the same series expansion method and obtained on a computer the first 10 eigenvalues and the corresponding eigenfunctions, but he had to use 50 significant figures. Today, by the use of most modern computers, however, it is possible to integrate Equation 6.100 numerically and obtain the eigenvalues and eigenfunctions up to any desired numerical accuracy. The first three eigenfunctions are shown in Figure 6.15. Once the eigenfunctions $R_n(\eta)$ are available, the general solution can be written as

$$\Theta(\xi, \eta) = \sum_{n=0}^{\infty} C_n R_n(\eta)e^{-\lambda_n^2 \xi}, \tag{6.103}$$

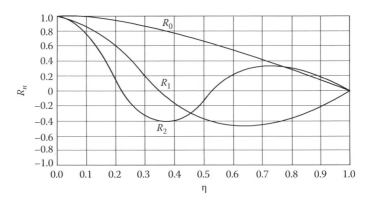

**FIGURE 6.15**
The first three eigenfunctions of Graetz solution.

where $C_n$ are the expansion coefficients. The use of the initial condition (6.95a) yields

$$1 = \sum_{n=0}^{\infty} C_n R_n(\eta). \tag{6.104}$$

Using the orthogonality relation (6.102a), we get

$$C_n = \frac{\displaystyle\int_0^1 \eta(1-\eta^2)R_n(\eta)\,d\eta}{\displaystyle\int_0^1 \eta(1-\eta^2)R_n^2(\eta)\,d\eta} = -\frac{2}{\lambda_n(\partial R_n/\partial \lambda_n)_{\eta=1}}, \tag{6.105}$$

where we have used Equation 6.102b and the following relation (see Problem 6.28):

$$\int_0^1 \eta(1-\eta^2)R_n(\eta)\,d\eta - \frac{1}{\lambda_n^2}\left(\frac{dR_n}{d\eta}\right)_{\eta=1}. \tag{6.106}$$

Hence, inserting the previous result for $C_n$ into Equation 6.103, we get

$$\Theta(\xi,\eta) = -2\sum_{n=0}^{\infty} \frac{e^{-\lambda_n^2 \xi}R_n(\eta)}{\lambda_n(\partial R_n/\partial \lambda_n)_{\eta=1}}. \tag{6.107}$$

The local heat-transfer coefficient is given by Equation 6.53b, that is,

$$h_x = \frac{k(\partial T/\partial r)_{r=r_0}}{T_w - T_m}, \tag{6.108}$$

which may be rewritten in terms of the dimensionless quantities as

$$h_x = -\frac{k}{r_0}\frac{1}{\Theta_m}\left(\frac{\partial \Theta}{\partial \eta}\right)_{\eta=1}. \tag{6.109}$$

Hence, the local Nusselt number is given by

$$Nu_x = \frac{2r_0 h_x}{k} = -\frac{2}{\Theta_m}\left(\frac{\partial \Theta}{\partial \eta}\right)_{\eta=1}. \tag{6.110}$$

The bulk temperature at any distance $x$ is obtained from Equation 6.49; the result is

$$\Theta_m = \frac{T_m - T_w}{T_i - T_w} = -8\sum_{n=0}^{\infty} \frac{e^{-\lambda_n^2 \xi}}{(\partial R_n/\partial \lambda_n)_{\eta=1}}\int_0^1 \eta(1-\eta^2)R_n(\eta)\,d\eta, \tag{6.111a}$$

and by the use of Equation 6.106, we get

$$\Theta_m = 8 \sum_{n=0}^{\infty} \frac{(dR_n/d\eta)_{\eta=1}}{\lambda_n^3 (\partial R_n/\partial \lambda_n)_{\eta=1}}. \qquad (6.111b)$$

From the temperature distribution, Equation 6.107, we have

$$\left(\frac{\partial \Theta}{\partial \eta}\right)_{\eta=1} = -2 \sum_{n=0}^{\infty} \frac{(dR_n/d\eta)_{\eta=1}}{\lambda_n (\partial R_n/\partial \lambda_n)_{\eta=1}} e^{-\lambda_n^2 \xi}. \qquad (6.112)$$

Thus, the Nusselt number becomes

$$Nu_x = \frac{\sum_{n=0}^{\infty} A_n e^{-\lambda_n^2 \xi}}{2 \sum_{n=0}^{\infty} \left(A_n/\lambda_n^2\right) e^{-\lambda_n^2 \xi}}, \qquad (6.113)$$

where

$$A_n = \frac{(dR_n/d\eta)_{\eta=1}}{\lambda_n (\partial R_n/\partial \lambda_n)_{\eta=1}}. \qquad (6.114)$$

Equation 6.113, which gives the variation of the Nusselt number in the thermal entrance region of a circular pipe with fully developed laminar flow under constant wall temperature boundary condition, can be evaluated at any $x$-location along the pipe by using the values of the constants $\lambda_n$, $C_n$, and $A_n [= -(1/2)C_n R_n'(1)]$ given by Brown [14] and Sellars et al. [19], where $R_n'(1) = (dR_n/d\eta)_{\eta=1}$. The eigenvalues $\lambda_n$ and the constants $C_n$ and $A_n$ of this problem as presented by Sellars et al. [19] are given in Table 6.2.

**TABLE 6.2**

First 10 Eigenvalues and Constants of the Graetz Problem for Laminar Flow in a Circular Pipe

| $n$ | $\lambda_n^2$ | $C_n$ | $A_n = -\dfrac{1}{2}C_n R_n'(1)$ |
|---|---|---|---|
| 0 | 7.1129 | +1.47989 | 0.7303 |
| 1 | 44.489 | −0.80345 | 0.53810 |
| 2 | 113.785 | +0.58732 | 0.460074 |
| 3 | 215.121 | −0.474993 | 0.413743 |
| 4 | 348.457 | +0.404448 | 0.381785 |
| 5 | 513.793 | −0.355345 | 0.357853 |
| 6 | 711.129 | +0.318858 | 0.338988 |
| 7 | 940.465 | −0.290488 | 0.323555 |
| 8 | 1201.8 | +0.267691 | 0.310596 |
| 9 | 1495.1 | −0.248895 | 0.29950 |

*Note:* Constant wall temperature [19].

**TABLE 6.3**

More Accurate Values of the
Eigenvalues and Constants of the
Graetz Problem for Laminar
Flow in a Circular Pipe

| $n$ | $\lambda_n$ | $C_n$ | $A_n$ |
|---|---|---|---|
| 0 | 2.7043644 | +1.46622 | 0.74879 |
| 1 | 6.679032 | −0.802476 | $S$ |
| 2 | 10.67338 | +0.587094 | 0.46288 |
| 3 | 14.67108 | −0.474897 | 0.41518 |
| 4 | 18.66987 | +0.404402 | 0.38237 |

*Note:* Constant wall temperature [22].

The solution presented by Sellars et al. [19] is valid for large $\lambda_n$. By employing a fairly rapidly converging series solution [21] of the Graetz problem, Lipkis [22] calculated to about five significant digits the lowest five values of the eigenvalues $\lambda_n$ and the constants $C_n$ and $A_n$. These values are sufficiently accurate for low values of $n$ so that better and more complete solutions could be obtained. These values are given in Table 6.3.

For large values of $(x/d)/RePr$ (>0.05), only the first term is of significance; then from Equation 6.113, with the value of $\lambda_0$ given in Table 6.3, we get

$$Nu_\infty = \frac{\lambda_0^2}{2} = \frac{(2.7043644)^2}{2} = 3.657,$$ (6.115)

which is the Nusselt number for fully developed temperature profile under constant wall temperature boundary condition.

Hausen [23] proposed the following correlation as representing the Graetz solution for constant wall temperature and parabolic velocity distribution in a pipe:

$$Nu = \frac{hd}{k} = 3.66 + \frac{0.0668(d/x)Pe}{1+0.04[(d/x)Pe]^{2/3}}.$$ (6.116)

This equation, in fact, gives the mean Nusselt number over a length $x$ of pipe. For large values of $(x/d)/Pe$, Equation 6.116 yields

$$Nu = 3.66.$$ (6.117)

The variations of the local and mean Nusselt numbers in the thermal entry region of a circular pipe with constant surface temperature (and constant wall heat flux) are shown in Figure 6.16 (see also Figure 6.17).

For small values of $(x/d)/RePr$, Equation 6.113 requires the use of a large number of terms. Sellars et al. [19] proposed the following relation for this case:

$$Nu_x = 1.3565\left(\frac{x/r_0}{Pe}\right)^{-1/3}, \quad \frac{x/r_0}{Pe} = \leq 0.01.$$ (6.118)

A numerical solution of this problem has also been given by Kays [1]. His findings are tabulated in Table 6.4.

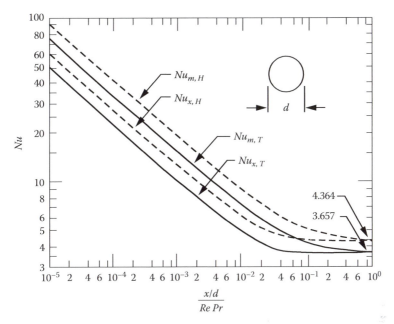

**FIGURE 6.16**
The local and mean Nusselt numbers in the thermal entrance region of a circular duct. (From Shah, R.K. and Bhatti, M.S., Laminar convective heat transfer in ducts, in *Handbook of Single-Phase Convective Heat Transfer*, S. Kakaç, R.K. Shah, and W. Aung, Eds., John Wiley & Sons, New York, 1987, Chapter 3; Shah, R.K. and London, A.L., *Laminar Flow Forced Convection in Ducts*, Supplement 1, *Advances in Heat Transfer*, Academic Press, New York, 1978.)

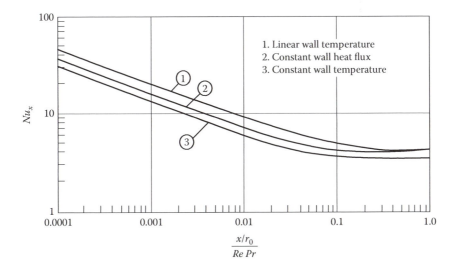

**FIGURE 6.17**
The local Nusselt numbers for laminar flow in a circular duct. (From Sellars, J.R. et al., *Trans. ASME*, 78, 441, 1956.)

**TABLE 6.4**

Numerical Solutions for the Local and Mean Nusselt Numbers in a Circular Duct with Constant Wall Temperature and Parabolic Velocity Profile

| $4(x/d)/Pe$ | $Nu_x$ | $Nu$ | $4(x/d)/Pe$ | $Nu_x$ | $Nu$ |
|---|---|---|---|---|---|
| 0.010 | 7.77 | 11.83 | 0.160 | 3.78 | 4.95 |
| 0.020 | 6.12 | 9.43 | 0.200 | 3.72 | 4.71 |
| 0.040 | 4.96 | 7.44 | 0.800 | 3.66 | 3.93 |
| 0.080 | 4.18 | 5.97 | $\infty$ | 3.66 | 3.66 |

*Note:* Thermal entrance region [1].

### 6.9.3 Extensions of the Graetz Problem

The classical Graetz problem has also been extended by Sellars et al. [19] for boundary conditions other than the constant wall temperature. The energy equation to be solved for the temperature distribution is

$$(1-\eta^2)\frac{\partial T}{\partial \xi} = \frac{1}{\eta}\frac{\partial}{\partial \eta}\left(\eta \frac{\partial T}{\partial \eta}\right),$$

(6.119)

where we have introduced

$$\xi = \frac{x/r_0}{Pe} \quad \text{and} \quad \eta = \frac{r}{r_0}.$$

(6.120a,b)

#### 6.9.3.1 Constant Wall Heat Flux

For this case, the inlet and boundary conditions are

$$T(0,\eta) = T_i,$$

(6.121a)

$$\frac{\partial T(\xi,1)}{\partial \eta} = \frac{q_w'' r_0}{k} = \text{constant},$$

(6.121b)

$$\frac{\partial T(\xi,0)}{\partial \eta} = 0,$$

(6.121c)

where $q_w''$ represents the prescribed constant wall heat flux. In this section, we will present the results given by Sellars et al. [19]: The local Nusselt number for the thermal entrance region is given by

$$Nu_x = \frac{1}{\dfrac{11}{48} + \dfrac{1}{2}\displaystyle\sum_{n=1}^{\infty}\left(e^{-\beta_n^2\xi}\big/\beta_n^4 R'\left(-\beta_n^2\right)\right)},$$

(6.122)

where the first three values of $\beta_n^2$ and $R'(-\beta_n^2)$ for a circular tube are given in Table 6.5.

**TABLE 6.5**

Values of $\beta_n^2$ and $R'(-\beta_n^2)$ in
Equation 6.122

| $n$ | $\beta_n^2$ | $-R'(-\beta_n^2)$ |
|---|---|---|
| 1 | 25.639 | $8.854 \times 10^{-3}$ |
| 2 | 84.624 | $2.062 \times 10^{-3}$ |
| 3 | 176.40 | $9.434 \times 10^{-4}$ |

*Source:* Sellars, J.R. et al., *Trans.
ASME*, 78, 441, 1956.

Equation 6.122 shows that the asymptotic value of the Nusselt number becomes

$$Nu_\infty = 4.364. \tag{6.123}$$

In fact, when $(x/r_0)Pe \geq 0.25$, $Nu \cong 4.364$. For small values of $(x/r_Q)/Pe$, the local Nusselt number can be approximated as

$$Nu_x = 1.6393 \left( \frac{x/r_0}{Pe} \right)^{1/3}, \quad \frac{x/r_0}{Pe} < 0.01. \tag{6.124}$$

The uniform heat-flux thermal entry region problem for laminar flow has also been solved by the method of separation of variables and the Sturm–Liouville theory by Siegel et al. [24], and the result for the local Nusselt number is given as

$$Nu_x = \left[ \frac{11}{48} - \frac{1}{2} \sum_{n=1}^{\infty} \frac{\exp\left(-\beta_n^2 \xi\right)}{A_n \beta_n^4} \right]^{-1}, \tag{6.125}$$

where $\beta_n^2$ and $A_n$ are given in Table 6.6. Table 6.7 shows the variation of $Nu_x$ from Equation 6.125 in the thermal entrance region.

For large values of $(x/r_0)/Pe$, Equation 6.125 gives $Nu_\infty = 4.364$ as expected. A numerical solution to this problem was also given by Kays [1]. His findings are summarized in Table 6.8.

**TABLE 6.6**

First Five Eigenvalues $\beta_n$ and
Constants $A_n$ in the Thermal
Entrance Solution (6.125) in a
Circular Pipe with $q_w'' = $ Constant

| $n$ | $\beta_n^2$ | $A_n$ |
|---|---|---|
| 1 | 25.68 | $7.630 \times 10^{-3}$ |
| 2 | 83.86 | $2.058 \times 10^{-3}$ |
| 3 | 174.2 | $0.901 \times 10^{-3}$ |
| 4 | 296.5 | $0.487 \times 10^{-3}$ |
| 5 | 450.9 | $0.297 \times 10^{-3}$ |

*Source:* Siegel, R. et al., *Appl. Sci.
Res.*, 7A, 386, 1958.

**TABLE 6.7**

Nusselt Numbers for the Circular
Tube ($q_w'' = $ Constant)

| $(x/r_0)/Pe$ | $Nu_x$ | $(x/r_0)/Pe$ | $Nu_x$ |
|---|---|---|---|
| 0 | | 0.020 | 6.14 |
| 0.002 | 12.00 | 0.040 | 5.19 |
| 0.004 | 9.93 | 0.100 | 4.51 |
| 0.010 | 7.49 | ∞ | 4.36 |

*Note:* Thermal entrance solution [24].

**TABLE 6.8**

Numerical Solution for the Case
of Laminar Flow in a Circular Tube
with Constant Wall Heat Flux and
Parabolic Velocity Profile

| $(x/r_0)/Pe$ | $Nu_x$ | $(x/r_0)/Pe$ | $Nu_x$ |
|---|---|---|---|
| 0.0025 | 13.81 | 0.050 | 5.11 |
| 0.005 | 10.40 | 0.100 | 4.59 |
| 0.010 | 8.06 | 0.200 | 4.45 |
| 0.020 | 6.56 | 0.400 | 4.40 |
| 0.030 | 5.76 | ∞ | 4.364 |
| 0.040 | 5.37 | | |

*Source:* Kays, W.M., *Trans. ASME*, 77, 1265,
1955.

### 6.9.3.2 Linear Wall Temperature Variation

For this case, the inlet and the boundary conditions are as follows:

$$T(0, \eta) = T_i, \tag{6.126a}$$

$$T(\xi, 1) = Tw(\xi) = T_i + C\xi, \tag{6.126b}$$

$$\frac{\partial T(\xi, 0)}{\partial \eta} = 0, \tag{6.126c}$$

where $C$ is a prescribed constant. The local Nusselt number is found to be [19]

$$Nu_x = \frac{(1/2) + 4\sum_{n=1}^{\infty}(C_n/2)\left(R_n'(1)/\lambda_n^4\right)e^{-\lambda_n^2\xi}}{(88/768) + 8\sum_{n=1}^{\infty}(C_n/2)\left(R_n'(1)/\lambda_n^4\right)e^{-\lambda_n^2\xi}}, \tag{6.127}$$

where the values of the pertinent quantities are given in Table 6.2. In this case, for small values of $(x/r_0)/Pe$, the local Nusselt number can be approximated by

$$Nu_x = 2.0348\left(\frac{x/r_0}{Pe}\right)^{-1/3}, \quad \frac{x/r_0}{Pe} < 0.01 \tag{6.128}$$

For far downstream the circular duct ($\xi \to \infty$), the asymptotic value of the Nusselt number becomes

$$Nu_\infty = 4.364. \tag{6.129}$$

In fact, when

$$\left(\frac{x}{r_0}\right)Pe \geq 0.5, \quad Nu \cong 4.364.$$

Figure 6.17 shows the variation of the local Nusselt number along the pipe for the previous three cases.

It is interesting to note that the asymptotic values given by Equations 6.123 and 6.129 are the same. But the thermal entrance length for the constant wall heat-flux case is one-half the thermal entrance length for the linear wall temperature case.

## 6.10 Laminar Flow Heat Transfer in the Combined Entrance Region of Circular Ducts

Heat-transfer phenomena in the combined entrance region of ducts have been examined widely due to their practical and theoretical importance. In the combined entrance region, simultaneous development of the hydrodynamic and thermal boundary layers has to be considered. This requires the knowledge of the development of the velocity profile in the entrance region, which is given in Reference 25. For integral methods of solution, profiles of various orders have been used in the literature. In the differential treatment, however, actual (exact or approximate) velocity components obtained for the entrance region had to be used.

Kays [1] solved the combined entrance region problem in a pipe for $Pr = 0.7$ by employing Langhaar's velocity profiles [26]. In his solution he neglected the effect of the radial velocity component and numerically integrated the following form of the energy equation:

$$\frac{u}{U_m}\frac{\partial \Theta}{\partial X} = \frac{\partial \Theta}{\partial \eta^2} + \frac{1}{\eta}\frac{\partial \Theta}{\partial \eta}, \tag{6.130}$$

where

$$X = \frac{2(x/r_0)}{Pe}$$

$$\eta = \frac{r}{r_0}$$

$$\Theta = \frac{T}{T_i}$$

For the case of uniform velocity profile at the pipe entrance, the velocity distribution in the entrance region is a function of $\eta$ and $X$ and has been given by Langhaar [26] as

$$\frac{u}{U_m} = \frac{I_0(\gamma) - I_0(\gamma\eta)}{I_2(\gamma)}, \tag{6.131}$$

where $I_0$ represents the Bessel function of the first kind of zero order (see Appendix C) and the quantity $\gamma$ is given by

$$\gamma = \psi\left(\frac{x/d}{Re}\right) = \psi(XPr), \tag{6.132}$$

which is tabulated in Reference 26.

The local Nusselt number as calculated from Equation 6.53 becomes

$$Nu_x = \frac{2(\partial\Theta(X,1)/\partial\eta)}{\Theta_w - \Theta_m}, \tag{6.133}$$

where

$$\Theta_m = 2\pi \int_0^1 \frac{u}{U_m} \Theta\eta\, d\eta \tag{6.134}$$

and

$$\Theta_w = \frac{T_w}{T_i}.$$

Integrating Equation 6.130 for the given boundary conditions yields $\Theta_w = \Theta(X,\eta)$. Kays presented solutions for the following cases obtained by numerical integration [1]:

1. Constant wall temperature, velocity, and temperature uniform at duct entrance
2. Constant wall-to-fluid temperature difference, velocity, and temperature uniform at duct entrance
3. Constant heat input per unit of tube length, velocity, and temperature uniform at duct entrance

The results of Kay's work are presented in Tables 6.9 through 6.11 for the previous three cases, respectively. The results for the variation of the local Nusselt numbers are also shown in Figure 6.18. In Figure 6.19, the mean Nusselt numbers for the constant wall temperature and constant temperature difference solutions are plotted.

The Nusselt number is infinitely large at the inlet and decreases rapidly in the direction of flow, asymptotically approaching a constant value as the temperature distribution becomes fully developed. This constant value is 3.66 when the wall temperature is uniform and 4.364 when the wall heat flux is constant or the wall temperature variation is linear.

**TABLE 6.9**

Variation of the Local and Average Nusselt Numbers
in the Combined Entrance Region

| 4(x/d)/Pe | $Nu_x$ | $Nu$ | 4(x/d)/Pe | $Nu_x$ | $Nu$ |
|-----------|--------|------|-----------|--------|------|
| 0.010 | 11.31 | 17.44 | 0.160 | 3.88 | 5.83 |
| 0.020 | 7.90 | 13.36 | 0.200 | 3.77 | 5.43 |
| 0.040 | 5.82 | 10.00 | 0.800 | 3.66 | 4.11 |
| 0.080 | 4.57 | 7.53 | ∞ | 3.66 | 3.66 |

*Note:* Constant wall temperature, Langhaar's velocity, and $Pr = 0.7$ [1].

**TABLE 6.10**

Variation of the Local and Average Nusselt Numbers
in the Combined Entrance Region

| 4(x/d)Pe | $Nu_x$ | $Nu$ | 4(x/d)Pe | $Nu_x$ | $Nu$ |
|----------|--------|------|----------|--------|------|
| 0.010 | 11.80 | 17.81 | 0.160 | 4.56 | 6.44 |
| 0.020 | 8.36 | 13.76 | 0.200 | 4.48 | 6.06 |
| 0.040 | 6.40 | 10.48 | 0.800 | 4.39 | 4.82 |
| 0.080 | 5.21 | 8.08 | ∞ | 4.364 | 4.364 |

*Note:* Constant wall temperature difference, Langhaar's velocity, and
$Pr = 0.7$ [1].

**TABLE 6.11**

Variation of the Local Nusselt Number
in the Combined Entrance Region

| 4(x/d)/Pe | $Nu_x$ | 4(x/d)/Pe | $Nu_x$ |
|-----------|--------|-----------|--------|
| 0.010 | 14.53 | 0.160 | 4.97 |
| 0.020 | 10.54 | 0.200 | 4.78 |
| 0.040 | 7.80 | 0.800 | 4.45 |
| 0.080 | 6.08 | ∞ | 4.364 |

*Note:* Constant heat input, Langhaar's velocity,
and $Pr = 0.7$ [1].

As seen in Figure 6.18, Kay's results predict that the asymptotic values are reached when $(x/d)/RePr \cong 0.05$. The thermal entrance length $L_t$, over which the temperature distribution becomes fully developed can therefore be estimated by $L_t/d \cong 0.05RePr$. On the other hand, the hydrodynamic entrance $L_h$, required for the velocity distribution to develop varies with the Reynolds number approximately as $L_h/d \cong 0.056Re$. Thus, for fluids of very small Prandtl numbers, the thermal entrance length would be very short compared with the hydrodynamic entrance length. The heat-transfer coefficient in this case can be estimated by assuming a uniform velocity distribution (i.e., slug flow assumption). For fluids of large Prandtl numbers, the thermal entrance length is much longer than the hydrodynamic entrance length. In this case, results for fully developed parabolic velocity distribution can be used to predict the heat-transfer coefficient.

The solutions presented so far are for fluids with constant thermophysical properties. When temperature differences are small, such an assumption would be quite satisfactory.

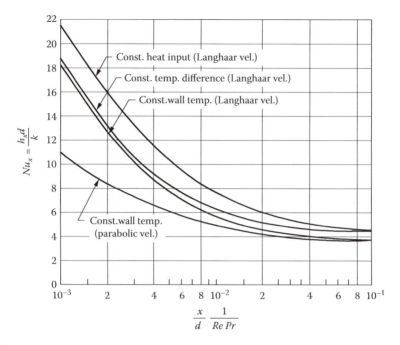

**FIGURE 6.18**
The local Nusselt numbers in the simultaneously developing region of a circular duct, $Pr = 0.7$. (From Kays, W.M., *Trans. ASME*, 77, 1265, 1955.)

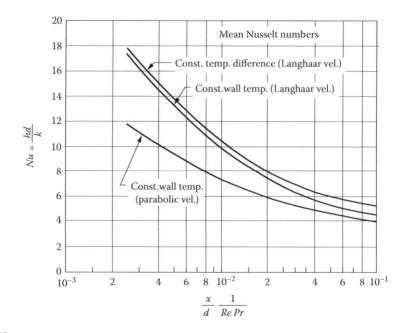

**FIGURE 6.19**
The mean Nusselt numbers in the simultaneously developing region of a circular duct, $Pr = 0.7$. (From Kays, W.M., *Trans. ASME*, 77, 1265, 1955.)

For large temperature differences, however, the fluid velocity distribution is influenced by the variation of the fluid properties with temperature as indicated in Figure 6.7. The effect of property variations on forced convection heat transfer will be deferred until Chapters 7 and 10.

The following empirical equations suggested by Hausen [23] can be used to approximate the solutions given in Reference 1. These results are more convenient to the practicing engineers.

*Constant wall temperature* (Langhaar's velocity profile)

$$Nu = 3.66 + \frac{0.104(RePr/(x/d))}{1 + 0.016(RePr/(x/d))^{0.8}}$$ (6.135)

*Constant temperature difference* (Langhaar's velocity profile)

$$Nu = 4.366 + \frac{0.10(RePr/(x/d))}{1 + 0.016(RePr/(x/d))^{0.8}}$$ (6.136)

*Constant heat input* (parabolic velocity profile)

$$Nu = 4.36 + \frac{0.023(RePr/(x/d))}{1 + 0.0012(RePr/(x/d))^{0.8}}$$ (6.137)

*Constant heat input* (Langhaar's velocity profile)

$$Nu = 4.36 + \frac{0.036(RePr/(x/d))}{1 + 0.0011(RePr/(x/d))^{0.8}}$$ (6.138)

It has been shown that for fluids in laminar flow in the Prandtl number range of gases, the assumption of a fully developed parabolic velocity distribution for heat-transfer calculations can result in a substantial underestimation of the Nusselt number if heat transfer starts at the pipe entrance.

Ulrichson and Schmitz [27] and Manohar [28] also obtained numerical solutions for the case of simultaneously developing velocity and temperature profiles in laminar flow in the entrance region of a circular tube for $Pr = 0.7$. Ulrichson and Schmitz used values for the axial velocity components taken from the work of Langhaar [26] and obtained the radial component from the continuity equation and Langhaar's profiles. Manohar's work was a numerical solution of the complete governing equations. These calculations were a refinement of the work of Kays [1], who had neglected the effect of the radial velocity component. The calculations of Ulrichson and Schmitz show a significant decrease in the local Nusselt number in the entrance region. For the constant wall heat-flux case, their results predict that the local Nusselt number close to the inlet is overestimated by about 6% if the radial velocity is neglected. On the other hand, their calculations for the constant wall temperature case show that the relative error in the local Nusselt number near the inlet

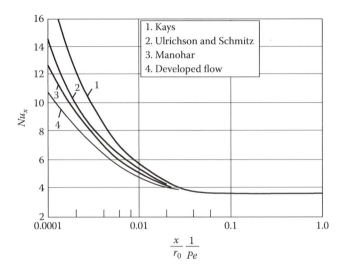

**FIGURE 6.20**
Variation of the local Nusselt number in the simultaneously developing region in a circular duct for constant wall temperature, $Pr = 0.7$. (From Manohar, R., *Int. J. Heat Mass Transfer*, 12, 15, 1969.)

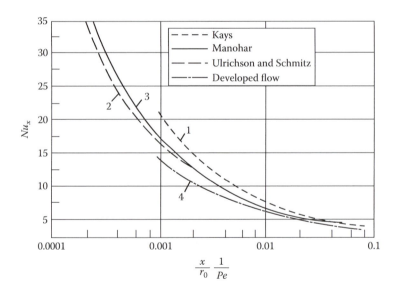

**FIGURE 6.21**
Variation of the local Nusselt number in the simultaneously developing region in a circular duct for constant wall heat flux, $Pr = 0.7$. (From Manohar, R., *Int. J. Heat Mass Transfer*, 12, 15, 1969.)

may be 15% larger than when the radial velocity components are taken to be zero. Figures 6.20 and 6.21 give the local Nusselt number from the solutions of Ulrichson, Schmitz, and Manohar, together with those of Kays, for both the constant wall temperature and the constant wall heat-flux cases. In the case of constant heat-flux case, the results of Ulrichson and Schmitz are in excellent agreement with those obtained by Manohar for $(x/r_0)/Pe > 0.008$.

Kakaç and Özgü [25] also presented solutions for incompressible steady laminar flow heat transfer in the combined entrance region of a circular tube for the cases of constant wall heat flux and constant wall temperature. They used the velocity profiles obtained by Sparrow et al. [29]. Figure 6.22 illustrates the variations of the local Nusselt number for the two cases considered in this work, and a comparison of the results of this work with those of References 1 and 27 is given in Table 6.12. It should be noted that the Reynolds number is based on diameter in Reference 25 and on radius in Reference 27.

Roy [30] made an integral treatment of the laminar heat transfer in the inlet of a tube for the constant heat-flux boundary condition with the assumption that the fluid enters the tube with a uniform velocity and temperature. He considered fluids of Prandtl numbers of 1, 10, 100, and 1000.

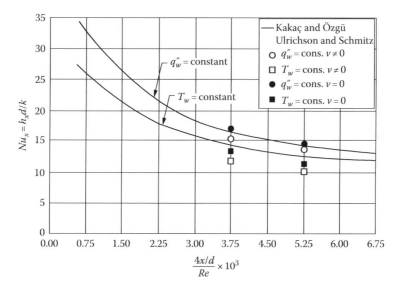

**FIGURE 6.22**
Variation of the local Nusselt numbers in the combined entrance region of a circular duct for $T_w$ constant and $q_w''$ = constant, $Pr = 0.7$. (From Sparrow, E.M. et al., *Phys. Fluids*, 7(3), 338, 1964.)

**TABLE 6.12**

Comparison of the Local Nusselt Numbers in the Combined Entrance Region of a Circular Tube

| $\frac{4x/d}{Re} \times 10^3$ | [25] | [27] $v \neq 0$ | [27] $v = 0$ | [1] |
|---|---|---|---|---|
| 3.75 | 16.72 | 15.75 | 17.24 | 19 |
| 4.50 | 15.58 | 14.75 | 15.8 | 17.5 |
| 5.25 | 14.58 | 14.0 | 14.56 | 16.3 |
| 6.00 | 13.92 | 13.0 | 13.92 | 15.5 |
| 3.75 | 14.58 | 12.0 | 13.44 | 16.0 |
| 4.50 | 13.5 | 11.0 | 12.4 | 14.5 |
| 5.25 | 12.92 | 10.21 | 11.52 | 13.2 |
| 6.00 | 12.5 | 9.68 | 10.90 | 12.5 |

*Note:* Constant wall heat flux, $Pr = 0.7$.

## 6.11  Laminar Convective Heat Transfer between Two Parallel Plates

This problem was first solved by Lévêque [31] in 1928. His method was essentially an extension of the method used by Boussinesq [32] in the flat plate case. Lévêque treated the problem as if the fluid were a solid slab moving between fixed heated plates under the following assumptions:

1. The $y$-components of the velocity, $v$, are everywhere zero.
2. $k/\rho c_p u$ is uniformly constant throughout the fluid.
3. $T_i$ is constant.
4. $T_w = T_i$ if $x < 0$, and $T_w$ = constant $\neq T_i$ if $x > 0$.
5. The temperature of the fluid in contact with the plates is the same as that of the plates.
6. Thermal conductivity is zero in the direction of fluid motion.
7. $T \to T_w$ as $x \to \infty$.

These are essentially the same assumptions made by Boussinesq in the case of single flat plate. Under these conditions, Lévêque solved the following form of the energy equation:

$$c_p \rho u \frac{\partial T}{\partial x} = k \frac{\partial^2 T}{\partial y^2}. \tag{6.139}$$

### 6.11.1  Cartesian Graetz Problem for Slug Flow with Constant Wall Temperature

All the assumptions introduced in Section 6.9 will also be used here. Rewriting the energy equation (6.139) for $u = U_m$ = constant, we get

$$\frac{\rho c_p U_m}{k} \frac{\partial T}{\partial x} = \frac{\partial^2 T}{\partial y^2}. \tag{6.140}$$

The inlet and boundary conditions are (see Figure 6.23)

$$T(0,y) = T_i, \tag{6.141a}$$

$$T(x,d) = T(x,-d) = T_w, \tag{6.141b}$$

$$\frac{\partial T(x,0)}{\partial y} = 0, \tag{6.141c}$$

where $2d$ is the distance between the plates.

We now introduce the following dimensionless variables:

$$\Theta = \frac{T - T_w}{T_i - T_w}, \quad \eta = \frac{y}{d}, \quad \xi = \frac{4x/d}{RePr}, \tag{6.142a,b,c}$$

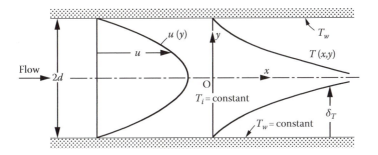

**FIGURE 6.23**
Thermal entrance region between two parallel walls.

where

$$Re = \frac{4dU_m\rho}{\mu} \quad \text{and} \quad Pr = \frac{\mu c_p}{k}. \tag{6.143a,b}$$

The energy equation (6.140) may now be rewritten in the following dimensionless form:

$$\frac{\partial \Theta}{\partial \xi} = \frac{\partial^2 \Theta}{\partial \eta^2}. \tag{6.144}$$

The inlet and the boundary conditions (6.141) become

$$\Theta(0,\eta) = 1, \tag{6.145a}$$

$$\Theta(\xi,1) = 0, \tag{6.145b}$$

$$\frac{\partial \Theta(\xi,0)}{\partial \eta} = 0. \tag{6.145c}$$

This problem may now be solved by the method of separation of variables, and accordingly we assume a solution of the form

$$\Theta(\xi,\eta) = X(\xi)Y(\eta). \tag{6.146}$$

Introducing Equation 6.146 into the energy equation (6.144) and the conditions (6.145) and also noting that $\Theta(\infty,\eta) = 0$, we obtain the following solution for $\Theta(\xi,\eta)$:

$$\Theta(\xi,\eta) = \sum_{n=0}^{\infty} A_n \cos \lambda_n \eta e^{-\lambda_n^2 \xi}, \tag{6.147}$$

where

$$\lambda_n = \frac{(2n+1)\pi}{2}, \quad n = 0,1,2,\ldots \tag{6.148}$$

The use of the nonhomogeneous inlet condition $\Theta(0,\eta) = 1$ gives

$$1 = \sum_{n=0}^{\infty} A_n \cos \lambda_n \eta \tag{6.149}$$

from which the expansion coefficients $A_n$ can be found [20]. The result is

$$A_n = 2\frac{(-1)^n}{\lambda_n}. \tag{6.150}$$

Hence, the temperature distribution becomes

$$\Theta(\xi, \eta) = 2\sum_{n=0}^{\infty} \frac{(-1)^n}{\lambda_n} \cos \lambda_n \eta e^{-\lambda_n^2 \xi}. \tag{6.151}$$

The dimensionless mean temperature $\Theta_m$ is calculated from Equation 6.50 that, in this case, reduces to

$$\Theta_m = \int_0^1 \Theta(\xi, \eta) \, d\eta. \tag{6.152}$$

Substituting Equation 6.150 into Equation 6.151, we get

$$\Theta_m = 2\sum_{n=0}^{\infty} \frac{e^{-\lambda_n^2 \xi}}{\lambda_n^2}. \tag{6.153}$$

The local heat-transfer coefficient is given by

$$h_x = \frac{k(\partial T/\partial y)_{y=d}}{T_w - T_m} = -\frac{k}{d} \frac{(\partial \Theta/\partial \eta)_{\eta=1}}{\Theta_m}. \tag{6.154}$$

Substituting Equations 6.151 and 6.152 into Equation 6.154, we obtain

$$h_x = \frac{(k/d)\sum_{n=0}^{\infty} e^{-\lambda_n^2 \xi}}{\sum_{n=0}^{\infty} \left(1/\lambda_n^2 e^{-\lambda_n^2 \xi}\right)} \tag{6.155}$$

or

$$Nu_x = \frac{4h_x d}{k} = 4\frac{\sum_{n=0}^{\infty} e^{-\lambda_n^2 \xi}}{\sum_{n=0}^{\infty} \left(1/\lambda_n^2\right) e^{-\lambda_n^2 \xi}} \tag{6.156}$$

which can also be written as

$$Nu_x = 4 \frac{\sum_{n=0}^{\infty} e^{-4\lambda_n^2 \frac{x/d}{RePr}}}{\sum_{n=0}^{\infty} \left(1/\lambda_n^2\right) e^{-4\lambda_n^2 \frac{x/d}{RePr}}}. \tag{6.157}$$

For large values of $(x/d)/RePr$, Equation 6.157 yields

$$Nu_{\infty} = \pi^2. \tag{6.158}$$

### 6.11.2 Cartesian Graetz Problem for Slug Flow with Constant Wall Heat Flux

If the same constant heat flux $q_w''$ is applied to both plates and the inlet temperature of the fluid is $T_i$, the formulation of the problems in term of $\Theta = T - T_i$ is given by

$$\frac{\partial \Theta}{\partial x} = \frac{k}{\rho c_p U_m} \frac{\partial^2 \Theta}{\partial y^2}, \tag{6.159}$$

$$\Theta(0,y) = 0, \tag{6.160a}$$

$$\frac{\partial \Theta(0,y)}{\partial y} = 0, \tag{6.160b}$$

$$\frac{\partial \Theta(x,d)}{\partial y} = \frac{q_w''}{k}, \tag{6.160c}$$

where $2d$ is the distance between the two plates.

Introducing an assumption of the form

$$\Theta(x,y) = \Psi(x,y) + \phi(y) + \chi(x), \tag{6.161}$$

the problem can be separated into the following simple problems:

$$\frac{d\chi}{dx} = \frac{k}{\rho c_p U_m} \frac{d^2\phi}{dy^2} \tag{6.162}$$

with

$$\frac{d\phi(0)}{dy} = 0, \tag{6.163a}$$

$$\frac{d\phi(d)}{dy} = \frac{q_w''}{k}, \tag{6.163b}$$

and

$$\frac{\partial \Psi}{\partial x} = \frac{k}{\rho c_p U_m} \frac{\partial^2 \Psi}{\partial y^2} \tag{6.164}$$

with

$$\Psi(0, y) = -\phi(y) - \chi(0), \tag{6.165a}$$

$$\frac{\partial \Psi(x, d)}{\partial y} = 0, \tag{6.165b}$$

$$\frac{\partial \Psi(x, d)}{\partial y} = 0. \tag{6.165c}$$

Since $X(x)$ and $\phi(y)$ can vary independently, Equation 6.162 should be equal to a constant, say $C$. Then, we obtain

$$\chi(x) = \frac{k}{\rho c_p U_m} Cx + c_1, \tag{6.166}$$

$$\phi(y) = \frac{1}{2} C y^2 + C_2 y + C_3. \tag{6.167}$$

From the boundary conditions (6.163), we conclude $C_2 = 0$ and $C = q''_w / kd$. Hence, the solutions (6.166) and (6.167) become

$$\chi(x) = \frac{q''_w}{\rho c_p U_m d} x + C_1, \tag{6.168}$$

$$\phi(y) = \frac{1}{2} \frac{q''_w}{kd} y^2 + C_3. \tag{6.169}$$

On the other hand, assuming a product solution in the form

$$\Psi(x, y) = Y(y)X(x), \tag{6.170}$$

we obtain the following solution for $\Psi(x, y)$ [20]:

$$\Psi(x, y) = A_0 + \sum_{n=1}^{\infty} A_n e^{(k/\rho U m c_p)\lambda_n^2 x} \cos \lambda_n y, \tag{6.171}$$

where

$$\lambda_n = \frac{n\pi}{d}, \quad n = 1, 2, 3, \ldots \tag{6.172}$$

The use of the inlet condition (6.165a) gives

$$\frac{q_w^n y^2}{2kd} = a_0 + \sum_{n=1}^{\infty} A_n \cos \lambda_n y, \tag{6.173}$$

where $a_0 = A_0 + C_1 + C_3$. From Equation 6.173, the coefficients $a_0 + A_n$ can be found [20], and the results are

$$a_0 = A_0 + C_1 + C_3 = -\frac{q_w^n d}{6k}, \tag{6.174a}$$

$$A_n = -(-1)^n \frac{2q_w^n d}{k(\lambda_n d)^2}. \tag{6.174b}$$

Thus, combining Equations 6.168, 6.169, and 6.173, the temperature distribution becomes

$$\Theta(x,y) = T(x,y) - T_i = \frac{q_w^n d}{k} \left[ \frac{k}{\rho c_p U_m} \frac{x}{d^2} + \frac{1}{2}\left(\frac{y}{d}\right)^2 - \frac{1}{6} - 2\sum_{n=1}^{\infty} \frac{(-1)^n}{(\lambda_n d)^2} e^{-(k/\rho c_p U_m)\lambda_n^2 x} \cos \lambda_n y \right]. \tag{6.175}$$

From Equation 6.175, we obtain the wall temperature as

$$T_w - T_i = \frac{q_w^n d}{k} \left[ \frac{1}{3} + \frac{k}{\rho c_p U_m} \frac{x}{d^2} - 2\sum_{n=1}^{\infty} \frac{1}{(\lambda_n d)^2} e^{-(k/\rho c_p U_m)\lambda_n^2 x} \right]. \tag{6.176}$$

Substituting Equation 6.175 into Equation 6.50, we obtain

$$T_m - T_i = \frac{q_w^n d}{k} \cdot \frac{k}{\rho c_p U_m} \frac{x}{d^2}. \tag{6.177}$$

Hence, we can write

$$T_m - T_w = -\frac{q_w^n d}{k} \left[ \frac{1}{3} - 2\sum_{n=1}^{\infty} \frac{1}{(\lambda_n d)^2} e^{-(k/\rho c_p U_m)\lambda_n^2 x} \right]. \tag{6.178}$$

The local heat-transfer coefficient is given by

$$h_x = \frac{q_w^n}{T_w - T_m}. \tag{6.179}$$

Substituting Equation 6.178 into Equation 6.179, we get

$$h_x = \frac{k}{d}\left[\frac{1}{3} - 2\sum_{n=1}^{\infty}\frac{1}{(\lambda_n d)^3}e - \frac{k}{\rho c_p U_m}\lambda_n^2 x\right]^{-1}$$

(6.180)

and the local Nusselt number becomes

$$Nu_x = \frac{4h_x d}{k} = 4\left[\frac{1}{3} - 2\sum_{n=1}^{\infty}\frac{1}{(\lambda_n d)^3}e^{-(k/\rho c_p U_m)\lambda_n^2 x}\right]^{-1}.$$

(6.181)

For large values of $x$, Equation 6.181 reduces to

$$Nu_\infty = 12.$$

### 6.11.3 Cartesian Graetz Problem for Parabolic Velocity Profile with Constant Wall Temperature

The energy equation to be solved in the thermal entrance region with parabolic velocity distribution between the two parallel walls separated by a distance $2d$ as shown in Figure 6.23 is given by

$$\frac{3}{2}U_m\left[1 - \left(\frac{y}{d}\right)^2\right]\frac{\partial T}{\partial x} = \alpha\frac{\partial^2 T}{\partial y^2},$$

(6.182)

where $U_m$ is the mean flow velocity. Introduce the following dimensionless quantities

$$\Theta = \frac{T - T_w}{T_i - T_w}, \quad \eta = \frac{y}{d}, \quad \text{and} \quad \xi = 4\frac{x/d}{RePr},$$

(6.183a,b,c)

where
$T_w$ is the temperature of the plates for $x \geq 0$
$T_i$ is the temperature of the fluid at $x = 0$

$$Re = \frac{4U_m d}{v}.$$

(6.184)

The energy equation (6.182) can then be rewritten in the following dimensionless form:

$$\frac{3}{2}(1 - \eta^2)\frac{\partial\Theta}{\partial\xi} = \frac{\partial^2\Theta}{\partial\eta^2}.$$

(6.185)

The inlet and boundary conditions for $\Theta(\xi,\eta)$ are given by

$$\Theta(0,\eta) = 1,$$

(6.186a)

$$\frac{\partial \Theta(\xi,0)}{\partial \eta} = 0, \tag{6.186b}$$

$$\Theta(\xi,1) = 0. \tag{6.186c}$$

Solution to Equation 6.185 under the conditions (6.186) is given by

$$\Theta(\xi,\eta) = \sum_{n=0}^{\infty} C_n Y_n(\eta) e^{-(2/3)\lambda_n^2 \xi}, \tag{6.187}$$

where $\lambda_n$ and $Y_n(\eta)$ are the eigenvalues and eigenfunctions of the following Sturm–Liouville-type eigenvalue problem:

$$\frac{d^2 Y}{d\eta^2} + \lambda^2 (1 - \eta^2) Y = 0, \tag{6.188}$$

$$\frac{dY(0)}{d\eta} = 0, \tag{6.189a}$$

$$Y(1) = 0. \tag{6.189b}$$

The previous eigenvalue problem was investigated in detail by a number of workers: Prins et al. [33] presented a series solution and were able to find numerically first three eigenvalues and eigenfunctions. Sellars et al. [19] determined the higher eigenvalues and obtained asymptotic formulas for the eigenfunctions using a matched asymptotic expansion technique. Recently, Mengüç [34] developed a modified series solution from which all the eigenvalues and eigenfunctions can be determined numerically on a computer. Here, we present in Table 6.13 the results of Sellars et al. to be used in Equation 6.187.

**TABLE 6.13**

First 10 Eigenvalues and the Constants of Cartesian Graetz Problem with Constant Wall Temperature

| $n$ | $\lambda_n^2$ | $C_n$ | $-A_n$ |
|---|---|---|---|
| 0 | 2.779 | +0.503 | 0.683 |
| 1 | 32.11 | −0.121 | 0.454 |
| 2 | 93.45 | +0.0648 | 0.380 |
| 3 | 186.9 | −0.0431 | 0.338 |
| 4 | 312.2 | +0.0319 | 0.311 |
| 5 | 469.6 | −0.0253 | 0.291 |
| 6 | 658.9 | +0.0207 | 0.274 |
| 7 | 880.3 | −0.0174 | 0.262 |
| 8 | 1134 | +0.0150 | 0.251 |
| 9 | 1419 | −0.0131 | 0.242 |

*Source:* Sellars, J.R. et al., *Trans. ASME*, 78, 441, 1956.

Using Equation 6.187, the local Nusselt number can be obtained as follows:

$$Nu_x = \frac{4h_x d}{k} = \frac{8}{3} \frac{\sum_{n=0}^{\infty} A_n e^{(-2/3)\lambda_n^2 \xi}}{\sum_{n=0}^{\infty} (1/\lambda_n^2) A_n e^{(-2/3)\lambda_n^2 \xi}}, \tag{6.190}$$

where

$$A_n = C_n \frac{dY_n(1)}{d\eta}. \tag{6.191}$$

Norris and Streid [35] proposed the following relations for laminar flow heat transfer in a parallel-plate channel for the case of constant wall temperatures:

$$Nu = \frac{hd_H}{k} = 1.85 \left( \frac{x/d_H}{Pe} \right)^{-1/3}, \quad \frac{x/d_H}{Pe} < 0.0005 \tag{6.192}$$

(*properties at arithmetic mean temperature*)

$$Nu = 1.85 \left( \frac{x/d_H}{Pe} \right)^{-1/3}, \quad \frac{x/d_H}{Pe} < 0.014 \tag{6.193}$$

(*properties at bulk mean temperature*)

$$Nu = 7.60, \quad \frac{x/d_H}{Pe} > 0.014, \tag{6.194}$$

where $d_H = 4d$ is the equivalent (i.e., hydraulic) diameter. For sufficiently small values of $Pe/(x/d_H)$, the theoretical results practically coincide with their asymptotic value so that

$$Nu = \frac{1}{2} \frac{Pe}{(x/d_H)}, \quad \frac{Pe}{(x/d_H)} < 6 \tag{6.195}$$

(*properties at arithmetic mean temperature*)

These relations, which were obtained by a theoretical study of laminar flow heat transfer between parallel plates, are also applicable to flat rectangular ducts when all sides are at the same uniform temperature.

## 6.12 Integral Method

Heat transfer in the combined and the thermal entrance regions of a rectangular duct can easily be studied by the use of the integral method. In this section, we present two solutions in the thermal entrance region of a flat rectangular duct by the integral method. The physical model consists of the following assumptions:

1. The flow is laminar.
2. All fluid properties are constant.

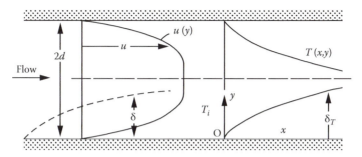

**FIGURE 6.24**
Development of the velocity and thermal boundary layers along the walls of a flat rectangular duct.

3. Both duct walls have the same uniform heat flux $q_w''$ imposed on them, or they are at the same uniform temperature $T_w$.

4. The viscous dissipation and the work of compression are both negligible.

5. There exists a thermal boundary layer of definite thickness $\delta_T$. Since the temperature of the fluid entering the duct differs from the temperature of the duct walls, there will be heat transfer to (or from) the fluid. The assumption will be made that the effect of the heat transfer plays a role only within the thermal boundary layer. The fluid outside the thermal boundary layer will be uninfluenced by the heat transfer and will therefore have a uniform temperature equal to the value $T_i$ at the thermal entrance ($x = 0$) of the duct.

6. The thermal boundary-layer thickness will be assumed zero at the entrance section at $x = 0$.

Figure 6.24 shows diagrammatically the developing velocity and thermal boundary layers. The velocity profile is assumed to be fully developed at $x = 0$. Therefore, the heated (or cooled) section starts from $x = 0$, and the thermal boundary layer grows in thickness along the length of the duct until it reaches the centerline where it meets the boundary layer from the other wall of the duct.

### 6.12.1 Constant Wall Heat Flux

The energy integral equation (5.26) would be applicable. Thus,

$$q_w'' = \frac{d}{dx}\left[\int_0^{\delta_T}\rho c_p u(T - T_i)dy\right],\tag{6.196}$$

which requires only that the conservation of energy be satisfied for the cross section as a whole. This is much less stringent requirement than that imposed by Equation 6.139.

Referring to Figure 6.24, the velocity profile to be used in Equation 6.196, from Equation 6.17a, can be written as

$$\frac{u}{U_c} = 2\left(\frac{y}{d}\right) - \left(\frac{y}{d}\right)^2.\tag{6.197}$$

For the temperature distribution in the thermal boundary layer, the following profile is chosen:

$$T - T_i = \frac{q_w'' \delta_T}{3k} \left[ 2 - 3\frac{y}{\delta_T} - \left(\frac{y}{\delta_T}\right)^3 \right], \quad 0 \le y \le \delta_T,$$

(6.198)

which is identical to Equation 5.94. Substituting Equations 6.197 and 6.198 into the energy integral equation (6.196), we obtain

$$q_w'' = \frac{d}{dx} \left[ \frac{\rho c_p U_c q_w''}{3k} \delta_T^2 \int_0^1 (2\eta\xi - \eta^2\xi^2)(2 - 3\eta + \eta^3) d\eta \right],$$

(6.199)

where

$$\eta = \frac{y}{\delta_T} \quad \text{and} \quad \xi = \frac{\delta_T}{d}.$$

(6.200a,b)

Performing the integral in Equation 6.199, we get

$$\frac{d\xi^3}{dx} = \frac{15}{2} \frac{k}{\rho c_p U_c d^2},$$

(6.201)

where we have neglected the higher-order terms in $\xi$ since $\xi < 1$. Solution of Equation 6.201 can be written as

$$\xi^3 = \frac{15}{2} \frac{k}{\rho c_p U_c d^2} x + C.$$

(6.202)

Since $\xi = 0$ at $x = 0$, we have $C = 0$. Hence, the solution for $\xi$ becomes

$$\xi = \sqrt[3]{\frac{80x/d_H}{RePr}},$$

(6.203)

where

$$d_H = 4d$$

$$Re = \frac{U_m d_H}{v}$$

$$U_m = \frac{2}{3} U_c$$

From Equation 6.198, we have

$$T_w - T_i = \frac{2q_w''}{3k} \delta_T.$$

(6.204a)

Since $\delta_T = \xi d$, Equation 6.204a can also be written as

$$T_w - T_i = \frac{2q_w''d}{3k}\sqrt[3]{\frac{80x/d_H}{RePr}}. \qquad (6.204b)$$

The heat-transfer coefficient based on $T_w - T_i$ is given by

$$h_x = \frac{q_w''}{T_w - T_i}. \qquad (6.205)$$

Substituting Equation 6.204b into Equation 6.205 yields

$$h_x = \frac{6k}{d_H}\sqrt[3]{\frac{RePr}{80x/d_H}}. \qquad (6.206)$$

Then, the Nusselt number becomes

$$Nu_x = \frac{hd_H}{k} = 6\sqrt[3]{\frac{RePr}{80x/d_H}}, \quad \text{(based on } T_w - T_i\text{)}. \qquad (6.207)$$

Using the definition of the bulk temperature as given by Equation 6.50, we obtain

$$T_m - T_i = \frac{4}{d_H}\int_0^{\delta_T}\frac{u}{U_m}(T - T_i)dy = \frac{4q_w''}{3kd_H}\delta_T^2\int_0^1 (2\eta\xi - \eta^2\xi^2)(2 - 3\eta + \eta^3)d\eta,$$

which yields

$$T_m - T_i = \frac{q_w''d_H}{2k}\left(\frac{\xi^3}{10} - \frac{\xi^4}{48}\right). \qquad (6.208)$$

Combining Equations 6.204a and 6.208, we get

$$T_w - T_m = \frac{q_w''d_H}{2k}\left[\frac{\xi}{3} - \frac{\xi^3}{10} + \frac{\xi^4}{48}\right]. \qquad (6.209)$$

Hence, the heat-transfer coefficient based on $T_w - T_m$ can be written as

$$h_x = \frac{q_w''}{T_w - T_m} = \frac{k}{d_H}\frac{2}{[(\xi/3) - ((\xi^3/10) + (\xi^4/48))]} \qquad (6.210)$$

and the Nusselt number becomes

$$Nu_x = \frac{hd_H}{k} = \frac{2}{[(\xi/3) - ((\xi^3/10) + (\xi^4/48))]}, \quad \text{(based on } T_w - T_m\text{)}. \qquad (6.211)$$

### 6.12.2 Constant Wall Temperature

The parallel-plate channel under consideration is the same as that shown in Figure 6.24. An incompressible fluid with constant thermophysical properties flows in laminar motion between the plates having a constant wall temperature boundary condition. The fully developed parabolic velocity profile is given by Equation 6.197. For the temperature profile in the thermal entrance region, we assume

$$\frac{T - T_i}{T_w - T_i} = 1 - \frac{1}{2}\frac{y}{\delta_T}\left[3 - \left(\frac{y}{\delta_T}\right)^2\right], \quad 0 \le y \le \delta_T, \tag{6.212}$$

which satisfies the assumptions listed in Section 5.5. Substituting the velocity and temperature distributions into the energy integral equation (5.29) and performing the integral, we obtain a differential equation for $\xi = \delta_T/d$, the solution of which yields

$$\xi = \sqrt[3]{\frac{120x/d_H}{RePr}}. \tag{6.213}$$

Using Equations 6.212 and 6.213, the heat-transfer coefficients and the Nusselt numbers based on $T_w - T_i$ and $T_w - T_m$ can be obtained. The results are (see Problem 6.34) as follows:

$$Nu_x = \frac{hd_H}{k} = 6\sqrt[3]{\frac{RePr}{120x/d_H}}, \quad \text{(based on } T_w - T_i\text{)}, \tag{6.214}$$

$$Nu_x = \frac{hd_H}{k} = \frac{6}{\xi[1 - (3/2)((\xi^2/5) + (\xi^3/24))]}, \quad \text{(based on } T_w - T_m\text{)}. \tag{6.215}$$

The variation of the Nusselt numbers $Nu_H$ and $Nu_T$ in the thermal entrance region of a circular duct of diameter $d$ for constant heat-flux and constant wall temperature boundary conditions calculated by the integral method is given in Table 6.14.

**TABLE 6.14**

Variation of the Nusselt Numbers in the Entrance Region of a Circular Duct

| $(x/d)/RePr \times 10^3$ | $Nu_H$ | $Nu_T$ |
|---|---|---|
| 0.2 | 39.47 | 32.23 |
| 2 | 18.30 | 15.07 |
| 4 | 14.47 | 11.88 |
| 6 | 12.60 | 10.32 |
| 8 | 11.42 | 9.338 |

**TABLE 6.15**

Nusselt Numbers for Fully Developed Condition
in Laminar Flow

| Geometry | Velocity Profile | Wall Condition | $Nu_\infty$ |
|---|---|---|---|
| Circular tube | Parabolic | $q_w''$ unif. | 4.36 |
| Circular tube | Parabolic | $T_w$ unif. | 3.66 |
| Circular tube | Slug | $q_w''$ unif. | 8.00 |
| Circular tube | Slug | $T_w$ unif. | 5.75 |
| Parallel plates | Parabolic | $q_w''$ unif. | 8.23 |
| Parallel plates | Parabolic | $T_w$ unif. | 7.60 |
| Triangular duct | Parabolic | $q_w''$ unif. | 3.00 |
| Triangular duct | Parabolic | $T_w$ unif. | 2.35 |

## 6.13 Asymptotic Values of Heat-Transfer Coefficients in Ducts

As it has been discussed in the previous sections, the heat-transfer coefficient decreases
with distance along a duct and asymptotically approaches a constant value. The asymptotic values of the Nusselt numbers for various geometries are summarized in Table 6.15
(also see Table 6.1).

## 6.14 Effect of Circumferential Heat-Flux Variation

In the previous examples, it was assumed that both the wall heat flux and the wall temperature are uniform around the tube periphery. But such circumferentially uniform thermal conditions are frequently not achieved in practice. Circumferential nonuniformities
may occur in well-established areas of application such as boilers, condenser, and heat-
exchanger tubes.

The case of fully developed laminar flow in a circular pipe with axially constant heat
input, but circumferentially varying heat flux, has been presented by Reynolds [36].

Under the assumptions of incompressible, steady, laminar, and fully developed flow in
a circular pipe with negligible viscous dissipation, constant properties, circumferentially
varying wall heat flux, and negligible axial conduction, the energy equation takes the following form:

$$\frac{1}{r}\frac{\partial}{\partial r}\left(r\frac{\partial T}{\partial r}\right) + \frac{1}{r^2}\frac{\partial^2 T}{\partial \phi^2} = \frac{u}{\alpha}\frac{\partial T}{\partial x} \tag{6.216}$$

If the temperature profile is also fully developed, then again it can be shown that

$$\frac{\partial T}{\partial x} = \frac{\partial T_m}{\partial x} = \text{constant.} \tag{6.217a}$$

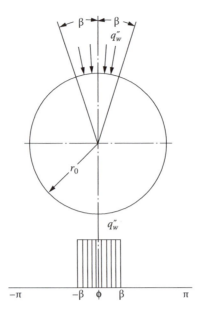

**FIGURE 6.25**
Circumferentially varying wall heat flux:

$$q_w'' = \text{constant}, \qquad -\beta < \phi < \beta$$
$$q_w'' = 0, \qquad \beta < \phi < 2\pi - \beta$$

Considering the variation of the wall heat flux as indicated in Figure 6.25, an energy balance over a differential volume element yields

$$\frac{dT_m}{dx} = \frac{2\beta q_w''}{\pi r_0 \rho U_m c_p}. \tag{6.217b}$$

Then the energy equation can be written as

$$\frac{\partial^2 \Theta}{\partial \eta^2} + \frac{1}{\eta} \frac{\partial \Theta}{\partial \eta} + \frac{1}{\eta^2} \frac{\partial^2 \Theta}{\partial \phi^2} = a(1 - \eta^2), \tag{6.218}$$

where we have introduced

$$\Theta(r, \phi) = T(r, \phi, x) - T_m(x),$$

$$\eta = \frac{r}{r_0},$$

$$a = \frac{4\beta r_0 q_w''}{\pi k},$$

Referring to Figure 6.25, the boundary conditions are as follows:

$$\frac{k}{r_0}\frac{\partial\Theta(1,\phi)}{\partial\eta} = q_w'', \quad -\beta \le \phi \le \beta; \quad \frac{\partial\Theta(\eta,0)}{\partial\phi} = 0, \tag{6.219a,b}$$

$$\frac{\partial\Theta(1,\phi)}{\partial\eta} = 0, \quad \beta \le \phi \le 2\pi - \beta; \quad \frac{\partial\Theta(\eta,\pi)}{\partial\phi} = 0. \tag{6.219c,d}$$

Under these conditions, Reynolds obtained the following expressions for the wall–bulk temperature difference and the Nusselt number:

$$\Theta_w(\phi) = T_w - T_m = \frac{4\beta r_0 q_w''}{\pi k}\frac{3}{16} - \frac{13}{48}\frac{\beta q_w'' r_0}{\pi k} + \sum_{n=1}^{\infty}\frac{2q_w'' r_0}{n^2 k\pi}\sin(n\beta)\cos(n\phi) \tag{6.220}$$

and

$$Nu(\phi) = \frac{2r_0 h}{k} = \frac{2r_0 q_w''}{k(T_w - T_m)} = \frac{1}{(11/48)(\beta/\pi) + \sum_{n=1}^{\infty}(1/n^2\pi)\sin(n\beta)\cos(n\phi)}. \tag{6.221}$$

When the heat flux is constant circumferentially ($\beta = \pi$), Equation 6.221 reduces to

$$Nu = \frac{48}{11} = 4.364, \tag{6.222}$$

which is the same result as Equation 6.123.

The solution previously presented can be used to obtain the solution in a tube heated at different rates over several segments of its circumference. Consider, for example, a tube heated as shown in Figure 6.26. Comparing this case with the case of Figure 6.25, we have

$$\beta = \frac{\Delta\Psi}{2}, \quad \phi' = \phi - \left(\Psi + \frac{\Delta\Psi}{2}\right). \tag{6.223a,b}$$

Hence, from Equation 6.221, the Nusselt number becomes

$$Nu(\phi, \Psi) = \frac{1}{(11/48)(\Delta\Psi/2\pi) + \sum_{n=1}^{\infty}(1/n^2\pi)\sin\left(n\Delta\Psi/2\right)\cos[n(\phi - \Psi - (\Delta\Psi/2))]}. \tag{6.224}$$

If an $m$th segment is heated at the rate $q_w''(\Psi_m)$ (and the remainder being insulated), then the solution given by Equation 6.224 is applicable, and since the energy equation (6.216) is linear, the sum of such solutions is also a solution. For a continuous variation of the circumferential heat flux, letting $\Delta\Psi$ be infinitesimally small $d\Psi$ and replacing the

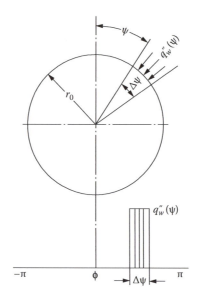

**FIGURE 6.26**
Circumferentially varying wall heat flux:

$$q_w'' = q_w''(\Psi), \qquad \Psi < \phi < \Psi + \Delta\Psi$$
$$q_w'' = 0, \qquad \Psi + \Delta\Psi < \phi < 2\pi + \Psi$$

summation over elemental segments by an integral around circumference, the Nusselt number can be obtained as

$$Nu(\phi) = \frac{q_w''(\phi)2r_0}{k\Theta_w(\phi)} = \frac{q_w''(\phi)}{\displaystyle\int_0^{2x} q_w''(\Psi)g(\phi,\Psi)(d\Psi/2\pi)}, \tag{6.225}$$

where

$$g(\phi,\Psi) = \frac{11}{48} + \sum_{n=1}^{\infty} \frac{\cos n(\phi - \Psi)}{n}. \tag{6.226}$$

For the case of constant heat-flux distribution around the circumference, Equation 6.225 reduces, as expected, to

$$Nu = \frac{48}{11} = 4.364. \tag{6.227}$$

Thus, first, a solution is obtained for the case of a tube with constant heat flux over a portion of its circumference, insulated over the remainder. From the solution of this case, a general solution to the axially constant but circumferentially varying wall heat-flux problem is obtained. The most important difficulty that arises in applying the previous results is to determine a simple mathematical expression for the circumferential wall heat-flux variation in applications. If the heat-flux distribution around the periphery, for example, is given by

$$q''(\phi) = q_0''(1 + b\cos\phi), \tag{6.228}$$

then, using Equation 6.220 together with Equation 6.223a,b, the wall temperature distribution can be obtained as

$$\Theta_w(\phi) = \int_0^{2\pi} q_0''(1 + b\cos\phi)\left(\frac{2r_0}{k}\right)\left[\frac{11}{48} + \sum_{n=1}^{\infty} \frac{\cos n(\phi - \Psi)}{n}\right]\frac{d\Psi}{2\pi} \tag{6.229}$$

or

$$\Theta_w(\phi) = q_0''\frac{2r_0}{k}\left(\frac{11}{48} + b\frac{\cos\phi}{2}\right). \tag{6.230}$$

The local Nusselt number can now be calculated from the given local heat flux and the calculated wall–bulk temperature difference as

$$Nu(\phi) = \frac{1 + b\cos\phi}{(11/48) + b(\cos\phi/2)}. \tag{6.231}$$

This result indicates a substantial variation of the heat-transfer coefficient around the tube periphery.

If $b = 0$, this case corresponds to a tube with constant wall heat-flux boundary condition, and the Nusselt number becomes equal to 4.364. If $b = 0.458$, then the Nusselt number becomes infinity at $\phi = \pi$. An infinite heat-transfer coefficient means that the surface temperature is the same as the fluid bulk temperature since the wall heat flux is always finite.

## 6.15 Heat Transfer in Annular Passages

An annular passage is a simple geometrical form used in practice for the purposes of heat transfer between two fluids. In double-pipe heat exchangers, for example, while one fluid flows in the inside tube, another one at different temperature flows in the annular space between the two tubes for heating or cooling purposes.

Although the number of works on annulus problems is quite limited, various limiting cases (circular pipes or parallel-plate channels) have received considerable attention.

In certain annulus problems, flow has been found to remain laminar for about 11 times the hydraulic diameter with a Reynolds number over 2900, with the transition occurring a little sooner on the inner wall than on the outer wall. Therefore, any investigation of heat transfer near the entrance of an annular duct should include considerations of laminar flow. The most widely used approach is to use Langhaar's solution for the velocity development.

The treatment of annulus problems is quite similar to that of the pipe problem. There is, however, an additional variable of radius ratio, and, since two surfaces are involved, there is also the possibility of many combinations of the boundary conditions.

Jacob and Ress [37] are among the first who calculated the temperature distribution in the thermal entrance region of an annular passage. The most complete analysis of heat transfer in the thermal entrance region of annular passages has been given by Reynolds et al. [38,39]. Worsoe-Schmidt [40] also presented a solution in the thermal entrance region.

Sugino [41] calculated the development of the velocity distribution in the hydrodynamic entrance region of an annular passage by applying Langhaar's method. Sparrow and Lin [42] presented a different but generalizable analysis for the calculation of the velocity distribution and pressure drop, and Manohar [43] gave a numerical solution again in the hydrodynamic entrance region. Murakawa [44] presented a numerical solution for heat transfer in the combined entrance region in an annular space between double pipes. Heaton et al. [45] solved the combined entrance region problem only for the constant heat-flux boundary condition. Yücel [46] and Kakaç and Yücel [47] gave a complete solution of the combined entrance region problem for Prandtl numbers of 0.01, 0.7, and 10.

In most heat-transfer applications [47,48], there are mainly two kinds of simple boundary conditions, namely, constant heat flux and constant temperature. Since there are two boundaries in an annulus, various combinations of these two boundary conditions are possible. But because of the linearity of the energy equation for constant-property flows, any axisymmetric surface temperature or surface heat-flux boundary condition can be satisfied by superposing one or more solutions of the so-called four fundamental problems. These fundamental problems are as follows:

1. *Fundamental problems of the first kind.* One surface at a constant temperature different from the inlet fluid temperature; the opposite surface temperature constant at the inlet fluid temperature. There are two solutions of the first kind, one for each of the two surfaces.

2. *Fundamental problems of the second kind.* One surface with a constant heat flux; the other surface insulated. There are also two solutions of the second kind.

3. *Fundamental problems of the third kind.* One surface at a constant temperature different from the inlet fluid temperature; the other surface insulated. There are two solutions of the third kind.

4. *Fundamental problems of the fourth kind.* One surface with a constant heat flux; the other surface at the inlet fluid temperature. There are two solutions of the fourth kind.

Therefore, with the eight solutions to these four fundamental problems, any specified heat flux or temperature on either surface varying in any desired manner in the flow direction can be readily handled.

The differential equations for the velocity and temperature fields in a steady, laminar, incompressible annulus flow with constant fluid properties and negligible axial conduction, internal energy generation, viscous energy dissipation, and axial rate of change of radial shear stress are given by

$$\frac{\partial u}{\partial x} + \frac{v}{r} + \frac{\partial v}{\partial r} = 0, \tag{6.232}$$

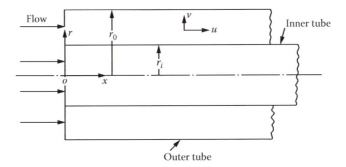

**FIGURE 6.27**
Coordinate system for the annular passage between two concentric tubes.

$$u\frac{\partial u}{\partial x}+v\frac{\partial v}{\partial r}=-\frac{1}{\rho}\frac{\partial p}{\partial x}+v\left(\frac{\partial^2 u}{\partial r^2}+\frac{1}{r}\frac{\partial u}{\partial r}\right),\tag{6.233}$$

$$u\frac{\partial u}{\partial x}+v\frac{\partial T}{\partial r}=\alpha\left(\frac{\partial^2 T}{\partial r^2}+\frac{1}{r}\frac{\partial T}{\partial r}\right).\tag{6.234}$$

The coordinate system for the model under consideration is shown in Figure 6.27.

With the simultaneous development of both velocity and temperature fields, the Prandtl number becomes a more significant parameter because its effect can no longer be included entirely in the dimensionless axial variable.

As an application, let us consider the second fundamental solutions for which the nomenclature is given in Figure 6.28. The developing temperature profiles for these solutions are obtained in terms of the dimensionless temperature

$$^{2}\psi^{i}=\frac{^{2}T^{i}-T_{e}}{q_{w}''d_{H}/2k}=\frac{^{2}\theta^{i}}{q_{w}''d_{H}/2k},\tag{6.235}$$

where
the left-hand superscript 2 indicates the second fundamental solution
the right-hand superscript $i$ identifies the wall that is heated
$q_{w}''$ is the heat flux on the respective surface

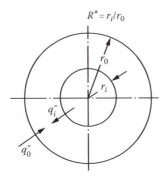

**FIGURE 6.28**
Nomenclature for the fundamental solutions of the second kind.

In addition, $T_w$ denotes the uniform fluid temperature at the entrance of the passage, and $d_H$ is the hydraulic diameter.

1. *Constant heat flux through inner surface and outer surface insulated.* The boundary conditions in this case are as follows:

$$T(0,r) = T_e,$$

(6.236)

$$\frac{\partial T(x,r_i)}{\partial r} = -\frac{q_i''}{k},$$

(6.237)

$$\frac{\partial T}{\partial r}(x,r_0) = 0,$$

(6.238)

$$T(\infty,r) = T_{fd}.$$

(6.239)

The solution to this problem can now be written in terms of $^2\Psi^i$ defined by Equation 6.35 as

$$^2\Theta^i = {}^2T^i - T_e = \frac{q_i'' d_H}{2k}\, {}^2\Psi^i.$$

(6.240)

2. *Constant heat flux through outer surfaces and inner surface insulated.* The *developing* temperature distribution in this case can be obtained in terms of $^2\Psi^i$ as

$$^2\Theta^0 = {}^2T^0 - T_e = \frac{q_0'' d_H}{2k}\, {}^2\Psi^0.$$

(6.241)

Superposition of these two solutions gives the general solution to the case of constant heat flux through both the inner and outer surfaces. This superposition can be formulated as

$$^2\Theta = {}^2T - T_e = \frac{q_i'' d_H}{2k}\, {}^2\Psi^i + \frac{q_0'' d_H}{2k}\, {}^2\Psi^0.$$

(6.242)

The Nusselt numbers for the previous two cases may be written as follows:

$$Nu = \frac{h d_H}{k} = \frac{2\left(q_w'' d_H / 2k\right)}{T_w - T_m},$$

(6.243)

where
$T_w$ is the temperature of the heated surface
$T_m$ is the local bulk temperature

Thus, for case (1), the Nusselt number defined as $^2Nu_{ii} = h_{ii}d_H/k$, where $h_{ii} = q_i/(T_i - T_m)$, may be written in terms of the fundamental solution variables as

$$^2Nu_{ii} = \frac{2\left(q_i''d_H/2k\right)}{^2T_1^i - {}^2T_m^i}. \tag{6.244}$$

By the use of Equation 6.240, we obtain

$$^2Nu_{ii} = \frac{2\left(q_i''d_H/2k\right)}{\left(q_i''d_H/2k\right){}^2\Psi_i^i - \left(q_i''d_H/2k\right){}^2\Psi_m^i} = \frac{2}{^2\Psi_i^i - {}^2\Psi_m^i}. \tag{6.245}$$

For the case (2), following the same procedure, the Nusselt number, defined as $^2Nu_{00} = h_{00}d_H/k$, becomes

$$^2Nu_{00} = \frac{2}{^2\Psi_0^0 - {}^2\Psi_m^0}. \tag{6.246}$$

By the superposition of the previous results, the method can readily be extended to the case of constant but different heat fluxes on each wall. In this case, the bulk and surface temperatures and the temperature gradients at the surface can be determined from Equation 6.242. Introducing these into Equation 6.243 and making the necessary simplifications, the Nusselt numbers at the inner and outer surfaces for a given heat-flux ratio become

$$^2Nu_i = \frac{2}{^2\Psi_i^i - {}^2\Psi_m^i + \left(q_0''/q_i''\right)\left(^2\Psi_i^0 - {}^i\Psi_m^0\right)} \tag{6.247}$$

and

$$^2Nu_0 = \frac{2}{^q\Psi_0^0 - {}^2\Psi_m^0 + \left(q_i''/q_0''\right)\left(^2\Psi_0^i - {}^i\Psi_m^i\right)}. \tag{6.248}$$

From Equations 6.245 through 6.248, the following simpler expressions can be obtained:

$$\frac{^2Nu_i}{^2Nu_{ii}} = \frac{1}{1 - \left(q_0''/q_i''\right){}^2C^i} \tag{6.249}$$

and

$$\frac{^2Nu_0}{^2Nu_{00}} = \frac{1}{1 - \left(q_i''/q_0''\right){}^2C^0}, \tag{6.250}$$

where

$$^2C^i = \frac{^2\Psi_m^0 - {}^2\Psi_i^0}{^2\Psi_i^i - {}^2\Psi_m^i} \quad \text{and} \quad ^2C^0 = \frac{^2\Psi_m^i - {}^2\Psi_0^i}{^2\Psi_0^0 - {}^2\Psi_m^0}, \tag{6.251a,b}$$

which are called influence coefficients. Therefore, the Nusselt numbers on the inner and outer surfaces for any flux ratio can be calculated from Equations 6.249 and 6.250. Detailed discussion on the Nusselt numbers and influence coefficients for all fundamental problems are given in References 46 and 47.

For a complete tabulation of solutions, the reader is referred to the original References 40,45–47. The results of one case with constant but different heat flux on each surface are tabulated in Tables 6.16 through 6.18. A circular tube and a parallel-plate channel represent two extremes of the annulus with $R^* = 0$ and $R^* = 1$, respectively. The Nusselt numbers are also plotted in References 46,47 as a function of dimensionless distance along the annular passage for four fundamental solutions at various values of $R^*$.

Reynolds et al. [39] obtained solutions in the thermal entrance region of circular tube annuli for all fundamental problems. A summary of their results is presented in Table 6.19.

The Nusselt numbers and the influence coefficients in circular tube annuli for fully developed velocity and temperature fields can be obtained as asymptotic values of the previous solutions. The results for the fundamental problem of the second kind under fully developed conditions are tabulated in Table 6.20.

The heat-transfer results presented thus far in Tables 6.16 through 6.20 give the solution only for the case where one wall is heated and the other wall insulated. The Nusselt numbers on the inner and outer surfaces for any heat-flux ratio can be calculated from

**TABLE 6.16**

Nusselt Numbers and Influence Coefficients in a Circular Tube Annulus for Fundamental Problems of the Second Kind

| $R^*(= r_i/r_0)$ | $X\left(= \dfrac{4x/d_H}{RePr}\right)$ | $^2C^i$ | $^2C^o$ | $^2Nu_{ii}$ | $^2Nu_{00}$ |
|---|---|---|---|---|---|
| 0.25 | 0.001 | 0.0355 | 0.0087 | 50.77 | 48.92 |
| | 0.002 | 0.0549 | 0.0132 | 37.80 | 36.06 |
| | 0.004 | 0.0845 | 0.0200 | 28.19 | 26.51 |
| | 0.01 | 0.1485 | 0.0341 | 19.265 | 17.59 |
| | 0.02 | 0.2290 | 0.0510 | 14.64 | 12.96 |
| | 0.04 | 0.3591 | 0.0770 | 11.39 | 9.722 |
| | 0.1 | 0.6251 | 0.1276 | 8.787 | 7.125 |
| | 0.2 | 0.6906 | 0.1537 | 7.922 | 6.081 |
| | 0.4 | 0.7864 | 0.1546 | 7.758 | 5.625 |
| | ∞ | 0.7932 | 0.1252 | 7.753 | 4.904 |
| 0.50 | 0.001 | 0.0289 | 0.0143 | 49.50 | 48.796 |
| | 0.002 | 0.0443 | 0.0218 | 36.52 | 35.855 |
| | 0.004 | 0.0674 | 0.0329 | 26.93 | 26.29 |
| | 0.01 | 0.1159 | 0.0559 | 18.02 | 17.38 |
| | 0.02 | 0.1747 | 0.0834 | 13.39 | 12.77 |
| | 0.04 | 0.2663 | 0.1258 | 10.13 | 9.557 |
| | 0.1 | 0.4484 | 0.2092 | 7.497 | 6.974 |
| | 0.2 | 0.5567 | 0.2550 | 6.497 | 5.926 |
| | 0.4 | 0.5807 | 0.2596 | 6.171 | 5.500 |
| | ∞ | 0.528 | 0.2154 | 6.181 | 5.036 |

*Note:* Combined thermal and hydrodynamic entrance region, $Pr = 0.01$ [46,47].

**TABLE 6.17**

Nusselt Numbers and Influence Coefficients in a Circular Tube Annulus for Fundamental Problems of the Second Kind

| $R^*(= r_i/r_0)$ | $X\left(=\dfrac{4x/d_H}{RePr}\right)$ | $^2C^i$ | $^2C^o$ | $^2Nu_{ii}$ | $^2Nu_{00}$ |
|---|---|---|---|---|---|
| 0.10 | 0.001 | 0.0377 | 0.0031 | 42.73 | 33.92 |
| | 0.002 | 0.0587 | 0.0044 | 33.24 | 24.30 |
| | 0.004 | 0.0944 | 0.0064 | 26.35 | 17.66 |
| | 0.01 | 0.1814 | 0.0108 | 20.10 | 11.89 |
| | 0.02 | 0.3060 | 0.0165 | 16.89 | 9.014 |
| | 0.04 | 0.5289 | 0.0256 | 14.63 | 7.044 |
| | 0.1 | 1.0023 | 0.0438 | 12.765 | 5.491 |
| | 0.2 | 1.2855 | 0.0538 | 12.043 | 4.969 |
| | 0.4 | 1.3705 | 0.0565 | 11.840 | 4.841 |
| | ∞ | 1.3835 | 0.0562 | 11.90 | 4.834 |
| 0.25 | 0.002 | 0.0433 | 0.0096 | 27.87 | 24.37 |
| | 0.004 | 0.0667 | 0.0140 | 21.16 | 17.72 |
| | 0.01 | 0.1209 | 0.0238 | 15.22 | 11.94 |
| | 0.04 | 0.3242 | 0.0565 | 10.19 | 7.10 |
| | 0.1 | 0.5904 | 0.0967 | 8.546 | 5.56 |
| | 0.2 | 0.7443 | 0.1189 | 7.931 | 5.046 |
| | 0.4 | 0.7897 | 0.1249 | 7.759 | 4.915 |
| | ∞ | 0.7932 | 0.125 | 7.735 | 4.904 |
| 0.50 | 0.001 | 0.0230 | 0.0112 | 35.52 | 34.05 |
| | 0.002 | 0.0336 | 0.0160 | 25.90 | 24.49 |
| | 0.004 | 0.0506 | 0.0235 | 19.24 | 17.86 |
| | 0.01 | 0.0887 | 0.0400 | 13.395 | 12.09 |
| | 0.02 | 0.1395 | 0.0612 | 10.50 | 9.23 |
| | 0.04 | 0.2252 | 0.0960 | 8.50 | 7.25 |
| | 0.1 | 0.3981 | 0.1652 | 6.924 | 5.713 |
| | 0.2 | 0.4979 | 0.2037 | 6.351 | 5.188 |
| | 0.4 | 0.5270 | 0.2147 | 6.190 | 5.044 |
| | ∞ | 0.5288 | 0.216 | 6.181 | 5.036 |

*Note:* Combined thermal and hydrodynamic entrance region, $Pr = 0.7$ [46,47].

Equations 6.249 and 6.250. For example, for a radius ratio of $R^* = 0.50$, under both hydrodynamically and thermally fully developed conditions, Equations 6.49 and 6.250 become

$$^2Nu_i = \frac{6.18}{1-0.528\left(q_0''/q_i''\right)} \quad \text{and} \quad ^2Nu_0 = \frac{5.04}{1-0.216\left(q_i''/q_0''\right)}. \tag{6.252a,b}$$

For the case of flow between two parallel plates, Equations 6.249 and 6.250 become identical:

$$^2Nu_i = {}^2Nu_0 = Nu = \frac{5.385}{1-0.346\left(q_0''/q_i''\right)}. \tag{6.253}$$

**TABLE 6.18**

Nusselt Numbers and Influence Coefficients in a Circular Tube Annulus for Fundamental Problems of the Second Kind

| $R^*(= r_i/r_0)$ | $X\left(=\dfrac{4x/d_H}{RePr}\right)$ | $^2C^i$ | $^2C^o$ | $^2Nu_{ii}$ | $^2Nu_{00}$ |
|---|---|---|---|---|---|
| 0.25 | 0.001 | 0.0242 | 0.0050 | 30.24 | 24.97 |
| | 0.002 | 0.0377 | 0.0074 | 23.64 | 18.63 |
| | 0.004 | 0.0604 | 0.0113 | 18.92 | 14.19 |
| | 0.01 | 0.1158 | 0.0205 | 14.53 | 10.30 |
| | 0.02 | 0.1928 | 0.0442 | 12.085 | 8.292 |
| | 0.04 | 0.3253 | 0.0542 | 10.226 | 6.811 |
| | 0.1 | 0.5944 | 0.0959 | 8.572 | 5.525 |
| | 0.2 | 0.7462 | 0.1187 | 7.934 | 5.043 |
| | 0.4 | 0.7899 | 0.1250 | 7.759 | 4.912 |
| | $\infty$ | 0.7932 | 0.125 | 7.753 | 4.904 |
| 0.50 | 0.001 | 0.0184 | 0.0085 | 27.546 | 25.40 |
| | 0.002 | 0.0280 | 0.0126 | 21.106 | 19.02 |
| | 0.004 | 0.0439 | 0.0193 | 16.54 | 14.56 |
| | 0.01 | 0.0823 | 0.0353 | 12.40 | 10.63 |
| | 0.02 | 0.1348 | 0.0570 | 10.14 | 8.58 |
| | 0.04 | 0.2237 | 0.0932 | 8.44 | 7.034 |
| | 0.1 | 0.4002 | 0.1644 | 6.934 | 5.695 |
| | 0.2 | 0.4985 | 0.2035 | 6.353 | 5.186 |
| | 0.4 | 0.5271 | 0.2147 | 6.190 | 5.044 |
| | $\infty$ | 0.5288 | 0.216 | 6.181 | 5.036 |

*Note:* Combined thermal and hydrodynamic entrance region, $Pr = 10$ [46,47].

If both sides of the channel are heated equally, that is, $q_0''/q_i'' = 1$, then $Nu_i = Nu_0 = 8.23$. If $q_0'' = 0$, which is the case where only one of the walls of the channel is heated, then the Nusselt number is $Nu_i = 5.385$

## Problems

**6.1** In Section 6.3.1, the steady and fully developed velocity distribution for laminar flow between two parallel walls, that is, Equation 6.17a or b, was obtained in the absence of body forces. Show that the same results would be obtained, say, if the momentum equation (6.8b) contained a body force in the form $fy = -\hat{j}g$, where $g$ represents the gravitational acceleration.

**6.2** Using the velocity distribution (6.28) for fully developed laminar flow in a circular pipe, obtain the expression (6.37) for the friction factor as defined by Equation 6.33.

**6.3** Under certain conditions, the temperature distribution in a flow between two parallel plates is governed by $k(d^2T/dy^2) = -\mu(du/dy)$. State clearly all assumptions underlying this differential equation.

**TABLE 6.19**

Nusselt Numbers and Influence Coefficients in a Circular
Tube Annulus for Fundamental Problems of the Second Kind

| $R^*(= r_i/r_0)$ | $X\left(=\dfrac{4x/d_H}{RePr}\right)$ | $^2C^i$ | $^2C^o$ | $^2Nu_{ii}$ | $^2Nu_{oo}$ |
|---|---|---|---|---|---|
| 0.05 | 0.004 | 0.1265 | 0.00255 | 33.2 | 13.4 |
|  | 0.02 | 0.460 | 0.00760 | 24.2 | 7.99 |
|  | 0.04 | 0.817 | 0.0125 | 21.5 | 6.58 |
|  | 0.20 | 2.13 | 0.0278 | 18.1 | 4.92 |
|  | 0.40 | 2.17 | 0.0293 | 17.8 | 4.80 |
|  | ∞ | 2.18 | 0.0294 | 17.8 | 4.79 |
| 0.10 | 0.004 | 0.0914 | 0.00491 | 25.1 | 13.5 |
|  | 0.02 | 0.311 | 0.0147 | 17.1 | 8.08 |
|  | 0.04 | 0.540 | 0.241 | 14.9 | 6.65 |
|  | 0.20 | 1.296 | 0.0531 | 12.1 | 4.96 |
|  | 0.40 | 1.38 | 0.0560 | 11.9 | 4.84 |
|  | ∞ | 1.38 | 0.0562 | 11.9 | 4.83 |
| 0.25 | 0.004 | 0.0605 | 0.01104 | 18.9 | 13.8 |
|  | 0.02 | 0.194 | 0.0328 | 12.1 | 8.28 |
|  | 0.04 | 0.325 | 0.6540 | 10.2 | 6.80 |
|  | 0.20 | 0.746 | 0.118 | 7.94 | 5.04 |
|  | 0.40 | 0.789 | 0.125 | 7.76 | 4.91 |
|  | ∞ | 0.793 | 0.125 | 7.75 | 4.90 |
| 0.50 | 0.004 | 0.0437 | 0.0189 | 16.4 | 14.2 |
|  | 0.02 | 0.1347 | 0.0570 | 10.1 | 8.55 |
|  | 0.04 | 0.224 | 0.0934 | 8.43 | 7.03 |
|  | 0.20 | 0.498 | 0.204 | 6.35 | 5.19 |
|  | 0.40 | 0.526 | 0.215 | 6.19 | 5.05 |
|  | ∞ | 0.528 | 0.216 | 6.18 | 5.04 |
| 1.00 | 0.001 | 0.01175 | 0.01175 | 23.5 | 23.5 |
|  | 0.01 | 0.0560 | 0.0560 | 11.2 | 11.2 |
|  | 0.04 | 0.1491 | 0.1491 | 7.49 | 7.49 |
|  | 0.20 | 0.327 | 0.327 | 5.55 | 5.55 |
|  | 0.50 | 0.346 | 0.346 | 5.39 | 5.39 |
|  | ∞ | 0.346 | 0.346 | 5.39 | 5.39 |

*Note:* Thermal entrance region [39].

**6.4** In a system composed of two large parallel plates, the upper plate is maintained at the uniform temperature $T_w$ and moves with a velocity $U_0$, while the lower plate is stationary and insulated. Find the steady-state temperature of the lower plate in terms of $T_w$, $U_0$, and the properties of the fluid between the plates. Neglect any pressure gradient in the flow. See Problems 2.12 and 2.15.

**6.5** Consider both hydrodynamically and thermally fully developed steady laminar flow of a constant-property fluid through a circular pipe of constant wall temperature. Obtain an expression for the Nusselt number by solving Equation 6.73 iteratively, first starting with the temperature distribution for the constant wall heat-flux case and then repeating the procedure until the Nusselt number reaches a limit.

**TABLE 6.20**

Nusselt Numbers and Influence Coefficients
in a Circular Tube Annulus for Fundamental
Problems of the Second Kind

| $R^*(= r/r_i)$ | $^2C^i$ | $^2C^o$ | $^2Nu_{ii}$ | $^2Nu_{oo}$ |
|---|---|---|---|---|
| 0 | $\infty$ | 0 | $\infty$ | 4.364 |
| 0.02 | 4.183 | 0.0121 | 32.7 | 4.73 |
| 0.05 | 2.18 | 0.0294 | 17.8 | 4.79 |
| 0.10 | 1.38 | 0.0562 | 11.91 | 4.83 |
| 0.25 | 0.793 | 0.125 | 7.75 | 4.90 |
| 0.50 | 0.528 | 0.216 | 6.18 | 5.04 |
| 1.00 | 0.346 | 0.346 | 5.39 | 5.39 |

*Note:* Fully developed velocity and temperature
profiles.

**6.6** Solve Problem 6.5 for flow between two parallel plates.

**6.7** Consider both hydrodynamically and thermally fully developed steady laminar flow
of a constant-property fluid through a circular pipe with a constant heat-flux condition maintained at the pipe wall. Neglecting axial conduction of heat and assuming
that the velocity profile can be approximated to be uniform across the entire flow area
of the pipe (i.e., slug flow), obtain an expression for the Nusselt number. Compare the
result with Equation 6.68 and explain the reason for the difference on a physical basis.

**6.8** Solve Problem 6.7 for flow between two parallel plates.

**6.9** Consider a fully developed steady laminar flow of a constant-property fluid through
a circular duct with a constant heat-flux condition maintained at the duct wall.
Neglecting axial conduction of heat, obtain an expression for the Nusselt number
with the effect of viscous dissipation included in the analysis. How does this frictional heating affect the Nusselt number?

**6.10** Solve Problem 6.9 for flow between two parallel plates.

**6.11** Under constant wall temperature boundary condition, for slug flow in a circular duct,
the asymptotic value of the Nusselt number is $Nu = 5.75$, whereas $Nu = 3.66$ when the
velocity distribution is parabolic. Give a physical explanation for this difference.

**6.12** The energy equation for steady fully developed laminar flow in a circular duct is
given, in the absence of viscous dissipation and for constant properties, by Equation
6.46. Under certain conditions, it is quite reasonable to assume that the fully developed velocity profile can be approximated by

$$\frac{u}{U_c} = 1 - \frac{r}{r_0},$$

where
  $U_c$ is the centerline velocity
  $r_0$ is the radius of the duct

For this type of flow, neglecting the axial conduction of heat, obtain an expression
for the Nusselt number for fully developed conditions under constant wall heat-flux
boundary condition.

**6.13** Solve Problem 6.12 by including the effect of viscous dissipation in the analysis. How does this frictional heating affect the Nusselt number?

**6.14** Derive Equation 6.47 by making an energy balance on an annular elemental control volume.

**6.15** Air at 1 atm and 20°C is to be heated at a rate of 0.04 kg/min in a circular pipe of 5 cm *ID* and 6 m length by maintaining a constant heat flux at the pipe wall.

**6.16** What is the required wall heat flux if the maximum local difference between the pipe wall and mean fluid temperatures is to be equal or less than 10°C?

    a.  Determine the exit air temperature.

**6.17** Consider the fully developed flow of a very viscous fluid in a circular pipe of radius $r_0$. Obtain an expression for the Nusselt number if the boundary condition is given as

$$\text{at } r = r_0 : \quad T = T_w < T_m,$$

where $T_m$ is the mean fluid temperature.

**6.18** Solve Problem 6.16 for the fully developed flow of very viscous fluid between two stationary parallel plates separated by a distance 2*d* for the boundary conditions that

$$\text{at } y = \mp d : \quad T - T_w < T_m.$$

**6.19** Solve Problem 6.17 for the following boundary conditions:

$$\text{at } y = -d : \quad T - T_{w_1}$$
$$\text{at } y = +d : \quad T - T_{w_2} < T_{w_1} < T_m.$$

**6.20** Consider steady flow of a constant-property fluid with a mean velocity of $U_m$ through a duct of constant cross-sectional area $A$ and perimeter $P$. The duct wall temperature is maintained constant $T_w$, while the fluid temperature at the inlet of the duct is $T_i$. Determine the variation of the mean fluid temperature $T_m$ with the distance $x$ along the length of the duct. Neglect axial conduction of heat and viscous dissipation, and assume that the variation of the local heat-transfer coefficient $h_x$ with $x$ is known.

**6.21** Crude oil, $c_p$ = 2094 J/(kg K), is to be heated at a flow rate of 800 kg/h from 20°C to 30°C in a 3/4 in. standard steel pipe (*OD* = 1.050 in., *ID* = 0.724 in.). If the pipe is jacketed with saturated steam at 100° C, what length of pipe will be required?

**6.22** Consider the flow of a fluid through a 4 in. *ID* pipe. It is given that at a location along the pipe, the velocity profile can be assumed to be uniform across the cross section of the pipe and the temperature of the fluid varies linearly from 30°C at the pipe wall to 0°C at the centerline. Also, the rate of heat transfer from the pipe wall to the fluid is given as 3.1 × 10⁴ W/m².

    a.  What is the value of the heat-transfer coefficient at this location?

    b.  Estimate the value (average) of the thermal conductivity of the fluid.

**6.23** Consider hydrodynamically fully developed steady laminar flow of a constant-property fluid through a circular pipe with a constant heat-flux condition maintained at the pipe wall. Neglect axial conduction of heat and assume that the velocity profile can be approximated to be uniform across the entire flow area of the pipe (i.e., slug flow).

Obtain an expression for the local Nusselt number in the thermal entrance region of the pipe. What is the asymptotic value of the Nusselt number for large values of the axial coordinate?

**6.24** Solve Problem 6.22 for parabolic velocity distribution in the pipe by neglecting the effect of viscous dissipation.

**6.25** Consider the steady and laminar flow of a constant-property liquid metal in a circular pipe of radius $r_0$. The liquid metal and the pipe wall are both at the same constant temperature $T_i$ for $x < 0$. Neglecting axial conduction of heat, obtain an expression for the local Nusselt number in the thermal entrance region of the pipe, if the pipe wall temperature for $x \geq 0$ is prescribed as

$$T(x, r_0) = T_w + Ax,$$

where $A$ is a given constant.

**6.26** Solve Problem 6.24 for steady and laminar flow of a constant-property fluid with hydrodynamically fully developed parabolic velocity distribution.

**6.27** Show that the eigenfunctions of the eigenvalue problem given by Equations 6.100 and 6.101a,b are orthogonal with respect to the weight function $\eta(1 - \eta^2)$ over the interval $0 < \eta| < 1$.

**6.28** Obtain the relation given by Equation 6.102b.

**6.29** Obtain the relation given by Equation 6.106.

**6.30** Consider hydrodynamically fully developed steady laminar flow of a constant-property fluid between two parallel plates separated by a distance $L$. Let the plates at $y = 0$ and $y = L$ be maintained at constant temperatures $T_{w_1}$ and $T_{w_2}$ respectively, for $x \geq 0$, and the fluid be at a constant temperature $T_i$ at $x = 0$. Assuming uniform velocity profile across the entire flow area (i.e., slug flow) and neglecting axial conduction of heat, (a) obtain an expression for the temperature distribution $T(x,y)$ in the thermal entrance region, and (b) determine the variation of the local Nusselt numbers at $y = 0$ and $y = L$ for $x \geq 0$.

**6.31** Solve Problem 6.29 for the situation where constant heat-flux boundary conditions are maintained at the two plate surfaces for $x \geq 0$, such that the heat flux to the fluid from the plate at $y = 0$ is twice the heat flux from the plate at $y = L$.

**6.32** Solve the problem defined by Equations 6.185 and 6.186a,b,c and obtain the expression (6.190) for the local Nusselt number.

**6.33** Consider an isothermal viscous fluid of temperature $T_0$ between two large parallel plates separated by a distance $L$. The upper plate moves with a constant velocity $U$ in the $x$-direction and is insulated. The lower plate is stationary and a step change in its temperature from $T_0$ to $T_w$ occurs for $x \geq 0$. Obtain, by a similarity solution, an expression for the local heat flux from the local plate to the fluid for those $x$-locations where $\delta_T \ll L$. Also, develop an expression for the local heat-transfer coefficient. Assume constant thermophysical properties and neglect viscous dissipation.

**6.34** Solve Problem 6.32 by the integral method by using a third-degree polynomial in $y$ for the temperature distribution in the thermal boundary layer, and develop an expression for the local heat-transfer coefficient for those $x$-locations where $\delta_T < L$. Compare and comment on the results obtained here and in Problem 6.32 for the local heat-transfer coefficient.

**6.35** Consider the steady and laminar flow of a constant-property fluid in the $x$-direction between two large parallel plates. The plates are maintained at the same constant temperature $T_w$ for $x \geq 0$ and the fluid temperature is $T_i$ at $x = 0$. Assuming slug flow, obtain an expression for the local Nusselt number by the integral method. How long would the thermal entrance length 5 be for the temperature distribution to be fully developed?

**6.36** Solve Problem 6.34 for hydrodynamically fully developed conditions with parabolic velocity $j$ profile.

**6.37** Solve Problem 6.34 for the situation where the same constant heat flux $q_w''$ is maintained from the plates to the fluid for $x \geq 0$.

**6.38** Evaluate the values of the local Nusselt number from Equation 6.113 at $(x/d)/(RePr) = 0.0025, 0.05, 0.2$, and 0.5 along the pipe.

**6.39** As an application of the results obtained in this chapter for laminar flows in circular ducts under constant wall temperature boundary condition, consider an oil heat exchanger, in which the $j$ length of the tubes is 200 diameters long. The Prandtl number of the oil is given as 100, and the $j$ Reynolds number of the flow in the tubes is 1000:

    a. Is it possible that the flow in the tubes in this heat exchanger can be assumed to be fully developed?

    b. Find the value of the local Nusselt number at the end of the tubes from Equation 6.113 by assuming hydrodynamically fully developed and thermally developing flow in the tubes.

    c. Assuming that the average Nusselt number is approximately twice the local value at the end of the tubes, estimate the error that could have been introduced if fully developed conditions were used.

**6.40** Consider the flow of a constant-property fluid at a mass flow rate $m$ in an electrically heated tube of diameter $d$ and length $L$. The heat flux to the fluid along the length of the tube is given as

$$q_w'' = q_0'' \sin \frac{\pi x}{L},$$

where $q_w''$ is a given constant. Determine the variation of the tube surface temperature along the length of the tube. Assume that the heat-transfer coefficient is constant and known.

**6.41** Consider the laminar flow of an oil inside a duct with a Reynolds number of 1000. The length of the duct is 2.5 m and the diameter is 2 cm. The duct is heated electrically by the use of its walls as an electrical resistance. Properties of the oil at the average oil temperature are $\rho = 870$ kg/m³, $c_p = 1959$ kJ/(kg K), $\mu = 0.004$ N s/m², and $k = 0.128$ W/(m K). Obtain the $j$ local Nusselt number at the end of the duct.

**6.42** Derive the energy integral equation for a circular duct. Write down the assumptions you made.

**6.43** Applying the energy integral equation of Problem 6.41, obtain an expression for the Nusselt number in the thermal entrance region of a circular duct.

**6.44** By the change of the dependent variable as $\Theta = F(\eta, \varphi) + a((\eta^2/4) - (\eta^4/16))$, solve the following:

    a. Simplify the differential equation (6.218) and obtain the boundary conditions (6.219) in terms of the new variable $\Theta$.

    b. Solve the resulting differential equation by the method of separation of variables to obtain the temperature distribution.

    c. By the use of the definition of the bulk temperature for an incompressible, constant-property fluid, obtain Equation 6.220.

**6.45** In laminar flow under fully developed conditions in a pipe, the circumferential variation of the heat flux around the pipe surface is given by the following expressions:

    a. $q''(\Psi) = q_0''(1 + b\sin\Psi)$

    b. $q''(\Psi) = q_0''[2 + a\cos\Psi + b\sin\Psi]$

Obtain expressions for the variation of Nusselt number for both cases and discuss the results.

**6.46** Find the heat fluxes and Nusselt numbers for a flow in an annulus with the inlet temperature $T_i$ outer wall maintained at $T_i$, over its entire length, and the inner wall at $T_w$ different from inlet fluid temperature. From these solutions, by the method of superposition obtain inner and outer Nusselt numbers for a given temperature ratio in terms of fundamental solutions and the influence coefficients.

**6.47** Suppose it is required to know the solution for the temperature distribution for an annulus with the inner wall subjected to a uniform and constant heat flux $q_i''$ and the outer wall maintained at $T_w$ from the inlet, where the inlet temperature is $T_i$. Obtain the solution as linear combination of a constant, one fundamental solution of the third kind and one fundamental solution of the fourth kind. Obtain inner and outer Nusselt numbers in terms of the fundamental solutions and the influence coefficients for a given ratio of $\theta_w/(q_i''d_H/2k)$.

---

# References

1. Kays, W. M., Numerical solutions for laminar-flow heat transfer in circular tubes, *Trans. ASME*, 77, 1265–1274, 1955.
2. Shah, R. K. and Bhatti, M. S., Laminar convective heat transfer in ducts, in *Handbook of Single-Phase Convective Heat Transfer*, S. Kakaç, R. K. Shah, and W. Aung (Eds.), John Wiley & Sons, New York, Chapter 3, 1987.
3. Zhi-qing, W., Study on correction coefficients of laminar and turbulent entrance region effect in round pipe, *Appl. Math. Mech.*, 3(3), 433–446, 1982.
4. Reynolds, O., On the experimental investigation of the circumstances which determine whether the motion of water will be direct or sinuous, and of the law of resistance in parallel channels, *Phil. Trans. R. Soc. London*, 174A, 935–982, 1883. Also, Papers on *Mechanical and Physical Subjects*, Vol. 1, University Press, Cambridge, England, 1990.
5. Bhatti, M. S. and Shah, R. K., Turbulent and transition flow convective heat transfer in ducts, in *Handbook of Single-Phase Convective Heat Transfer*, S. Kakaç, R. K. Shah, and W. Aung (Eds.), John-Wiley & Sons, Chapter 4, 1987.
6. Schlichting, H., *Boundary-Layer Theory* (translated into English by J. Kestin), 7th edn., McGraw-Hill, New York, 1979.

7. Moody, L. F., Friction factors for pipe flow, *Trans. ASME*, 66, 671–684, 1944.
8. Shah, R. K. and London, A. L., *Laminar Flow Forced Convection in Ducts*, Supplement 1, *Advances in Heat Transfer*, Academic Press, New York, 1978.
9. McAdams, W. H., *Heat Transmission*, 3rd edn., McGraw-Hill, New York, 1954.
10. Shah, R. K., Fully developed laminar flow forced convection in channels, in *Low Reynolds Number Flow Heat Exchanger*, S. Kakaç, R. K. Shah, and A. E. Bergles (Eds.), Hemisphere Publishing Co., New York, Chapter 4, pp. 75–108, 1983.
11. Graetz, L. Über die Wärmeleitungs fähigkeit von Flüssigkeiten, *Annalen der Physik and Chemie.*, 18, 79–94, 1883; 25, 337–357, 1885.
12. Nusselt, W., Die Abhängigkeit der Wärmeübergangszahl von der Rohrlänge, *Z Ver. Deut. Ing.*, 54, 1154–1158, 1910.
13. Jakob, M., *Heat Transfer*, Vol. 1, John Wiley & Sons, New York, 1956.
14. Brown, G. M., Heat or mass transfer in a fluid in laminar flow in a circular or flat conduit, *AIChE J.*, 6, 179–183, 1960.
15. Clark, S. H. and Kays, W. M., Laminar flow forced convection in rectangular ducts, *Trans. ASME*, 75, 859–866, 1953.
16. Cess, R. D. and Shaffer, E. C., Heat transfer to laminar flow between parallel plates with a prescribed wall heat flux, *Appl. Sci. Res.*, 8A, 339–344, 1959.
17. Cess, R. D. and Shaffer, E. C., Heat transfer to laminar flow between parallel plates with a prescribed wall heat flux, *Appl. Sci. Res., Sec. A*, 8, 64, 1959.
18. Cess, R. D. and Shaffer, E. C., Summary of laminar heat transfer between parallel plates with unsymmetrical wall temperature, *J. Aerosp. Sci.*, 26(8), 548, 1960.
19. Sellars, J. R., Tribus, M., and Klein, J. S., Heat transfer to laminar flow in a round tube or flat conduit—The Graetz problem extended, *Trans. ASME*, 78, 441–448, 1956.
20. Yener, Y. and Kakaç, S., *Heat Conduction*, 4th edn., Taylor & Francis Group, New York, 2008.
21. Abramowitz, M., On the solution of differential equation occurring in the problem of heat convection laminar flow through a tube, *J. Math. Phys.*, 32, 184–187, 1953.
22. Lipkis, R. P., Heat transfer to an incompressible fluid in laminar motion, M.S. thesis, University of California, Los Angeles, CA, 1954.
23. Hausen, H., Darstellung des Wärmeüberganges in Rohren durch verallgemeinerte Potenzbeziehungen, *Z Ver deutsch. Ing. Beih. Verfahrenstech.*, 4, 91–98, 1943.
24. Siegel, R., Sparrow, E. M., and Hallman, T. M., Steady laminar heat transfer in a circular tube with prescribed wall heat flux, *Appl. Sci. Res.*, 7A, 386–392, 1958.
25. Kakaç, S. and Özgü, M. R., Analysis of laminar flow forces convection heat transfer in the entrance region of a circular pipe, *Wärme-und Stoffübertragung*, 2, 240–245, 1969.
26. Langhaar, H. L., Steady flow in the transition length of a straight tube, *J. Appl. Mech., Trans. ASME*, 9, 4–55, 1942.
27. Ulrichson, D. L. and Schmitz, R. A., Laminar-flow heat transfer in the entrance region of circular tubes, *Int. J. Heat Mass Transfer*, 8, 253–258, 1965.
28. Manohar, R., Analysis of laminar-flow heat transfer in the entrance region of circular tubes, *Int. J. Heat Mass Transfer*, 12, 15–22, 1969.
29. Sparrow, E. M., Lin, S. H., and Lundgren, T. S., Flow developments in the hydrodynamic entrance region of tubes and ducts, *Phys. Fluids*, 7(3), 338–347, 1964.
30. Roy, D. N., Laminar heat transfer in the inlet of a uniformly heated tube, *Trans. ASME, J. Heat Transfer*, 84, 425, 1965.
31. Lévêque, M. A., Les Lois de la Transmission de Chaleur par Convection, *Annales des Mines, Memoires, Ser. 12*, 13, 201–229, 305–362, 381–415, 1928.
32. Boussinesq, J., Calcul du Poirior Refroidissant des Courants Fluids, *Math. Pures Appl.*, 60, 285, 1905.
33. Prins, J. A., Mulder, J., and Schenk, J., Heat transfer in laminar flow between parallel plates, *Appl. Sci. Res.*, A2, 431–438, 1956.
34. Mengüç, P., Heat transfer in a radiating laminar flow between two parallel plates, MS thesis, Middle East Technical University, Ankara, Turkey, 1980.

35. Norris, R. H. and Streid, D. D., Laminar flow heat-transfer coefficient for ducts, *Trans. ASME*, 62, 525–533, 1940.
36. Reynolds, W. C., Heat transfer to fully developed laminar flow in a circular tube with arbitrary circumferential heat flux, *J. Heat Transfer*, 82, 108–112, 1960.
37. Jacob, M. and Ress, K. A., Heat transfer to a fluid in laminar flow through an annular space, *Trans. AIChE*, 37, 619, 1941.
38. Reynolds, W. C., Lunderg, R. E., and McCuen, P. A., Heat transfer in annular passages. General formulation of the problem for arbitrarily prescribed wall temperature or heat flux, *Int. J. Heat Mass Transfer*, 6, 483–493, 1963.
39. Reynolds, W. C., Lunderg, R. E., and McCuen, P. A., Heat transfer in annular passages. Hydrodynamically developed laminar flow with arbitrarily prescribed wall temperature or heat flux, *Int. J. Heat Mass Transfer*, 6, 495–529, 1963.
40. Worsoe-Schmidt, P. M., Heat transfer in the thermal entrance region of circular tubes and annular passages with fully developed laminar flow, *Int. J. Heat Transfer*, 10, 541–551, 1967.
41. Sugino, E., Velocity distribution and pressure drop in the inlet of a pipe with annular space, *Bull. JSME*, 5(20), 651–655, 1962.
42. Sparrow, E. M. and Lin, S. M., Developing laminar flow and pressure drop in the entrance region of annular duct, *J. Basic Eng., Trans. ASME*, 86, 827–834, 1964.
43. Manohar, R., An exact analysis of laminar flow in the entrance region of an annular duct, *Zeitschrift fur Angewandte Math. und Mech.*, 45, 171–176, 1964.
44. Murakawa, K., Heat transfer in entry length of double pipes, *Int. J. Heat Mass Transfer*, 2, 240–251, 1961.
45. Heaton, H. S., Reynolds, W. C., and Kays, W. M., Heat transfer in annular passages: Simultaneous development of velocity and temperature fields in laminar flow, *Int. J. Heat Mass Transfer*, 7, 763–781, 1964.
46. Yücel, O., Laminar flow heat transfer in the combined entrance region of concentric circular annulus, M.S. thesis, Middle East Technical University, Ankara, Turkey, 1972.
47. Kakaç, S. and Yücel, O., Laminar flow heat transfer in an annulus with simultaneous development of velocity and temperature fields, TUBITAK, ISITEK Report No. 19, Ankara, Turkey, 1974.
48. Bejan, A., *Convection Heat Transfer*, 2nd edn., John Wiley & Sons, New York, 1995.

# 7

# Forced Convection in Turbulent Flow

## Nomenclature

| | |
|---|---|
| $A$ | constant, Equation 7.46 |
| $A^+$ | van Driest damping constant, Equation 7.38 |
| $B$ | constant, Equation 7.46 |
| $C_f$ | skin friction coefficient $= 2\tau_w/\rho U_\infty^2$ |
| $c_p$ | specific heat at constant pressure, J/(kg K) |
| $D(y)$ | effective thermal diffusivity $= \alpha + \varepsilon_h$, m²/s |
| $d$ | diameter of a circular duct or distance between parallel plates, m |
| $d_e$ | equivalent diameter, m |
| $f$ | fanning friction factor $= 2\tau_w/\rho U_m^2$ |
| $h$ | heat-transfer coefficient, W/(m² K) |
| $K(y)$ | effective thermal conductivity $= k + \rho c_p \varepsilon_h$, W/(m K) |
| $k$ | thermal conductivity, W/(m K) |
| $k_t$ | turbulent conductivity $= \rho c_p \varepsilon_h$ |
| $L$ | reference length, m |
| $l$ | mixing length, m |
| $Nu$ | Nusselt number $= hd/k$, $hL/k$ |
| $Pe$ | Péclet number $= RePr$ |
| $Pr$ | Prandtl number $= c_p \mu/k$ |
| $Pr_t$ | turbulent Prandtl number $= \varepsilon_m/\varepsilon_h$ |
| $p$ | pressure, Pa |
| $\Delta p$ | static pressure drop, Pa |
| $q''$ | heat flux, W/m² |
| $Re$ | Reynolds number based on length $L$ or $d$, $= \rho U_\infty L/\mu$, $\rho U_m d/\mu$ |
| $r$ | radius, radial coordinate, m |
| $St$ | Stanton number $= Nu/(RePr)$ |
| $T$ | temperature, °C, K |
| $t$ | time, s |
| $U$ | mean or free-stream velocity, m/s |
| $u$ | velocity component in $x$-direction, m/s |
| $u^+$ | nondimensional velocity $= u/(\tau_w/\rho)^{1/2}$ |
| $V$ | velocity vector, m/s |
| $v$ | velocity component in $y$-direction |
| $x$ | rectangular coordinate, m |
| $Y$ | dimensionless distance $= y/d$, m |

| | |
|---|---|
| $y$ | rectangular coordinate normal to surface, m |
| $y^+$ | nondimensional distance from surface $= (\tau_w/\rho)^{1/2}y/v$ |
| $w$ | velocity component in $z$-direction, m/s |

## Greek Symbols

| | |
|---|---|
| $\alpha$ | thermal diffusivity $= k/\rho c$, m$^2$/s |
| $\delta$ | boundary-layer thickness, m |
| $\varepsilon_h$ | thermal eddy diffusivity, m$^2$/s |
| $\varepsilon_m$ | momentum eddy diffusivity, m$^2$/s |
| $\kappa$ | von Karman mixing-length constant |
| $\mu$ | dynamic viscosity, Pa s |
| $\mu_e$ | effective viscosity $= \mu + \rho\varepsilon_m$, Pa s |
| $\mu_t$ | turbulent viscosity $= \rho\varepsilon_m$, Pa s |
| $v$ | kinematic viscosity, m$^2$/s |
| $\rho$ | density, kg/m$^3$ |
| $\tau$ | shear stress, Pa |

## Subscripts

| | |
|---|---|
| $b$ | buffer layer |
| $c$ | center |
| $fp$ | flat plate |
| $i$ | inner or inlet |
| $l$ | laminar sublayer |
| $m$ | mean value |
| $o$ | outer |
| $q$ | constant heat flux |
| $T$ | constant wall temperature |
| $t$ | turbulent quantity |
| $x$ | local value |
| $w$ | wall condition |
| $\infty$ | free-stream conditions |

## 7.1 Introduction

Most flows occurring in practical applications are turbulent. Turbulent flow is characterized by disorderly displacement of individual volumes of fluid within the flow. Velocity, temperature, pressure, and other properties change continuously in time at every point of a turbulent flow. The governing equations describing the conservation of mass, momentum, and energy, which were described in Chapter 2, are valid for turbulent as well as for laminar flows. It must be noted, however, that the quantities such as velocity, pressure, and temperature in these equations are instantaneous values. In turbulent flow, the instantaneous values always vary with time, and the variations are completely random with irregular fluctuations about mean values. This leads to seemingly insurmountable difficulties in the solution of these equations in most turbulent flow problems. That is, the Navier–Stokes and energy equations could be applied to the case of turbulent flow if we could follow out all irregular fluctuations in velocity, pressure, temperature, etc. In any case, for most engineering purposes, we are only interested in the mean values. Hence, it becomes necessary to take the statistical nature of turbulence into account in turbulent flow.

The methods of analysis of heat transfer in turbulent flow are by no means complete at present, because the nature of the mechanism of turbulence is very complex. However, many analytical procedures with various turbulence models and empirical correlations have been proposed by various investigators. In turbulent flow, the fluctuations that are superimposed on the principle motion are so complex that mathematical treatment seems impossible. In order to attack the problems involving turbulence, analytically or numerically, it is convenient to define mean and fluctuating components (conventional Reynolds decomposition) of velocity, pressure, temperature, density, etc., such that

$$
\begin{aligned}
u &= \bar{u} + u', & p &= \bar{p} + p' \\
v &= \bar{v} + v', & T &= \bar{T} + T'. \\
w &= \bar{w} + w', & \rho &= \bar{\rho} + \rho'
\end{aligned}
\tag{7.1}
$$

The velocity component $u$, for example, in turbulent flow at a fixed location varies as a function of time as shown in Figure 7.1. The fluctuating component $u'$ can be seen from this figure to have both positive and negative values, and the mean value $\bar{u}$ is defined as

$$
\bar{u} \equiv \frac{1}{\Delta t} \int_{t_0}^{t_0 + \Delta t} u \, dt,
\tag{7.2}
$$

where $\Delta t$ is sufficiently large compared to the period of the random fluctuations associated with the turbulence, to obtain a true average, that is, large enough for recording the turbulent fluctuations but sufficiently small for the velocity $u$ to be unaffected by the external disturbances on the flow. Hence, it becomes clear that

$$
\bar{u}' = \int_{t_0}^{t_0 + \Delta t} u' \, dt \equiv 0,
\tag{7.3}
$$

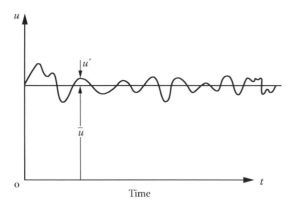

**FIGURE 7.1**
Variation of the velocity component $u$ with time.

because during the period $\Delta t$ all the fluctuating components of velocity cancel each other. Therefore, one can describe the field of real velocities as a field of averaged velocities in time and the superimposed field of fluctuations. The same is true for temperature, pressure, and density fields and with other dependent variables.

Turbulent flow is called steady when the mean values $\bar{u}$, $\bar{v}$, $\bar{w}$, $\bar{p}$, $\bar{T}$, etc., do not change with time. This is reasonable; experiments show that instruments with relatively long response times (i.e., a thermocouple), when placed in a turbulent stream, show readings that are entirely stable with time. This means that the values of the velocities and temperatures in a turbulent flow remain constant on the average when taken over a sufficiently long time interval in steady turbulent flow. Although the actual values continuously deviate from the average value, the mean values remain constant over certain time interval $\Delta t$. Therefore, the time average of all quantities describing the fluctuations is equal to zero.

In some cases, the properties such as density, viscosity, specific heat, and thermal conductivity may also be variable, but the fluctuations of the fluid properties will be neglected.

## 7.2 Governing Equations with Steady Turbulent Flow

The continuity equation, equations of motion and the energy equation derived in Chapter 2, must be satisfied by the instantaneous velocity components, pressure, and temperature. We shall consider a steady, 2D turbulent flow with constant properties, meaning that $\bar{w} = 0$, $\bar{u} = \bar{u}(x, y)$, and $\bar{v} = \bar{v}(x, y)$. Even in this case, $u' = u'(x, y, z, t)$, $v' = v'(x, y, z, t)$, and $w' \neq 0$. We shall however assume that

$$u' = u'(x, y, t),$$

$$v' = v'(x, y, t), \tag{7.4}$$

$$w' = 0.$$

With the earlier assumptions, inserting Equations 7.1 into Equations 2.14, 2.33, and 2.57, we can obtain the continuity, momentum, and energy equations for turbulent flow. In this derivation, the mean behavior rather than instantaneous behavior of turbulent flow will be considered. After inserting the mean and fluctuating components into the governing equations, the equations will be averaged over long time periods.

### 7.2.1 Continuity Equation

The continuity equation written in the Cartesian coordinate system for the instantaneous velocity components $u$ and $v$ from Equation 2.14 is

$$\frac{\partial u}{\partial x} + \frac{\partial v}{\partial y} = 0. \tag{7.5}$$

Substituting Equations 7.1 into Equation 7.5, we obtain

$$\frac{\partial \bar{u}}{\partial x} + \frac{\partial \bar{v}}{\partial y} + \frac{\partial u'}{\partial x} + \frac{\partial v'}{\partial y} = 0. \tag{7.6}$$

When this equation is averaged over a small time interval $\Delta t$ by using the definitions (7.2) and (7.3), the time average of the fluctuating components becomes zero and the average quantities $\bar{u}$ and $\bar{v}$ remain the same. Thus, Equation 7.6 reduces to

$$\frac{\partial \bar{u}}{\partial x} + \frac{\partial \bar{v}}{\partial y} = 0, \tag{7.7}$$

and hence

$$\frac{\partial u'}{\partial x} + \frac{\partial v'}{\partial y} = 0. \tag{7.8}$$

Therefore, the time-averaged velocity components and the fluctuating components each satisfy the incompressible equation of continuity as in the laminar flow.

## 7.2.2 Momentum Equations

Substituting the velocity components and pressure into Equation 2.33a, for the $x$-momentum equation, we get

$$\frac{\partial u'}{\partial t} + \left(\bar{u} + u'\right)\frac{\partial}{\partial x}\left(\bar{u} + u'\right) + \left(\bar{v} + v'\right)\frac{\partial}{\partial y}\left(\bar{u} + u'\right) = -\frac{1}{\rho}\left(\frac{\partial \bar{p}}{\partial x} + \frac{\partial p'}{\partial x}\right) + v\left(\frac{\partial^2 \bar{u}}{\partial x^2} + \frac{\partial^2 \bar{u}}{\partial y^2} + \frac{\partial^2 u'}{\partial x^2} + \frac{\partial^2 u'}{\partial y^2}\right), \tag{7.9}$$

where we have neglected the body force in the $x$-direction. An analogous expression may be set up for the $y$-direction. This equation will now be time averaged.

Before taking the time average of Equation 7.9 over the time interval $\Delta t$, we shall adopt the following rules for averaging in accordance with Reynolds:

1. A second averaging with respect to the same variable does not change the value of the quantity obtained after the first averaging, that is, if

$$\bar{u} = \frac{1}{\Delta t}\int_{t_0}^{t_0 + \Delta t} u \, dt,$$

then

$$\bar{\bar{u}} = \bar{u}. \tag{7.10}$$

2. The average of a product of an averaged quantity and actual quantity is equal to the product of averaged quantities, that is,

$$\overline{\bar{u}\,u} = \bar{u}\,\bar{u}. \tag{7.11}$$

3. The average of the product of two averaged quantities is equal to the product of averaged quantities, that is,

$$\overline{\overline{u}\,\overline{u}} = \overline{u}\,\overline{u}. \tag{7.12}$$

4. The average of the sum of two averaged quantities is equal to the sum of the averaged quantities, that is,

$$\overline{\overline{u}+\overline{u}} = \overline{u}+\overline{u}. \tag{7.13}$$

5.

$$\overline{\int u\,dx} = \int \overline{u}\,dx \quad \text{and} \quad \overline{\frac{\partial u}{\partial x}} = \frac{\partial \overline{u}}{\partial x}. \tag{7.14}$$

Considering these relationships, we find

$$\overline{\overline{u}\frac{\partial \overline{u}}{\partial x}} = \overline{u}\frac{\partial \overline{u}}{\partial x}, \quad \overline{\frac{\partial^2 \overline{u}}{\partial x^2}} = \frac{\partial^2 \overline{u}}{\partial x^2},$$

$$\overline{\overline{v}\frac{\partial \overline{u}}{\partial y}} = \overline{v}\frac{\partial \overline{u}}{\partial y}, \quad \overline{\frac{\partial^2 \overline{u}}{\partial y^2}} = \frac{\partial^2 \overline{u}}{\partial y^2}, \tag{7.15}$$

$$\frac{1}{\rho}\overline{\frac{\partial \overline{p}}{\partial x}} = \frac{1}{\rho}\frac{\partial \overline{p}}{\partial x}.$$

Accordingly, we get

$$\overline{\frac{\partial^2 u'}{\partial x^2}} = \frac{1}{\Delta t}\int_{t_0}^{t_0+\Delta t}\frac{\partial^2 u'}{\partial x^2}\,dt = \frac{1}{\Delta t}\frac{\partial^2}{\partial x^2}\int_{t_0}^{t_0+\Delta t}u'\,dt = 0. \tag{7.16}$$

Similarly,

$$\overline{\frac{\partial u'}{\partial t}} = \overline{u'\frac{\partial \overline{u}}{\partial x}} = \overline{v'\frac{\partial \overline{u}}{\partial x}} = 0. \tag{7.17}$$

Taking now the time average of each term of Equation 7.9, in view of the earlier expressions, we obtain

$$\overline{u}\frac{\partial \overline{u}}{\partial x} + \overline{v}\frac{\partial \overline{u}}{\partial y} + \overline{u'\frac{\partial u'}{\partial x}} + \overline{v'\frac{\partial u'}{\partial y}} = -\frac{1}{\rho}\frac{\partial \overline{p}}{\partial x} + v\left(\frac{\partial^2 \overline{u}}{\partial x^2} + \frac{\partial^2 \overline{u}}{\partial y^2}\right). \tag{7.18}$$

Considering the following relations,

$$\overline{u'\frac{\partial u'}{\partial x}} = \frac{1}{2}\frac{\partial}{\partial x}\overline{(u')^2}, \tag{7.19a}$$

$$\overline{v'\frac{\partial u'}{\partial y}} = \frac{\partial}{\partial y}\overline{(u'v')} + \frac{1}{2}\frac{\partial}{\partial x}\overline{(u')^2}, \tag{7.19b}$$

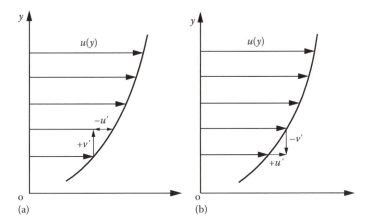

**FIGURE 7.2**
Fluctuating velocity components.

Equation 7.18 may also be written as

$$\bar{u}\frac{\partial \bar{u}}{\partial x}+\bar{v}\frac{\partial \bar{u}}{\partial y}=-\frac{1}{\rho}\frac{\partial \bar{p}}{\partial x}+\frac{1}{\rho}\left[\mu\left(\frac{\partial^2 \bar{u}}{\partial x^2}+\frac{\partial^2 \bar{u}}{\partial y^2}\right)-\frac{\partial}{\partial x}\left(\rho\overline{u'^2}\right)-\frac{\partial}{\partial y}\left(\rho\overline{u'v'}\right)\right]. \tag{7.20}$$

Figure 7.2a presumes a positive velocity fluctuating component $v'$ by eddy movement; the effect of $v' > 0$ will be to decrease the longitudinal velocity $u$ by causing $u' < 0$. The opposite will occur in the case of Figure 7.2b. Figure 7.2 shows that on the average a positive $v'$ is always associated with a negative $u'$ and a negative $v'$ is associated with a positive $u'$. Therefore, the time average $\overline{u'v'}$ always has the negative sign and is not zero.

Similarly, for the $y$-component of the momentum equation, we get

$$\bar{u}\frac{\partial \bar{v}}{\partial x}+\bar{v}\frac{\partial \bar{v}}{\partial y}=-\frac{1}{\rho}\frac{\partial \bar{p}}{\partial y}+\frac{1}{\rho}\left[\mu\left(\frac{\partial^2 \bar{v}}{\partial x^2}+\frac{\partial^2 \bar{v}}{\partial y^2}\right)-\frac{\partial}{\partial x}\left(\rho\overline{u'v'}\right)-\frac{\partial}{\partial y}\left(\rho\overline{v'^2}\right)\right]. \tag{7.21}$$

Equations 7.20 and 7.21 are the well-known Reynolds equations of motion or the Reynolds-averaged equations. The $x$-component of the momentum equation shows that in averaged motion, the fluctuations of velocity lead to the appearance of the terms like $-\rho\overline{u'v'}$ and $-\rho u'2$. The first of these is a shear stress $\tau_{yx}$ and the latter is a normal stress $\sigma_x$. These are called the Reynolds stresses and they are analogous to the viscous stress terms. The Reynolds stresses are turbulent stress terms and express the transfer of momentum by large (macroscopic) volumes of fluid displaced as a result of velocity fluctuations in the flow.

Applying Prandtl's order of magnitude analysis in the same way as we did for laminar flow, the hydrodynamic boundary-layer equations for the 2D incompressible turbulent flow become

$$\frac{\partial \bar{u}}{\partial x}+\frac{\partial \bar{v}}{\partial y}=0, \tag{7.22}$$

$$\bar{u}\frac{\partial \bar{u}}{\partial x}+\bar{v}\frac{\partial \bar{u}}{\partial y}=-\frac{1}{\rho}\frac{d\bar{p}}{dx}+\frac{1}{\rho}\left[\frac{\partial}{\partial y}\left(\mu\frac{\partial \bar{u}}{\partial y}-\rho\overline{u'v'}\right)\right]. \tag{7.23}$$

The boundary-layer momentum equation in the $y$-direction gives $\partial \bar{p}/\partial y = 0$, that is, $\bar{p}$ in the boundary layer over a flat surface is a function of $x$ only. The first term inside the bracket on the right-hand side of Equation 7.23 represents viscous shear stress proportional to the gradient of the averaged velocity, and the second term is the turbulent shear stress caused by momentum transfer due to velocity fluctuations.

### 7.2.3 Energy Equation

In nonisothermal turbulent flows, the fluctuations of velocity result also in fluctuations of temperature.

Neglecting the viscous dissipation term in the energy equation (2.57) and substituting the velocity and temperature components from Equation 7.1, we get, after averaging each term in the energy equation over a time interval $\Delta t$ and applying Prandtl's order of magnitude analysis, the following Reynolds form of the energy equation for a steady 2D turbulent flow:

$$\bar{u}\frac{\partial \bar{T}}{\partial x} + \bar{v}\frac{\partial \bar{T}}{\partial y} = \frac{1}{\rho c_p}\left[\frac{\partial}{\partial y}\left(k\frac{\partial \bar{T}}{\partial y} - \rho c_p \overline{u'T'}\right)\right]. \tag{7.24}$$

It should be noted that in Equation 7.24, heat diffusion only in the $y$-direction is assumed. As seen, there is one additional term in the energy equation.

For 3D constant-property flows, the general energy equation becomes

$$\frac{D\bar{T}}{Dt} = \frac{k}{\rho c_p}\nabla^2\bar{T}\left[\frac{\partial}{\partial x}\left(\overline{u'T'}\right) + \frac{\partial}{\partial y}\left(\overline{v'T'}\right) + \frac{\partial}{\partial z}\left(\overline{w'T'}\right)\right] \tag{7.25}$$

or

$$\frac{D\bar{T}}{Dt} = \alpha\nabla^2\bar{T} - \nabla\left(\overline{v'T'}\right), \tag{7.26}$$

where $\mathbf{V'} = \hat{\mathbf{i}}u' + \hat{\mathbf{j}}v' + \hat{\mathbf{k}}w'$. These additional terms involving the fluctuating components represent the transfer of heat by turbulent convection, and they are analogous to the Reynolds stresses in the equation of motion.

The first term on the right-hand side of Equation 7.24 represents the transfer of heat by molecular conduction within the fluid, and the second term represents the process of turbulent or eddy transfer (convective transfer). Transfer processes in a turbulent flow are controlled by two mechanisms: molecular and turbulent (convective). Therefore, property fluctuations in a turbulent system produce apparent (total) stresses and fluxes. Total shearing stress in the $x$-direction and the heat flux in the $y$-direction can be written as follows:

$$\tau = \mu\frac{\partial \bar{u}}{\partial y} - \rho\overline{u'v'}. \tag{7.27}$$

Here, the first term is the viscous stress, and the second term is the turbulent stress:

$$q'' = -\left(k\frac{\partial \bar{T}}{\partial y} - \rho c_p\overline{v'T'}\right). \tag{7.28}$$

The first term of this sum is the heat flux due to conduction, while the second term represents heat convection by eddies. Equations 7.27 and 7.28 may be regarded as the rate equations for turbulent flow systems. Each flux consists of a molecular diffusion flux (as in laminar flow) and an eddy flux. The eddy momentum flux $\rho\overline{u'v'}$ in Equation 7.27 can be interpreted as a turbulent shear stress in the same manner that the molecular shear stress was interpreted. The eddy diffusivity of momentum $\varepsilon_m$ for turbulent shear stress was defined by Boussinesq [1] as

$$-\overline{u'v'} = \varepsilon_m \frac{\partial \overline{u}}{\partial y}. \tag{7.29}$$

Then, the total shear stress (apparent shear stress) in the $x$-direction can be written in the following way:

$$\tau = (\mu + \rho\varepsilon_m) \frac{\partial \overline{u}}{\partial y}. \tag{7.30}$$

If it is assumed that the principal direction of flow is parallel to the $x$-axis, that is, $\overline{u} = \overline{u}(y)$ and $\overline{v} = 0$, then we have

$$\tau_{yx} = -\rho\overline{u'v'}.$$

The quantity $\rho\varepsilon_m$ is usually interpreted as an eddy viscosity (turbulent viscosity) analogous to the molecular viscosity $\mu$, but whereas the latter is a fluid property, the former is a parameter of fluid motion.

In a similar way, the enthalpy flux $\rho c_p\overline{v'T'}$ in Equation 7.24 suggests the concept of an eddy diffusivity $\varepsilon_h$ for energy transfer, which was defined by Boussinesq as

$$-\overline{v'T'} = \varepsilon_h \frac{\partial \overline{T}}{\partial y}. \tag{7.31}$$

Then, the total heat flux (apparent heat flux) in the $y$-direction becomes

$$q'' = -(k + \rho c_p\varepsilon_h) \frac{\partial \overline{T}}{\partial y}. \tag{7.32}$$

The product $\rho c_p\varepsilon_h$ has the units of conductivity and is called eddy conductivity or turbulent conductivity, and again it is not only a fluid property but a flow parameter.

These two coefficients are, in general, different from each other, because the mechanisms for momentum and thermal energy transport are not identical. But for a given turbulent system, $\varepsilon_m$ and $\varepsilon_h$ might be expected to be closely related.

Therefore, substituting the definitions of $\varepsilon_m$ and $\varepsilon_h$ into the boundary-layer equations (7.23) and (7.24), the momentum equation in the $x$-direction and the energy equation for turbulent boundary-layer flows can be written in the following forms:

*Momentum equation*

$$\overline{u} \frac{\partial \overline{u}}{\partial x} + \overline{v} \frac{\partial \overline{u}}{\partial y} = -\frac{1}{\rho} \frac{\partial \overline{p}}{\partial x} + \frac{1}{\rho} \frac{\partial}{\partial y} \left[ (\mu + \rho\varepsilon_m) \frac{\partial \overline{u}}{\partial y} \right], \tag{7.33}$$

where $\mu_e = \mu + \rho\varepsilon_m$ (effective viscosity).

*Energy equation*

$$\bar{u}\frac{\partial \bar{T}}{\partial x} + \bar{v}\frac{\partial \bar{T}}{\partial y} = \frac{1}{\rho c_p}\frac{\partial}{\partial y}\left[(k + \rho c_p \varepsilon_h)\frac{\partial \bar{T}}{\partial y}\right], \tag{7.34a}$$

which can also be written as

$$\bar{u}\frac{\partial \bar{T}}{\partial x} + \bar{v}\frac{\partial \bar{T}}{\partial y} = \frac{1}{\rho c_p}\frac{\partial}{\partial y}\left[K(y)\frac{\partial \bar{T}}{\partial y}\right], \tag{7.34b}$$

where $K(y) = k + \rho c_p \varepsilon_h$ (effective thermal conductivity) or

$$\bar{u}\frac{\partial \bar{T}}{\partial x} + \bar{v}\frac{\partial \bar{T}}{\partial y} = \frac{\partial}{\partial y}\left[D(y)\frac{\partial \bar{T}}{\partial y}\right], \tag{7.34c}$$

where $D(y) = \alpha + \varepsilon_h$ (effective thermal diffusivity). Strictly speaking, if the flow pattern is independent of $x$, then $K$ and $D$ are functions of $y$ alone. Equations 7.34 can also be written in the following vectorial forms:

$$\frac{D\bar{T}}{Dt} = \alpha\,\text{div}\left[\left(1 + \frac{\varepsilon_h}{\alpha}\right)\text{grad}\,\bar{T}\right] \tag{7.34d}$$

or

$$\frac{D\bar{T}}{Dt} = \alpha\,\text{div}\left[\left(1 + \frac{k_t}{k}\right)\text{grad}\,\bar{T}\right], \tag{7.34e}$$

where
   $k_t = \rho c_p \varepsilon_h$ (turbulent conductivity)
   $k$ is the molecular conductivity

By making use of the effective viscosity and effective thermal conductivity concepts, Equations 7.33 and 7.34 can be derived by writing a force and momentum balance and a thermal energy balance on a control volume: consider a 2D turbulent flow system of heat and momentum transfer to a fluid flowing in a channel as illustrated in Figure 7.3. We assume that the fluid is incompressible and the flow is 2D and steady with negligible heat conduction in the direction of flow and that the thermophysical properties remain constant throughout the field. Referring to Figure 7.4a, we consider the forces acting on an elemental control volume $\Delta x$ and $\Delta y$ by unity in the $z$-direction. It is assumed that there are no pressure gradients in the direction perpendicular to flow direction and the viscous forces in the $y$-direction are negligible. Equating the sum of the viscous shear and pressure forces to the net rate of momentum transfer in the $x$-direction, we obtain Equation 7.33.

In Figure 7.4b, the rate of heat flow into the element by means of transport process (convection) and rate of heat conduction in the $y$-direction are shown. Writing the energy balance with the quantities shown in Figure 7.4b yields Equation 7.34b.

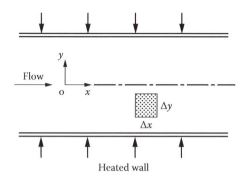

**FIGURE 7.3**
A 2D turbulent flow system.

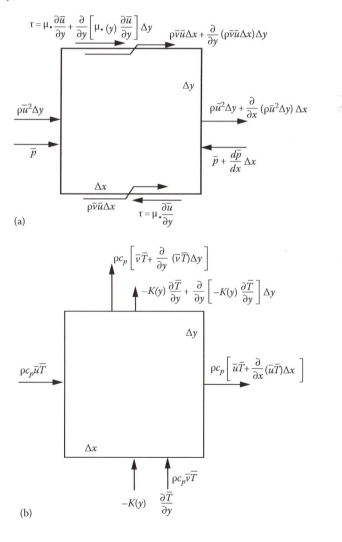

**FIGURE 7.4**
Elemental control volume for force and momentum, and energy balances in 2D turbulent flow. (a) The forces acting on an elemental control volume and (b) the rate of heat flow into the element by means of transport process (convection) and rate of heat conduction in the y-direction.

The earlier form of the energy equation takes into account the dependence of thermal conductivity on temperature. It is also noted that in these equations, as in the general equations, the Reynolds stresses and fluxes are additive to the corresponding molecular terms and that the energy equation is linear.

The following dimensionless number is known as the *turbulent Prandtl number*:

$$Pr_t = \frac{\varepsilon_m}{\varepsilon_h}. \tag{7.35}$$

Turbulent flow boundary-layer equations (Equations 7.22, 7.33, and 7.34) show that there are five unknowns ($\bar{u}$, $\bar{v}$, $\bar{T}$, $\varepsilon_m$, and $\varepsilon_h$) with three equations. The expressions for eddy diffusivity of momentum, $\varepsilon_m$, and eddy diffusivity of heat, $\varepsilon_h$, are obtained from the existing turbulence models, which will be discussed in the next section. When the velocity profiles and pressure distribution are known by measurements, then $\varepsilon_m$ can be calculated as a function of the coordinates and possibly of the Reynolds and Prandtl numbers. Knowing the turbulent Prandtl number and from this introducing $\varepsilon_h$ together with the velocity profiles, Equation 7.34a can be solved for the temperature field.

In the discussions that follow in this and the following chapters, the use of the over-bar notation for a time-averaged quantity will be discontinued since all the dependent variables in turbulent flow are time averages.

## 7.3 Turbulence Models

In order to obtain solutions to the Reynolds equations, it becomes necessary to evaluate the turbulent stresses and heat fluxes appearing in these equations that characterize turbulence modeling. The literature on turbulence models is extensive. None of them are both general and accurate. Only a few of the most commonly used models will be described in this section.

Turbulence models to close the Reynolds equations can be divided into two categories according to whether or not the Boussinesq assumption is used. Models using the Boussinesq assumption are referred to as *turbulent-viscosity* models. Most models currently employed in engineering calculations are of this type. Models without Boussinesq assumption are known as the *Reynolds-stress* or *stress-equation* models.

The other common classification of models is according to the number of supplementary partial differential equations, which must be solved in order to supply the modeling parameters. This number ranges from 0 for the simplest algebraic models to 12 for the most complex of the Reynolds-stress models [2].

For the solution of a boundary-layer form of the conservation equation, Equations 7.23 and 7.24, the modeling task is to find expressions for $-\rho\overline{u'v'}$ and $-\rho c_p \overline{v'T'}$. A large number of convective problems can still be solved through the use of the boundary-layer equations. Common models have been discussed in Reference 3. The description here follows the chapter of Pletcher in Reference 3.

*Zero-Equation Models*: These models are also called *algebraic turbulence models*, which utilize the Boussinesq assumption. One of the most successful models of this type was suggested by Prandtl [4,5], which is called the mixing-length model:

$$\mu_t = \rho\varepsilon_m = \rho l^2 \left( \frac{\partial u}{\partial y} \right) \tag{7.36}$$

or from the definition of momentum eddy diffusivity (Equation 7.29)

$$-\overline{u'v'} = \varepsilon_m \frac{\partial u}{\partial y} = l^2 \left( \frac{\partial u}{\partial y} \right)^2,$$
(7.37)

where *l* is the so-called *mixing length* that can be thought of as a transverse distance over which particles maintain their original momentum and is approximately of the order of a mean free path for the collision or mixing of globules of fluid.

The evaluation of the mixing length *l* depends on the type of flow. For internal or external flows along a solid surface, good results are obtained by evaluating *l* according to the following relation [6]:

$$l_i = \kappa y \left( 1 - e^{-y^+/A^+} \right),$$
(7.38)

where

$$y^+ = \frac{y(\tau_w/\rho_w)^{1/2}}{\nu_w}.$$
(7.39)

$\kappa$ is the von Karman constant, usually taken as 0.41, and $A^+$ is the damping constant, most commonly evaluated as 26. The mixing length *l* is evaluated by the use of Equation 7.38 in the inner region closest to the solid boundaries. When $l_i$ predicted by Equation 7.38 first exceeds $l_o$, then Equation 7.37 is used for the outer region with the following relation:

$$l_o = C_1 \delta,$$
(7.40)

where
   the constant $C_1$ is usually assigned a value of 0.089
   $\delta$ is the velocity boundary-layer thickness

The quantity in parentheses in Equation 7.38 is the van Driest damping function [6] and is the most common expression used to bridge the gap smoothly between the fully turbulent region where $l = \kappa y$ and the viscous (laminar) sublayer where $l \to 0$. Therefore, for the fully turbulent region, if one assumes that the eddy diffusivity of momentum $\varepsilon_m$ is zero within the laminar sublayer, then $\varepsilon_m$ can be written as

$$\varepsilon_m = \kappa^2 y^2 \left( \frac{\partial u}{\partial y} \right).$$
(7.41)

An alternative treatment to Equation 7.40 is also used to evaluate the turbulent viscosity in the outer region [7]. The exponential expression of Equation 7.38 has been modified to take into account the effects of property variations, pressure gradients, blowing, and surface roughness [7]. Equation 7.38 gives the mixing length for the inner region (law-of-the-wall zone) of the turbulent flow ($y^+ \leq 400$), and Equation 7.40 produces the outer *wakelike* region. Therefore, Equations 7.38 and 7.40 give a two-region turbulence model.

The Reynolds heat-flux term $\rho c_p \overline{v'T'}$ is usually handled in algebraic models defining a turbulent conductivity ($\rho c_p \overline{v'T'} = -k_t \partial T / \partial y$) or eddy thermal diffusivity $\varepsilon_h$ as defined by Equation 7.31. In most of the algebraic turbulence models, the turbulent Prandtl number is taken as constant near 1; mostly, $Pr_t = 0.9$.

By the use of the turbulent Prandtl number, the turbulent heat flux can also be expressed as

$$-\rho c_p \overline{v'T'} = \frac{c_p \mu_t}{Pr_t} \frac{\partial T}{\partial y}. \tag{7.42}$$

The Boussinesq-type approximation can be extended to other components of the temperature gradient; one can evaluate $-\rho c_p \overline{u'T'}$ as

$$-\rho c_p \overline{u'T'} = \frac{c_p \mu_t}{Pr_t} \frac{\partial T}{\partial x}. \tag{7.43}$$

Therefore, using Equations 7.37 and 7.38 or an equivalent model, Equations 7.22 and 7.23 can be integrated numerically to determine the flow field $u, v$.

### 7.3.1 Eddy Diffusivity of Heat and Momentum

The simplest approach to the solution of Equations 7.34 is to make some reasonable assumptions; one of these is the assumption concerning the turbulent Prandtl number ($Pr_t = \varepsilon_m/\varepsilon_h$). The first statement of this type was made by Reynolds [8] who reached the conclusion that $Pr_t = 1$, which indicates that in a fully turbulent field, both momentum and heat are transferred at the same rate as a result of the motion of eddies. Details of this argument will be discussed in Section 7.6.

Measurements by Corcoran et al. [9] have indicated that the turbulent Prandtl number is not constant, but in a boundary-layer type of flow, it depends on the Reynolds number and on the distance from the wall for a specified fluid. Their results are given in Table 7.1.

**TABLE 7.1**

Eddy Diffusivity of Momentum and Heat

| $Y = (y/d) \times 10^2$ | $\varepsilon_m \times 10^3$ (ft²/s) | $\varepsilon_h \times 10^3$ (ft²/s) | $\varepsilon_m/\varepsilon_h$ |
|---|---|---|---|
| 2 | 0.2 | 0.3 | 0.5 |
| 4 | 0.83 | 0.93 | 0.56 |
| 6 | 1.55 | 1.63 | 0.67 |
| 8 | 2.26 | 2.50 | 0.78 |
| 10 | 2.84 | 3.27 | 0.82 |
| 15 | 3.79 | 4.72 | 0.81 |
| 20 | 4.35 | 5.60 | 0.78 |
| 25 | 4.66 | 5.99 | 0.78 |
| 30 | 4.64 | 5.84 | 0.79 |
| 35 | 4.39 | 5.38 | 0.81 |
| 40 | 4.00 | 4.90 | 0.82 |
| 45 | 3.70 | 4.52 | 0.83 |
| 50 | | 4.34 | |

*Source:* Corcoran, W. H. et al., *Ind. Eng. Chem.*, 44, 410, 1952.
*Note:* $Re = 17,100$; $Pr = 0.73$.

They have also obtained sufficient data to establish experimentally the point values of thermal flux, temperature, velocity, and shear stresses for each set of conditions. Their velocity profile agrees well with the universal velocity profile of the other investigations. These experimental values can be used in the numerical investigations of the steady and time-dependent turbulent forced convection heat transfer.

One of the first proposals for a modification of the Reynolds analogy was made by Jenkins [10] who calculated the diffusivity ratio $\varepsilon_h/\varepsilon_m$ based on a very simple eddy model. The result is an expression for the diffusivity ratio, $\varepsilon_h/\varepsilon_m$, in terms of mixing length, transverse velocity of the eddy, and the Prandtl number. The values predicted by Jenkins for the Prandtl numbers of the order of unity are known to be in error by 10%–20%. Jenkins' results indicate that the ratio is less than unity for the Prandtl numbers around 0.7, but measurements seem to indicate a ratio somewhat in excess of unity.

Rohsenow and Cohen [11] derived an expression for $\varepsilon_h/\varepsilon_m$. The validity of their result is questionable in that it leads to infinite values of $\varepsilon_h/\varepsilon_m$ as the Prandtl number increases and, furthermore, no dependence on the Reynolds number emerges from the analysis.

Sleicher and Tribus [12] analyzed turbulent flow in circular tubes using Jenkins' analysis adjusted by a multiplying factor at the single Prandtl number of 0.7.

Kays and Leung [13] employed Jenkins' curves with a smaller bumping factor and obtained results that are in better agreement with subsequent experiments for the Prandtl number of 0.7 and are not bad at low Prandtl numbers. In the analysis made by Reynolds [14], the Prandtl numbers of 0.7 and below the Jenkins' ratios were multiplied by 1.15 to estimate $\varepsilon_h/\varepsilon_m$.

There have been several attempts to provide the analysis of the variation of the turbulent Prandtl number. One of the analyses is due to Azer and Chao [15]. It is based on a modification of Prandtl's mixing-length hypothesis. It was assumed that there is a continuous change of momentum and energy during the flight of the eddies. Two expressions giving the ratio of eddy diffusivities of heat and momentum were obtained for a fully developed circular duct flow. One of the expressions is for the Prandtl numbers ranging from 0.6 to 15, and the other one is for liquid metals.

Various investigators [16–18] have suggested various forms of variation for $\varepsilon_h/\varepsilon_m$. It is therefore clear that the question of the turbulent Prandtl number is still wide open. At the present time, no unified and consistent picture emerges.

In the constant-property analysis, it is assumed that the properties of the fluid are constant throughout the fluid. The compressibility effects are excluded from most of the analysis, and it is therefore permissible to assume that the density $\rho$ remains constant. The remaining properties of interest are the absolute viscosity $\mu$, the kinematic viscosity $\nu$, the thermal conductivity $k$, the thermal diffusivity $\alpha = k/c_p\rho$, the Prandtl number, and the specific heat $c_p$. Variation of these properties for gases with temperature is given in Table 7.2.

**TABLE 7.2**

Variation of Gas Properties with Temperature

| Gases | $\mu_{100}/\mu_{20}$ | $\nu_{100}/\nu_{20}$ | $k_{100}/k_{20}$ | $\alpha_{100}/\alpha_{20}$ | $c_{p100}/c_{p20}$ | $Pr_{100}/Pr_{20}$ |
|---|---|---|---|---|---|---|
| Air | 1.2 | 1.5 | 1.2 | 1.5 | 1.0 | 1.0 |
| Hydrogen | 1.2 | 1.5 | 1.2 | 1.6 | 1.0 | 0.96 |
| Helium | 1.2 | 1.5 | 1.1 | 1.4 | 1.0 | 1.0 |

*Note:* Ratio of property at 100°C to that at 20°C at 1 atm.

It is useful to note that temperature differences that correspond to a given heat flux are smaller in turbulent than in laminar convection and, consequently, in practical problems, the assumption of constant properties is better justified in turbulent than in laminar convection.

To take the effects of wall roughness, transpiration, strong pressure gradients, etc., into account, adjustments or extensions must be made to the simplest form of algebraic turbulence models; then, predictions in agreement with experimental results can be obtained. Such adjustments are given in References 6,19–23.

There are several more complex turbulence models that permit the variation of the model parameter $\mu_t$ with the primary flow direction in a manner determined by the solution to an ordinary differential equation. These models are called *one-half equation*, *one-equation*, and *one and one-half* models. The details of these models are given in Reference 3.

One of the most frequently used two-equation model is the $k$–$\varepsilon$ model first proposed by Harlow and Nakayama [24]. The description of this model can be found in References 25,26. Numerous two-equation models have been suggested in the literature [27–29].

## 7.4 Velocity Distribution in Turbulent Flow

Experimental observations show that the flow field near a solid surface can be divided into three regions as shown in Figure 7.5. In a very thin layer (up to A–A), viscous forces predominate and the fluid motion is laminar (the laminar sublayer or viscous layer). Next to this layer is a region in which the motion may be either streamlined or turbulent at any instant. It serves as a transition region or buffer layer between the laminar sublayer and the fully developed turbulent region.

The velocity is zero at the wall and increases linearly with distance from the wall in the laminar sublayer. As a result of turbulent mixing, which then becomes active, the velocity increases less rapidly in the buffer layer. In the turbulent region, the velocity increases only in proportion to the 1/7 power of the distance from the wall as compared to the first power in the laminar sublayer. Investigations of the velocity distribution in circular ducts with turbulent flows also indicate the presence of three zones.

It has been possible to describe the results in the form of a generalized distribution curve as shown in Figure 7.6. This representation is due to a combination of theoretical and

**FIGURE 7.5**
Three regions in turbulent boundary layer.

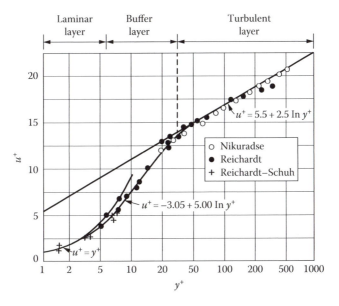

**FIGURE 7.6**
Generalized velocity distribution in a fully developed turbulent flow. (From Schlichting, H., *Boundary Layer Theory*, translated by J. Kestin, 4th edn., McGraw-Hill, New York, 1960; Reichardt, H., *Arclt. Ges. Wärmetech.*, 2, 129, 1951; Reichardt, H., *Z. Angew. Math. Mech.*, 20, 297, 1940; Reichardt, H., *Z. Angew. Math. Mech.*, 24, 268, 1944.)

empirical efforts to correlate the measured data. The ordinate $u^+$ in Figure 7.6 is the ratio of the velocity $u$ at any distance from the wall to the so-called friction velocity $\sqrt{\tau_w/\rho}$, that is,

$$u^+ = \frac{u}{\sqrt{\tau_w/\rho}}, \tag{7.44}$$

where $\tau_w$ is the shear stress at the wall. The abscissa $y^+$ is a dimensionless distance defined as

$$y^+ = y\frac{\sqrt{\tau_w/\rho}}{\nu}. \tag{7.45}$$

Figure 7.6 is a semilogarithmic diagram of measured turbulent velocity profiles in the neighborhood of the wall flow. In this way of plotting, the velocity profiles can be represented approximately as a single curve, which is independent of the Reynolds number. This profile is, therefore, called the *profile* profile. On Figure 7.6, $y^+ < 5$ corresponds to the laminar sublayer with heat and momentum transfer by molecular processes, $5 < y^+ < 30$ is the buffer zone with molecular and eddy transfer, and $y^+ > 30$ is the turbulent zone with eddy transfer being dominant.

Prandtl [30] suggested a semiempirical expression for the universal velocity profile in the following form:

$$u^+ = A \ln y^+ + B, \tag{7.46}$$

which is commonly used to describe the universal velocity profile. The experimental measurements indicate that in the turbulent region, $A \cong 2.5$ and $B \cong 5.5$. In the formulation of *the law of the wall*, the velocity distribution in the laminar sublayer is taken as

$$u^+ = y^+.$$
(7.47)

von Karman [38] introduced a buffer layer, postulating in it a velocity distribution as

$$u^+ = 5.0 \ln y^+ - 3.05,$$
(7.48)

which provides a smooth rather than abrupt transition between laminar sublayer and turbulent zone. Equation 7.46 gives a straight line in the diagram and represents the experimental values for dimensionless distance greater than 30. Note that the constants in von Karman's approximate expression for the velocity distribution in the buffer zone are chosen to give continuity both in the velocity and the velocity gradient at the edge of the laminar sublayer where $y^+ = 5$. At the junction between the buffer zone and the turbulent zone, however, these values give continuity only in the velocity and not in the velocity gradient. This outer edge of the buffer zone, specified by $y^+ = 30$, however, has no real significance.

A representative list of formulas proposed for the law of the wall by different authors has been reviewed in References 31–34 and is summarized here in Table 7.3. The main motivation for the alternative improved models was the desire to smooth out the transition from one expression to the other in continuity with experimental findings and also to simplify the derivations.

## 7.5 Friction Factors for Turbulent Flow

It is shown in Section 6.4 that for geometrically similar systems, the friction factor is a function of the Reynolds number and the relative roughness, Equation 6.31. All friction data for turbulent flow in circular ducts have been correlated by plotting the friction factor versus the Reynolds number for geometrically similar systems. Extensive data on various fluids and various sizes of ducts have been obtained since the Reynolds experiments. In 1913, Blasius [35] analyzed all the data and obtained a correlation between the friction factor and the Reynolds number and suggested the following correlation for the friction factor:

$$f = 0.079 Re^{-1/4},$$
(7.49)

which may be used to predict the friction factor for turbulent flow in smooth circular ducts for values of the Reynolds number up to $10^5$.

The shear stress $\tau_w$ at the wall and the pressure drop under these conditions are given by

$$\tau_w = f \frac{1}{2} \rho U_m^2 = 0.0395 \rho v^{1/4} d^{-1/4} U_m^{7/4}$$
(7.50)

and

$$\Delta p = 4f \frac{L}{d} \frac{\rho U_m^2}{2} = 0.158 L \rho v^{1/4} d^{-5/4} U_m^{7/4}.$$
(7.51)

**TABLE 7.3**

Various Analytical Expressions Proposed for the Universal Velocity Profile (Law of the Wall)

| Form of $u^+(y^+)$ | $1 + \dfrac{\mu_t}{\mu} = \varepsilon^+ = dy^+/du^+$ | Range | Reference |
|---|---|---|---|
| $u^+ = y^+$ | $\varepsilon^+ = 1$ | $0 \le y^+ \le 11.5$ | Prandtl [30] |
| $u^+ = 2.5\ln y^+ + 5.5$ | $\varepsilon^+ = 0.4y^+$ | $y^+ > 11.5$ | |
| $u^+ = y^+$ | $\varepsilon^+ = 1$ | $0 \le y^+ \le 5$ | von Karman [38] |
| $u^+ = 5\ln y^+ - 3.05$ | $\varepsilon^+ = 0.2y^+$ | $5 \le y^+ \le 30$ | |
| $u^+ = 2.5\ln y^+ + 5.5$ | $\varepsilon^+ = 0.4y^+$ | $y^+ > 30$ | |
| $u^+ = 14.53\tanh\,(y^+/14.53)$ | $\varepsilon^+ = 1 + \sinh^2(y^+/14.53)$ | $0 \le y^+ \le 27.5$ | Rannie [39] |
| $u^+ = 2.5\ln y^+ + 5.5$ | $\varepsilon^+ = 0.4y^+$ | $y^+ > 27.5$ | |
| $\dfrac{du^+}{dy^+} = \dfrac{2}{1 + \{1 + 4\kappa^2 y^{+2}[1 - \exp(-y^+/A^+)]^2\}^{1/2}}$ <br> $k = 0.4,\ A^+ = 26$ | $\varepsilon^+ = 1 + \{1 + 4\kappa^2 y^{+2}[1 - \exp(-y^+/A^+)]^2\}^{1/2}$ | all $y^+$ | van Driest [6] |
| $u^+ = 2.5\ln(1 + 0.4y^+) + 7.8[1 - \exp(-y^+/11)$ <br> $- (y^+/11)\exp(-0.33y^+)]$ | $\dfrac{1}{\varepsilon^+} = \dfrac{1}{1 + 0.4y^+} + 7.8[(1/11)\,\exp(-y^+/11)]$ <br> $- (1/11)\,\exp(-0.33y^+) + 0.03y^+\,\exp(-0.33y^+)$ | all $y^+$ | Reichardt [31] |
| $\dfrac{du^+}{dy^+} = \dfrac{1}{1 + n^2 u^+ y^+[1 - \exp(-n^2 u^+ y^+)]}$ <br> $n = 0.124,\ u^+ = 2.78\ln y^+ + 3.8$ | $\varepsilon^+ = 1 + n^2 u^+ y^+[1 - \exp(-n^2 u^+ y^+)]$ <br> $\varepsilon^+ = y^+/2.78$ | $0 \le y^+ \le 26$ <br> $y > 26$ | Deissler [40] |
| $y^+ = u^+ + A\left[ \exp Bu^+ - 1 - Bu^+ - \dfrac{1}{2}(Bu^+)^2 - \dfrac{1}{6}(Bu^+)^3 - \dfrac{1}{24}(Bu^+)^4 \right]$ <br> (The last term in $u + 4$ may be omitted.) | $\varepsilon^+ = 1 + AB\left[ \exp Bu^+ - 1 - Bu^+ - \dfrac{1}{2}(Bu^+)^2 - \dfrac{1}{6}(Bu^+)^3 \right]$ <br> (The last term in $u + 3$ may be omitted.) | all $y^+$ <br> $A = 0.1108$ <br> $B = 0.4$ | Spalding [41] |

*Source:* Kestin, J. and Richardson, P.D., *Int. J. Heat Mass Transfer*, 6, 147, 1963.

For higher values of the Reynolds number, beyond $10^5$, Nikuradse [36] obtained the following relationship between $f$ and $Re$:

$$\frac{1}{\sqrt{f}} = 4.0 \log\left(Re\sqrt{f}\right) - 0.40. \tag{7.52}$$

von Karman [38] derived a theoretical expression, using the universal velocity distribution for turbulent flow:

$$\frac{1}{\sqrt{f}} = 4.06 \log\left(Re\sqrt{f}\right) - 0.60. \tag{7.53}$$

A simpler expression between the Reynolds number and the friction factor is given by

$$f = 0.046(Re)^{-0.2}, \quad 3 \times 10^4 \le Re \le 10^6. \tag{7.54}$$

It should be noted that these expressions apply to smooth circular ducts. Commercial ducts are usually quite rough. Moody [42] has presented friction factor charts for the determination of friction factors in either smooth or rough circular ducts. A complete frictional plot is given by Moody's chart (Figure 6.6).

## 7.6 Analogies between Heat and Momentum Transfer

The principal mechanism of heat and momentum transfer in turbulent flow in regions away from the wall is the crosswise mixing of fluid particles due to eddies. Therefore, analogies between heat and momentum can be established to determine the heat-transfer coefficient from the knowledge of the friction factor. The first and the simplest of such relations was developed by Reynolds [43,44]. In this section, we first give the derivation of the Reynolds analogy and then introduce the other more refined analogies.

### 7.6.1 Reynolds Analogy

In 1874, Osborne Reynolds published a remarkable paper [8] entitled "On the Extent and Acting of the Heating Surface for Steam Boilers." In this paper, Reynolds suggested that momentum and heat in a fluid are transferred in the same way. He concluded that in geometrically similar systems, a simple proportionality relation must exist between fluid friction and heat transfer. Suppose, for example, a flow parallel to the $x$-axis in which the mean velocity of fluid is a function of the $y$-coordinate only. It is known from Reynolds' previous investigations that the shear stress occurring in an arbitrary plane perpendicular to the $y$-axis is equal to the total momentum transferred by the molecular and turbulent exchange and is given by Equation 7.27.

    If the temperature $T$ of the fluid varies with $y$-coordinate, the heat flow in the $y$-direction, that is, the amount of heat transferred by unit area across the same plane, can be expressed in a similar way. It consists of two parts: the molecular heat conduction and the turbulent heat transfer due to the fluctuations of velocity and temperature. The first contribution is equal to $-k\,dT/dy$, and the second part is equal to the mean value of the product $\rho c_p \overline{v'T'}$. Hence, the total heat flow is given by Equation 7.28.

The analogy suggested by Reynolds amounts to the following statement: If the mean value of $\overline{v'u'}$ is expressed in the form of $\varepsilon du/dy$, then the mean value of $\overline{v'T'}$ is also equal to $\varepsilon dT/dy$ where $\varepsilon$ has the same numerical value in the two expressions and is called the coefficient of the turbulent exchange.

Introducing the kinematic viscosity $\nu = \mu/\rho$ and the thermometric conductivity (diffusivity) $\alpha = k/\rho c_p$, one obtains the following fundamental equations from Equations 7.30 and 7.32:

$$\frac{\tau}{\rho} = (\nu + \varepsilon)\frac{du}{dy}, \tag{7.55}$$

$$\frac{q''}{\rho c_p} = -(\alpha + \varepsilon)\frac{dT}{dy}. \tag{7.56}$$

It follows from Equations 7.55 and 7.56 that direct proportionality between shearing stress and heat transfer can be expected if

    a. $\nu$ and $\alpha$ are negligible compared to $\varepsilon$ or

    b. $\nu = \alpha$

In the case of turbulent flow in ducts or around submerged bodies, the condition (a) is generally satisfied with the exception of a relatively narrow range near solid walls.

Consider two planes $y = y_1$ and $y = y_2$ in the turbulent region, and assume first that $\tau$ and $q$ are independent of $y$. Then, under the earlier conditions (a) and (b), it follows by integration from Equations 7.55 and 7.56 that

$$\frac{(u_2 - u_1)}{\tau} = \frac{c_p(T_1 - T_2)}{q''}. \tag{7.57}$$

This is the analytical expression of the Reynolds analogy. It is evident that the same relation prevails if $\tau$ and $q''$ vary with $y$ in the same way, that is, $q/\tau$ = constant. In most cases, for example, in a flow in a rectangular or a circular duct, the variations of $\tau$ and $q''$ are not exactly identical. Between parallel walls or in a circular duct, $\tau$ varies linearly with the distance from the wall, whereas the variation of $q''$ is determined by the balance between the heat carried by the fluid in the flow direction and the heat flow across the stream and therefore depends on the velocity distribution. However, the influence of the deviation between the distributions of $\tau$ and $q''$ is not very significant, and Equation 7.57 can be generally applied in the fully developed turbulent region.

Condition (b) is approximately satisfied by gases, since the kinematic viscosity and thermal diffusivity of gases are of the same order of magnitude. However, for liquids, $\nu$ is larger than $\alpha$. The ratio $Pr = \nu/\alpha$ is for some liquids as high as $Pr = 200$. Hence, the Reynolds analogy cannot be applied directly to the heat transfer between solids and liquids.

One can write the relation (7.57) between wall conditions and mean bulk conditions to get (Figure 7.7):

$$\frac{U_m}{\tau_w} = \frac{c_p(T_w - T_m)}{q''_w}. \tag{7.58}$$

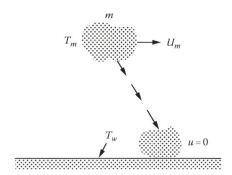

**FIGURE 7.7**
Simple model of the Reynolds analogy.

The heat-transfer coefficient is defined by

$$q_w'' = h(T_w - T_\infty),$$ (7.59)

and in general,

$$\tau_w = f \frac{1}{2} \rho U_m^2$$ (7.60)

is used to define the friction coefficient; the choice of the reference velocity is more or less arbitrary. For flow through a pipe, the centerline velocity or the average velocity over the cross section may be chosen. Substituting the expressions for $t_w$ and $q_w''$ in Equation 7.58 gives

$$St = \frac{h}{\rho c_p U_m} = \frac{Nu}{RePr} = \frac{f}{2},$$ (7.61)

which is called the Reynolds analogy. It relates the heat-transfer rate to the frictional loss. The foregoing treatment is consistent with the assumption that temperature and velocity profiles are similar and the heat and momentum are transported at the same rate.

Consider now the energy and the $x$-momentum equations, that is, Equations 7.33 and 7.34c. These equations are similar if

1. $\alpha = \nu$,

2. $\varepsilon_h = \varepsilon_m$, (7.62a,b,c)

3. $\dfrac{\partial p}{\partial x} = 0$.

For flow over a flat plate with constant free-stream velocity, the pressure gradient $\partial p / \partial x$ is equal to zero. Hence, the shape of two functions $T$ and $u$ will be similar if the boundary conditions are also similar.

If we consider pipe flow, then $\partial p / \partial x \neq 0$. Therefore, in general, these two equations are not similar. But for a fully developed turbulent flow, $dp/dx$ (pressure gradient along the pipe) is almost constant, that is, independent of $x$ and $y$. For a fully developed temperature distribution, $\partial T / \partial x$ is also constant (provided that the wall heat flux is constant).

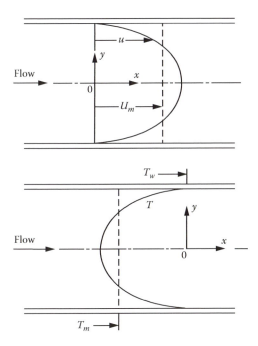

**FIGURE 7.8**
Fully developed velocity and temperature profiles between two parallel plates.

As illustrated in Figure 7.8, in turbulent flow, the velocity distribution is flat, and therefore the term $u(\partial T/\partial x)$ in Equation 7.34c may be assumed to be constant. On the other hand, for fully developed conditions, the momentum equation (7.33) and the energy equation (7.34c) become

$$\frac{\partial}{\partial y}\left[(\nu+\varepsilon_m)\frac{\partial u}{\partial y}\right]=\frac{1}{\rho}\frac{dp}{dx}, \tag{7.63}$$

$$\frac{\partial}{\partial y}\left[(\alpha+\varepsilon_h)\frac{\partial T}{\partial y}\right]=u\frac{\partial T}{\partial x}. \tag{7.64}$$

Therefore, for fully developed turbulent channel flows, these equations are similar if the earlier conditions (7.62a,b) are fulfilled. Hence, the shapes of the two functions $T$ and $u$ will be similar if the boundary conditions are similar (Figure 7.8). Thus, it follows that

$$\frac{U_m}{(du/dy)_w}=\frac{T_w-T_m}{(dT/dy)_w}, \tag{7.65}$$

which is the same result as Equation 7.58.

It should be noted that the picture that leads to the Reynolds analogy is oversimplified, not only in the mechanism postulated for turbulent flow but also because no account is taken of the existence of any laminar sublayer. In spite of the rather drastic simplifications in the analysis, this analogy gives quite good agreement with experimental results for gases.

### 7.6.2 Prandtl–Taylor Analogy

Taylor [46] and Prandtl [4,48] took into account approximately the influence of fluid flow behavior at the wall on heat transfer, and the improved analogy is called Prandtl–Taylor analogy.

In the Prandtl–Taylor analogy, the previous method is extended to include the effect of a laminar-flow region adjacent to the wall, and thus the flow field is now pictured as being divided into two zones as illustrated in Figure 7.9, that is, a laminar sublayer is assumed to extend from the wall to a thickness $\delta_L$. In this region, only viscous shearing stresses are supposed to exist, and heat transfer is maintained by the mechanism of conduction alone. Outside the laminar sublayer, the flow is assumed to be entirely turbulent, and the molecular coefficients $\mu$ and $k$ can be neglected. In this outer region, stresses arise through the transfer of momentum by the turbulent motion of small particles of fluid, while heat is similarly transferred by the mechanism of turbulent convection alone.

In the laminar sublayer, assuming a linear variation of velocity and temperature, we get

$$\tau = \mu \frac{du}{dy} = \mu \frac{u_L}{\delta_L} \tag{7.66}$$

and

$$q'' = -k \frac{dT}{dy} = -\frac{k(T_L - T_w)}{\delta_L}. \tag{7.67}$$

Hence,

$$\frac{q''}{\tau} = -\frac{k(T_L - T_w)}{\mu u_L} = \frac{c_p(T_L - T_w)}{Pr\, u_L}. \tag{7.68}$$

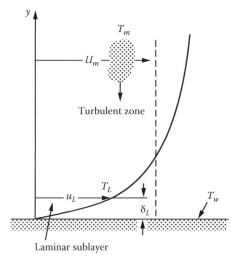

**FIGURE 7.9**
Velocity distribution for the Prandtl–Taylor analogy.

In the turbulent zone, assuming that the Reynolds analogy can be applied for movement of particles from a region with average properties $U_m$ and $T_m$ to the edge of laminar sublayer where the velocity is $u_L$ and temperature is $T_L$, then Equation 7.57 becomes

$$\frac{q''}{\tau} = -\frac{c_p(T_m - T_L)}{U_m - u_L}. \tag{7.69}$$

Prandtl introduced the assumption that the ratio $q''/\tau$ remains constant across the whole width of the boundary layer. Thus,

$$\frac{q''_w}{\tau_w} = -\frac{c_p(T_m - T_L)}{U_m - u_L}. \tag{7.70}$$

Hence, eliminating $T_L$ between Equations 7.68 and 7.70, we obtain

$$\frac{q''_w}{\tau_w}[(U_m - u_L) + Pr\, u_L] = -c_p(T_m - T_w) \tag{7.71a}$$

or

$$\frac{q''_w}{\rho U_m c_p(T_w - T_m)} = \frac{\tau_w}{\rho U_m^2} \frac{1}{1 + (u_L/U_m)(Pr - 1)}, \tag{7.71b}$$

which can also be written as

$$St = \frac{1}{2}\frac{f}{1 + (u_L/U_m)(Pr - 1)} \tag{7.72}$$

or

$$Nu = \frac{1}{2}f\frac{RePr}{1 + (u_L/U_m)(Pr - 1)}. \tag{7.73}$$

This equation was derived independently by Prandtl in 1910 [4] and Taylor in 1916 [46]. In the special case of the Prandtl number being numerically equal to 1.0, this reduces to the Reynolds analogy.

In the case of turbulent flow in a circular duct, the ration of the velocity at the outer edge of the laminar sublayer to the velocity $U_m$ is given by

$$\frac{u_L}{U_m} = 5\sqrt{\frac{\tau_w}{\rho U_m^2}} = 5\sqrt{\frac{1}{2}f}. \tag{7.74}$$

With this approximation, Equation 7.72 becomes

$$St = \frac{(1/2)f}{1 + 5\sqrt{(f/2)}(Pr - 1)}. \tag{7.75}$$

Introducing the value of $f = 0.079(Re)^{-1/4}$ for a fully developed turbulent flow in smooth circular ducts as given by Equation 7.49, we obtain

$$St = \frac{0.0395 Re^{-1/4}}{1 + 0.994 Re^{-1/8}(Pr - 1)}.$$ (7.76)

The local friction coefficient, $f$, for a flat plate when the boundary layer is turbulent from the leading edge onward is $f = 0.0592 Re_x^{-0.2}$. Introducing this into Equation 7.72, the local value of Stanton number may also be obtained for a flat plate.

The ratio $u_L/U_m$ is most often replaced by its empirical value given by Hoffman [49]:

$$\frac{u_L}{U_m} = 1.5 Re^{-1/8} Pr^{-1/6}.$$ (7.77)

Therefore, substituting Equation 7.77 into Equation 7.72, the Prandtl–Taylor analogy becomes

$$St = \frac{(1/2)f}{1 + 1.5 Re^{-1/8} Pr^{-1/6}(Pr - 1)}.$$ (7.78)

### 7.6.3 von Karman Analogy

In deriving the Prandtl–Taylor analogy, it was assumed that the boundary layer could be divided into two regions. von Karman [38] improved this analogy by considering three regions in the velocity profile as shown in Figure 7.10 and derived a similar relation between the heat-transfer and friction coefficients.

The linear temperature drop in the laminar sublayer is represented by

$$q'' = q_w'' = -k\frac{dT}{dy}.$$ (7.79)

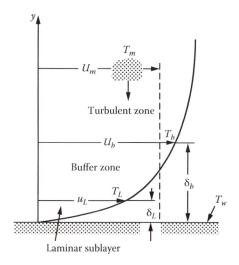

**FIGURE 7.10**
Velocity distribution for von Karman analogy.

Integrating Equation 7.79 from $y^+ = 0$ to $y^+ = 5$ gives

$$T_L - T_w = -\frac{5 \, v q_w''}{k\sqrt{\tau_w/\rho}} = -\frac{5Pr}{\rho c_p}\sqrt{\frac{\rho}{\tau_w}} q_w''. \tag{7.80}$$

In the buffer zone with the assumption that $q''$ and $\tau$ are both constants, Equations 7.26 and 7.28 become

$$\tau_w = \rho(v + \varepsilon_m)\frac{du}{dy}, \tag{7.81}$$

$$q_w'' = -\rho c_p(\alpha + \varepsilon_h)\frac{dT}{dy}. \tag{7.82}$$

In addition, von Karman assumed that $Pr_t = 1$ or $\varepsilon_m = \varepsilon_h$. From the velocity profile (Equation 7.48) for the buffer zone, we have

$$\frac{du^+}{dy^+} = \frac{5}{y^+} \quad \text{or} \quad \frac{du}{dy} = \frac{\tau_w}{\rho v}\frac{5}{y^+}. \tag{7.83}$$

Therefore, from Equation 7.81, we obtain

$$\varepsilon = v\left(\frac{y^+}{5} - 1\right), \tag{7.84}$$

and then substituting into Equation 7.82, we get

$$\frac{dT}{dy^+} = \frac{q_w''}{-\rho c_p}\sqrt{\frac{\rho}{\tau_w}}\frac{1}{1/Pr + (y^+/5 - 1)}. \tag{7.85}$$

Integrating Equation 7.85 from $y^+ = 5$ to $y^+ = 30$ gives

$$T_b - T_L = -\frac{q_w''}{\rho c_p}\frac{5}{\sqrt{\tau_w/\rho}}\ln(5Pr + 1). \tag{7.86}$$

In the turbulent zone, extending beyond $y^+ = 30$, we can assume that the Reynolds analogy applies for the movement of particles of fluid having mean field properties $U_m$ and $T_m$ up to the edge of the buffer zone, and hence from Equation 7.57, we have

$$\frac{q_w''}{\tau_w} = -\frac{c_p(T_m - T_b)}{U_m - u_b}, \tag{7.87}$$

from which we obtain

$$T_m - T_b = -\frac{q_w''}{c_p \tau_w}(U_m - u_b) = -\frac{q_w''}{\rho c_p}\frac{U_m^+ - u_b^+}{\sqrt{\tau_w/\rho}}. \tag{7.88}$$

For the velocity at the outer edge of the buffer zone, we have

$$u_b^+ - u_L^+ = 5.0 \ln \frac{30}{5} = 5.0 \ln 6,$$
(7.89)

$$u_b^+ = 5(1 + \ln 6),$$
(7.90)

$$u_b = 5\sqrt{\frac{\tau_w}{\rho}}(1 + \ln 6).$$
(7.91)

Substituting for $u_b^+$ into Equation 7.88 and noting that

$$\sqrt{\frac{\tau_w}{\rho}} = U_m \sqrt{\frac{f}{2}},$$

we get

$$T_m - T_b = -\frac{q_w''}{\rho c_p \sqrt{\tau_w/\rho}}\left[\frac{2}{f} - 5(1 + \ln 6)\right].$$
(7.92)

Adding the three temperature differences from Equations 7.80, 7.86, and 7.92 gives the overall temperature difference between the wall and the end of the turbulent zone

$$\Delta T = T_w - T_m = \frac{q_w''}{\rho c_p}\sqrt{\frac{\rho}{\tau_w}}\left[\frac{2}{f} + 5(Pr-1) + 5\ln\left(\frac{5Pr+1}{6}\right)\right].$$
(7.93)

The dimensionless heat-transfer coefficient becomes

$$St = \frac{q_w''}{\rho U_m c_p \Delta T} = \frac{(1/2)f}{1 + 5\sqrt{\frac{f}{2}}\left[(Pr-1) + \ln((5Pr+1)/6)\right]}.$$
(7.94)

This equation is valid for flat plates as well as for ducts. Therefore, it depends on the selection of friction coefficient, $f$. For a smooth flat surface $f = 0.0592\,(Re_x)^{-0.2}$ and for smooth ducts, either Equation 7.49 or Equation 7.54 can be used. One may also select friction factor correlations from Table 9.5.

The expressions for heat transfer obtained by von Karman and Prandtl–Taylor are applicable for constant physical properties over the range of 0.7 to 10–20 for the Prandtl number. They neglected turbulent heat transfer in a viscous sublayer, which leads to essential error when $Pr$ is large; they also neglected heat transfer by conduction in the turbulent core, which is not true for low Prandtl numbers.

Colburn [50] suggested that most experimental data can be approximated by simply introducing a correction term in the Reynolds analogy to allow for the effect of the Prandtl number variation. His suggestion led to the following relation:

$$St_x Pr^{2/3} = \frac{f}{2} \quad \text{(Colburn analogy)}.$$
(7.95)

In all preceding derivations, we have assumed that the turbulent Prandtl number $Pr_t = 1$. It is, however, known from measurements that the value of $Pr_t$ differs from unity. Rubesin [51] and van Driest [6,52] modified the equations due to Prandtl–Taylor and von Karman to include $Pr_t$.

The analogy relations between the rate of heat transfer and friction in turbulent flow are of great practical importance since they can be used for arbitrary turbulent flows.

### Example 7.1

Compare heat-transfer coefficients for water flowing at an average temperature of 40°C and at a velocity of 0.5 m/s in a 2.54 cm diameter duct using (1) the Prandtl–Taylor analogy, (2) von Karman analogy, and (3) Colburn analogy.

### Solution

The properties of water at 60°C, from Appendix B, are

$$\rho = 992.2 \, \text{kg/m}^3$$

$$c_p = 4178 \, \text{J/(kg K)}$$

$$\nu = 0.658 \times 10^{-6} \, \text{m}^2/\text{s}$$

$$Pr = 4.35$$

The Reynolds number is

$$Re = \frac{U_m d}{\nu} = \frac{0.5 \times 2.54 \times 10^{-2}}{0.658 \times 10^{-6}} = 1.93 \times 10^4.$$

Hence, the flow is turbulent. Assuming a smooth duct, friction factor Equation 7.50 is applicable:

$$f = 0.079 Re^{-1/4}.$$

1. *Prandtl–Taylor analogy*: To calculate the heat-transfer coefficient, we use Equation 7.79:

$$St = \frac{0.0395 Re^{-1/4}}{1 + 1.5 Re^{-1/8} Pr^{-1/6} (Pr - 1)}$$

$$= \frac{0.0395 \times (1.93 \times 10^4)^{-1/4}}{1 + 1.5 \times (1.93 \times 10^4)^{-1/8} \times (4.35)^{-1/6} \times (4.35 - 1)}$$

$$= 0.00156.$$

Hence, the heat-transfer coefficient is

$$h = \rho c_p U_m St = 992.2 \times 4178 \times 0.5 \times 0.00156$$

$$= 323 \, \text{W/(m}^2 \, \text{K)}.$$

2. *von Karman analogy*: We use Equation 7.94:

$$St = \frac{0.0395 Re^{-1/4}}{1 + 5\sqrt{(f/2)[(Pr - 1) + \ln((5Pr + 1)/6)]}}$$

$$= \frac{0.0395(1.93 \times 10^{-4})^{-1/4}}{1 + 5\sqrt{0.0395} \times (1.93 \times 10^4)^{-1/8} \times [(4.35 - 1) + \ln((5 \times 4.35 + 1)/6)]}$$

$$= 0.00126.$$

Hence, the heat-transfer coefficient is

$$h = \rho c_p U_m St = 992.2 \times 4178 \times 0.5 \times 0.00126$$

$$= 2612 \, \text{W} / (\text{m}^2 \, \text{K}).$$

3. *Colburn analogy*: We use Equation 7.95:

$$St Pr^{2/3} = \frac{f}{2},$$

which gives

$$St = \frac{f}{2} Pr^{-2/3} = 0.0395 Re^{-1/4} Pr^{-2/3}$$

$$= 0.0395 \times (1.93 \times 10^4)^{-1/4} \times (4.35)^{-2/3}$$

$$= 0.0001258.$$

Hence, the heat-transfer coefficient is

$$h = \rho c_p U_m St = 992.2 \times 4178 \times 0.5 \times 0.001258$$

$$= 2607 \, \text{W} / (\text{m}^2 \, \text{K}).$$

## 7.7 Further Analogies in Turbulent Flow

### 7.7.1 Turbulent Flow through Circular Tubes

An extension of von Karman's analysis of turbulent heat transfer to fluids in enclosed conduits has been given by Boelter et al. [53]. In von Karman's ideal system, the resistances to heat flow from a solid boundary to a fluid consist of laminar sublayer, a buffer layer, and a turbulent core. Thermal resistances of each of the fluid layers are calculated from the universal velocity profile, which relates $u^+$ and $y^+$. From the analysis of the previous section, Table 7.4 can be prepared. It is assumed that (1) all the physical properties are independent of temperature; (2) the shear stress and heat flux are constant both in laminar and buffer layers, and in turbulent zone, they vary linearly to zero at the center ($y = 0$ corresponds

**TABLE 7.4**

Heat Flux, Shear Stress, and Temperature Drop in Turbulent Flow

| Regions of Velocity Profile | Universal Velocity Distribution | Shear Stress | Heat Flux | Temperature Drop | Resistance |
|---|---|---|---|---|---|
| Laminar | $u^+ = y^+$ | $\dfrac{\tau}{\rho} = \dfrac{\tau_w}{\rho} = v\dfrac{du}{dy}$ | $\dfrac{q''}{\rho c_p} = \dfrac{q_w''}{\rho c_p} = -\alpha\dfrac{dT}{dy}$ | $\Delta T = T_w - T_L = \dfrac{5q_w'' v}{k\sqrt{\tau_w/\rho}}$ | $\dfrac{1}{Nu_1} = \dfrac{5}{Re}\sqrt{\dfrac{2}{f}}$ |
| Sublayer | $0 < y^+ < 5$ | | | | |
| Buffer layer | $u^+ = -3.05 + 5\ln y^+$ $5 \le y^+ \le 30$ | $\dfrac{\tau}{\rho} = \dfrac{\tau_w}{\rho} = (v+\varepsilon)\dfrac{du}{dy}$ | $\dfrac{q''}{\rho c_p} = \dfrac{q_w''}{\rho c_p} = -\left(\dfrac{v}{Pr}+\varepsilon\right)\dfrac{dT}{dy}$ | $\Delta T = T_b - T_L$ $= \dfrac{5q_w''}{\rho c_p\sqrt{\tau_w/\rho}}\ln(5Pr+1)$ | $\dfrac{1}{Nu_2} = \dfrac{5}{RePr}\sqrt{\dfrac{2}{f}}\ln[5Pr+1]$ |
| Turbulent core | $u^+ = 5.5 + 2.5\ln y^+$ $y^+ \ge 30$ | $\dfrac{\tau}{\rho} = \dfrac{\tau_w}{\rho}\left(1-\dfrac{y}{r_0}\right)$ $= \varepsilon\dfrac{du}{dy}$ | $\dfrac{q''}{\rho c_p} = \dfrac{q_w''}{\rho c_p}\left(r-\dfrac{y}{r_0}\right)$ $= -\varepsilon\dfrac{dT}{dy}$ | $\Delta T = T_b - T_c$ $= \dfrac{2.5q_w''}{\rho c_p\sqrt{\tau_w/\rho}}\ln\left(\dfrac{Re}{60}\sqrt{\dfrac{f}{2}}\right)$ | $\dfrac{1}{Nu_3} = \dfrac{2.5}{RePr}\sqrt{\dfrac{2}{f}}\ln\left(\dfrac{Re}{60}\sqrt{\dfrac{f}{2}}\right)$ |

*Note:* $y$ is the distance from the wall.

to the solid boundary); and (3) the velocity gradient (*du/dy*) is not zero at the center of the pipe. This is a well-known weakness of all present-day logarithmic expressions for the velocity distribution in turbulent flow.

From Table 7.4, the total resistance to heat transfer is

$$\frac{1}{Nu} = \sum_{n=1}^{3} \frac{1}{Nu_n}$$

$$= \frac{5}{Re}\sqrt{\frac{2}{f}} + \frac{5}{RePr}\sqrt{\frac{2}{f}}\ln[5Pr+1] + \frac{2.5}{RePr}\ln\left(\frac{Re}{60}\sqrt{\frac{f}{2}}\right)$$

$$= \frac{5}{RePr}\sqrt{\frac{2}{f}}\left[Pr + \ln(5Pr+1) + 0.5\ln\left(\frac{Re}{60}\sqrt{\frac{f}{2}}\right)\right]. \tag{7.96}$$

Hence, the Stanton number becomes

$$St = \frac{Nu}{RePr} = \frac{\sqrt{f/2}}{5\left[Pr + \ln(1+5Pr) + 0.5\ln\left((Re/60)\sqrt{f/2}\right)\right]}. \tag{7.97}$$

This expression of Stanton number is based upon the temperature difference between the pipe wall and the center of the pipe. To obtain the Stanton number based on mean temperature difference, the right side of Equation 7.97 should be divided by the ratio of $(\Delta T)_{mean}/(\Delta T)_{max}$, which can be calculated from the velocity and temperature distributions, and thus Equation 7.97 becomes

$$St = \frac{Nu}{RePr} = \frac{\sqrt{f/2}((\Delta T)_{max}/(\Delta T)_{mean})}{5\left[Pr + \ln(1+5Pr) + 0.5\ln\left((Re/60)\sqrt{f/2}\right)\right]}. \tag{7.98}$$

The plot of $(\Delta T)_{mean}/(\Delta T)_{max}$ for various values of the Reynolds and Prandtl numbers is given in Figure 7.11 as given in Reference 48. It is shown that for $Pr = 1$, the following relation results in

$$St = \frac{Nu}{RePr} \cong \frac{f}{2}. \tag{7.99}$$

The analogy between heat and momentum transfer has been extended to molten metals by Martinelli [54]. He solved for the heat-transfer problem during turbulent flow in a circular duct at larger distances from the entrance section (fully developed conditions) under the uniform wall temperature boundary condition. The hydrodynamic and thermal equations in his work can be written as

$$\frac{1}{\rho}\frac{dp}{dx} = \frac{1}{r}\frac{\partial}{\partial r}\left[r(\varepsilon_m + v)\frac{\partial u}{\partial r}\right], \tag{7.100}$$

$$u\frac{dT}{dx} = \frac{1}{r}\frac{\partial}{\partial r}\left[r(\varepsilon_h + \alpha)\frac{\partial T}{\partial r}\right]. \tag{7.101}$$

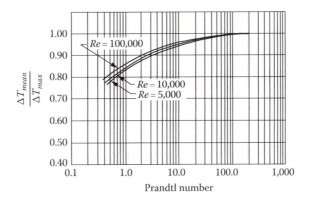

**FIGURE 7.11**
Ratio of mean to maximum temperature difference. (From Boelter, L.M.K. et al., *Trans. ASME*, 63, 447, 1941.)

These two equations must be solved simultaneously to obtain the solution to heat-transfer problem. If $\tau$ and $q''$ vary linearly from the wall, Equations 7.100 and 7.101 can be reduced to the following more simple forms:

$$\frac{\tau_w}{\rho}\left(1 - \frac{y}{r_0}\right) = (\varepsilon_m + \nu)\frac{\partial u}{\partial y}, \tag{7.102}$$

$$\frac{q''_w}{c_p\rho}\left(1 - \frac{y}{r_0}\right) = -(\varepsilon_h + \alpha)\frac{\partial T}{\partial y}, \tag{7.103}$$

where $y = r_0 - r$. To obtain the temperature as a function of $y$, Equations 7.102 and 7.103 have to be solved simultaneously.

The flow field is subdivided into three layers of interest: (1) laminar sublayer, (2) buffer layer, and (3) turbulent core. In the laminar sublayer region, both $\varepsilon_m$ and $\varepsilon_h$ are assumed to be zero. Besides, the shear and heat-transfer rates are assumed to be invariant with $y$, that is, $(1 = y/r_0) = 1.0$. Then, by integrating Equation 7.103, we obtain

$$T_w - T = \left(\frac{q''_w}{c_p\rho\sqrt{\tau_w/\rho}}\right)Pr\, y^+ \quad \text{(laminar sublayer)}. \tag{7.104}$$

The expressions for $\varepsilon_m$ in the buffer layer and turbulent core are obtained from Equation 7.102 and the velocity distribution given in Figure 7.6. The results are

$$\varepsilon_m = \frac{\sqrt{\tau_w/\rho}(1 - y/r_0)}{5} - \nu \quad \text{(in the buffer zone)}, \tag{7.105}$$

$$\varepsilon_m = \frac{\sqrt{\tau_w/\rho}(1 - y/r_0)y}{2.50} \quad \text{(in the turbulent core)}. \tag{7.106}$$

Then, assuming $\varepsilon_h = \zeta\varepsilon_m$, Equation 7.103 can be integrated to obtain temperature distribution in two steps giving

$$T_L - T = \frac{5q_w''}{\zeta c_p\rho\sqrt{\tau_w/\rho}}\ln\left[1 + \zeta Pr\left(\frac{y^+}{5} - 1\right)\right] \quad \text{(in the buffer zone)} \tag{7.107}$$

and

$$T_b - T = \frac{2.5q_w''}{\zeta c_p\rho\sqrt{\tau_w/\rho}}\ln\frac{y^+}{30} \quad \text{(in the turbulent core).} \tag{7.108}$$

In the buffer zone, $(1 - y/r_0) = 1$ was also assumed. In the turbulent zone, $\nu$ and $\alpha$ were assumed negligible. The elimination of the intermediate temperatures $T_L$ and $T_b$ from the earlier equations and the use of equation $\sqrt{\tau_w/\rho} = U_m\sqrt{f/2}$ give the relation for the temperature drop between the wall and the center of the pipe:

$$T_w - T_c = \frac{5q_w''}{\zeta c_p\rho\sqrt{\tau_w/\rho}}\left[\zeta Pr + \ln(1 + 5\zeta Pr) + 0.5\ln(Re/60)\sqrt{f/2}\right]. \tag{7.109}$$

The equations for the dimensionless temperature distribution derived by Martinelli are as follows:

1. Laminar sublayer $(0 < y^+ < 5)$

$$\frac{T_w - T}{T_w - T_c} = \frac{\zeta Pr(y/y_1)}{\left[\zeta Pr + \ln(1 + 5\zeta Pr) + 0.5\ln(Re/60)\sqrt{f/2}\right]}, \tag{7.110}$$

   where $y_1$ is the value of $y$ at $y^+ = 5$.
2. Buffer layer $(5 < y^+ < 30)$

$$\frac{T_w - T}{T_w - T_c} = \frac{\zeta Pr + \ln[1 + \zeta Pr((y/y_1) - 1)]}{\left[\zeta Pr + \ln(1 + 5\zeta Pr) + 0.5\ln(Re/60)\sqrt{f/2}\right]}. \tag{7.111}$$

3. Turbulent core $(y^+ > 30)$

$$\frac{T_w - T}{T_w - T_c} = \frac{\zeta Pr + \ln(1 + 5\zeta Pr) + 0.5\ln(Re/60)\sqrt{(f/2)(y/r_0)}}{[\zeta Pr + \ln(1 + 5\zeta Pr) + 0.5\ln(Re/60)\sqrt{f/2}]}, \tag{7.112}$$

   where $T_c$ is the tube centerline temperature.

The Nusselt number can be written as

$$Nu = \left[\frac{\partial[(T_w - T)/(T_w - T_m)]}{\partial(y/d)}\right]_w, \tag{7.113}$$

which can be obtained by differentiating Equation 7.110. Then, the Stanton number is readily evaluated. The result is

$$St = \frac{Nu}{RePr} = \frac{\zeta\sqrt{f/2}((T_w - T_c)/(T_w - T_m))}{5\left[\zeta Pr + \ln(1 + 5\zeta Pr) + 0.5\ln(Re/60)\sqrt{f/2}\right]}. \tag{7.114}$$

When the magnitude of the Prandtl number is very low as in the case of liquid metals, the thermal diffusivity $\alpha$ cannot be neglected in comparison with the eddy diffusivity $\varepsilon_h$ even at high values of the Reynolds number. Martinelli [54] also studied this case and obtained the temperature distribution in the turbulent core. The equation to be integrated is

$$T_b - T = 2.5\frac{q''_w}{\zeta c_p \rho \sqrt{\tau_w/\rho}} \int_{30}^{y^+} \frac{(1 - (y/r_0))dy^+}{(1 - (y/r_0))y^+ + (2.50/\zeta Pr)}, \tag{7.115}$$

and the result of integration is

$$T_b - T = \frac{1.25q''_w}{\zeta c_p \rho \sqrt{\tau_w/\rho}} \ln\left[\frac{5\lambda + (y/r_0)(1 - (y/r_0))}{5\lambda + (y_2/r_0)(1 - (y_2/r_0))}\right]$$

$$+ \frac{1.25q''_w}{\sqrt{1 + 20\lambda}} \ln\left[\frac{((2y/r_0) - 1) + \sqrt{1 + 20\lambda}((2y_2/r_0) - 1) - \sqrt{1 + 20\lambda}}{((2y/r_0) - 1) - \sqrt{1 + 20\lambda}((2y_2/r_0) - 1) + \sqrt{1 + 20\lambda}}\right], \tag{7.116}$$

where $y_2$ is the value of $y$ at $y^+ = 30$.

$$\lambda = \frac{1}{\zeta Pr Re\sqrt{f/2}}$$

Temperature distributions for different values of the Reynolds and Prandtl numbers can be prepared from Equations 7.104, 7.107, and 7.108, or 7.116. Equation 7.116 can be used for values $Pr < 1$ and Equation 7.108 for $Pr > 1$. Figure 7.12 shows the curves for $Re = 10,000$.

In order to simplify calculations of the Nusselt number and to bring out the importance of molecular conduction at low values of Prandtl number, a factor $F$ is defined as the ratio of the total thermal resistances of the turbulent core due to molecular and eddy conduction to the thermal resistance of the turbulent core due to eddy conduction only (Equation 7.116 divided by Equation 7.108 with $r = r_0$). The values of $F$ are given in Figure 7.13.

Therefore, when the molecular conduction in turbulent core is considered, Equations 7.109 through 7.112 change only by the factor $F$. Multiplying the third term in the brackets by factor $F$, the equation for the Stanton number becomes

$$St = \frac{Nu}{RePr} = \frac{\zeta\sqrt{(f/2)((T_w - T_c)/(T_w - T_m))}}{5\left[\zeta Pr + \ln(1 + 5\zeta Pr) + 0.5F\ln(Re/60)\sqrt{f/2}\right]}. \tag{7.117}$$

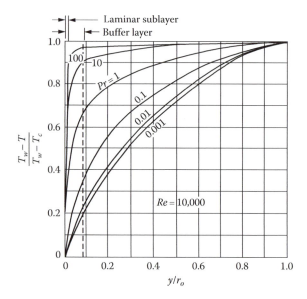

**FIGURE 7.12**
Radial temperature distributions in a tube for turbulent flow. (From Martinelli, R.C., *Trans. ASME*, 69, 947, 1947.)

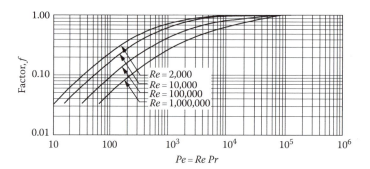

**FIGURE 7.13**
*F* values. (From Martinelli, R.C., *Trans. ASME*, 69, 947, 1947.)

Dimensionless bulk temperature is given by

$$\frac{T_w - T_m}{T_w - T_c} = \frac{\int_0^{r_o} (u/U_{max})((T_w - T)/(T_w - T_c))r\,dr}{\int_0^{r_o} (u/U_{max})r\,dr}. \tag{7.118}$$

This calculation was performed by Martinelli [54], and the results are shown in Figure 7.14. The variation of the Nusselt number calculated from Equation 7.117 is shown in Figure 7.15 as a function of the Reynolds and Prandtl numbers. It is observed that for low values of the Prandtl number, the Nusselt number no longer varies with 0.8 power of the Reynolds number. In particular, for values of Prandtl number less than 0.01, the Nusselt number is almost independent of the Reynolds number until the value of $Re \cong 10^5$ is reached. This

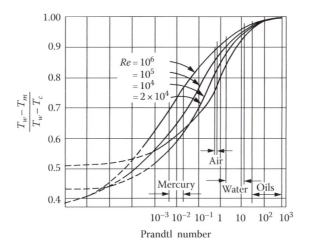

**FIGURE 7.14**
Ratio of mean to maximum temperature difference. (From Martinelli, R.C., *Trans. ASME*, 69, 947, 1947.)

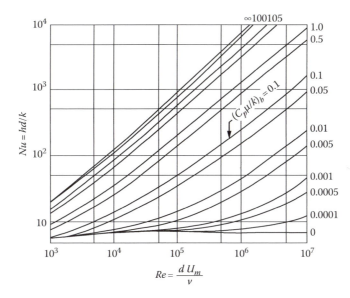

**FIGURE 7.15**
Variation of the Nusselt number with the Reynolds and Prandtl numbers ($\varepsilon_m = \varepsilon_h$). (From Martinelli, R.C., *Trans. ASME*, 69, 947, 1947.)

is a result of the molecular conduction in the turbulent core, which makes the effect of increasing turbulence less important in increasing the unit thermal conductance. But the ratio of $L/d$ to attain a fully developed profile can be considerably large for fluids with a low Prandtl number.

### 7.7.2 Turbulent Flow between Two Parallel Plates

It can be shown that the equation for the Stanton number for heat transfer to a fluid flowing between two parallel flat plates, a distance $d$ apart, is identical to Equation 7.117. Certain

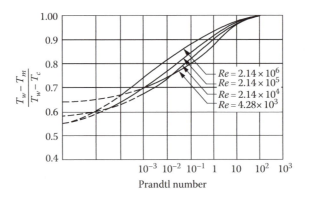

**FIGURE 7.16**
Ratio of mean to maximum temperature difference as a function of the Reynolds and Prandtl number for turbulent heat transfer between flat plates ($d_e = 2d$). (From Martinelli, R.C., *Trans. ASME*, 69, 947, 1947.)

terms in this equation, however, must be evaluated in a different manner. As a good approximation, the friction factor for flow between flat plates can be determined by using the equivalent diameter of the flat plates ($d_e = 2d$) in the Reynolds number and obtaining $f$ from appropriate curves or from correlations for circular ducts. The equation for the Stanton number can be written by assuming $\zeta = \varepsilon_h/\varepsilon_m = 1$ as

$$St = \frac{Nu}{RePr} = \frac{\sqrt{f/2}((T_w - T_c)/(T_w - T_m))}{5\left[Pr + \ln(1 + 5Pr) + 0.5F\ln(Re/120)\sqrt{f/2}\right]}, \tag{7.119}$$

where the mean mixed fluid temperature is given in Figure 7.16 and the value of $F$ is determined from Figure 7.13 at a value of

$$Pe = \frac{(Re)_{fp}Pr}{2}\sqrt{\frac{f_{fp}}{f}}, \tag{7.120}$$

$$Re = \frac{(Re)_{fp}}{2}\sqrt{\frac{f_{fp}}{f}}, \tag{7.121}$$

where subscript *fp* stands for flat plates.

The variation of the Nusselt number as calculated from Equation 7.119 is shown in Figure 7.17 as a function of the Reynolds and Prandtl numbers.

## 7.8 Turbulent Heat Transfer in a Circular Duct with Variable Circumferential Heat Flux

The problem of turbulent flow heat transfer in a circular duct with fully developed velocity and temperature profiles under variable circumferential heat-flux boundary condition has been analyzed by Reynolds [14]. But wall heat flux is assumed to be invariant in the flow

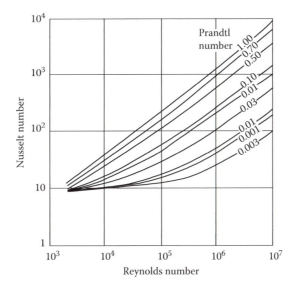

**FIGURE 7.17**
Variation of the Nusselt number with the Reynolds and Prandtl numbers ($\varepsilon_m = \varepsilon_h$) for turbulent heat transfer between flat plates ($d_e = 2d$). (From Martinelli, R.C., *Trans. ASME*, 69, 947, 1947.)

direction. It is also assumed that the diffusivities for heat in the radial and circumferential directions are identical. Under these simplifying assumptions, the governing differential energy equation is given by

$$\frac{1}{r}\frac{\partial}{\partial r}\left[r(\alpha + \varepsilon_h)\frac{\partial T}{\partial r}\right] + \frac{1}{r}\frac{\partial}{\partial \phi}\left[(\alpha + \varepsilon_h)\frac{\partial T}{\partial \phi}\right] = u\frac{\partial T}{\partial x}. \tag{7.122}$$

Under the fully developed conditions, we have

$$\frac{\partial T}{\partial x} = \frac{\partial T_m}{\partial x} = \text{constant}.$$

The prescribed heat-flux boundary condition is

$$k\left(\frac{\partial T}{\partial r}\right)_{r_0} = q_w''(\phi) \quad \text{(prescribed)}, \tag{7.123}$$

which can be represented in the form

$$q_w''(\phi) = q_0'' + F(\phi), \tag{7.124}$$

where $F(\phi)$ is the heat-flux variation about the mean value. It is assumed that $F(\phi)$ has a Fourier expansion.

The local wall temperature corresponding to the arbitrarily prescribed heat flux (7.124) is obtained as

$$T_w(\phi, x) - T_m(x) = \frac{r_0}{k}\left[S_0 q_0 + \sum_{n=1}^{\infty} S_n(a_n \sin n\phi + b_n \cos n\phi)\right],$$ (7.125)

where $S_n$ is called the wall temperature function

$$S_0 = \frac{2}{Nu_0}$$

$$Nu_0 = \left(\frac{q_0''}{\Delta T}\right)\left(\frac{2r_0}{k}\right).$$ (7.126)

It is seen that once the values of $S_n$ have been determined, one can calculate the temperature difference for any prescribed circumferential heat flux.

For laminar flow, it is found that

$$S_n = \frac{1}{n}, \quad n > 0; \quad S_0 = \frac{2}{Nu_0} = \frac{11}{24} = 0.458.$$ (7.127)

The values of $S_n(Re, Pr)$ for turbulent flow were computed for the first five harmonics for a series of the Reynolds and Prandtl numbers, and the results are summarized in Table 7.5. For the complete tabulation, see Reference 14. In the numerical solution, an expression for the eddy diffusivity for momentum suggested by Cess [56] was employed.

**Example 7.2**

As an illustration, let us consider a fluid ($Pr = 3$) flow in a channel at $Re = 30,000$ and suppose that the circumferential heat-flux distribution is given by

$$q''(\phi) = q_0''(1 + 0.5\cos\phi).$$

**Solution**
By the use of Equation 7.125, we have

$$T_w(\phi, x) - T_m(x) = \frac{r_0}{k}\left[S_0 q_0'' + 0.5 q_0'' \, S_1 \cos\phi\right]$$

or

$$\frac{T_w - T_m}{q_0'' \, r_0/k} = S_0 + 0.5 S_1 \cos\phi.$$

For $Pr = 3$ and $Re = 30,000$, from Table 7.5, we get

$$S_0 = 0.0134 \quad \text{and} \quad S_1 = 0.0178.$$

**TABLE 7.5**

Circumferential Heat-Flux Functions $S_n(Re,Pr)$

| Pr | n | 10⁴ | 3 × 10⁴ | 10⁵ | 3 × 10⁵ | 10⁶ |
|---|---|---|---|---|---|---|
| | | | | **Re** | | |
| 0.001 | 0 | 0.318 | 0.302 | 0.293 | 0.282 | 0.246 |
| | 1 | 1.000 | 1.000 | 0.990 | 0.974 | 0.901 |
| | 2 | 0.500 | 0.500 | 0.499 | 0.491 | 0.496 |
| | 3 | 0.333 | 0.333 | 0.333 | 0.329 | 0.320 |
| | 4 | 0.250 | 0.250 | 0.250 | 0.248 | 0.244 |
| | 5 | 0.200 | 0.200 | 0.200 | 0.199 | 0.196 |
| 0.003 | 0 | 0.318 | 0.302 | 0.282 | 0.246 | 0.156 |
| | 1 | 0.999 | 0.994 | 0.957 | 0.831 | 0.473 |
| | 2 | 0.500 | 0.498 | 0.484 | 0.435 | 0.279 |
| | 3 | 0.333 | 0.332 | 0.325 | 0.299 | 0.203 |
| | 4 | 0.250 | 0.249 | 0.245 | 0.229 | 0.170 |
| | 5 | 0.200 | 0.200 | 0.197 | 0.186 | 0.145 |
| 0.01 | 0 | 0.311 | 0.286 | 0.224 | 0.141 | 0.0655 |
| | 1 | 0.991 | 0.952 | 0.733 | 0.409 | 0.161 |
| | 2 | 0.473 | 0.483 | 0.397 | 0.246 | 0.109 |
| | 3 | 0.332 | 0.325 | 0.279 | 0.186 | 0.0894 |
| | 4 | 0.249 | 0.245 | 0.217 | 0.153 | 0.0784 |
| | 5 | 0.199 | 0.197 | 0.178 | 0.132 | 0.0710 |
| 0.03 | 0 | 0.290 | 0.220 | 0.126 | 0.0618 | 0.0248 |
| | 1 | 0.0923 | 0.699 | 0.348 | 0.145 | 0.0535 |
| | 2 | 2.473 | 0.383 | 0.214 | 0.0986 | 0.0402 |
| | 3 | 0.302 | 0.272 | 0.165 | 0.0816 | 0.0353 |
| | 4 | 0.243 | 0.213 | 0.138 | 0.0720 | 0.0326 |
| | 5 | 0.195 | 0.176 | 0.120 | 0.0654 | 0.0307 |
| 0.7 | 0 | 0.0631 | 0.0283 | 0.0112 | 0.00465 | 0.00174 |
| | 1 | 0.121 | 0.0490 | 0.0180 | 0.00721 | 0.00275 |
| | 2 | 0.0900 | 0.0378 | 0.0141 | 0.00578 | 0.00226 |
| | 3 | 0.0784 | 0.0336 | 0.0127 | 0.00525 | 0.00209 |
| | 4 | 0.0716 | 0.0313 | 0.0119 | 0.00469 | 0.00199 |
| | 5 | 0.0668 | 0.0297 | 0.0114 | 0.04477 | 0.00193 |
| 3 | 0 | 0.0325 | 0.0134 | 0.00495 | 0.00194 | 0.03690 |
| | 1 | 0.0448 | 0.0178 | 0.00629 | 0.00246 | 0.03902 |
| | 2 | 0.0379 | 0.0151 | 0.00540 | 0.00213 | 0.03791 |
| | 3 | 0.0353 | 0.0142 | 0.00508 | 0.00201 | 0.03751 |
| | 4 | 0.0338 | 0.0137 | 0.00490 | 0.00194 | 0.03728 |
| | 5 | 0.0327 | 0.0133 | 0.00479 | 0.00190 | 0.03714 |
| 10 | 0 | 0.0201 | 0.00806 | 0.00290 | 0.00111 | 0.03383 |
| | 1 | 0.0239 | 0.00931 | 0.00322 | 0.00123 | 0.03438 |
| | 2 | 0.0218 | 0.00853 | 0.00296 | 0.00113 | 0.03405 |
| | 3 | 0.0211 | 0.00824 | 0.00286 | 0.00110 | 0.03393 |
| | 4 | 0.0206 | 0.00808 | 0.00281 | 0.00108 | 0.03386 |
| | 5 | 0.0203 | 0.00797 | 0.00277 | 0.00107 | 0.03382 |

*Source:* Reynolds, W.C., *Int. J. Heat Mass Transfer*, 6, 445, 1962.

Then,

$$\frac{T_w - T_m}{q_0'' r_0 / k} = 0.0134 + 0.5 \times 0.0178 \cos\phi$$

$$= 0.0134(1 + 0.664 \cos\phi).$$

If we simply use the local heat flux, together with the uniform heat-flux Nusselt number, to estimate the peripheral temperature variation, we obtain

$$\frac{T_w - T_m}{q_0'' r_0 / k} = 0.0134(1 + 0.5 \cos\phi).$$

The importance of the function $S_n$ on heat-transfer coefficient is evident. It is found that for a given heat-flux distribution, the circumferential effects will be more pronounced in turbulent flow with low Prandtl number than in laminar flow.

The Nusselt number around the periphery will be

$$Nu(\phi) = \frac{1 + 0.5 \cos\phi}{(1/Nu_0) + ((0.5 S_1 / 2)) \cos\phi}.$$

If $Pr = 3$ and $Re = 3 \times 10^4$, from Table 7.10 $Nu_0 = 149$, and from Table 7.5 $S_1 = 0.0178$. Inserting these values into the $Nu(\phi)$ expression, we see that Nusselt will vary between 134.4.

---

## 7.9  Turbulent Heat Transfer in Annular Passages

This problem is the extension of the analysis made by Kays and Leung that we have discussed in Chapter 6 for the same geometry in laminar flow. The turbulent flow problem is two orders of magnitude more complex than its laminar-flow counterpart because the Reynolds and Prandtl numbers become parameters, and it is further difficult because of the incomplete knowledge of the details of the turbulent heat transport mechanism. Kays and Leung [13] solved this problem for the fundamental solutions of the second kind. In this work for four annulus radius ratios, 0.192, 0.255, 0.376, and 0.500, the fundamental solutions of the second kind are developed for air ($Pr = 0.76$) entirely from experimental data. An asymptotic solution is then developed analytically (velocity and temperature profiles fully developed) for the Prandtl numbers from 0 to $10^3$, the Reynolds numbers from $10^4$ to $10^6$, and the radius ratios from 0.1 to 1.0. Under the conditions of steady hydrodynamically fully developed turbulent flow with constant fluid properties, negligible axial conduction, and axially symmetric heating, the energy differential equation can be written as follows:

$$\frac{\partial}{\partial r}\left[ r(\varepsilon_h + \alpha)\frac{\partial T}{\partial r} \right] = ru(r)\frac{\partial T}{\partial x}. \tag{7.128}$$

There are two solutions:

1. Constant heat flux through inner surface, outer surface is insulated.
2. Constant heat flux through outer surface, inner surface is insulated.

The superposition of these two solutions gives the general solution to the case of asymmetric heating. It can be shown that (see Section 6.15 for derivation and nomenclature), for a given heat-flux ratio, the Nusselt number for inner and outer tubes can be evaluated from

$$\frac{^2Nu_i}{^2Nu_{ii}} = \frac{1}{1-\left(q_0''/q_i''\right)^2C^i} \tag{7.129}$$

and

$$\frac{^2Nu_o}{^2Nu_{oo}} = \frac{1}{1-\left(q_0''/q''_i\right)^2C^0}, \tag{7.130}$$

where

$$^2C^i = \frac{^2\Psi_m^0 - {}^2\Psi_i^0}{^2\Psi_i^i - {}^2\Psi_m^i}, \quad ^2C^i = \frac{^2\Psi_m^0 - {}^2\Psi_i^0}{^2\Psi_i^i - {}^2\Psi_m^i}. \tag{7.131}$$

Kays and Leung solved the fully developed turbulent flow energy equation, Equation 7.128, for the second fundamental solution. The results for the Nusselt numbers and influence coefficients in circular duct annulus, parallel plates, and circular ducts are given in Tables 7.6 through 7.10. For the complete tabulation of solutions, the reader is referred to the original Reference 13.

## 7.10 Effect of Boundary Conditions on Heat Transfer

When the heat transfer across the boundaries of a continuum cannot be prescribed, it may be assumed to be proportional to the temperature difference between the boundaries and the ambient, $q'' = h(T_w - T_\infty)$.

The heat-transfer coefficient between a plate, circular or noncircular duct wall, and the fluid flowing over or through it depends both on the flow characteristics and on the thermal boundary conditions. The latter may be called the secondary effects on heat-transfer coefficient. Secondary effects associated with the manner in which heat is added to the fluid are (1) the magnitude of the heat flux and (2) the distribution of heat flux over the heat-transfer surface.

Secondary effects on heat-transfer coefficient have been the subject of considerable interest. Hall and Khan [57] presented the experimental results for forced convection heat transfer in the combined entrance region of a circular duct; the results were obtained by using both the boundary conditions of uniform heat flux and uniform wall temperature (Figure 7.18). A significant difference that is attributable to the different boundary conditions was observed for the Reynolds numbers below about $Re = 3 \times 10^4$, but at higher Reynolds numbers, the difference rapidly diminishes.

The effect of these two boundary conditions was also investigated by utilizing an eigenvalue formulation for the thermal entrance region in circular ducts by Siegel and

**TABLE 7.6**

Nusselt Numbers and Influence Coefficients for a Fully Developed Turbulent Flow in a Circular Duct Annulus for Fundamental Solutions of the Second Kind where $R^* = 0.10$

| Re | $10^4$ | | $3 \times 10^4$ | | $10^5$ | | $3 \times 10^5$ | | $10^6$ | |
|---|---|---|---|---|---|---|---|---|---|---|
| | | | | *Heating from outer tube* | | | | | | |
| Pr | $Nu_\infty$ | $^2C^0$ | $Nu_{00}$ | $^2C^0$ | $Nu_{00}$ | $^2C^0$ | $Nu_{00}$ | $^2C^0$ | $Nu_{00}$ | $^2C^0$ |
| 0 | 6.00 | 0.077 | 6.12 | 0.079 | 6.32 | 0.081 | 6.50 | 0.084 | 6.68 | 0.085 |
| 0.001 | 6.00 | 0.077 | 6.12 | 0.079 | 6.40 | 0.082 | 6.60 | 0.082 | 7.20 | 0.082 |
| 0.003 | 6.00 | 0.077 | 6.24 | 0.081 | 6.55 | 0.083 | 7.34 | 0.082 | 10.8 | 0.071 |
| 0.01 | 6.13 | 0.076 | 6.50 | 0.081 | 7.80 | 0.077 | 12.1 | 0.067 | 26.4 | 0.052 |
| 0.03 | 6.45 | 0.076 | 7.95 | 0.075 | 13.7 | 0.065 | 28.2 | 0.051 | 71.8 | 0.036 |
| 0.5 | 24.8 | 0.039 | 53.4 | 0.032 | 134 | 0.028 | 320 | 0.025 | 860 | 0.022 |
| 0.7 | 29.8 | 0.032 | 66.0 | 0.028 | 167 | 0.024 | 409 | 0.022 | 1,100 | 0.020 |
| 1.0 | 36.5 | 0.026 | 81.8 | 0.023 | 212 | 0.021 | 520 | 0.019 | 1,430 | 0.017 |
| 3 | 61.5 | 0.013 | 147 | 0.013 | 395 | 0.012 | 1,000 | 0.012 | 2,830 | 0.011 |
| 10 | 99.2 | 0.006 | 246 | 0.006 | 685 | 0.006 | 1,780 | 0.006 | 5,200 | 0.006 |
| 30 | 143 | 0.003 | 360 | 0.003 | 1,030 | 0.003 | 2,720 | 0.003 | 8,030 | 0.003 |
| 100 | 205 | 0.002 | 525 | 0.002 | 1,500 | 0.002 | 4,030 | 0.002 | 12,100 | 0.002 |
| 1,000 | 378 | – | 980 | – | 2,850 | – | 7,600 | – | 23,000 | – |
| | | | | *Heating from inner tube* | | | | | | |
| | $Nu_{ii}$ | $^2C^1$ | $Nu_{ii}$ | $^2C^1$ | $Nu_{ii}$ | $^2C^1$ | $Nu_{ii}$ | $^2C^1$ | $Nu^{ii}$ | $^2C^1$ |
| 0 | 11.5 | 1.475 | 11.5 | 1.502 | 11.5 | 1.500 | 11.5 | 1.460 | 11.6 | 1.477 |
| 0.001 | 11.5 | 1.475 | 11.5 | 1.502 | 11.5 | 1.480 | 11.7 | 1.462 | 12.3 | 1.410 |
| 0.003 | 11.5 | 1.475 | 11.5 | 1.475 | 11.7 | 1.473 | 12.6 | 1.391 | 17.0 | 1.124 |
| 0.01 | 11.8 | 1.482 | 11.8 | 1.442 | 13.5 | 1.323 | 19.4 | 1.090 | 39.0 | 0.760 |
| 0.03 | 12.5 | 1.472 | 14.1 | 1.330 | 21.8 | 1.027 | 42.0 | 0.760 | 103 | 0.526 |
| 0.5 | 40.8 | 0.632 | 81.0 | 0.486 | 191 | 0.394 | 443 | 0.339 | 1,160 | 0.294 |
| 0.7 | 48.5 | 0.512 | 98.0 | 0.407 | 235 | 0.338 | 550 | 0.292 | 1,510 | 0.269 |
| 1.0 | 58.5 | 0.412 | 120 | 0.338 | 292 | 0.286 | 700 | 0.256 | 1,910 | 0.232 |
| 3 | 93.5 | 0.202 | 206 | 0.175 | 535 | 0.162 | 1,300 | 0.152 | 3,720 | 0.148 |
| 10 | 140 | 0.089 | 328 | 0.081 | 890 | 0.078 | 2,300 | 0.078 | 6,700 | 0.077 |
| 30 | 195 | 0.041 | 478 | 0.039 | 1,320 | 0.038 | 3,470 | 0.038 | 10,300 | 0.040 |
| 100 | 272 | 0.017 | 673 | 0.015 | 1,910 | 0.015 | 5,030 | 0.016 | 15,200 | 0.018 |
| 1,000 | 486 | 0.004 | 1,240 | 0.003 | 3,600 | 0.003 | 9,600 | 0.004 | 28,700 | 0.004 |

*Source:* Kays, W.M. and Leung, E.Y., *Int. J. Heat Mass Transfer*, 6, 537, 1963.

Sparrow [58]. Their findings for $Re = 10,000$ and $Pr = 0.7$ are given in Table 7.11. The percentage difference in Table 7.11 is defined as

$$\frac{\left(Nu_q - Nu_T\right)}{Nu_T},$$

where
  $Nu_q$ is the Nusselt number for the constant heat-flux boundary condition
  $Nu_T$ is the Nusselt number for the constant wall temperature boundary condition

**TABLE 7.7**

Nusselt Numbers and Influence Coefficients for a Fully Developed Turbulent Flow in a Circular Duct Annulus for Fundamental Solutions of the Second Kind where $R^* = 0.50$

| Re | $10^4$ | | $3 \times 10^4$ | | $10^5$ | | $3 \times 10^5$ | | $10^6$ | |
|---|---|---|---|---|---|---|---|---|---|---|
| | *Heating from outer tube* | | | | | | | | | |
| Pr | $Nu_{00}$ | $^2C^0$ | $Nu_{00}$ | $^2C^0$ | $Nu_{00}$ | $^2C^0$ | $Nu_{00}$ | $^2C^0$ | $Nu_{00}$ | $^2C^0$ |
| 0 | 5.66 | 0.281 | 5.78 | 0.294 | 5.80 | 0.296 | 5.83 | 0.302 | 5.95 | 0.310 |
| 0.001 | 5.66 | 0.281 | 5.78 | 0.294 | 5.80 | 0.296 | 5.92 | 0.302 | 6.40 | 0.304 |
| 0.003 | 5.66 | 0.281 | 5.78 | 0.294 | 5.85 | 0.294 | 6.45 | 0.301 | 9.00 | 0.278 |
| 0.01 | 5.73 | 0.281 | 5.88 | 0.289 | 6.80 | 0.289 | 10.3 | 0.264 | 22.6 | 0.217 |
| 0.03 | 6.03 | 0.279 | 7.05 | 0.284 | 11.6 | 0.258 | 24.4 | 0.214 | 64.0 | 0.163 |
| 0.5 | 22.6 | 0.162 | 49.8 | 0.142 | 125 | 0.123 | 298 | 0.111 | 795 | 0.098 |
| 0.7 | 28.3 | 0.137 | 62.0 | 0.119 | 158 | 0.107 | 380 | 0.097 | 1,040 | 0.090 |
| 1.0 | 34.8 | 0.111 | 78.0 | 0.101 | 200 | 0.092 | 490 | 0.085 | 1,340 | 0.078 |
| 3 | 60.5 | 0.059 | 144 | 0.058 | 384 | 0.055 | 960 | 0.054 | 2,730 | 0.052 |
| 10 | 100 | 0.028 | 246 | 0.028 | 680 | 0.028 | 1,750 | 0.028 | 5,030 | 0.028 |
| 30 | 143 | 0.013 | 365 | 0.013 | 1,030 | 0.014 | 2,700 | 0.014 | 8,000 | 0.015 |
| 100 | 207 | 0.006 | 530 | 0.006 | 1,500 | 0.006 | 4,000 | 0.006 | 12,000 | 0.006 |
| 1,000 | 387 | 0.001 | 990 | 0.001 | 2,830 | 0.001 | 7,600 | 0.001 | 23,000 | 0.001 |
| | *Heating from inner tube* | | | | | | | | | |
| | $Nu_{ii}$ | $^2C^1$ | $Nu_{ii}$ | $^2C^1$ | $Nu_{ii}$ | $^2C^1$ | $Nu_{ii}$ | $^2C^1$ | $Nu_{ii}$ | $^2C^1$ |
| 0 | 6.28 | 0.620 | 6.30 | 0.632 | 6.30 | 0.651 | 6.30 | 0.659 | 6.30 | 0.654 |
| 0.001 | 6.28 | 0.620 | 6.30 | 0.632 | 6.30 | 0.651 | 6.40 | 0.659 | 6.75 | 0.644 |
| 0.003 | 6.28 | 0.620 | 6.30 | 0.632 | 6.40 | 0.656 | 6.85 | 0.637 | 9.40 | 0.585 |
| 0.01 | 6.37 | 0.622 | 6.45 | 0.636 | 7.30 | 0.623 | 10.8 | 0.540 | 23.2 | 0.427 |
| 0.03 | 6.75 | 0.627 | 7.53 | 0.598 | 12.0 | 0.533 | 24.8 | 0.430 | 65.5 | 0.333 |
| 0.5 | 24.6 | 0.343 | 52.0 | 0.292 | 130 | 0.253 | 310 | 0.229 | 835 | 0.208 |
| 0.7 | 30.9 | 0.300 | 66.0 | 0.258 | 166 | 0.225 | 400 | 0.206 | 1,080 | 0.185 |
| 1.0 | 38.2 | 0.247 | 83.5 | 0.218 | 212 | 0.208 | 520 | 0.183 | 1,420 | 0.170 |
| 3 | 66.8 | 0.219 | 152 | 0.121 | 402 | 0.115 | 1,010 | 0.114 | 2,870 | 0.111 |
| 10 | 106 | 0.059 | 260 | 0.059 | 715 | 0.059 | 1,850 | 0.059 | 5,400 | 0.061 |
| 30 | 153 | 0.028 | 386 | 0.027 | 1,080 | 0.028 | 2,850 | 0.031 | 8,400 | 0.032 |
| 100 | 220 | 0.006 | 558 | 0.006 | 1,600 | 0.006 | 4,250 | 0.007 | 12,600 | 0.007 |
| 1,000 | 408 | 0.002 | 1,040 | 0.002 | 3,000 | 0.002 | 8,000 | 0.002 | 24,000 | 0.002 |

*Source:* Kays, W.M. and Leung, E.Y., *Int. J. Heat Mass Transfer*, 6, 537, 1963.

Kakaç and Paykoç [59] also analyzed numerically the turbulent forced convection heat transfer between two parallel plates to examine the effects of uniform wall temperature and uniform wall heat-flux boundary conditions in both the fully developed and thermal entrance regions. They used experimental values of eddy diffusivity of heat and velocity distribution [9]. The variations of the Nusselt number with $x/d_e$ (where $d_e = 2d$, $d$ is the distance between plates) for the aforementioned two boundary conditions are given in Figure 7.19. The findings of Deissler [60] are also indicated in Figure 7.19.

**TABLE 7.8**

Nusselt Numbers and Influence Coefficients for a Fully Developed Turbulent Flow in a Circular Duct Annulus for Fundamental Solutions of the Second Kind where $R^* = 0.80$

| Re | $10^4$ | | $3 \times 10^4$ | | $10^5$ | | $3 \times 10^5$ | | $10^6$ | |
|---|---|---|---|---|---|---|---|---|---|---|
| | | | | *Heating from outer tube* | | | | | | |
| Pr | $Nu_{00}$ | $^2C^0$ | $Nu_{00}$ | $^2C^0$ | $Nu^{00}$ | $^2C^0$ | $Nu_{00}$ | $^2C^0$ | $Nu_{00}$ | $^2C^0$ |
| 0 | 5.65 | 0.379 | 5.70 | 0.386 | 5.75 | 0.398 | 5.80 | 0.407 | 5.85 | 0.409 |
| 0.001 | 5.65 | 0.379 | 5.70 | 0.386 | 5.75 | 0.398 | 5.88 | 0.406 | 6.25 | 0.407 |
| 0.003 | 5.65 | 0.379 | 5.70 | 0.386 | 5.84 | 0.397 | 6.35 | 0.407 | 8.80 | 0.374 |
| 0.01 | 5.75 | 0.381 | 5.85 | 0.386 | 6.72 | 0.390 | 9.95 | 0.361 | 21.0 | 0.286 |
| 0.03 | 6.10 | 0.388 | 6.90 | 0.380 | 11.1 | 0.319 | 23.2 | 0.290 | 62.0 | 0.216 |
| 0.5 | 22.4 | 0.225 | 48.0 | 0.191 | 121 | 0.169 | 292 | 0.153 | 790 | 0.136 |
| 0.7 | 28.0 | 0.192 | 61.0 | 0.166 | 156 | 0.150 | 378 | 0.136 | 1,020 | 0.122 |
| 1.0 | 34.8 | 0.159 | 76.5 | 0.141 | 197 | 0.129 | 483 | 0.120 | 1,330 | 0.111 |
| 3 | 61.3 | 0.083 | 142 | 0.079 | 382 | 0.078 | 960 | 0.076 | 2,730 | 0.073 |
| 10 | 100 | 0.039 | 243 | 0.039 | 670 | 0.039 | 1,740 | 0.040 | 5,050 | 0.040 |
| 30 | 146 | 0.019 | 365 | 0.019 | 1,040 | 0.020 | 2,720 | 0.021 | 8,000 | 0.022 |
| 100 | 209 | 0.008 | 533 | 0.008 | 1,500 | 0.009 | 4,000 | 0.009 | 12,000 | 0.101 |
| 1,000 | 385 | 0.002 | 1,000 | 0.002 | 2,870 | 0.002 | 7,720 | 0.002 | 23,000 | 0.002 |
| | | | | *Heating from inner tube* | | | | | | |
| | $Nu_{ii}$ | $^2C^1$ | $Nu_{ii}$ | $^2C^1$ | $Nu_{ii}$ | $^2C^1$ | $Nu_{ii}$ | $^2C^1$ | $Nu_{ii}$ | $^2C^1$ |
| 0 | 5.87 | 0.489 | 5.90 | 0.505 | 5.92 | 0.515 | 5.95 | 0.525 | 5.97 | 0.528 |
| 0.001 | 5.87 | 0.489 | 5.90 | 0.505 | 5.92 | 0.515 | 6.00 | 0.518 | 6.33 | 0.516 |
| 0.003 | 5.87 | 0.489 | 5.90 | 0.505 | 6.03 | 0.485 | 6.40 | 0.504 | 8.80 | 0.468 |
| 0.01 | 5.95 | 0.485 | 6.07 | 0.506 | 6.80 | 0.493 | 10.0 | 0.452 | 21.7 | 0.382 |
| 0.03 | 6.20 | 0.478 | 7.05 | 0.485 | 11.4 | 0.445 | 23.0 | 0.357 | 61.0 | 0.276 |
| 0.5 | 22.9 | 0.268 | 49.5 | 0.250 | 123 | 0.214 | 296 | 0.193 | 800 | 0.174 |
| 0.7 | 28.5 | 0.244 | 62.3 | 0.212 | 157 | 0.186 | 384 | 0.172 | 1,050 | 0.160 |
| 1.0 | 35.5 | 0.200 | 78.3 | 0.181 | 202 | 0.166 | 492 | 0.154 | 1,350 | 0.140 |
| 3 | 63.0 | 0.108 | 145 | 0.102 | 386 | 0.097 | 973 | 0.096 | 2,750 | 0.093 |
| 10 | 102 | 0.051 | 248 | 0.051 | 693 | 0.052 | 1,790 | 0.051 | 5,150 | 0.051 |
| 30 | 147 | 0.027 | 370 | 0.027 | 1,050 | 0.028 | 2,750 | 0.029 | 8,100 | 0.030 |
| 100 | 215 | 0.010 | 540 | 0.010 | 1,540 | 0.010 | 4,050 | 0.011 | 12,000 | 0.012 |
| 1,000 | 393 | 0.002 | 1,000 | 0.002 | 2,890 | 0.002 | 7,700 | 0.002 | 23,000 | 0.002 |

*Source:* Kays, W.M. and Leung, E.Y., *Int. J. Heat Mass Transfer*, 6, 537, 1963.

Turbulent heat transfer in the thermal entrance region of a circular duct with uniform heat flux has been studied using a method similar to the Graetz formulation for laminar thermal entry region problem [55]. It is assumed that the fluid enters the pipe with a uniform temperature and a fully developed turbulent velocity profile. In the course of flow through the heated pipe, the temperature profile will change until, at some distance from the entrance, a fully developed shape is substantially achieved. The thermal entrance length is that length required for the local heat-transfer coefficient to approach

**TABLE 7.9**

Nusselt Numbers and Influence Coefficients for a Fully Developed Turbulent Flow between Parallel Plates for Fundamental Solutions of the Second Kind

| Re | $10^4$ | | $3 \times 10^4$ | | $10^5$ | | $3 \times 10^5$ | | $10^6$ | |
|---|---|---|---|---|---|---|---|---|---|---|
| *Parallel plates* | | | | | | | | | | |
| Pr | Nu | $^2C$ | Nu | $^2C$ | Nu | $^2C$ | Nu | $^2C$ | Nu | $^2C$ |
| 0 | 5.70 | 0.428 | 5.78 | 0.445 | 5.80 | 0.456 | 5.80 | 0.460 | 5.80 | 0.468 |
| 0.001 | 5.70 | 0.428 | 5.78 | 0.445 | 5.80 | 0.456 | 5.88 | 0.460 | 6.23 | 0.460 |
| 0.003 | 5.70 | 0.428 | 5.80 | 0.445 | 5.90 | 0.450 | 6.32 | 0.450 | 8.62 | 0.422 |
| 0.01 | 5.80 | 0.428 | 5.92 | 0.445 | 6.70 | 0.440 | 9.80 | 0.407 | 21.5 | 0.333 |
| 0.03 | 6.10 | 0.428 | 6.90 | 0.428 | 11.0 | 0.390 | 23.0 | 0.330 | 61.2 | 0.255 |
| 0.5 | 22.5 | 0.256 | 47.8 | 0.222 | 120 | 0.193 | 290 | 0.174 | 780 | 0.157 |
| 0.7 | 27.8 | 0.220 | 61.2 | 0.192 | 155 | 0.170 | 378 | 0.156 | 1,030 | 0.142 |
| 1.0 | 35.0 | 0.182 | 76.8 | 0.162 | 197 | 0.148 | 486 | 0.138 | 1,340 | 0.128 |
| 3 | 60.8 | 0.095 | 142 | 0.092 | 380 | 0.089 | 966 | 0.087 | 2,700 | 0.084 |
| 10 | 101 | 0.045 | 241 | 0.045 | 680 | 0.045 | 1,760 | 0.045 | 5,080 | 0.046 |
| 30 | 147 | 0.021 | 367 | 0.022 | 1,030 | 0.022 | 2,720 | 0.023 | 8,000 | 0.024 |
| 100 | 210 | 0.009 | 514 | 0.009 | 1,520 | 0.010 | 4,030 | 0.010 | 12,000 | 0.011 |
| 1,000 | 390 | 0.002 | 997 | 0.002 | 2,880 | 0.002 | 7,650 | 0.002 | 23,000 | 0.002 |

*Source:* Kays, W.M. and Leung, E.Y., *Int. J. Heat Mass Transfer*, 6, 537, 1963.
*Note:* $R^* = 1.00$.

**TABLE 7.10**

Nusselt Numbers and Influence Coefficients for a Fully Developed Turbulent Flow in a Circular Duct for Fundamental Solutions of the Second Kind

| Re | $10^4$ | $3 \times 10^4$ | $10^5$ | $3 \times 10^5$ | $10^6$ |
|---|---|---|---|---|---|
| *Circular tube* | | | | | |
| Pr | Nu | Nu | Nu | Nu | Nu |
| 0 | 6.30 | 6.64 | 6.84 | 6.95 | 7.06 |
| 0.001 | 6.30 | 6.64 | 6.84 | 7.08 | 8.12 |
| 0.003 | 6.30 | 6.64 | 7.10 | 8.14 | 12.8 |
| 0.01 | 6.43 | 7.00 | 8.90 | 14.2 | 30.5 |
| 0.03 | 6.90 | 9.10 | 15.9 | 32.4 | 80.5 |
| 0.5 | 26.3 | 57.3 | 142 | 340 | 895 |
| 0.7 | 31.7 | 70.7 | 178 | 430 | 1,150 |
| 1.0 | 37.8 | 86.0 | 222 | 543 | 1,470 |
| 3 | 61.5 | 149 | 404 | 1,030 | 2,900 |
| 10 | 99.8 | 248 | 690 | 1,810 | 5,220 |
| 30 | 141 | 362 | 1,030 | 2,750 | 8,060 |
| 100 | 205 | 522 | 1,510 | 4,030 | 12,000 |
| 1,000 | 380 | 975 | 2,830 | 7,600 | 22,600 |

*Source:* Kays, W.M. and Leung, E.Y., *Int. J. Heat Mass Transfer*, 6, 537, 1963.
*Note:* $R^* = 0$.

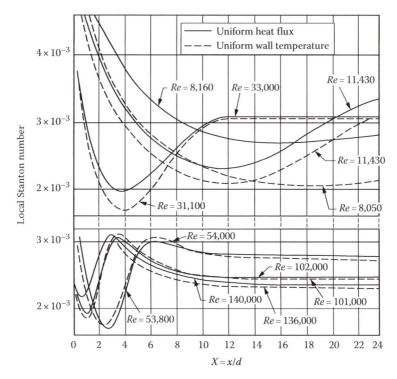

**FIGURE 7.18**
Stanton number distributions, $A^a$. (From Hall, W.B. and Khan, S.A., *J. Mech. Eng. Sci.*, 6, 250, 1964.)

**TABLE 7.11**

Effect of Boundary Conditions

| x/d | $Nu_q$ | $Nu_T$ | % Difference |
|-----|--------|--------|--------------|
| 2   | 42.83  | 39.28  | 9.0          |
| 5   | 36.9   | 34.68  | 6.4          |
| 10  | 34.15  | 32.44  | 5.3          |
| 20  | 32.72  | 31.32  | 4.5          |
| 30  | 32.42  | 31.11  | 4.2          |
| ∞   | 32.32  | 31.06  | 4.1          |

*Source:* Siegel, R. and Sparrow, E.M., *ASME J. Heat Transfer*, 82, 152, 1960.
*Note:* $Re = 10^4$; $Pr = 0.7$.

its asymptotic value within a few percent (e.g., 5%). Energy equation (7.128) has been solved under the following inlet and boundary conditions for the function $T(x,r)$:

$$T(0,r) = T_i,$$

$$\frac{\partial T(x,r_0)}{\partial r} = \frac{q''_w}{k},$$

(7.132a,b,c)

$$\frac{\partial T(x,0)}{\partial r} = 0.$$

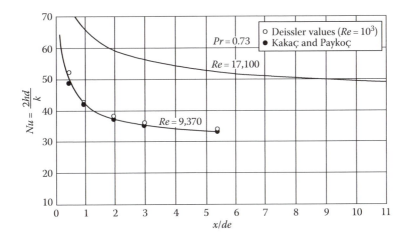

**FIGURE 7.19**
Variation of the Nusselt number with $x/d_e$. Uniform wall heat flux. (From Kakaç, S. and Paykoç, S., *METU J. Pure Appl. Sci.*, 1(1), 27, 1968.)

Defining $\Theta = (T - T_i)/(q_w'' r_0/k)$, the entry and fully developed solutions are separated as

$$\Theta = \Theta_1 + \Theta_2,$$

where
  $\Theta_1$ is the fully developed solution
  $\Theta_2$ is simply the remainder of the temperature distribution after $\Theta_i$ is subtracted from $\Theta$

Evidently, $\Theta_2$ will approach zero for large values of $x$. The complete solution of the problem is obtained as

$$\Theta(r^+, x^+) = \frac{4}{RePr} x^+ + G(r^+) + \sum_{n=0}^{\infty} C_n \phi_n(r^+) \exp\left[-\frac{2\beta_n^2}{Re} x^+\right], \tag{7.133}$$

where
  $G(r^+)$ is the fully developed temperature profile
  $\phi_n$ and $\beta_n^2$ are the eigenfunctions and eigenvalues associated with the developing temperature profile

The local heat-transfer coefficient and the Nusselt number are defined in the usual way, and the result is

$$Nu_x = \frac{2}{G(r_0^+) + \sum_{n=0}^{\infty} C_n \phi_n(r^+) \exp\left[-\left(4\beta_n^2/Re\right)(x/d)\right]}. \tag{7.134}$$

The fully developed Nusselt number, denoted by $Nu_\infty$, is found by evaluating Equation 7.134 for large values of $x/d$. So we obtain

$$Nu_\infty = \frac{2}{G(r_0^+)}. \tag{7.135}$$

The numerical solutions of the problem were carried out for the Prandtl numbers of 0.7, 10, and 100, and the fully developed results are approximated by

$$Nu_\infty = 0.0245Re^{0.77}, \quad Pr = 0.7$$

$$Nu_\infty = 0.0387Re^{0.85}, \quad Pr = 10 \quad . \qquad\qquad (7.136a,b,c)$$

$$Nu_\infty = 0.0611Re^{0.88}, \quad Pr = 100$$

By combining Equations 7.134 and 7.135, the following expression for the entrance region Nusselt number is obtained:

$$\frac{Nu_\infty}{Nu_\infty} = \frac{1}{1+\sum_{n=0}^{\infty} A_n \exp\left[-\left(4\beta_n^2/Re\right)(x/d)\right]}. \qquad (7.137)$$

To evaluate Equation 7.137, $A_n$ and the eigenvalues $\beta_n^2$ are tabulated in Tables 7.12 and 7.13, and the values of $x/d$ corresponding to $Nu/Nu_\infty = 1.05$ are given in Reference 55.

Hatton and Quarmby [62] presented the solutions to the heat-transfer problem with turbulent flow between parallel plates with heating on one side only, that is, uniform temperature on one wall, the other wall is insulated, and uniform heat input on one wall, the other insulated. The methods used are essentially the same as those given by Sparrow et al. [61], Sleicher and Tribus [12], and Siegel and Sparrow [58]. The differential equation that must be solved for a fully developed turbulent flow ($v = 0$) between parallel plates is Equation 7.34c. But in these solutions, the velocity and eddy diffusivity variations due to Deissler are used, and eddy diffusivities of heat and momentum are taken as equal. The solutions can be obtained with the same procedure as applied in the laminar flow, that is, the separation variables are first used, and the solutions are put in the same form as for the laminar-flow counterpart. Hatton and Quarmby obtained the

**TABLE 7.12**

Eigenvalues $\beta_n^2$

| Pr | $\beta_0^2$ | $\beta_1^2$ | $\beta_2^2$ | $\beta_3^2$ | $\beta_4^2$ | $\beta_5^2$ |
|---|---|---|---|---|---|---|
| | | | *Re = 50,000* | | | |
| 0.7 | 0 | 1,390 | 3,737 | 7,054 | 11,350 | 16,630 |
| 10 | 0 | 1,368 | 3,658 | 6,864 | 10,980 | 15,990 |
| 100 | 0 | 1,367 | 3,653 | 6,849 | 10,950 | 15,920 |
| | | | *Re = 100,000* | | | |
| 0.7 | 0 | 2,541 | 6,811 | 12,830 | 20,600 | 30,150 |
| 10 | 0 | 2,520 | 6,732 | 12,640 | 20,230 | 59,520 |
| 100 | 0 | 2,158 | 6,726 | 12,620 | 20,200 | 29,470 |
| | | | *Re = 500,000* | | | |
| 0.7 | 0 | 10,640 | 28,420 | 53,370 | 85,510 | 124,800 |
| 10 | 0 | 10,620 | 28,340 | 53,170 | 85,130 | 124,200 |
| 100 | 0 | 10,620 | 28,330 | 53,160 | 85,110 | 124,200 |

*Source:* Sparrow, E.M. et al., *Appl. Sci. Res., Ser. A*, 7, 37, 1957.

**TABLE 7.13**

Coefficient $A_n$

| Pr | $-A_0$ | $-A_1$ | $-A_2$ | $-A_3$ | $-A_4$ | $-A_5$ |
|---|---|---|---|---|---|---|
| | | | *Re = 50,000* | | | |
| 0.7 | 0 | 0.1785 | 0.09402 | 0.06502 | 0.05008 | 0.04177 |
| 10 | 0 | 0.05055 | 0.02945 | 0.02311 | 0.02106 | 0.02113 |
| 100 | 0 | 0.01150 | 0.007287 | 0.00633 | 0.00675 | 0.00821 |
| | | | *Re = 100,000* | | | |
| 0.7 | 0 | 0.1669 | 0.08858 | 0.06093 | 0.04633 | 0.03770 |
| 10 | 0 | 0.04876 | 0.02726 | 0.02013 | 0.01669 | 0.01506 |
| 100 | 0 | 0.01112 | 0.006479 | 0.00487 | 0.00505 | 0.00446 |
| | | | *Re = 500,000* | | | |
| 0.7 | 0 | 0.1435 | 0.07764 | 0.05386 | 0.04121 | 0.03341 |
| 10 | 0 | 0.04612 | 0.02531 | 0.01786 | 0.01393 | 0.01154 |
| 100 | 0 | 0.01080 | 0.005962 | 0.004232 | 0.003328 | 0.002783 |

*Source:* Sparrow, E.M. et al., *Appl. Sci. Res., Ser. A*, 7, 37, 1957.

following Nusselt number expression (using the heat-transfer coefficient based on the wall to bulk temperature difference) for uniform temperature on one wall, the other wall being insulated:

$$Nu = \frac{4Pr\sum_{n=1}^{\infty} C_n Y'_{ni} \exp\left(-8\lambda_n^2 x^+ / Re\right)}{\sum_{n=1}^{\infty} \left(C_n Y'_{ni}/\lambda_n^2\right)\exp\left(-8\lambda_n^2 x^+ / Re\right)}, \tag{7.138}$$

where $x^+ = x/d_e$. The eigenvalues and constants in these solutions were obtained by the use of computers. Table 7.14 gives the values of $\lambda_n$ and $C_n Y'_{ni}$ for the Reynolds and three Prandtl numbers.

The result of constant heat input on one wall, the other insulated, is obtained as

$$Nu = \frac{1}{G_i\left[1 - \sum_{n=1}^{\infty} C_n \exp(-8\lambda_n^2 x^+ / Re)\right]}, \tag{7.139}$$

where $G_i$ is the difference between wall and mean temperatures on the heat input side for the fully developed profile. Table 7.15 gives values of $\lambda_n$ and $C_n$, which may be used in Equation 7.139. Compare this table with Table 7.9. Hatton and Quarmby also extended these two basic solutions to certain axial variations by using superposition; the results are obtained for unequal uniform heat fluxes on each side of the passage.

In some problems, for example, the cooling of a nuclear reactor, the heat-flux variation in the direction of flow of the coolant is specified and is far from being constant. For such a case, it is necessary to have some method of assessing the consequent variation of heat-transfer coefficient.

**TABLE 7.14**

Eigenvalues and Constants for Turbulent Flow; Thermal Entry Length;
Uniform Temperature on One Side, the Other Side Insulated

| | Pr = 0.1 | | Pr = 1.0 | | Pr = 10 | |
|---|---|---|---|---|---|---|
| $n$ | $\lambda_n$ | $C_n Y_{ni}$ | $\lambda_n$ | $C_n Y_{ni}$ | $\lambda_n$ | $C_n Y_{ni}$ |
| | | | *Re* = 7,096 | | | |
| 1 | 4.57381 | 0.147757 | 2.55361 | 0.049787 | 1.38538 | 0.0151622 |
| 2 | 14.0447 | 0.105989 | 2.06643 | 0.0172367 | 7.81528 | 0.00144318 |
| 3 | 23.7382 | 0.081551 | 16.0894 | 0.009226 | 14.5916 | 0.0007255 |
| 4 | 33.4205 | 0.068169 | 22.8363 | 0.0074637 | 20.6899 | 0.0006865 |
| 5 | 42.9915 | 0.063112 | 29.1253 | 0.007433 | 25.9532 | 0.0008020 |
| 6 | 52.5218 | 0.060519 | 35.2911 | 0.0077921 | 30.9724 | 0.0009383 |
| 7 | 62.0815 | 0.057585 | 41.4927 | 0.0079426 | 36.0476 | 0.0010268 |
| | | | *Re* = 73,612 | | | |
| 1 | 8.62619 | 0.0741824 | 6.08550 | 0.0390701 | 3.60242 | 0.0139752 |
| 2 | 27.6631 | 0.040843 | 23.4587 | 0.0101968 | 21.5889 | 0.0011034 |
| 3 | 47.3420 | 0.027441 | 42.0244 | 0.0052399 | 40.2887 | 0.0004807 |
| 4 | 66.5868 | 0.023149 | 52.6688 | 0.004032 | 57.7009 | 0.0003519 |
| 5 | 85.4110 | 0.022076 | 76.6308 | 0.0035303 | 74.3562 | 0.0002944 |
| 6 | 104.234 | 0.0211419 | 93.6654 | 0.0030976 | 91.1214 | 0.0002537 |
| 7 | 123.185 | 0.0199186 | 110.869 | 0.002764 | 108.022 | 0.0002309 |
| | | | *Re* = 494,576 | | | |
| 1 | 16.8765 | 0.053976 | 12.7438 | 0.0318024 | 8.04710 | 0.0128536 |
| 2 | 57.8632 | 0.021774 | 52.9425 | 0.006566 | 49.9012 | 0.0009328 |
| 3 | 101.125 | 0.012881 | 95.9135 | 0.0032389 | 93.2850 | 0.0004012 |
| 4 | 142.559 | 0.010684 | 136.230 | 0.0024958 | 133.390 | 0.0002894 |
| 5 | 182.613 | 0.0101542 | 174.867 | 0.0021720 | 171.765 | 0.0002337 |
| 6 | 222.795 | 0.0096531 | 213.784 | 0.0018972 | 210.545 | 0.0001917 |
| 7 | 263.294 | 0.009272 | 253.016 | 0.0017133 | 249.663 | 0.0001646 |

*Source:* Hatton, A.P. and Quarmby, A., *Int. J. Heat Mass Transfer*, 6, 903, 1963.

Kays and Nicoll [63] and Burgoyne [64] have measured the effect on heat-transfer coefficient of linearly increasing and decreasing heat fluxes along a pipe to air through it.

Hall and Price [65] studied the uniform heat flux, exponentially increasing heat fluxes along the pipe and finally a sinusoidal variation of heat flux for a turbulent flow in a pipe. The experimental results for two different exponential heat-flux distributions are shown in Figure 7.20 together with the theoretical predictions based on the results of the uniform heat input experiment.

It should be noted that Stanton number tends to a constant value as $X$ (= $x/d$) increases. The case of sinusoidal heat flux distribution is of particular interest in nuclear reactor cooling because the distribution of heat generation along a fuel element is usually sinusoidal. Experimental results and the theoretical prediction are shown in Figure 7.21.

Hall and Price also made an analysis to find the effect of an arbitrary heat-flux distribution over the length of a circular duct from the results of uniform heat flux by the principle of superposition. This was first extensively used by Duhamel [66].

**TABLE 7.15**

Eigenvalues and Constants for Turbulent Flow; Thermal Entry Length; Uniform
Temperature on One Side, the Other Side Insulated

| | $Pr = 0.1$ | | $Pr = 1.0$ | | $Pr = 10$ | |
|---|---|---|---|---|---|---|
| | | | $Re = 7{,}096$ | | | |
| $G_i$ | 0.2141052 | | 0.7387098 | | 0.02581698 | |
| $G_o$ | 0.0768648 | | −0.0148205 | | 0.0018105 | |
| $n$ | $\lambda_n$ | $c_n$ | $\lambda_n$ | $c_n$ | $\lambda_n$ | $c_n$ |
| 1 | 10.5851 | 0.453048 | 7.80291 | 0.263048 | 7.46092 | 0.090942 |
| 2 | 20.5723 | 0.140911 | 15.0006 | 0.104846 | 14.2449 | 0.048317 |
| 3 | 30.3202 | 0.076073 | 21.6133 | 0.078378 | 20.1805 | 0.053720 |
| 4 | 39.7985 | 0.050089 | 27.6359 | 0.066777 | 25.1755 | 0.065565 |
| 5 | 49.2541 | 0.035389 | 33.5781 | 0.054367 | 29.9135 | 0.069065 |
| 6 | 58.7887 | 0.026398 | 39.6470 | 0.042935 | 34.7616 | 0.064725 |
| 7 | 68.2729 | 0.020616 | 45.7495 | 0.033627 | 39.6840 | 0.057708 |
| | | | $Re = 73{,}612$ | | | |
| $G_i$ | 0.0623753 | | 0.0131940 | | 0.0038308 | |
| $G_o$ | 0.0181794 | | 0.0020704 | | 0.0002212 | |
| $n$ | $\lambda_n$ | $c_n$ | $\lambda_n$ | $c_n$ | $\lambda_n$ | $c_n$ |
| 1 | 22.0430 | 0.376900 | 20.9064 | 0.205371 | 20.7885 | 0.072623 |
| 2 | 42.2838 | 0.133795 | 39.9055 | 0.080378 | 39.6525 | 0.029597 |
| 3 | 61.3268 | 0.080998 | 57.4954 | 0.054798 | 57.0367 | 0.021372 |
| 4 | 79.7950 | 0.055965 | 74.2859 | 0.042367 | 73.6425 | 0.0177233 |
| 5 | 98.4660 | 0.040181 | 91.2552 | 0.033770 | 90.3637 | 0.0154379 |
| 6 | 117.3824 | 0.030419 | 108.3859 | 0.028689 | 107.1940 | 0.0145074 |
| 7 | 136.2531 | 0.024265 | 125.3590 | 0.025626 | 123.7957 | 0.0146026 |
| | | | $Re = 494{,}576$ | | | |
| $G_i$ | 0.0623753 | | 0.0131940 | | 0.0038308 | |
| $G_o$ | 0.0037636 | | −0.0003836 | | −0.00003578 | |
| $n$ | $\lambda_n$ | $c_n$ | $\lambda_n$ | $c_n$ | $\lambda_n$ | $c_n$ |
| 1 | 48.7477 | 0.294157 | 48.2497 | 0.167790 | 48.1997 | 0.062193 |
| 2 | 93.1388 | 0.122429 | 92.0672 | 0.066072 | 91.9585 | 0.026276 |
| 3 | 134.036 | 0.073979 | 132.2351 | 0.045081 | 132.0501 | 0.018122 |
| 4 | 173.328 | 0.054151 | 170.6821 | 0.034155 | 170.4069 | 0.013909 |
| 5 | 213.136 | 0.040150 | 209.593 | 0.026287 | 209.2205 | 0.010878 |
| 6 | 253.358 | 0.031384 | 248.853 | 0.021096 | 248.3750 | 0.008909 |
| 7 | 293.277 | 0.025240 | 287.759 | 0.017469 | 287.1742 | 0.007569 |

*Source:* Hatton, A.P. and Quarmby, A., *Int. J. Heat Mass Transfer*, 6, 903, 1963.

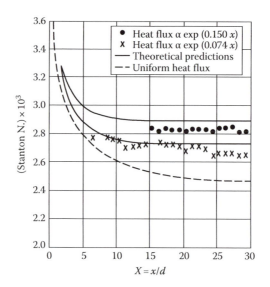

**FIGURE 7.20**
Experimental results for exponential heat-flux distribution compared with mean curve for uniform heat flux (*x* distance from start of heating). (From Hall, W.B. and Price, P.H., The effect of a longitudinally varying wall heat flux on the heat transfer coefficient for turbulent flow in a pipe, *International Heat Transfer Conference, ASME IMechE*, Boulder, CO, 1961.)

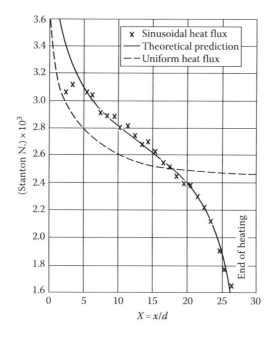

**FIGURE 7.21**
Experimental results for sinusoidal heat-flux distribution compared with mean curve for uniform heat flux (*x* distance from start of heating). (From Hall, W.B. and Price, P.H., The effect of a longitudinally varying wall heat flux on the heat transfer coefficient for turbulent flow in a pipe, *International Heat Transfer Conference, ASME IMechE*, Boulder, CO, 1961.)

### 7.10.1 Constant Heat-Transfer Coefficient Boundary Condition

If a constant heat-transfer coefficient to a prescribed ambient temperature is given, the boundary condition can be written as

$$\frac{\partial T}{\partial y} + h(T - T_\infty) = 0. \tag{7.140}$$

Then, zero heat flux corresponds to $h = 0$, and zero wall temperature corresponds to $h = \infty$. The constant $h$ case should thus lie between these two.

While investigating the effects of various forms of boundary conditions along a pipe [65], it was discovered that a wall heat flux that varies exponentially along the channel yields a longitudinally uniform heat-transfer coefficient in the fully developed thermal region. In another work, it was shown analytically that constant heat flux and constant wall temperature boundary conditions are special cases of the exponential heat-flux boundary condition [67].

In another analysis [68], the constant heat-transfer boundary condition along the direction of flow is investigated by Kakaç and Price [68] both theoretically and numerically for the fully developed flow between two parallel plates.

Let us consider a fully developed turbulent flow through a channel, subject to the limitations noted before. The energy equation is

$$u \frac{\partial T}{\partial x} = \frac{\partial}{\partial y}\left[\left(\frac{k}{\rho c_p} + \varepsilon_h\right)\right]\frac{\partial T}{\partial y}. \tag{7.141}$$

Defining $X = 2x/d/RePr$, $Y = y/d$, and $(1 + \varepsilon_h/\alpha) = G(Y)$, Equation 7.141 becomes

$$\frac{u}{U_m} \frac{\partial T}{\partial X} = \frac{\partial}{\partial Y}\left[G(Y)\frac{\partial T}{\partial Y}\right]. \tag{7.142}$$

Assuming a product solution in the form

$$T(X, Y) = R(X)\phi(Y) \tag{7.143}$$

and introducing it into Equation 7.141, we get

$$\frac{1}{\phi} \frac{d}{dY}\left[G(Y)\frac{d\phi}{dY}\right] - \frac{u}{U_m}\frac{1}{R}\frac{dR}{dX} = 0. \tag{7.144}$$

From Equation 7.144, we have two equations:

$$\frac{1}{R}\frac{dR}{dX} = \lambda \tag{7.145}$$

and

$$\frac{d}{dY}\left[G(Y)\frac{d\phi}{dY}\right] - \frac{u}{U_m}\lambda\phi = 0. \tag{7.146}$$

The solution of Equation 7.145 is simply

$$R(X) = Ae^{\lambda x}. \tag{7.147}$$

So that we may write for the solution as

$$T(X,Y) = Ce^{\lambda X}\phi(Y), \tag{7.148}$$

the function $\phi(Y)$ must be determined by Equation 7.146 with appropriate boundary conditions. From the temperature distribution (7.148), the heat flux at the wall, $q_w''$, is

$$q_w'' = -\frac{k}{d}Ce^{\lambda x}\left(\frac{d\phi}{dY}\right)_w. \tag{7.149}$$

It is of interest to note from Equation 7.149 that the heat flux also changes as $e^{\lambda}x$ and is proportional to $(d\phi/dY)_w$. Introducing the local temperature $T_w$ (at $y = 0$) into Equation 7.148, we get

$$T_w = Ce^{\lambda X}\phi(0) \tag{7.150}$$

and

$$T_w - T = C\left[\phi(0) - \phi(Y)\right]e^{\lambda X}. \tag{7.151}$$

From Equation 7.148, the mean bulk temperature $T_b$ at $y = b$ is

$$T_b = Ce^{\lambda X}\phi(b). \tag{7.152}$$

Then, the Nusselt number becomes

$$Nu = \frac{2hd}{k} = \frac{2q_w''}{(T_w - T_b)}\frac{d}{k} = -\frac{2(d\phi/dY)_w}{\phi(0) - \phi(b)} = \text{constant}. \tag{7.153}$$

The Nusselt number is constant for a given value of $\lambda$. Thus, the wall heat flux that varies exponentially in the direction of flow gives a uniform heat-transfer coefficient, or if one imposes a constant heat-transfer coefficient boundary condition, an exponentially varying wall heat flux in space is obtained.

As a special case, assuming a slug flow, Equation 7.146 becomes

$$\frac{d^2\phi}{dY^2} - \gamma^2\phi = 0, \tag{7.154}$$

where $y^2 = \lambda$. The solution of the Equation 7.154 is then simply

$$\phi(Y) = C_1 \cos h\gamma Y + C_2 \sin h\gamma Y. \tag{7.155}$$

The most usual boundary conditions for a channel with equal heat fluxes through the two walls at $Y = 0$ and $Y = 1/2$ are

$$\text{at } Y = 1/2; \quad \frac{d\phi(Y)}{dY} = 0,$$
$$\text{at } Y = 0; \quad \phi(Y) = 1. \tag{7.156a,b}$$

The former comes from the consideration of symmetry, and the latter is arbitrary. From the boundary conditions (7.156), we get

$$C_1 = 0 \quad \text{and} \quad C_2 = -\frac{\sin h(\gamma/2)}{\cos h(\gamma/2)}.$$

Then,

$$\phi(Y) = \cos h(\gamma Y) - \frac{\sin h(\gamma/2)}{\cos h(\gamma/2)} \sin h(\gamma Y)$$

and

$$T(X,Y) = C\left( \cos h(\gamma Y) - \frac{\sin h(\gamma/2)}{\cos h(\gamma/2)} \sin h(\gamma Y) e^{\lambda X} \right). \tag{7.157}$$

Hence,

$$T_w = C e^{\lambda X}. \tag{7.158}$$

By the use of temperature distribution (7.157), for bulk temperature, we get

$$T_b = C\frac{1}{r}\left( \sin h\,\gamma - \frac{\sin h(\gamma/2)}{\cos h(\gamma/2)} \cos h\gamma + \frac{\sin h(\gamma/2)}{\cos h(\gamma/2)} \right) e^{\lambda X}. \tag{7.159}$$

Then,

$$T_w - T_b = C e^{\lambda X}\left[ 1 - \frac{1}{r}\left( \sinh\,\gamma - \frac{\sinh(\gamma/2)}{\cosh(\gamma/2)} \cosh\gamma \right) + \frac{\sinh(\gamma/2)}{\cosh(\gamma/2)} \right] \tag{7.160}$$

and

$$\left( \frac{d\phi}{dY} \right)_w = -\gamma \frac{\sinh(\gamma/2)}{\cosh(\gamma/2)}. \tag{7.161}$$

Substituting Equations 7.160 and 7.161 into Equation 7.153, the Nusselt number becomes

$$Nu = \frac{2\gamma}{(2/\gamma) - \coth(\gamma/2)}. \tag{7.162}$$

This is a general expression for the Nusselt number from which the surface heat-transfer coefficient can be calculated. For small values of $\gamma$, Equation 7.162 can be written in the following form:

$$Nu = \frac{-2\gamma}{(2/\gamma) - \dfrac{1 + (1/2!)(\gamma/4)^2 + \cdots}{(\gamma/2) + (1/3!)(\gamma/2)^3 + \cdots}} \tag{7.163}$$

$$Nu = \frac{2hd}{k} \rightarrow 12. \tag{7.164}$$

This corresponds to constant heat-flux boundary condition for slug flow. It can be shown that the result of Equation 7.164 does not depend on the sign of the $\gamma$.

Therefore, if one imposes a constant heat-transfer coefficient boundary condition on the walls of a channel through which the flow is fully developed, an exponentially varying wall heat flux along the channel is obtained, or the heat flux that varies exponentially in the direction of flow gives a uniform heat-transfer coefficient. Constant heat-flux boundary condition is a special case of exponential heat-flux boundary condition. For the small values of the exponent, it approaches to the case of constant wall heat flux. Similar analysis also applies to laminar flow.

## 7.11 Turbulent Flow on a Flat Plate

Measurements of the velocity distribution in turbulent boundary layers indicate that the velocity distribution may be approximated by a power law as

$$\frac{u}{U_\infty} = \left(\frac{y}{\delta}\right)^{1/n}, \tag{7.165}$$

where
$\quad n$ is a positive number
$\quad \delta = \delta(x)$ denotes the boundary-layer thickness

In applying the approximate method to the calculation of turbulent boundary layers, something more than the velocity distribution must be assumed since the shearing stress at the wall cannot be calculated from the earlier velocity distribution. This additional information is a relation between the wall shearing stress and other local properties of the boundary layer. The equations for shearing stress at the wall and velocity distribution are often taken from the circular duct:

$$\frac{\tau_w}{\rho U_\infty^2} = 0.0225 \left(\frac{U_\infty \delta}{v}\right)^{-1/4}, \tag{7.166}$$

$$\frac{u}{U_\infty} = \left(\frac{y}{\delta}\right)^{1/7}, \tag{7.167}$$

for the Reynolds numbers near $10^7$. With these relations, the boundary-layer thickness may be arrived at by integrating momentum equation across the boundary layer. The result is

$$\frac{\delta}{x} = \frac{0.37}{\left(U_\infty x/v\right)^{1/5}} = 0.37\,Re_x^{-1/5}. \tag{7.168}$$

The boundary-layer thickness is seen to increase with the power $x^{4/5}$ of the distance, whereas in laminar flow, we have $\delta \sim x^{1/2}$. The corresponding drag coefficient is

$$F_D = \frac{0.074}{\left(U_\infty L/v\right)^{1/5}}. \tag{7.169}$$

The skin friction coefficient $C_f = 2\tau_w/\rho U_\infty^2$ can also be calculated from the momentum equation. When a laminar boundary layer exists over the forward portion of the plate, a correction must be made to the preceding equation. The amount of this correction depends on the value of the critical Reynolds number at which transition to turbulence occurs. Schlichting [37] gives

$$F_D = \frac{0.074A}{\left(U_\infty L/v\right)^{1/5}} - \frac{A}{\left(U_\infty L/v\right)}, \quad 5\times10^5 < Re_{cr} < 10^7, \tag{7.170}$$

where $A$ depends on the critical Reynolds number and is tabulated in the following:

| $(U_\infty L/v)_{cr}$ | $3\times10^5$ | $5\times10^5$ | $10^6$ | $3\times10^6$ |
|---|---|---|---|---|
| $A$ | 1050 | 1700 | 3300 | 8700 |

For a critical Reynolds number of $8.5 \times 10^4$, Eckert and Drake [69] give $A = 300$.

The heat-transfer properties of turbulent boundary layers are often derived by means of the Colburn analogy (Equation 7.96) (modified Reynolds analogy), which relates the heat-transfer coefficient to skin friction:

$$\frac{h}{\rho c_p U_\infty}Pr^{2/3} = \frac{1}{2}\left[\frac{2\tau_w}{\rho U_\infty^2}\right] = \frac{\tau_w}{\rho U_\infty^2}. \tag{7.171}$$

The heat-transfer coefficient is obtained by using Equations 7.166 and 7.171:

$$\frac{h_x}{\rho c_p U_\infty}Pr^{2/3} = 0.023\left(\frac{U_\infty \delta}{v}\right)^{-1/4}. \tag{7.172}$$

From Equation 7.168, we get

$$\frac{U_\infty \delta}{v} = 0.37\left(\frac{U_\infty x}{v}\right)^{4/5}.$$

Therefore,

$$\frac{h_x}{\rho c_p U_\infty} Pr^{2/3} = 0.0296 \left( \frac{U_\infty x}{v} \right)^{-1/5}. \tag{7.173}$$

The local Nusselt number is then given by

$$Nu_x = \frac{h_x x}{k} = 0.0296 Pr^{1/3} Re_x^{4/5}. \tag{7.174}$$

Somewhat different results are obtained if other analogies are used.
The average heat-transfer coefficient for a plate of length $L$ is

$$h = \frac{1}{L} \int_0^L h_x dx = \frac{5}{4} h_L. \tag{7.175}$$

The average Nusselt number for the entire plate is

$$Nu = \frac{hL}{k} = 0.037 Pr^{1/3} Re_L^{4/5}. \tag{7.176}$$

When the forward portion of the plate is covered with a laminar boundary layer, a correction to the preceding equation is necessary. In this case, an average heat-transfer coefficient for the turbulent flow must be obtained and combined with the average coefficient for laminar region:

$$h = \frac{\int_0^{x_{cr}} h_x dx + \int_{x_{cr}}^L h_x dx}{L}. \tag{7.177}$$

Then, the average Nusselt number becomes

$$Nu = 0.037 Pr^{1/3} \left( Re_L^{4/5} - 23,550 \right), \tag{7.178}$$

for a critical Reynolds number of $5 \times 10^5$, and

$$Nu = 0.037 Pr^{1/3} \left( Re_L^{4/5} - 4325 \right), \tag{7.179}$$

for a critical Reynolds number of $10^5$.
A semiempirical integral analysis of the turbulent boundary layer on a flat plate with unheated starting section has been carried out in Reference 70. It is found that the correction factor by which the isothermal correlation must be multiplied to give local coefficients downstream of the discontinuity in wall temperature is

$$\left[ 1 - \left( \frac{x_0}{x} \right)^{9/10} \right]^{-1/9}. \tag{7.180}$$

The data to verify these results and comparisons with other analyses are also included in Reference 70.

## Example 7.3

Air at 15°C and atmospheric pressure flows over a flat plate at a speed of 6 m/s. The plate is heated over its entire length to a temperature of 105°C. The length of the plate is 3 m. Calculate the average heat-transfer coefficient over the length of the plate.

### Solution

Let us first calculate the transition point. The properties should be evaluated at film temperature, $T_f$:

$$T_f = \frac{105 + 15}{2} = 60°C.$$

The properties of air at 60°C, from Appendix B, are

$$v = 18.9 \times 10^{-6} \, m^2/s$$

$$k = 0.0285 \, W/(m\,K).$$

$$Pr = 0.709$$

For a critical Reynolds number of $5 \times 10^5$, the critical length is

$$x_{cr} = \frac{5 \times 10^5 \times 18.9 \times 10^{-6}}{6} = 1.575 \, m.$$

Therefore, the forward portion of the plate is covered with a laminar boundary layer, and the rest of the flow over the plate is turbulent. Hence, to calculate the average heat-transfer coefficient over the plate, we use Equation 7.178. At $x = 3$ m,

$$Re_L = \frac{U_\infty L}{v} = \frac{6 \times 3}{18.9 \times 10^{-6}} = 9.524 \times 10^5.$$

Thus,

$$Nu = 0.036 Pr^{1/3} \left( Re_L^{4/5} - 23,100 \right)$$

$$= 0.036(0.709)^{1/3}[(9.524 \times 10^5)^{4/5} - 23,100],$$

$$= 559.9$$

which yields

$$h = Nu \frac{k}{L} = 559.9 \times \frac{0.0285}{3} = 5.281 \, W/(m^2\,K).$$

(Compare this result with the result of Example 4.2.)

The literature on turbulent forced convection is very extensive. In this chapter, an attempt has been made to include fundamental equations, concepts, and basics of turbulent forced convection. Turbulent forced convection has been extensively reviewed in Chapter 4 of Reference 4. For turbulence modeling and the numerical computation of turbulent flows, References 26,65–68 are highly recommended.

## Problems

**7.1** Taking a control volume and applying the general laws together with the use of appropriate particular laws, obtain Equations 7.33 and 7.34a.

**7.2** Assume that the velocity distribution in the turbulent core for tube flow may be represented by

$$\frac{u}{U_c} = \left(1 - \frac{r}{r_0}\right)^{1/7},$$

where
$U_c$ is the velocity at the center
$r_0$ is the tube radius

The velocity in the laminar sublayer may be assumed to vary linearly with the radius. Using the friction factor given by Equation 7.49, derive an equation for the thickness of the laminar sublayer. For this problem, the average flow velocity may be calculated using only the turbulent velocity distribution.

**7.3** Using the velocity profile in Problem 7.2, obtain an expression for the eddy diffusivity of momentum as a function of radius.

**7.4** Estimate the temperature rise for water flowing in a heat-exchanger tube 3 m long, 1 in. OD with a 16 BWG (ID = 8.870 in.) wall. The water flows at the rate of 60 L/min and enters the tube at a temperature of 18°C. Neglect entrance effects and use the properties of water evaluated at the arithmetic average bulk temperature. The tube wall temperature is 95°C (use the Reynolds, Prandtl, and von Karman analogies).

**7.5** Show that the ratio $q''/\tau$ remains exactly constant across the whole width of the boundary layer for the flat plate.

**7.6** Using Hoffman's empirical value (Equation 7.77) for $u_L/U_m$, obtain an expression for Stanton number for a fully developed turbulent flow through smooth circular ducts.

**7.7** By the use of the following analogies, plot the variation of the Nusselt numbers with the Reynolds number for $Pr = 0.01$, 1, and 10 in the case of turbulent flow heat transfer in a circular duct flow:

1. Reynolds
2. Prandtl–Taylor
3. von Karman

**7.8** Develop an analogy between $St$ and $f$ similar to von Karman's analogy, by approximating the velocity distribution in the turbulent core as $u^+ = 8.74(y^+)^{1/7}$.

Calculate the heat-transfer coefficient for the conditions given in Example 7.1 and compare it with the other results obtained in Example 7.1.

**7.9** Determine the thickness of the laminar sublayer for a fully developed turbulent flow of water at 40°C ($v = 0.658 \times 10^{-6}$ m²/s) through a 2.54 cm diameter duct with a mean velocity of 0.5 m/s.

**7.10** Derive the total resistance given in Equation 7.96.

**7.11** Obtain Equations 7.107 through 7.109.

**7.12** Calculate the Martinelli factor $F$ as described in Section 7.7 for a smooth tube assuming $\zeta = \varepsilon_h/\varepsilon_m = 1$, $Pr = 0.1$, and $Re = 10^4$.

**7.13** Calculate the value of $h$ in Example 7.1 using Martinelli equation (7.117). How does it compare with the values from Example 7.1?

**7.14** A thin smooth metal circular duct 5 cm inside diameter and 60 cm long is perfectly insulated against heat losses to its surroundings. Water is forced through this tube at a velocity of 0.5 cm/s. The average temperature of water inside the circular duct is 40°C:

   a.  What must be the rate of heat generation (by electrical or other means) within the duct material in order to maintain the average duct wall temperature at 60°C?

   b.  What is the power that must be generated in kW?

**7.15** At some section of a cooled circular duct, having an inner diameter of 5 cm air is at 150°C and 1 atm and flows with a velocity of 8 m/s. The surface temperature of the duct is maintained at 40°C. If the static pressure drop per meter length of duct is $1.6 \times 10^{-4}$ bar, calculate approximately the temperature drop of the air over the next meter of duct.

**7.16** A thin symmetrical profile is tested in a wind tunnel where the free-stream velocity is 20 m/s and air temperature is 20°C. A total drag force of 100 kg$_f$ is measured. If the profile temperature is raised to 30°C and the free-stream temperature is lowered to 10°C but all other conditions remain the same, calculate the total heat-transfer rate.

**7.17** Consider a fully developed turbulent flow through a circular duct of radius $r_0$. Let the velocity distribution be specified as

$$u^+ = 5.5 + 2.5 \ln y^+,$$

with $y$ being the distance measured from the pipe wall. Develop an expression for the friction factor $f$ defined as

$$f = \frac{\tau_w}{(1/2)\rho U_m^2}.$$

**7.18** Show that thermal resistance from the wall to any point $y$, located in the turbulent core, is given by

$$\frac{5}{RePr}\sqrt{\frac{2}{f}}\left[ Pr + \ln(1+5Pr) + 0.5 \ \ln\frac{Re}{60}\frac{y}{r_0}\sqrt{\frac{f}{2}} \right].$$

Obtain an expression for the resistance to the center of the pipe and then the $\Delta T/\Delta T_{max}$ ratio.

**7.19** Show that the Martinelli factor $F$ is given by the following expression:

$$F = \frac{\ln\left[\dfrac{5\lambda}{5\lambda + \dfrac{y_2}{r_0}\left(1 - \dfrac{y_2}{r_0}\right)}\right] + \dfrac{1}{\sqrt{1+20\lambda}}\ln\left[\dfrac{1+\sqrt{1+20\lambda}}{1-\sqrt{1+20\lambda}}\dfrac{\left(\dfrac{2y_2}{r_0} - 1\right) - \sqrt{1+20\lambda}}{\left(\dfrac{2y_2}{r_0} - 1\right) + \sqrt{1+20\lambda}}\right]}{2\ln\dfrac{Re}{60}\sqrt{\dfrac{f}{2}}}.$$

**7.20** Let us consider a liquid metal flow ($Pr = 0.01$) in a reactor channel at $Re = 3 \times 10^4$ and suppose that the circumferential heat-flux distribution is

$$q''(\phi) = q_0''(1 + 0.40 \sin \phi).$$

Calculate $Nu(\phi)$ and $Nu_0$, and comment on the effect of circumferential variation of heat flux on the local Nusselt number and local temperature difference.

**7.21** The surface temperature of a thin plate located parallel to an air stream is 90°C. The main stream velocity is 60 m/s and its temperature is 0°C. The plate is 60 cm long and 45 cm wide. The width is along the direction of the main stream. Assume that the flow in the boundary layer changes abruptly from laminar to turbulent at a transition Reynolds number of $Re = 4 \times 10^5$. Neglect the end effect of the plate:

a. Find the average heat-transfer coefficients in the laminar and the turbulent regions.

b. Calculate the rate of heat transfer for the entire plate, considering both sides.

c. Plot curves for the local heat-transfer coefficient along the plate.

**7.22** In a 2D turbulent boundary-layer flow of a constant-property fluid over a flat plate, show that

$$\frac{\delta}{x} = 0.37(Re_x)^{-1/5}.$$

**7.23** Show that the average heat-transfer coefficient for turbulent flow over a flat plate when the forward portion of the plate is covered with a laminar boundary layer is given by

$$Nu = 0.036 Pr^{1/3} \left[ Re_L^{4/5} - 4200 \right],$$

if $Re_{cr} = 105$.

**7.24** In turbulent boundary layer on an isothermal flat surface at temperature $T_w$ (assuming that fluid properties are constants, no viscous dissipation, no body forces, and zero pressure gradient) under steady-state conditions, the boundary-layer equations can be written as

$$\frac{\partial u}{\partial x} + \frac{\partial v}{\partial y} = 0$$

$$u \frac{\partial u}{\partial x} + v \frac{\partial u}{\partial y} = \frac{\partial}{\partial y} \left[ (v + \varepsilon_m) \frac{\partial u}{\partial y} \right].$$

$$u \frac{\partial T}{\partial x} + v \frac{\partial T}{\partial y} = \frac{\partial}{\partial y} \left[ (\alpha + \varepsilon_h) \frac{\partial T}{\partial y} \right]$$

Let $U_\infty$ and $T_\infty$ be the free-stream velocity and temperature, respectively. If $Pr = Pr_t = 1$, then without solving the preceding equations show that

a. $St_x = (1/2)C_{fx}$.

b. The ratio $q''/\tau$ remains constant across the whole width of the boundary layer. What is this constant?

# References

1. Boussinesq, J., Essai sur la theorie des equanx courantes, *Mem. Presentes Acad. Sci. Paris*, 23, 46, 1877.
2. Donaldson, C. and Rosenbaum, H., Calculation of turbulent shear flows through closure of the Reynolds equations by invariant modeling, Aeronautical Research Associations Princeton Report, 127, 1968.
3. Pletcher, R. H., External flow forced convection, in *Handbook of Single-Phase Convective Heat Transfer*, S. Kakaç, R. K. Shah, and W. Aung (Eds.), John Wiley & Sons, New York, Chapter 2, 1987.
4. Prandtl, L., Eine beziehung zwischen warmeaüstausch und strömungswiderstand der flussiskeiten, *Phys. Z*, 11, 1072–1078, 1910.
5. Prandtl, L., *Essentials of Fluid Dynamics*, Blackie & Son, London, U.K., 1969, p. 117; in German, Führer durch die Strömung-slehre, Vieweg, Braunschweig, Germany, 1949.
6. van Driest, E. R., On turbulent flow near a wall, *J. Aerosp. Sci.*, 23, 1007–1011, 1956.
7. Cebeci, T. and Smith, A. M. O., *Analysis of Turbulent Boundary Layers*, Academic Press, New York, 1974.
8. Reynolds, O., On the extent and action of the heating surface for steam boilers, *Proc. Manchester Lit. Phil. Soc*, 14, 7–12, 1874.
9. Corcoran, W. H., Page, F. Jr., Schlinger, W. G., Breaux, D. K., and Sage, B. H., Temperature gradients in turbulent gas streams, *Ind. Eng. Chem.*, 44, 410–430, 1952.
10. Jenkins, R., Variation of Eddy conductivity with Prandtl modulus and its use in prediction of turbulent heat transfer, *Heat Transfer Fluid Mech. Inst.*, pp. 147–158, 1951.
11. Rohsenow, W. M. and Cohen, L. S., M.I.T. Heat Transfer Laboratory Report, June, 1960.
12. Sleicher, C. A. and Tribus, M., Heat transfer in a pipe with turbulent flow and arbitrary wall temperature distribution, *Trans. ASME*, 79, 765, 1957.
13. Kays, W. M. and Leung, E. Y., Heat transfer in annular passages: Hydrodynamically developed turbulent flow with arbitrarily prescribed heat flux, *Int. J. Heat Mass Transfer*, 6, 537–557, 1963.
14. Reynolds, W. C., Turbulent heat transfer in a circular tube with variable circumferential heat flux, *Int. J. Heat Mass Transfer*, 6, 445–454, 1963.
15. Azer, N. Z. and Chao, B. T., A mechanism of turbulent heat transfer in liquid metals, *Int. J. Heat Mass Transfer*, 1, 121–138, 1960.
16. Lykoudis, P. S. and Touloukian, Y. S., Analytical study of heat transfer in liquid metals, *Trans. ASME*, 80, 653–666, 1958.
17. Deissler, R. G., Heat transfer and fluid friction for fully developed turbulent flow of air and supercritical water with variable fluid properties, *Trans. ASME*, 76, 73–85, 1954.
18. Tyldesley, J. R. and Silver, R. S., The prediction of the transport properties of a turbulent fluid, *Int. J. Heat Mass Transfer*, 11, 1325–1340, 1968.
19. Bushnell, D. M., Cary, A. M. Jr., and Harris, J. E., Calculation methods for compressible turbulent boundary layers, von Karman Institute for Fluid Dynamics, Lecture Series 86 on Compressible Turbulent Boundary Layers, Rhode Saint Genese, Belgium, Vol. 2, 1976.
20. Adams, J. C. Jr. and Hodge, B. K., The calculation of compressible, transitional, turbulent, and relaminarizational boundary layers over smooth and rough surfaces using an extended mixing length hypothesis, *Proceedings of the AIAA 10th Fluid and Plasma Dynamics Conference*, Albuquerque, NM, AIAA Paper 77–682, 1977.
21. Christoph, G. H. and Pletcher, R. H., Prediction of rough-wall skin-friction and heat transfer, *AIAA J.*, 21(4), 509–515, 1983.
22. Pletcher, R. H., Prediction of transpired turbulent boundary layers, *J. Heat Transfer*, 96, 89–94, 1974.
23. Kays, W. M. and Moffat, R. J., The behavior of transpired turbulent boundary layers, in *Studies in Convection: Theory, Measurement, and Applications*, B. E. Launder (Ed.), Vol. 1, Academic Press, New York, pp. 223–319, 1975.

24. Harlow, F. H. and Nakayama, P. I., Transport of turbulence energy decay rate, Los Alamos Scientific Laboratory Report LA-3584, Los Alamos, NM, 1968.

25. Jones, W. R. and Launder, B. E., The prediction of laminarization with a two-equation model of turbulence, *Int. J. Heat Mass Transfer*, 15, 301–314, 1972.

26. Launder, B. E. and Spalding, D. B., The numerical computation of turbulent flows, *Comput. Meth. Appl. Mech. Eng.*, 3, 269–289, 1974.

27. Ng, K. H. and Spalding, D. B., Turbulence model for boundary layers near walls, *Phys. Fluids*, 15, 20–30, 1972.

28. Wilcox, D. C. and Traci, R. M., A complete model of turbulence. *AIAA Guidance and Control Conference*, Paper 76–351, San Diego, CA, 1976.

29. Saffman, P. G. and Wilcox, D. C, Turbulence model prediction for turbulent boundary layers, *AIAA J.*, 12, 541–546, 1974.

30. Prandtl, L., The mechanics of viscous fluids, in Aerodynamic Theory III, W.F. Durand (Ed.), vol. 3, pp. 34–208, Julius Springer, Berlin, 1935.

31. Kestin, J. and Richardson, P. D., Heat transfer across Turbulent, incompressible boundary layers, *Int. J. Heat Mass Transfer*, 6, 147–189, 1963.

32. Reichardt, H., Die Grundlagen des Tuibulenten Warmeüberganges, *Arclt. Ges. Wärmetech.*, 2, 129–142, 1951.

33. Reichardt, H., Heat transfer through turbulent friction layers, N.A.C. A TM 1047, 1943, translation of Die Warmeü-bertragung in Tuibulenten Reibungs-schichten, *Z. Angew. Math. Mech.*, 20, 297–328, 1940.

34. Reichardt, H., Impuls-and wärmeaustausch in Freier turblenz, *Z. Angew. Math. Mech.*, 24, 268, 1944.

35. Blasius, H., Des Aehnlichkeitsgesetz bei Reibungsvorgangen in Flussigkeiten, *VDI Mitt. Forschungsarb.*, 131, 1–39, 1913.

36. Nikuradse, J., Stromungsgesetz in rauhren rohren, *vDI Forschungshefte*, p. 361, 1933.

37. Schlichting, H., *Boundary Layer Theory*, translated by J. Kestin, Fourth Edition, McGraw-Hill, New York, 1960.

38. von Karman, Th., The analogy between fluid friction and heat transfer, *Trans. Am. Soc. Mech. Eng.*, 61, 705–710, 1939; Mechanische Ahnlichkeit und Turblenz, Ges. der Wiss. zu Gott, Nachrichten, *Matk-Phys. Kl*, 58–76, 1930.

39. Rannie, W. D., Heat transfer in turbulent shear flow, *J. Aerosp. Sci.*, 23, 485–489, 1956.

40. Deissler, R. G., Analysis of turbulent heat transfer, mass transfer and friction in smooth tubes at high Prandtl and Schmidt numbers, N.A.C.A. TN 3145, 1954: also Rep. 1210, 1955.

41. Spalding, D. B., A single formula for the "Law of the Wall", *J Appl. Mech.*, 28, 455–457, 1961.

42. Moody, F. F., Friction factors for pipe flow, *Trans. ASME*, 66, 671, 1944.

43. Reynolds, O., *Scientific Papers*, Cambridge University Press, London, U.K., Vol. 1, 1901.

44. Reynolds, O., On the dynamical theory of incompressible viscous fluids and the determination of the criterion, *Philos. Trans. R. Soc. London Ser. A*, 186, 123–164, 1895.

45. Reynolds, O., On the Extent and Action of the Heating Surface for Steam Boilers, *Proc. Manchester Lit. Phil. Soc*, 14, 7–12, 1874.

46. Taylor, G. I., Conditions at the surface of a body exposed to the wind, British Advisory Committee for Aeronautics Reports and Memoranda No. 272, Vol. H, pp. 423–429, 1916.

47. Prandtl, L., Eine Beziehung Zwischen Warmeaustausch und Strömungswiderstand der Flüssigkeit, *Phys. Zeits*, 11, 1072–1078, 1910.

48. Prandtl, L., Bemeikung über den Warmeübergang im Rohr, *Phys. Z.*, 29, 487–489, 1928.

49. Hoffman, E., Der Warmeübergang bei der Strömung im Rohr, *Z. Ges. Kalte-Ind*, 44, 99–107, 1937.

50. Colburn, A. P., A method of correlating forced convection heat transfer data and comparison with fluid friction, *Trans. AIChE*, 29, 174, 1933.

51. Rubesin, M. W., A modified Reynolds analogy for compressible turbulent boundary layer on a flat plate, NACA TN 2917, 1953.

52. van Driest, E. R., The turbulent boundary layer with variable Prandtl number, *Fifty Years of Boundary Layer Research*, Braunschweig, Germany, pp. 257–271, 1955.

53. Boelter, L. M. K., Martinelli, R. C., and Jonassen, F., Remarks on the analogy between heat transfer and momentum transfer, *Trans. ASME,* 63, 447–456, 1941.

54. Martinelli, R. C., Heat transfer to molten metals, *Trans. ASME,* 69, 947–959, 1947.

55. Reynolds, W. C., Turbulent heat transfer in a circular tube with variable circumferential heat flux, *Int. J. Heat Mass Transfer,* 6, 445–454, 1962.

56. Cess, R. D., *A Survey of the Literature on Heat Transfer in Turbulent Tube Flow,* School of Engineering and Mines, University of Pittsburgh, Pittsburgh, PA, 1958.

57. Hall, W. B. and Khan, S. A., Experimental investigation into the effect of the thermal boundary condition on heat transfer in the entrance region of a pipe, *J. Mech. Eng. Sci.,* 6, 250–256, 1964.

58. Siegel, R. and Sparrow, E. M., Comparison of turbulent heat-transfer results for uniform wall heat flux and uniform wall temperature, *ASME J. Heat Transfer,* 82, 152, 1960.

59. Kakaç, S. and Paykoç, S., Analysis of turbulent forced convection heat transfer between parallel plates, *METU J. Pure Appl. Sci.,* 1(1), 27–47, 1968.

60. Deissler, R. G., Turbulent heat transfer and friction in the entrance regions of smooth passages, *Trans. ASME,* 77, 1221, 1955.

61. Sparrow, E. M., Hallman, T. M., and Siegel, R., Turbulent heat transfer in the thermal entrance region of a pipe with uniform heat flux, *Appl. Sci. Res., Ser. A,* 7, 37–52, 1957.

62. Hatton, A. P. and Quarmby, A., Shorter communications, comments on the effect of axially varying and unsymmetrical boundary conditions on heat transfer with turbulent flow between parallel plates, *Int. J. Heat Mass Transfer,* 7, 1143–1144, 1964.

63. Kays, W. M. and Nicoll, W. B., The influence of non-uniform heat flux on the convection conductances in a nuclear reactor, Technical Report No. 33, Department of Mechanical Engineering, Stanford University, Stanford, CA, 1957.

64. Burgoyne, T., The effect of a variable heat flux on the heat transfer coefficient, Technical Note No. IGR-TN/W.306UKAEA, 1956.

65. Hall, W. B. and Price, P. H., The effect of a longitudinally varying wall heat flux on the heat transfer coefficient for turbulent flow in a pipe, *International Heat Transfer Conference,* ASME IMechE, Boulder, CO, 1961.

66. Duhamel, M., Memoire sur la method generate relative au mouvement de la chaleur dans les corps sondes plange dans desmilieux dont la temperature varie avec le temps, *J. Ecole Polytechnique, Paris,* 14, Cahier 22, 20, 1833.

67. Hall, W. B., Jackson, J. D., and Price, P. H., Note on forced convection in a pipe having a heat flux which varies exponentially along its length, *J. Mech. Eng. Sci.,* 5(1), 48–53, 1963.

68. Kakaç, S. and Price, P. H., On the constant heat transfer coefficient boundary condition on forced convection heat transfer in a channel, *METU J. Pure Appl. Sci.,* 2(3), 239–262, 1969.

69. Eckert, E. R. G. and Drake, R. M. Jr., *Analysis of Heat and Mass Transfer,* McGraw-Hill, New York, pp. 364–368, 1972.

70. Reynolds, W. C., Kays, W. M., and Kline, S. J., Heat transfer in the turbulent incompressible boundary layer with constant wall temperature, Final Report, Part n, Contract NAW-6494, Stanford University, Stanford, CA, 1957.

71. Anderson, D., Tannehill, J. C., and Pletcher, R. H., *Computational Fluid Mechanics and Heat Transfer,* Hemisphere Publishing Corporation, Washington, DC, 1984.

72. Launder, B. E. and Spalding, D. B., *Mathematical Models of Turbulence,* Academic Press, New York, 1972.

73. Launder, B. E. and Spalding, D. B., The numerical computation of Turbulent flows, *Comput. Methods Appl. Mech. Eng.,* 3, 269–289, 1974.

74. Rodi, W., Examples of turbulence models for incompressible flows, *AIAA J.,* 20, 872–879, 1982.

75. Kakaç, S. and Spalding, D. B., *Turbulent Forced Convection in Channels and Bundles,* Vols. 1 and 2, Hemisphere Publishing Corporation, Washington, DC, 1979.

# 8

# Unsteady Forced Convection in Ducts

## Nomenclature

| | |
|---|---|
| $A$ | dimensionless temperature amplitude function |
| $a^*$ | fluid-to-wall thermal capacitance ratio, $(\rho c_p)_f L/(\rho c)_w \ell$ |
| $Bi$ | Biot number, $hL/k$ |
| $c_f$ | fluid specific heat at constant pressure, J/(kg K) |
| $c_w$ | wall specific heat, J/(kg K) |
| $D$ | diameter $= 2r_0$, m |
| $D_h$ | hydraulic diameter $= 4L$, m |
| $D(y)$ | effective diffusivity $= \alpha + \varepsilon_h$ |
| $h$ | heat-transfer coefficient |
| $i$ | imaginary unit $= \sqrt{-1}$ |
| $J_0, J_1$ | Bessel functions of the first kind and of zero and first orders |
| $k$ | fluid thermal conductivity, W/(m K) |
| $L$ | half the distance between parallel plates; distance between parallel plates, m |
| $\ell$ | thickness of the wall |
| $P$ | pressure, Pa |
| $P_n$ | eigenfunction |
| $Pr$ | Prandtl number $= \mu c_p/k = v/\alpha$ |
| $Q_n$ | eigenfunction |
| $q''$ | heat flux, W/m$^2$ |
| $Re$ | Reynolds number $= \rho u D_h/\mu$ |
| $R_n$ | eigenfunction |
| $r$ | radial coordinate, m |
| $r_0$ | tube radius, m |
| $r_0^+$ | dimensionless radius $= r_0\sqrt{\tau_w/\rho}\,/v$ |
| $T$ | temperature, °C, K |
| $T_0$ | cycle mean temperature, °C, K |
| $t$ | time, s |
| $t^+$ | dimensionless time $= vt/r_0^2$ |
| $t_s$ | steady-state time, s |
| $U_m$ | mean velocity, m/s |
| $u$ | velocity component in $x$-direction, m/s |
| $u^+$ | dimensionless velocity $= u/\sqrt{\tau w/\rho}$ |
| $x$ | distance parallel to flow direction along the duct, m |

$x^+$        dimensionless $x$-coordinate $= x/D$
$y$          transverse distance in parallel-plate channels, m
$y^+$        dimensionless variable $= r_0^+ - r^+$

## Greek Symbols

$\alpha$     thermal diffusivity $= k/\rho c_p$, m²/s
$\alpha_n$   eigenvalues
$\beta$      inlet frequency, 1/s
$\beta_n$    eigenvalues
$\Delta T(y)$  inlet temperature amplitude profile
$\varepsilon_h$  eddy diffusivity of heat, m²/s
$\varepsilon_m$  eddy diffusivity of momentum, m²/s
$\eta$       dimensionless $r$ coordinate $= r/r_0$
$\theta$     dimensionless temperature, defined by Equation 8.4e or 8.31
$\lambda_n$  eigenvalues
$\mu$        fluid dynamic viscosity, Pa s
$\nu$        kinematic viscosity, m²/s
$\xi$        dimensionless axial coordinate $= (2x/D)/(RePr)$ or $(x/D_h)(D_h/L)^2(RePr)$
$\rho_f$     fluid density, kg/m³
$\rho_w$     density of wall material, kg/m³
$\tau$       dimensionless time $= \alpha t/r_0^2$
$\tau_s$     dimensionless steady-state times $= \nu t_s/r_0^2$ or $\alpha t_s/L^2$
$\Omega$     dimensionless inlet frequency $= 2\pi\beta L^2/\alpha$

## Subscripts

$b$          bulk value
$i$          inlet condition
$m$          mean condition
$s$          steady-state condition
$w$          wall condition
$\infty$     environment condition

## 8.1  Introduction

The understanding and evaluation of steady or unsteady flows with transient forced convection have recently become more important in connection with the precise control of modern high-performance heat-transfer systems. The increasingly greater use of automatic control devices, for instance, for the accurate regulation of fluid systems involving heat-exchange equipment, has stimulated a greater interest in transient heat-transfer phenomena. This is an especially important case when the positive control of industrial heat exchangers or the nuclear reactor systems must be assured during power changes. In addition, accurate prediction of the transient response of thermal systems is also highly important for the understanding of such adverse effects as reduced thermal performance and severe thermal stresses that they can produce, with eventual mechanical failure.

Problems such as start-up, shutdown, power surge, and pump failure have motivated investigations of thermal response of internal flows to step changes in thermal and hydrodynamic conditions. In the flow channels of nuclear reactors, heat generation can vary along the length of the channel walls due to nonuniformities in the neutron flux or spatial variations in fuel loading. Such problems have promoted investigations of thermal response of flows in channels and rod bundles to prescribed timewise variation of wall heat flux, wall temperature, or internal heat generation.

The literature on unsteady forced convection is small but growing. Some of the important contributions are listed in the references at the end of this chapter.

This chapter is mainly concerned with the study of the heat-transfer behavior of duct flows with thermal transients. Two geometries that are commonly encountered in practice will be considered for the analyses, namely, the parallel-plate channel and the circular duct. The flow in the parallel-plate channel or the circular duct may be laminar or turbulent.

The general problem of transient forced convection heat transfer can be stated as follows: The temperature distribution is to be determined in the system at an arbitrary instant of time, given the following:

1. The inlet temperature distribution as an arbitrary function of time and space.
2. The initial temperature distribution for $x \geq 0$ as an arbitrary function of time and space.
3. A prescribed boundary condition that may take many forms. Some possible forms are described as follows:
    a. A prescribed temperature distribution or a heat-flux distribution may in some way be enforced on the boundaries of the system, and this distribution may furthermore be constant or variable with time and/or space.
    b. A constant heat-transfer coefficient to a prescribed ambient temperature.
4. A time-dependent velocity distribution, that is, unsteady flow. In case of flow in the cooling channels of a nuclear reactor, if a pump failure, for example, occurs, the control system will simultaneously shut down the power. Hence, the decrease in the flow rate will be accompanied by a transient in the heat being transferred at the channel walls.

## 8.2 Transient Laminar Forced Convection in Ducts

In this section, first, a short review of the literature on transient laminar forced convection in parallel-plate channels and circular ducts is given, and then, some fundamental problems are discussed.

One of the earliest works on transient laminar forced convection was the calculation of transient temperatures in pipes and heat exchangers by Dusinberre [1], who presented explicit iteration formulas and numerical computation guides. The case of a compressible fluid flowing through an insulated tube with an exponential or step-function fluid inlet temperature was analyzed by Rizika [2]. The transient conditions for a compressible fluid flowing through a heat exchanger were also partially analyzed by Deissler in Reference 3. Rizika in Reference 4 extended his previous work in Reference 3 to obtain the transient solutions due to thermal lags in flowing incompressible fluid systems. Both the case of an incompressible fluid with a step-function temperature input flowing through a circular tube and the case of an incompressible fluid flowing in a simple heat exchanger were examined in detail. An example that demonstrates the transient condition at the exit of a simple (condensing-steam-water) heat exchanger was also presented.

Sparrow and Siegel [5] made an analysis of transient laminar forced convection in the thermal entrance region of circular tubes. They first determined the thermal responses to step changes both in wall temperature and in wall heat flux, using an integral formulation of the energy equation in connection with the method of characteristics. Then, using the linearity of the energy equation, they generalized die step-function results for arbitrary time-dependent boundary conditions and expressed their results in the form of integrals that can easily be evaluated for particular applications. Siegel and Sparrow in Reference 6 also made a similar analysis in the thermal entrance region of flat ducts.

Siegel in Reference 7 investigated laminar slug flow in a circular duct and a parallel-plate channel where the walls were given a step change in heat flux or, alternately, a step change in temperature. Siegel [8] also investigated laminar forced convection both in a circular duct and in a parallel-plate channel with arbitrary time variations in wall temperature. The velocity distributions in both cases were assumed to be steady and fully developed. First, the analyses were done for a step change in wall temperature, and then the results were generalized for arbitrary time variations. The method used was not an exact solution, but involved an approximation. The validity of the approximation was tested by comparing the results with the exact ones available for part of the solution, and good agreement was obtained. It was also demonstrated that the slug flow assumption does, in fact, lead to the essential physical behavior of the systems considered, although the numerical results were somewhat in error.

Perlmutter and Siegel [9] studied transient heat transfer with unsteady laminar flow between two parallel plates. The transients were caused by simultaneously changing the fluid pumping pressure and either the wall temperature or the wall heat flux. During the solution, the time-dependent slug flow simplification was made. Within this limitation, exact solutions for the fluid temperature distribution were obtained for a step change in wall temperature or wall heat flux with a simultaneous step change in the pumping pressure. Then, using superposition, solutions for more involved situations were developed. Perlmutter and Siegel in Reference 10 analyzed transient heat transfer in a 2D unsteady incompressible laminar duct flow between two parallel plates with a step change in wall temperature. Some results were also presented for the case where the transient heat conduction through the bounding walls was taken into consideration. Siegel and Perlmutter [11] also made an analysis of unsteady incompressible laminar forced convection between two parallel plates with the wall heat flux varying with both time and axial position. The flow velocity was assumed constant over the channel cross section (slug flow assumption), but allowed to vary with time. General analytical expressions were derived that could be used for computing the transient heat transfer in a channel with the wall heat flux varying with time and axial position.

Siegel [12] analyzed laminar forced convection between two parallel plates for slug flow conditions by including the heat capacity of the walls and with the wall heating assumed variable with axial position and time. The walls were assumed to be sufficiently thin or highly conducting so that the temperature variations across the wall thicknesses could be neglected. The method used involved coupling the heat-transfer behavior within the fluid to that in the wall and solving the resultant equations together with the energy equation. After the introduction of a general method of solution, some illustrative examples were considered. These included uniform wall heating varying sinusoidally in time and heating varying sinusoidally with axial distance and exponentially in time.

References 13–16 are a series of papers on the dynamic response of heat exchangers having time-varying internal heat sources. In these papers, theoretical results are also compared with experiments. In addition, Kardas [17] studied heat transfer from parallel flat plates to fluids flowing between them with an inlet temperature varying with time. He presented an analytical solution of the unidirectional regenerator problem.

Namatame [18] presented a modified quasi-steady solution for the transient temperature response of an annular slug flow in which the dependence of the surface temperatures upon axial position was taken into consideration.

Campo and Yoshimura [19] made a theoretical study to describe the influence of randomly varying ambient temperatures on the heat-transfer performance of a fully developed flow through a parallel-plate channel. In the analysis, the energy equation was simplified by the use of a lumped formulation in the transverse direction of the channel, and the time-dependent fluid temperature in the axial direction was obtained.

Lin and Shih [20] studied the unsteady thermal entrance heat transfer for fully developed laminar flow of power-law non-Newtonian fluids in pipes and plate slits with step change in surface temperature by the so-called instant-local similarity method that uses the concept of the extended Lévêque method by restricting the solution to large Graetz numbers and converting the energy equation to a boundary-layer-type equation. The effects of the flow index, viscous dissipation, and the Graetz number on the heat-transfer rate were demonstrated with numerical solutions.

Sucec [21] presented an improved quasi-steady approach which took both the effect of thermal history and of the thermal energy storage capacity of the fluid into approximate account in problems of transient conjugated laminar forced convection. The method was applied to two problems in a parallel-plate channel in which the finite-thermal capacity walls and fluid were both at constant temperature initially, when the transient was initiated by either a step change in fluid inlet temperature or a sinusoidal variation with time. Exact solutions were given for slug flow and for a linear velocity profile.

By the use of finite-difference numerical schemes, Lin et al. [22,23] solved the transient 2D energy equation for various flow conditions. The first paper studied the thermal entrance heat transfer in laminar pipe flows subjected to a step change in ambient temperature. The transient thermal entrance laminar flow heat transfer resulting from a step change in both the pressure gradient and the inlet temperature was studied in the second paper. Lin et al. [24] also studied heat transfer in the thermal entrance region of laminar pipe flows resulting from a step change in the inlet temperature, when coupled with the unsteady temperature variations in the surrounding enclosure, representing a refrigerator cabinet. The Nusselt number, fluid bulk mean temperature, and pipe wall temperature were presented over wide ranges of the parameters involved.

Chen et al. [25] gave a direct numerical solution for transient laminar heat transfer inside a circular duct subjected to a step change in either wall temperature or heat flux.

Cotta et al. [26] solved the slug flow problem considered in References 27,28 for both circular tubes and parallel-plate channels by developing a new approach for the solution of the associated complex eigenvalue problem and provided accurate results for the related eigenvalues; this approach, however, was not extended to parabolic flow situations. A second-order accurate finite-difference scheme was also developed for transient forced convection resulting from a step change in inlet temperature [29]. Cotta and Özisik [30] presented an analytical solution to periodic laminar forced convection inside ducts by making use of the generalized integral transform technique, completely avoiding the complex eigenvalue problem, and solving instead the resulting coupled system of complex ordinary differential equations. An experimental investigation of transient forced convection in ducts with a timewise variation of inlet temperature is described in References 31–34.

A theoretical and experimental study of laminar forced convection in the thermal entrance region of a rectangular duct, subjected to a sinusoidally varying inlet temperature, is presented in Reference 35. In this work, which is an extension of the theoretical work given in Reference 30, a general boundary condition of the fifth kind that accounts for both external convection and duct wall thermal capacitance effects is considered, and an analytical solution is obtained by extending the generalized integral transform technique [36]. The results thus obtained are compared with experimental findings to verify the validity of the proposed model. Numerical results are also presented for various combinations of the relevant parameters, such as the dimensionless inlet frequency, Biot number, fluid-to-wall thermal capacitance ratio, and nonuniform periodic inlet temperature profile. The effects of these parameters on the bulk and center temperature amplitudes and phase lags are critically examined.

An extensive review of the work on transient laminar forced convection in ducts has also been given by Kakaç and Yener [37], Yener and Kakaç [38], Li [39], and Li and Kakaç [40].

## 8.2.1 Transient Laminar Forced Convection in Circular Ducts with Step Change in Wall Temperature

Consider a circular duct as shown in Figure 8.1. A steady and fully developed laminar flow passes through the duct in the $x$-direction. The duct wall and fluid are initially isothermal at temperature $T_i$. Let the temperature of the duct wall for $x \geq 0$ be instantaneously changed (say at time $t = 0$) to a new value $T_w$ and be maintained at this value for all times thereafter.

The transient temperature distribution $T(x,r,t)$ in the duct for $x \geq 0$ and times $t > 0$ will satisfy the unsteady energy equation for fully developed laminar flow in a circular duct; that is,

$$\frac{\partial T}{\partial t} + u \frac{\partial T}{\partial x} = \alpha \frac{1}{r} \frac{\partial}{\partial r} \left( r \frac{\partial T}{\partial r} \right),$$ (8.1)

where the fluid properties have been assumed to be constant and viscous dissipation and axial heat conduction have been neglected. The initial, inlet, and boundary conditions are given by

$$T(x,r,0) = T_i, \quad T(0,r,t) = T_i,$$ (8.2a,b)

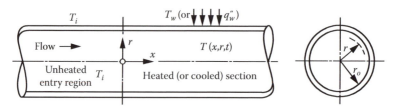

**FIGURE 8.1**
Coordinate system for circular duct geometry.

$$\left(\frac{\partial T}{\partial r}\right)_{r=0} = 0, \quad T(x, r_0, t) = T_w. \tag{8.2c,d}$$

In the following sections, two solutions of the foregoing problem are presented for slug flow and parabolic velocity distribution in the circular duct.

### 8.2.1.1 Solution for Slug Flow

If the velocity distribution $u$ in Equation 8.1 is assumed to be uniform over the duct cross section and equal to the mean flow velocity $U_m$, then the formulation of the problem can be rewritten in the following dimensionless forms:

$$\frac{\partial \theta}{\partial \tau} + \frac{1}{2}\frac{\partial \theta}{\partial \xi} = \frac{1}{\eta}\frac{\partial}{\partial \eta}\left(\eta \frac{\partial \theta}{\partial \eta}\right), \tag{8.3}$$

$$\theta(\xi, \eta, 0) = 1, \quad \theta(0, \eta, \tau) = 1, \tag{8.4a,b}$$

$$\left(\frac{\partial \theta}{\partial \eta}\right)_{\eta=0} = 0, \quad \theta(\xi, 1, \tau) = 0. \tag{8.4c,d}$$

Here

$$\theta = \frac{T - T_w}{T_i - T_w}, \quad \eta = \frac{r}{r_0}, \quad \xi = \frac{2x/D}{RePr}, \quad \tau = \frac{\alpha t}{r_0^2}, \tag{8.4e,f,g,h}$$

with

$$Re = \frac{U_m D}{\nu}, \quad Pr = \frac{\nu}{\alpha}, \quad D = 2r_0, \tag{8.4i,j,k}$$

where
  $\alpha$ is the thermal diffusivity
  $\nu$ is the kinematic viscosity of the fluid

The method of solution as described in Reference 7 is as follows: The fluid that was at $x = 0$ when the transient began does not reach a position $x$ until a time $x/U_m$ has elapsed. Thus, in the region where $x/U_m \geq t$, the heat-transfer process will not be influenced by the inlet condition. Consequently, in this region, the fluid at any cross section undergoes the same transient heating process as that at any other cross section, with the effect of heat convection being zero. Therefore, when $t \leq x/U_m$ (i.e., $\tau \leq 2\xi$), the solution at $x$ is governed by

$$\frac{\partial \theta}{\partial \tau} = \frac{1}{\eta}\frac{\partial}{\partial \eta}\left(\eta \frac{\partial \theta}{\partial \eta}\right), \tag{8.5}$$

with the following initial and boundary conditions:

$$\theta(\eta, \tau = 0) = 1, \tag{8.6a}$$

$$\left(\frac{\partial\theta}{\partial\eta}\right)_{\eta=0} = 0, \quad \theta(\eta = 1, \tau) = 0. \tag{8.6b,c}$$

The solution of this problem can readily be found to be [41]

$$\theta = 2\sum_{n=1}^{\infty} e^{-\lambda_n^2 \tau} \frac{J_0(\lambda_n \eta)}{\lambda_n J_1(\lambda_n)}, \quad \tau \le 2\xi, \tag{8.7}$$

where
$\lambda_n$ are the positive zeroes of $J_0(\lambda) = 0$
$J_0$ and $J_1$ are the Bessel functions of the first kind and of zero and first orders, respectively

Now consider the steady-state solution, which is governed by

$$\frac{1}{2}\frac{\partial\theta}{\partial\xi} = \frac{1}{\eta}\frac{\partial}{\partial\eta}\left(\eta\frac{\partial\theta}{\partial\eta}\right), \tag{8.8}$$

with

$$\theta(\xi = 0, \eta) = 1, \tag{8.9a}$$

$$\left(\frac{\partial\theta}{\partial\eta}\right)_{\eta=0} = 0, \quad \theta(\xi, \eta = 1) = 0. \tag{8.9b,c}$$

Since the differential equations (8.5) and (8.8) and the conditions (8.6a,b,c) and (8.9a,b,c) are the same in $\tau$ and $2\xi$, the steady-state solution will have the same form as Equation 8.7; that is,

$$\theta = 2\sum_{n=1}^{\infty} e^{-2\lambda_n^2 \xi} \frac{J_0(\lambda_n \eta)}{\lambda_n J_1(\lambda_n)}. \tag{8.10}$$

This result matches the initial transient solution, Equation 8.7, when $\xi = 1/2\tau$, and therefore, it must be the solution for $\tau \ge 2\xi$. Thus, Equations 8.7 and 8.10 together represent the complete solution of the problem for all times $t > 0$. Furthermore, these results show that the steady-state temperature distribution at a particular cross section is reached over the time period that it takes for the fluid to flow from the entrance to that particular section, that is, $x/U_m$.

### 8.2.1.2 Solution for Parabolic Velocity Distribution

For fully developed steady laminar flow in a circular duct of radius $r_0$, the velocity distribution is given by

$$\frac{u}{U_m} = 2\left[1 - \left(\frac{r}{r_0}\right)^2\right], \tag{8.11}$$

where $U_m$ is the mean flow velocity. With this parabolic profile, the energy equation (8.1) can be rewritten in dimensionless form as

$$\frac{\partial \theta}{\partial \tau} + (1 - \eta^2) \frac{\partial \theta}{\partial \xi} = \frac{1}{\eta} \frac{\partial}{\partial \eta} \left( \eta \frac{\partial \theta}{\partial \eta} \right), \tag{8.12}$$

where the definitions of various dimensionless quantities are the same as in the previous case. Furthermore, the same dimensionless initial, inlet, and the boundary conditions, that is, Equations 8.4, also apply to this case.

To develop the general solution, we now first consider the steady-state solution.

*8.2.1.2.1 Steady-State Solution*

Under steady-state conditions, the energy equation (8.12) reduces to

$$(1 - \eta^2) \frac{\partial \theta_s}{\partial \xi} = \frac{1}{\eta} \frac{\partial}{\partial \eta} \left( \eta \frac{\partial \theta_s}{\partial \eta} \right), \tag{8.13}$$

with the following inlet and boundary conditions:

$$\theta_s(0, \eta) = 1, \tag{8.14a}$$

$$\left( \frac{\partial \theta_s}{\partial \eta} \right)_{\eta=0} = 0, \quad \theta_s(\xi, 1) = 0. \tag{8.14b,c}$$

The problem consisting of the differential equation (8.13) and the conditions (8.14) is the classical Graetz problem, and its solution is given in Chapter 6 as

$$\theta_s(\xi, \eta) = \sum_{n=0}^{\infty} C_n e^{-\lambda_n^2 \xi} R_n(\eta), \tag{8.15}$$

where
$\lambda_n$ are the eigenvalues
$R_n(\eta)$ are the corresponding eigenfunctions of the following eigenvalue problem:

$$\frac{d^2 R^2}{d\eta^2} + \frac{1}{\eta} \frac{dR_n}{d\eta} + \lambda_n^2 (1 - \eta^2) R_n = 0, \tag{8.16a}$$

$$\frac{dR_n(0)}{d\eta} = 0, \quad R_n(1) = 0, \tag{8.16b,c}$$

and the constants $C_n$ are given by

$$C_n = -\frac{2}{\lambda_n (\partial R_n / \partial \lambda_n)_{\eta=1}}. \tag{8.17}$$

*8.2.1.2.2 Transient Solution*

Let the transient solution be given by a series expansion about the steady-state solution as

$$\theta(\xi, \eta, \tau) = \sum_{n=0}^{\infty} C_n F_n(\xi, \tau) R_n(\eta). \tag{8.18}$$

For large times, the functions $F_n$ should, by comparison with Equation 8.15, converge to

$$F_n = e^{-\lambda_n^2 \xi}. \tag{8.19}$$

As suggested by Siegel [8], the following approximation is now introduced: multiply Equation 8.12 by $\eta$ and then integrate from 0 to 1 to obtain

$$\int_0^1 \eta \frac{\partial \theta}{\partial \tau} d\eta + \int_0^1 \eta(1-\eta^2) \frac{\partial \theta}{\partial \xi} d\eta = \left(\frac{\partial \theta}{\partial \eta}\right)_{\eta=1}. \tag{8.20}$$

Substitution of Equation 8.18 into Equation 8.20 yields

$$\frac{\partial F_n}{\partial \tau} \int_0^1 \eta R_n(\eta) d\eta - \frac{1}{\lambda_n^2} \frac{\partial F_n}{\partial \xi} \frac{dR_n(l)}{d\eta} - F_n \frac{dR_n(l)}{d\eta} = 0, \quad n = 0, 1, 2, \ldots, \tag{8.21}$$

where the following relation has also been used (see Chapter 6):

$$\int_0^1 \eta(1-\eta^2) R_n(\eta) = \frac{1}{\lambda_n^2} \frac{dR_n(l)}{d\eta}. \tag{8.22}$$

The differential equation (8.21) can readily be solved by the method of characteristics to obtain $F_n(\xi,\tau)$, and the result for $\theta(\xi,\eta,\tau)$ is then given by [8]

$$\theta(\xi, \eta, \tau) = \frac{T - T_w}{T_i - T_w} = \sum_{n=0}^{\infty} C_n R_n(\eta) \begin{Bmatrix} e^{-\beta_n \tau}, & \tau \le a_n \xi \\ e^{-\lambda_n^2 \xi}, & \tau \ge a_n \xi \end{Bmatrix}, \tag{8.23}$$

where

$$a_n = \frac{\lambda_n^2}{\beta_n}$$

$$\beta_n = -\frac{dR_n(l)d\eta}{\int_0^1 \eta R_n(\eta) d\eta}. \tag{8.24}$$

The solution (8.23) satisfies all the required conditions of the problem, converges exactly to the steady-state solution for large times, and is approximate to the extent that it satisfies an integrated form of the energy equation, that is, Equation 8.20.

The wall heat flux can be evaluated from Fourier's law,

$$q_w'' = k \left( \frac{\partial T}{\partial r} \right)_{r=r_0} ,$$

(8.25)

which yields

$$\frac{q_w'' r_0}{K(T_w - T_i)} = -\sum_{n=0}^{\infty} C_n \frac{dR_n(l)}{d\eta} \begin{cases} e^{-\beta_n \tau}, & \tau \leq a_n \xi \\ e^{-\lambda_n^2 \xi}, & \tau \geq a_n \xi \end{cases} ,$$

(8.26)

where $q_w''$ is defined as the rate of heat transfer per unit surface area to the fluid at the wall. Figure 8.2 gives, together with the results of Chen et al. [25] by finite differences, the dimensionless wall heat flux from Equation 8.26 versus $\tau$ for various values of $\xi$ as computed by Siegel [8].

A quantity of practical importance is the time required to reach the steady-state conditions at any location $x$. The steady-state time, $\tau_s = \alpha t_s / r_0^2$, is given in Figure 8.3 as a function of $\xi$, where $\tau$ is defined as the time required for the local heat flux to approach within 5% of the value reached for infinite time. Two lines are also drawn in Figure 8.3. The line $\tau_s = \xi$ represents a lower bound on the steady-state time and is determined by the fact that the heat-transfer process cannot be stabilized at a location $x$ until a time of at least $x/U_{max}$ has elapsed. The upper line is obtained from the slug flow solution, which gives a steady-state time of $t_s = x/U_m$ or $\tau_s = 2\xi$. As Figure 8.3 shows, the slug flow solution underestimates the steady-state times for small values of $\xi$ and overestimates them for large $\xi$. Physically, for small values of $\xi$, the establishment of a steady state depends on the convection process in the thin thermal boundary layers near the wall, where the velocities are smaller than indicated by the slug flow approximation, and accordingly the slug flow solution yields lower

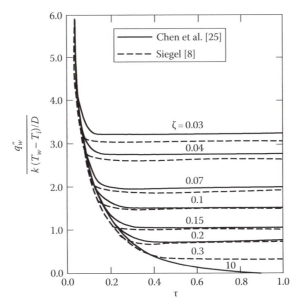

**FIGURE 8.2**
Transient variation in wall heat flux following a step change in wall temperature for fully developed steady laminar flow in a circular duct.

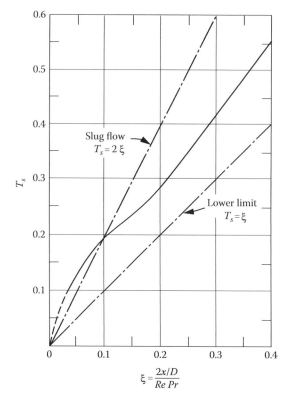

$$\xi = \frac{2x/D}{Re\ Pr}$$

**FIGURE 8.3**

Time to reach steady state after a step change in wall temperature for fully developed steady laminar flow in a circular duct. (From Siegel, R., *Trans. ASME, J. Appl. Mech.*, 82E, 241, 1960.)

steady-state times. On the other hand, for large values of $\xi$, heat has already penetrated all the way across the tube, and the fluid temperature near the wall is already close to the temperature of the tube wall. In this region, the establishment of a steady state is evidently more dependent on the velocities in the central portion of the tube cross section, which are, in fact, higher than the slug flow velocity. The steady-state times are therefore overestimated by the slug flow solution.

The transient heat-transfer problem considered here has also been solved in Reference 42 using an integral method. However, in this reference, only the thermal entrance region is considered and the results do not extend far down the tube. On the other hand, although the present series solutions, Equations 8.23 and 8.26, are valid for the entire length of the tube, many terms are required in the calculations for regions very close to the tube entrance. Thus, the results presented here and those of Reference 5 can be used simultaneously to obtain information for all positions along the tube length.

### 8.2.2 Transient Laminar Forced Convection in Circular Ducts with Arbitrary Time Variations in Wall Temperature

In the previous section, results describing the transient behavior following a step change in wall temperature have been presented. Since the energy equation (8.1) is linear, by using a superposition technique, these results can be generalized to apply for arbitrary time variations in wall temperature.

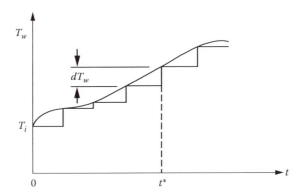

**FIGURE 8.4**
Representation of an arbitrary time-dependent wall temperature.

Let $T_w(t)$ represent the wall temperature as some arbitrary function of time. At any instant, $T_w$ is spatially uniform over the duct wall. As illustrated in Figure 8.4, this arbitrary wall temperature variation can be visualized as a series of differential steps. The effects of those steps can then be superposed to determine the response for an arbitrary variation in $T_w$.

First, consider a system isothermal at $T_i$, and let the tube wall be given a step change $dT_w$ in temperature at time $t^*$. Then, from the result (8.26), the wall heat-flux response to this differential step will be given by

$$dq_w'' = -\frac{k}{r_0} \sum_{n=0}^{\infty} C_n \frac{dR_n(l)}{d\eta} \begin{cases} e^{-\beta_n(\tau-\tau^*)}, & 0 < (\tau-\tau^*) \le a_n\xi \\ e^{-\lambda_n^2\xi}, & \tau-\tau^* \ge a_n\xi \end{cases} dT_w, \tag{8.27}$$

where
$\tau^* = t^*/r_0^2$

all other parameters are as defined in the previous section

As explained in Reference 8, when this result is integrated over an arbitrary wall temperature variation, the variation in wall heat flux $q_w''$ is obtained as

$$\frac{q_w''(\xi,\tau)r_0}{k} = -\sum_{n=0}^{N-1} C_n \frac{dR_n(l)}{d\eta} \left[ e^{-\lambda_n^2\xi} \{T_w(\tau-a_n\xi) - T_i\} - \beta_n \int_{\tau-a_n\xi}^{\tau} e^{-\beta_n(\tau-\tau^*)} \{T_w(\tau^*) - T_i\} d\tau^* \right]$$

$$+ \sum_{n=N}^{\infty} \beta_n C_n \frac{dR_n(l)}{d\eta} \int_0^{\tau} e^{-\beta_n(\tau-\tau^*)} \{T_w(\tau^*) - T_i\} d\tau^*, \tag{8.28}$$

where, for a given $\tau$, the value of $N$ is found from the relation

$$a_{N-1}\xi < \tau \le a_N\xi, \tag{8.29}$$

and for $N = 0$ the first summation is defined as

$$\sum_{n=0}^{-1} \equiv 0, \tag{8.30}$$

and also $a_{-1} \equiv 0$. In evaluating the heat flux at a particular location $\xi = \xi_i$ for early times, $\tau$ will be less than all of the $a_n \xi_i$, and therefore, only the second summation from $n = 0$ to $n = \infty$ is needed. For later times, more and more terms are needed from the first summation. For very large times, however, only the first summation needs to be considered.

If the transient starts from an already established initial steady-state heat-transfer situation in which the wall is at a uniform temperature $T_w$ different from the inlet fluid temperature $T_i$, then the heat-transfer behavior can be determined from the previous results by first letting the system go through an initial transient process with a step change in wall temperature from $T_i$ to $T_w$ and keeping the wall temperature at $T_w$ until the steady state is reached. Then the specified transient is initiated, and the results for this part of the computation yield the desired response from the initial steady-state heat-transfer condition.

### 8.2.3 Transient Laminar Forced Convection in Circular Ducts with Step Change in Wall Heat Flux

Attention is again directed to a hydrodynamically fully developed steady laminar flow through a circular duct of radius $r_0$ (see Figure 8.1), where the tube wall and the fluid are initially isothermal at $T_i$, but the duct wall is subjected to a constant heat flux $q_w''$ for $x \geq 0$ and for times $t > 0$. In this case, the energy equation in dimensionless form is also given by Equation 8.12; however, the dimensionless temperature $\theta(\xi, \eta, \tau)$ is defined as

$$\theta = \frac{T - T_i}{q_w'' r_0 / k}, \tag{8.31}$$

whereas the definitions of the other dimensionless quantities are as before. The inlet and the boundary conditions are then given accordingly by

$$\theta(\xi, \eta, 0) = 0, \quad \theta(\xi, \eta, \tau) = 0, \tag{8.32a,b}$$

$$\left(\frac{\partial \theta}{\partial \eta}\right)_{\eta=0} = 0, \quad \left(\frac{\partial \theta}{\partial \eta}\right)_{\eta-1} = 1. \tag{8.32c,d}$$

Chen et al. [25] solved this problem numerically by the finite-difference method, and Figure 8.5 shows their results for the transient wall temperature distribution versus the dimensionless time $\tau$ for various values of the dimensionless axial distance $\xi$.

### 8.2.4 Transient Laminar Forced Convection in a Parallel-Plate Channel with Step Change in Wall Temperature

Consideration is now given to a parallel-plate channel as shown in Figure 8.6. A steady and fully developed laminar flow passes through the channel in the $x$-direction. The channel

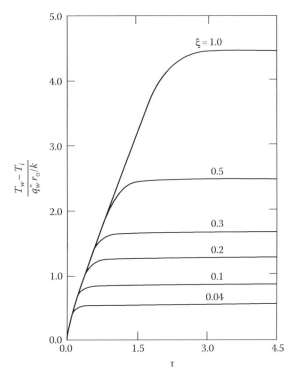

**FIGURE 8.5**
Transient variation in the wall temperature following a step change in the wall heat flux for fully developed steady laminar flow through a circular duct of radius $r_0$. (From Chen, S. C. et al., *J. Heat Transfer*, 105, 922, 1983.)

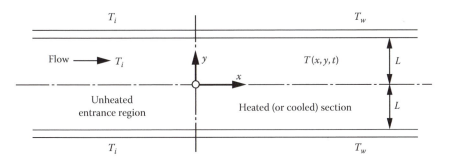

**FIGURE 8.6**
Coordinate system for parallel-plate channel geometry.

walls and the fluid are initially isothermal at temperature $T_i$. The temperature of the channel walls is suddenly changed at $t = 0$ to a new value $T_w$ and maintained at this value for all times thereafter.

The transient temperature distribution $T(x,y,t)$ in the channel for $x \geq 0$ and $t > 0$ will satisfy

$$\frac{\partial T}{\partial t} + u \frac{\partial T}{\partial x} = \frac{\partial^2 T}{\partial y^2}, \tag{8.33}$$

with the following initial, inlet, and boundary conditions:

$$T(x,y,0) = T_i, \quad T(0,y,t) = T_i, \tag{8.34a,b}$$

$$\left(\frac{\partial T}{\partial y}\right)_{y=0} = 0, \quad T(x,L,t) = T_w, \tag{8.34c,d}$$

where the fluid properties have been considered constant and viscous dissipation and axial heat conduction have been neglected. In addition, the velocity distribution is given by

$$\frac{u}{U_m} = \frac{3}{2}\left[1 - \left(\frac{y}{L}\right)^2\right], \tag{8.35}$$

where $U_m$ is the mean flow velocity in the channel.

This problem was first solved by Siegel and Sparrow [6], who obtained an approximate solution in the thermal entrance region using an integral formulation of the energy equation (8.33), together with the use of the method of characteristics. Later, Siegel [8] developed a solution by a method similar to the one discussed in Section 8.2.1 for laminar flow through a circular duct with the parabolic velocity distribution, Equation 8.11. However, instead of using the eigenfunctions of the corresponding Graetz problem in his expressions of the steady-state and transient temperature distributions, he employed the eigenfunctions that would result in the solution of the same problem with the slug flow assumption. Figures 8.7 and 8.8 show his results for the local transient heat flux and the steady-state times $\tau_s = \alpha t_s/L^2$ as a function of distance along the channel length, respectively. In these two figures, the Reynolds number, $Re$, is defined in terms of the hydraulic diameter $D_h = 4L$, where $L$ is the half distance between the plates.

### 8.2.5 Transient Laminar Forced Convection in a Parallel-Plate Channel with Unsteady Flow

Consider again the same parallel-plate channel shown in Figure 8.6. In this section, the transient heat-transfer phenomena that occur in this channel when there are simultaneous changes in fluid pumping pressure and wall heating conditions are discussed. Let there be a hydrodynamic entrance region, so that the flow is always fully developed for $x > 0$. Therefore, in the fully developed region, the velocity distribution, although time-dependent, does not vary with the axial position along the channel. Furthermore, if the fluid is assumed incompressible, then the velocity distribution for $x > 0$ will be governed by

$$\frac{\partial u}{\partial t} = -\frac{1}{\rho}\frac{\partial p}{\partial x} + v\frac{\partial^2 u}{\partial y^2}, \tag{8.36}$$

where $\partial p/\partial x$ is a function of time only.

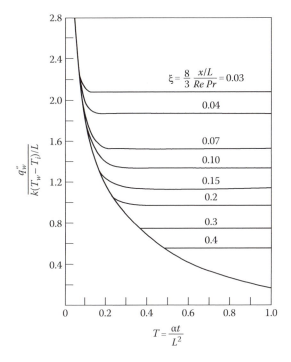

**FIGURE 8.7**

Transient variation in wall heat flux following a step change in wall temperature for fully developed steady laminar flow in a parallel-plate channel. (From Siegel, R., *Trans. ASME, J. Appl. Mech.*, 82E, 241, 1960.)

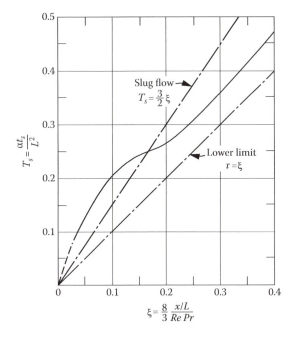

**FIGURE 8.8**

Time to reach steady state after a step change in wall temperature for fully developed steady laminar flow in a parallel-plate channel. (From Siegel, R., *Trans. ASME, J. Appl. Mech.*, 82E, 241, 1960.)

### 8.2.5.1 Step Change in Both Wall Temperature and Pressure Gradient from an Unhealed Initial Condition

Initially, let the flow through the channel be steady with a mean velocity $U_1$ and isothermal at temperature $T_i$. At time $t = 0$, the pressure gradient is abruptly changed so that the fluid velocity undergoes a transient to a new mean value $U_2$. At $t = 0$, when the pressure gradient is changed, the temperature of the bounding walls is also changed to a new value $T_w$ and maintained at this value for $t > 0$.

If the fluid properties are assumed as constants and viscous dissipation and axial conduction are neglected, then the unsteady temperature distribution $T(x,y,t)$ in the channel for $x \geq 0$ and $t > 0$ will satisfy Equation 8.33 together with the conditions of Equations 8.34a,b,c,d. However, the velocity distribution $u$ in Equation 8.33 will be given by the solution of Equation 8.36.

Perlmutter and Siegel [10] obtained an analytical solution of this problem by expanding the transient temperature distribution in a series in the same form as the steady-state solution of the problem. They evaluated the expansion coefficients in their time and axial-coordinate dependence by restricting the expansion to satisfy an integrated form of the energy equation (8.33) and then solving resulting partial differential equation by the method of characteristics. Once they obtained the transient temperature distribution, they calculated the variation of the heat flux $q_w''$ to the fluid at the channel walls from Fourier's law. Figure 8.9 gives their calculations for the variation of the wall heat flux with time for a fluid with $Pr = 0.7$ and for the special case where $U_1 = 0$, that is, initially there is no flow and both the channel and the fluid are isothermal at $T_i$. In this figure, the Reynolds number, $Re$,

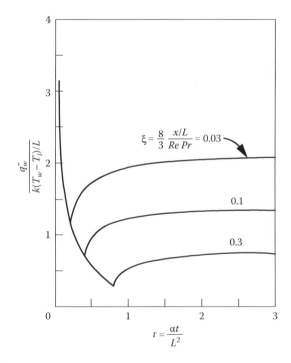

**FIGURE 8.9**
Transient variation in wall heat flux following a step change in pressure gradient and wall temperature. $Pr = 0.7$; $U_1 = 0$. (From Perlmutter, M. and Siegel, R., *Int. J. Heat Mass Transfer*, 3, 94, 1961.)

is defined in terms of the hydraulic diameter $D_h = 4L$ and the mean velocity $U_2$. As seen in Figure 8.9, at each axial location the wall heat flux goes through a minimum and then increases toward the constant steady-state value. The reason for the initial decrease in $q_w''$ is that after the initiation of the transient, the heat transfer at any location proceeds as if the channel were of infinite extent until the fluid particles that were at $x = 0$ at the beginning of the transient reach that location. For this early part of the process, the convective term drops out of the energy equation, and the transient temperature distribution becomes independent of the axial direction.

### 8.2.5.2 Step Change in Pressure Gradient Only, with Initial Steady Heating

Now consider the situation where the transient is caused by a sudden change in the pumping pressure when there is a steady-state heat-transfer process in the channel with the inlet temperature $T_i$ and the temperature of the walls $T_w$. By following an analysis similar to the one in the previous case, Perlmutter and Siegel [10] also developed a solution to this problem, and Figure 8.10 gives their results for the variation of the wall heat flux $q_w''$ for a fluid with $Pr = 0.7$ and for the special case where $U_1 = 0$. As Figure 8.10 shows, since there is no flow and the axial conduction was neglected, there is no heat transfer during the early transient period. After the fluid particles with the mean velocity $U_2$ reach a specific location, then heat transfer begins at that location and the wall heat flux rises toward its steady-state value.

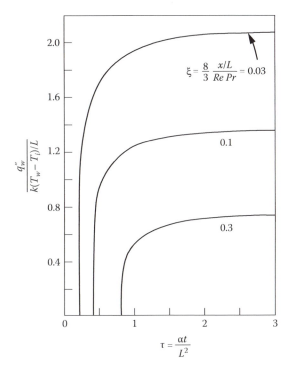

**FIGURE 8.10**
Transient variation in wall heat flux following a step change in pressure gradient with initial steady heating. $Pr = 0.7$; $U_1 = 0$. (From Perlmutter, M. and Siegel, R., *Int. J. Heat Mass Transfer*, 3, 94, 1961.)

### 8.2.5.3 *Step Change in Both Pressure Gradient and Wall Temperature with Initial Steady Heating*

Consider a more general case where the pressure gradient and the wall temperature are suddenly changed to new values when initially there is a steady-state heat-transfer process in the channel with nonzero flow velocity. The resulting transient can be evaluated by a superposition of the solutions to the previous two simpler cases. The details of this superposition are explained in Reference 10, and they will not be repeated here.

## 8.3 Transient Turbulent Forced Convection in Ducts

The literature on transient turbulent forced convection is sparse. Abbrecht and Churchill [43] presented the results of an experimental investigation of heat transfer in the thermal entrance region following a step change in wall temperature in fully developed turbulent flow in a tube. Radial and longitudinal temperature gradients, radial heat fluxes, and eddy diffusivities for heat and momentum were computed from the measurements.

Sparrow and Siegel [44] investigated transient turbulent heat transfer in the thermal entrance region of a tube whose wall temperature varies arbitrarily with time. As a first step, the heat-transfer response to a step jump in wall temperature was analyzed, and then this was generalized by a superposition technique to apply to arbitrary time variations. Use of the generalized results was illustrated by the application to the case where the wall temperature variation was linear with time. The method used permitted the heat-transfer coefficient to vary with time and position in accordance with the energy conservation principle.

Kakaç [45] analyzed transient heat transfer in incompressible turbulent flow between two parallel plates for a step jump in wall heat flux or wall temperature. The variations of the fluid velocity and effective diffusivity over the channel cross section were taken into account. It was assumed that the velocity profile was fully developed throughout the length of the channel. The thermal response of the system was obtained by solving the energy equation for air on a digital computer. The Nusselt number was presented, in the form of graphs, as a function of time and space. A method was also discussed to obtain the velocity distribution from the distribution of the turbulent eddy diffusivity of momentum.

Kakaç in Reference 46 presented a general closed-form solution to the transient energy equation under boundary conditions of zero wall temperature or zero heat flux for the decay of the inlet and initial temperature distributions in an incompressible turbulent flow between two parallel plates. However, the eigenfunctions and the corresponding eigenvalues were left to be determined to complete the solution.

Gartner [47] analyzed the unsteady convective heat transfer in a hydrodynamically stabilized steady turbulent flow of a viscous incompressible fluid in a concentric annulus with the wall heat flux varying with time. The formulation permitted the heat-transfer coefficient to vary with time and position. The energy equation was solved by using the method of superposition and separating variables by finite integral transforms. The use of the generalized results was illustrated by application to the case where the wall heat flux varies exponentially with time.

Kawamura in Reference 48 examined the variation of the heat-transfer coefficient experimentally in a steady turbulent flow through a circular tube cooled by water and heated stepwise with time. In addition, a numerical analysis was made for the same configuration, the results of which agreed well with the experimental findings. Furthermore, an analytical expression for the variation of heat-transfer coefficient was obtained. The time required for the heat-transfer coefficient to reach its steady-state value was also evaluated.

Other important contributions in the field of transient turbulent forced convection are given in the references of the papers cited herein.

### 8.3.1 Transient Turbulent Forced Convection in Circular Ducts

Consider the circular tube shown in Figure 8.1. Let the flow through this tube now be steady, turbulent, and fully developed and the tube wall and fluid be initially isothermal at $T_i$. Assume that the tube wall is given an instantaneous step in temperature (say at time $t = 0$) to reach a new value $T_w$ and maintained at $T_w$ for all times thereafter. The starting point in the analysis is the unsteady energy equation for fully developed turbulent flow in a circular tube:

$$\frac{\partial T}{\partial t} + u \frac{\partial T}{\partial x} = \frac{1}{r} \frac{\partial}{\partial r} \left[ r(\alpha + \varepsilon_h) \frac{\partial T}{\partial t} \right]. \tag{8.37}$$

The initial, inlet, and boundary conditions are given by

$$T(x,r,0) = T_i, \quad T(0,r,t) = T_i, \tag{8.38a,b}$$

$$\left( \frac{\partial T}{\partial r} \right)_{r=0} = 0, \quad T(x,r_0,t) = T_w. \tag{8.38c,d}$$

Following an approach similar to the one discussed in Section 8.2.1, Sparrow and Siegel [44] first obtained the steady-state solution of this problem and then expanded the transient solution about the steady-state conditions, which was only required to satisfy the integrated form of the energy equation. The following is their result for the transient temperature distribution:

$$\frac{T - T_w}{T_i - T_w} = \sum_{n=1}^{\infty} C_n F_n(x^+, t^+) R_n(r^+), \tag{8.39}$$

where

$$F_n = \begin{cases} \exp\left[ \dfrac{\left(r_0^+\right)^3 dR_n\left(r_0^+\right)/dr^+}{pr \displaystyle\int_0^{r_0^+} r^+ R_n dr^+} \right], & t^+ \leq a_n x^+ \\[30pt] \exp\left[ -\dfrac{4\beta_n^2}{Re} x^+ \right], & t^+ \geq a_n x^+ \end{cases} \tag{8.40}$$

Here, $\beta_n$ and $R_n(r^+)$ are the eigenvalues and eigenfunctions, respectively, of the following eigenvalue problem:

$$\frac{d}{dr^+}\left(r^+\gamma\frac{dR_n}{dr^+}\right)+\left(\frac{2\beta_n^2}{Re}\frac{r^+}{r_0^+}u^+\right)R_n=0, \tag{8.41a}$$

$$\frac{dR_n(0)}{dr^+}=0, \quad R_n\left(r_0^+\right)=0, \tag{8.41b,c}$$

and

$$C_n=\frac{\int_0^{r_0^+}r^+u^+R_n\,dr^+}{\int_0^{r_0^+}r^+u^+R_n^2\,dr^+}. \tag{8.42}$$

Various quantities in the previous equations are defined as follows:

$$x^+=\frac{x}{2r_0}, \quad r^+=r\frac{\sqrt{\tau_w/\rho}}{v}, \quad r_0^+=r_0\frac{\sqrt{\tau_w/\rho}}{v}, \quad u^+=\frac{u}{\sqrt{\tau_w/\rho}}, \quad t^+=\frac{vt}{r_0^2},$$

where $\tau_w$ is the wall shear stress

$$\gamma=\frac{\alpha+\varepsilon_h}{v}$$

Once the temperature distribution (8.39) is available, the wall heat-flux variation for the entire transient period can be calculated from Fourier's law, and the result is

$$\frac{q_w''r_0}{(T_w-T_i)k}=-r_0^+\sum_{n=0}^{\infty}C_nF_n(x^+,t^+)\frac{dR_n\left(r_0^+\right)}{dr^+}. \tag{8.43}$$

Sparrow and Siegel [44] evaluated Equation 8.43 for several combinations of Reynolds number and Prandtl number. Figure 8.11 is a representative result from Reference 44, where the heat-transfer responses at various positions ranging from $x/D = 2$ to $x/D = 100$ are given for $Pr = 0.7$ and $Re = 100,000$. At any position along the tube, initially the heat transfer is only by pure diffusion and follows the envelope curve, decreasing with increasing time. Then, at a certain time, for example, at $t^+ = 0.00078$ for $x/D = 20$, convection begins to act and the curve breaks away from the pure-diffusion envelope, with the heat transfer continuing to decrease until the horizontal steady state is reached. Sparrow and Siegel [44] in their numerical evaluations used the following correlations given in Reference 3:

$$\frac{du^+}{dy^+}=\left\{1+(0.124)^2u^+y^+\left[1-e^{-(0.124)^2u^+y^+}\right]\right\}^{-1}, \quad 0\le y^+\le 26, \tag{8.44}$$

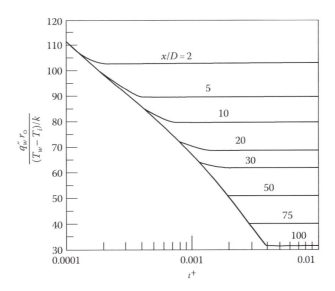

**FIGURE 8.11**
Wall heat-flux response to a step change in wall temperature for fully developed turbulent flow through a circular duct. $Pr = 0.73$; $Re = 10^5$. (From Sparrow, E. M. and Siegel, R., *Trans. ASME, J. Heat Transfer*, 82C, 170, 1960.)

$$u^+ = \frac{1}{0.36}\ln\left(\frac{y^+}{26}\right)+12.8493, \quad y^+ \geq 26, \tag{8.45}$$

where $y^+ = r_0^+ - r^+$. The total diffusivity was evaluated from Reference 49

$$\gamma = \frac{1}{Pr}+(0.124)^2 u^+ y^+\left[1-e^{-(0.124)^2 u^+ y^+}\right], \quad 0 \leq y^+ \leq 26, \tag{8.46}$$

$$\gamma = \frac{1}{Pr}+0.36 y^+\left(1-\frac{y^+}{y_0^+}\right)-1, \quad y^+ > 26. \tag{8.47}$$

The value of $\gamma$ at $y^+ = 26$ was taken as the average of Equations 8.46 and 8.47. The −1 appearing on the right-hand side of Equation 8.47 was retained for $26 < y^+ < r_0^+/2$ and deleted for larger values of $y^+$.

The steady-state times, $t_s$, defined as the time period required for the heat transfer to come to within 5% of the steady-state value were also calculated as a function of position by Sparrow and Siegel [44]. Their results are given here in Figure 8.12. As this figure shows, the steady-state time decreases as the Prandtl number increases, but by no more than a factor of 3 for this Prandtl number range. The Reynolds number also has a significant effect on the steady-state times, which are approximately in inverse proportion to the Reynolds number. Also appearing in this figure are two straight dashed lines corresponding to the time $x/U_m$, which approximately represent the time at which the heat-transfer process at any position $x$ begins to be influenced by the convection of fluid from the tube entrance.

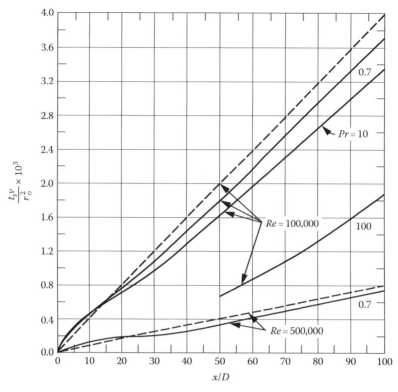

**FIGURE 8.12**
Steady-state times after a step change in wall temperature for fully developed turbulent flow through a circular duct. (From Sparrow, E. M. and Siegel, R., *Trans. ASME, J. Heat Transfer*, 82C, 170, 1960.)

This simple relation $t_s = x/U_m$, or $\tau_s = 4(x/D)/Re$ in dimensionless form, gives a fairly good estimate for a Prandtl number around unity, but tends to overestimate the steady-state times as the Prandtl number increases. However, for the purpose of providing an order-of-magnitude estimate, $x/U_m$ appears to be useful.

### 8.3.2 Transient Turbulent Forced Convection in a Parallel-Plate Channel

Kakaç [45] made a numerical analysis by finite differences of transient forced convection for a hydrodynamically fully developed incompressible steady turbulent flow in a parallel-plate channel when there is a step change in wall heat flux or wall temperature. He used experimentally determined values for the eddy diffusivities of momentum and heat in his calculations and presented the variation of the Nusselt number as a function of time and axial position along the channel. Figures 8.13 and 8.14 show two of his calculations for the local Nusselt number for air ($Pr = 0.73$) at two locations along the channel and for two Reynolds numbers with step changers in wall temperature and in wall heat flux, respectively. In these figures, $q_w''$ is the wall heat flux to the fluid, $T_b$ is the fluid bulk mean temperature, and $D_h = 4L$, where $L$ is the half distance between the channel walls.

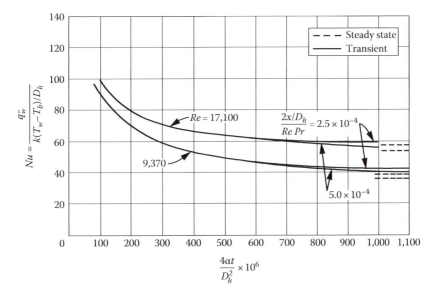

**FIGURE 8.13**
Transient Nusselt numbers for a step change in wall temperature for fully developed turbulent flow in a parallel-plate channel. $Pr = 0.73$. (From Kakaç, S., *Warme Stoffübertragung*, 1, 169, 1968.)

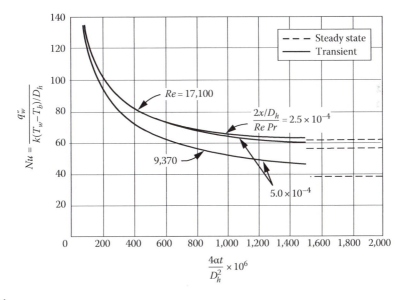

**FIGURE 8.14**
Transient Nusselt numbers for a step change in wall heat flux for fully developed turbulent flow in a parallel-plate channel. $Pr = 0.73$. (From Kakaç, S., *Warme Stoffübertragung*, 1, 169, 1968.)

## 8.4 Analysis of Transient Forced Convection for Timewise Variation of Inlet Temperature

The periodic thermal response of duct flows to imposed cyclic variations in thermal conditions has also been investigated. Sparrow and De Farias [27] made an analysis of unsteady laminar heat transfer in a parallel-plate channel with periodically varying inlet temperature. The midplane of each wall was considered insulated, and the wall temperature was dynamically determined by a balance of heat transfer and energy storage. In the analytical formulation, the commonly used quasi-steady assumption was lifted in favor of the local application of the energy equation, the solution of which involved an eigenvalue problem with complex eigenvalues and eigenfunctions. Numerical evaluation of the analytical results provided the time and space dependence of the wall and bulk temperatures and of the Nusselt number. In addition, results for the overall performance of the channel as a heat exchanger were presented in terms of the energy carried across the exit cross section relative to that carried across the entrance section. For comparison purposes, results for the overall performance were also derived by using the quasi-steady model. It was found that for a range of operating conditions the quasi-steady model was able to give accurate performance predictions, especially when it was used in conjunction with spatially varying heat-transfer coefficients.

Kakaç and Yener [28] obtained an exact solution to the transient energy equation for laminar slug flow of an incompressible fluid in a parallel-plate channel with time-varying inlet temperature. The results were confirmed experimentally by the frequency response method for a limited range of Reynolds number.

Acker and Fourcher [50] studied the laminar flow in a storage unit in thermal periodic regime. The energy equations were solved simultaneously both for the wall and for the fluid flow between two parallel plates by Laplace transforms with the slug flow assumption and when there is a sinusoidal variation in the inlet fluid temperature.

Sucec and Sawant [51] made a study of the unsteady, conjugated laminar forced convection in a parallel-plate channel with periodically varying inlet fluid temperature. They obtained the wall and the fluid bulk temperatures as a function of distance along the channel and of time for a sinusoidal inlet temperature variation, the channel walls being adiabatic on their outside surfaces and communicating thermally with the fluid across their inside surfaces.

Kim and Özisik [52] discussed the turbulent forced convection inside a parallel-plate channel with periodically varying inlet temperature under constant wall temperature boundary condition.

Cotta and Özisik [30], Kakaç et al. [31–33,35], Ding [34], and Li [39] also studied unsteady forced convection in ducts theoretically and experimentally for periodic variations of inlet temperature under various boundary conditions. Some of these solutions are discussed in the following sections.

### 8.4.1 Heat Transfer in Laminar Slug Flow through a Parallel-Plate Channel with Periodic Variation of Inlet Temperature

The parallel-plate channel under consideration is shown in Figure 8.15. The fluid entering the heated section has a temperature that is spatially uniform across the entrance section but varies sinusoidally with time as

$$T(0, y, t) = T_0 + (\Delta T)_0 \sin 2\pi \beta t, \tag{8.48}$$

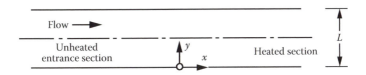

**FIGURE 8.15**
Coordinate system for parallel-plate channel.

where
$T_0$ is the cycle mean temperature
$(\Delta T)_0$ is the amplitude
$\beta$ is the frequency of the inlet temperature variation

The following idealizations are made in the analysis:

1. Flow between the plates is steady, fully developed, and laminar.
2. Viscous dissipation is negligible.
3. Axial conduction is negligible with respect to bulk transport in the $x$-direction. This is a reasonable assumption when the Peclet number exceeds 100.
4. Fluid properties are constant.
5. Thermal resistance of the channel walls is negligible.

The starting point of the analysis is again the unsteady energy equation for a fully developed laminar flow in a parallel-plate channel, which can be written as

$$\frac{\partial \theta}{\partial t} + u \frac{\partial \theta}{\partial x} = \alpha \frac{\partial^2 \theta}{\partial y^2}, \tag{8.49}$$

where

$$\theta(x,y,t) = \frac{T(x,y,t) - T_0}{(\Delta T)_0}, \tag{8.50}$$

with the following inlet and boundary conditions:

$$\theta(0,y,t) = \sin 2\pi\beta t, \tag{8.51a}$$

$$\left(\frac{\partial \theta}{\partial y}\right)_{y=0} = 0, \quad \left(k\frac{\partial \theta}{\partial y} + h\theta\right)_{y=L} = f(x), \tag{8.51b,c}$$

where the function $f(x)$ is given in Table 8.1 for various boundary conditions at $y = L$. One obtains the temperature boundary condition at $y = L$ by setting $k = 0$ and $h = 1$ and the heat-flux boundary condition by setting $h = 0$. When $h$ and $k$ are finite, Equation 8.51c means that the boundary at $y = L$ is losing heat by convection to the environment at temperature $T_\infty(x)$.

The foregoing problem can be separated into two as follows:

$$\theta(x,y,t) = \theta_1(x,y) + \theta_2(x,y,t), \tag{8.52}$$

**TABLE 8.1**

Function $f(x)$

| Boundary Condition at $y = L$ | Function $f(x)$ |
|---|---|
| First kind ($k = 0, h = 1$) | $\dfrac{T_w(x) - T_0}{(\Delta T)_0}$ |
| Second kind ($h = 0$) | $\dfrac{q_w''(x)}{(\Delta T)_0}$ |
| Third kind ($k$ and $h$ finite) | $h\dfrac{T_\infty - T_0}{(\Delta T)_0}$ |

where $\theta_1(x,y)$ and $\theta_2(x,y,t)$ are solutions of the following problems:

$$u\frac{\partial \theta_1}{\partial x} = \alpha\frac{\partial^2 \theta_1}{\partial y^2}, \tag{8.53a}$$

with

$$\theta_1(0, y) = 0, \tag{8.53b}$$

$$\left(\frac{\partial \theta_1}{\partial y}\right)_{y=0} = 0, \quad \left(k\frac{\partial \theta_1}{\partial y} + h\theta_1\right)_{y=L} = f(x), \tag{8.53c,d}$$

and

$$\frac{\partial \theta_2}{\partial t} + u\frac{\partial \theta_2}{\partial x} = \alpha\frac{\partial^2 \theta_2}{\partial y^2}, \tag{8.54a}$$

with

$$\theta_2(0, y, t) = \sin 2\pi\beta t, \tag{8.54b}$$

$$\left(\frac{\partial \theta_2}{\partial y}\right)_{y=0} = 0, \quad \left(k\frac{\partial \theta_2}{\partial y} + h\theta_2\right)_{y=L} = 0 \tag{8.54c,d}$$

To simplify the method of analysis, the velocity profile $u$ across the entire flow area of the channel will be taken constant (i.e., slug flow idealization).

### 8.4.1.1 Solution for $\theta_1(x,y)$

The solution of the problem given by Equations 8.53 can be written as [28]

$$\theta_1(x, y) = \sum_{n=1}^{\infty} \frac{\cos \lambda_n y}{N_n} \int_0^x e^{-(\alpha\lambda_n^2/u)(x-x')} A_n(x')dx', \tag{8.55}$$

**TABLE 8.2**

Eigenvalues

| Boundary Condition at $y = L$ | $\lambda_n$ |
|---|---|
| First kind ($k = 0$, $h = 1$) | $\lambda_n = \dfrac{2n-1}{L}\dfrac{\pi}{2}$, $\quad n = 1, 2, \ldots$ |
| Second kind ($h = 0$) | $\lambda_n = \dfrac{n-1}{L}\pi$, $\quad n = 1, 2, \ldots$ |
| Third kind ($k$ and $h$ finite) | Positive roots of $\lambda_n \tan \lambda_n L = h/k$ |

**TABLE 8.3**

Function $A_n(x)$ in Equation 8.55

| Boundary Condition at $y = L$ | $A_n(x)$ |
|---|---|
| First kind | $\dfrac{\alpha \lambda_n}{u} f(x) \sin \lambda_n$ |
| Second and third kinds | $\dfrac{\alpha}{uk} f(x) \cos \lambda_n$ |

where

$\lambda_n$ and $A_n(x)$ are given in Tables 8.2 and 8.3

$N_n$ is defined by

$$N_n = \frac{L}{2} + \frac{1}{4\lambda_n} \sin 2L\lambda_n. \tag{8.56}$$

### 8.4.1.2 Solution for $\theta_2(x,y,t)$

The solution of the problem given by Equations 8.54 can be written as [28]

$$\theta_2(x,y,t) = \begin{cases} \sin\left[2\pi\beta\left(t - \dfrac{x}{u}\right)\right] \displaystyle\sum_{n=1}^{\infty} \dfrac{\cos\lambda_n y \sin\lambda_n L}{\lambda_n N_n} e^{-\left(\alpha\lambda_n^2/u\right)x}, & h \neq 0 \\[4mm] \sin\left[2\pi\beta\left(t - \dfrac{x}{u}\right)\right], & h = 0 \end{cases} \tag{8.57}$$

When the boundary condition for $\theta(x,y,t)$ at $y = L$ is homogeneous, that is, when $f(x) = 0$, then $\theta_1(x,y)$ becomes identically zero, and in that case,

$$\theta(x,y,t) = \theta_2(x,y,t). \tag{8.58}$$

When the temperature or the convection boundary condition is homogeneous, the walls lose heat in such a way that each mode of $\theta_2(x,y,t)$ decays exponentially along the duct and this decay is inversely proportional to the velocity $u$. Therefore, as the velocity $u$ is increased, the rate of decay decreases. It is also seen that the phase lag along the tube is linear with the slope $2\pi\beta/u$, and as the velocity $u$ is increased, this slope decreases.

When the heat-flux boundary condition is homogeneous, there will be no heat conduction in the $y$-direction. Since the axial diffusion of heat has already been neglected, the amplitude of $\theta_2(x,r,t)$ remains constant. The phase lag, however, is the same as in the other two cases because of the convention in the $x$-direction.

### 8.4.2  Heat Transfer in Laminar Flow through a Parallel-Plate Channel with Periodic Variation of Inlet Temperature

We now consider transient laminar forced convection inside a parallel-plate channel with hydrodynamically fully developed flow subjected to periodic variations in the inlet temperature. The geometry for the analysis is again as shown in Figure 8.15. Neglecting axial conduction of heat and free convection effects and assuming constant thermophysical properties for the fluid, the energy equation governing the diffusion in the $y$-direction and the convection in the $x$-direction can be written as

$$\frac{\partial T}{\partial t} + u(y)\frac{\partial T}{\partial x} = \alpha\frac{\partial^2 T}{\partial y^2}, \quad \text{for } 0 < y < L, x > 0, t > 0, \tag{8.59a}$$

with the inlet and boundary conditions given, respectively, by

$$T(0,y,t) = T_\infty + \Delta T(y)e^{i2\pi\beta t}, \quad \text{for } 0 < y < L, \tag{8.59b}$$

$$\left(\frac{\partial T}{\partial y}\right)_{y=0} = 0, \quad \text{for } x > 0. \tag{8.59c}$$

On the duct wall at $y = L$, a boundary condition of the fifth kind that accounts for both external convection and wall heat capacitance effects is imposed as

$$h(T - T_\infty) + k\frac{\partial T}{\partial y} + (\rho c)_w \ell\frac{\partial T}{\partial t} = 0, \quad \text{at } y = L, \quad \text{for } x > 0, \tag{8.59d}$$

where
  $h$ is the heat-transfer coefficient between the outer surface of the duct wall and the environment that is at temperature $T_\infty$
  $\ell$ represents the thickness of the wall and is considered to be thin enough to neglect the variation of the wall temperature in the transverse direction

In addition, $(\rho c)_w$ is the thermal capacitance of the wall.
  Kakaç et al. [33,35] obtained an analytical solution to this problem for periodic thermal response by following the method introduced in Reference 30. They presented results for the variations of the dimensionless bulk and centerline fluid temperatures along the duct for various values of the Biot number, $Bi = hL/k$; the fluid-to-wall thermal capacitance ratio, $\alpha^* = (\rho c_p)_f L/(\rho c)_w \ell$; and the dimensionless inlet frequency, $\Omega = 2\pi\beta L^2/\alpha$. Figures 8.16 and 8.17 illustrate the effects of the Biot number on the amplitude and phase lag of bulk temperatures as functions of the dimensionless axial distance with $a^* = \infty$ (negligible wall thermal capacitance) and $\Omega = 0.1$. It is clearly seen that for $Bi \geq 0$, the results will not differ

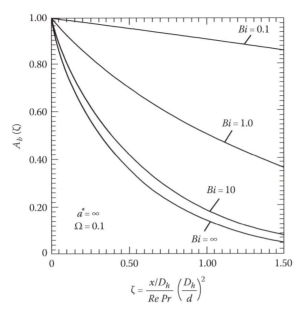

**FIGURE 8.16**
Amplitudes of dimensionless bulk temperature along the duct for various values of Biot number ($a^* = \infty$, $\Omega = 0.1$).

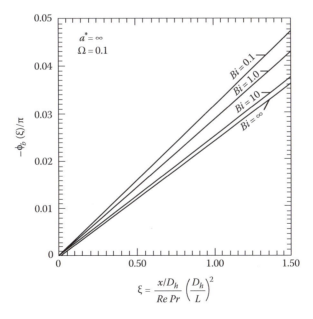

**FIGURE 8.17**
Phase lag of dimensionless bulk temperature along the duct for various values of Biot number ($a^* = \infty$, $\Omega = 0.1$).

significantly from those for $Bi = \infty$, especially for phase lags. Therefore, under experimental air flow situations when $Bi$ is expected to be large, its precise determination might be of limited relevance.

Figures 8.18 and 8.19 illustrate the effects of the fluid-to-wall thermal capacity ratios on the bulk temperature amplitudes and phase lags for $Bi = 10$ and $\Omega = 0.1$. As seen, only for large wall capacitances (or small $a^*$), the storage of heat at the wall itself will be of some importance to the fluid bulk temperature evolution along the duct. The effect will be much more significant for the phase lags. It should be noted that although amplitudes are practically unchanged for $a^* \geq 0.1$, some time phase lags are already present in the fluid temperature due to the presence of the wall.

The effect of the dimensionless frequency, $\Omega$, on the variation of the amplitude and phase lag along the channel is presented in Figures 8.20 and 8.21 for $Bi = 10$ and $a^* = 1.0$. The deviations in amplitudes for different frequencies are practically unnoticeable within the range of $\Omega$ studied. The phase lags, however, experience a more significant change and demonstrate an almost linear relationship with the frequency. Also, the four curves in Figure 8.21 practically collapse to one when the phase lags are divided by the corresponding value of $\Omega$. This behavior allows one to characterize the thermal response of a system for most practical purposes, subjected to different excitation frequencies, by a single set of results, namely, a set of amplitudes and normalized phase lags in terms of the duct length.

An experimental technique to study the decay of sinusoidal thermal inlet conditions for laminar (as well as for turbulent) forced convection in a parallel-plate channel is discussed in References 31,32,34. Figure 8.22 shows some typical experimental results for the decay of the amplitude of the centerline temperature along the channel in the thermal entrance region at a fixed Reynolds number $Re \cong 650$ and for different inlet frequencies.

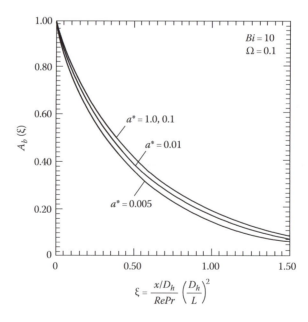

**FIGURE 8.18**
Amplitudes of dimensionless bulk temperature along the duct for various values of fluid-to-wall thermal capacitance ratio ($Bi = 10$, $\Omega = 0.1$).

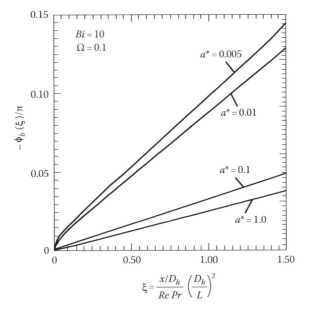

**FIGURE 8.19**
Phase lag of dimensionless bulk temperature along the duct for various values of fluid-to-wall thermal capacitance ratio ($Bi = 10$, $\Omega = 0.1$).

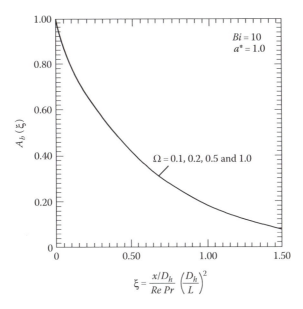

**FIGURE 8.20**
Amplitudes of dimensionless bulk temperature along the duct for various values of dimensionless inlet frequencies ($Bi = 10$, $a^* = 1.0$).

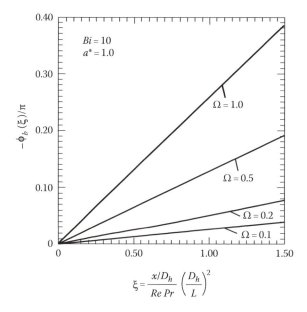

**FIGURE 8.21**
Phase lag of dimensionless bulk temperature along the duct for various values of dimensionless inlet frequencies ($Bi = 10$, $a^* = 1.0$).

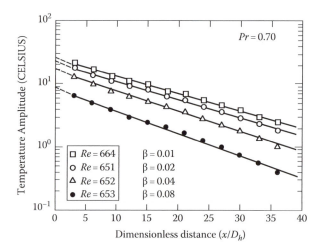

**FIGURE 8.22**
Variation of centerline temperature amplitude along the duct for $Re \cong 650$ and $\beta \cong 0.01$–$0.08$ Hz.

The analytical solutions presented in References 33,35 indicate that the fluid temperature response to periodic variation of the inlet temperature is a superposition of a series of periodic modes, each of which decays exponentially with distance along the duct, as is the case with the solution (8.57) for slug flow assumption. At a given inlet frequency, the higher modes decay so fast that ultimately only the basic mode remains in the solution. Figure 8.22 demonstrates this effect experimentally by exhibiting linear variations of the centerline temperature amplitudes on semilog scale along the major portion of the channel, except in regions very close to the inlet.

### 8.4.3 General Solution to the Transient Forced Convection Energy Equation for Timewise Variation of the Inlet Temperature

In this section, following the work of Kakaç [46], formal solutions for the decay of a periodically varying inlet temperature in a fully developed turbulent flow between two parallel plates with linear and homogeneous boundary conditions are given.

Consider a steady, fully developed turbulent flow through a parallel-plate channel whose walls are separated by a distance $L$ as shown in Figure 8.15. Neglecting axial diffusion and viscous dissipation and assuming constant fluid properties, the energy equation governing the conduction (in the $y$-direction) and the convection (in the $x$-direction) can be written as

$$\frac{\partial T}{\partial t} + u(y)\frac{\partial T}{\partial x} = \frac{\partial}{\partial y}\left(D(y)\frac{\partial T}{\partial y}\right),\tag{8.60}$$

where
$u(y)$ is the fully developed velocity profile
$D(y)$ is the effective diffusivity, which is assumed to be a function of $y$ only

Suppose that the system satisfying Equation 8.60 is subject to a periodic inlet condition of the form

$$T(0,y,t) = e^{i2\pi\beta t}, \quad i = \sqrt{-1}\tag{8.61}$$

and two linear homogeneous boundary conditions of the following general forms:

$$\left(a_1 T + b_1 \frac{\partial T}{\partial y}\right)_{y=0} = 0,\tag{8.62a}$$

$$\left(a_2 T + b_2 \frac{\partial T}{\partial y}\right)_{y=L} = 0,\tag{8.62b}$$

where $a_i$ and $b_i$, $i = 1, 2$ are given real constants. If a periodic solution of the form

$$T(x,y,t) = e^{i2\pi\beta t}X(x)Y(y)\tag{8.63}$$

for the decay of the inlet condition of Equation 8.60 is assumed, then it can be shown that the solution for $T(x,y,t)$ is given by Reference 46

$$T(x,y,t) = \sum_{n=1}^{\infty} c_n e^{-\alpha_n x}[P_n(y)\cos 2\pi(\beta t - \delta_n x) - Q_n(y)\sin 2\pi(\beta t - \delta_n x)]$$

$$+ i\sum_{n=1}^{\infty} c_n e^{-\alpha_n x}[P_n(y)\sin 2\pi(\beta t - \delta_n x)Q_n(y)\cos 2\pi(\beta t - \delta_n x)],\tag{8.64}$$

where

$P_n(y)$ and $Q_n(y)$ are the eigenfunctions

$\alpha_n$ are the eigenvalues of the following coupled eigenvalue problem

$$\frac{d}{dy}\left(D\frac{dP_n}{dy}\right) = -\alpha_n u P_n + (\delta_n u - \beta)Q_n, \tag{8.65a}$$

$$\frac{d}{dy}\left(D\frac{dQ_n}{dy}\right) = -\alpha_n u Q_n - (\delta_n u - \beta)P_n, \tag{8.65b}$$

with

$$a_1 P_n(0) + b_1 \frac{dP_n(0)}{dy} = 0, \quad a_1 Q_n(0) + b_1 \frac{dQ_n(0)}{dy} = 0, \tag{8.66a,b}$$

$$a_2 P_n(L) + b_2 \frac{dP_n(0)}{dy} = 0, \quad a_2 Q_n(L) + b_2 \frac{dQ_n(L)}{dy} = 0, \tag{8.66c,d}$$

and

$$\delta_n = \beta \frac{\displaystyle\int_0^L \left(P_n^2 + Q_n^2\right)dy}{\displaystyle\int_0^L u\left(P_n^2 + Q_n^2\right)dy}. \tag{8.67}$$

As an example, let the system satisfying Equation 8.60 be subjected to a periodic inlet condition given by

$$T(0,y,t) = T_m + (\Delta T)_0 \sin 2\pi\beta t \tag{8.68}$$

and two linear homogeneous boundary conditions of the following forms:

$$\left(\frac{\partial T}{\partial y}\right)_{y=0} = 0 \quad \text{and} \quad T(x,L) = T_w \tag{8.69a,b}$$

The solution will then be given by

$$\frac{T - T_m}{(\Delta T)_0} = \theta_1(x,y) + \sum_{n=1}^{\infty} C_n e^{-\alpha_n x} \sqrt{P_n^2(y) + Q_n^2(y)} \sin 2\pi(\beta t - \delta_n x + \varepsilon_n), \tag{8.70}$$

where

$$\varepsilon_n = \tan^{-1}\frac{Q_n(y)}{P_n(y)}, \tag{8.71}$$

$\theta_1(x,y)$ satisfies the following problem:

$$u\frac{\partial\theta_1}{\partial x} = \frac{\partial}{\partial y}\left(D\frac{\partial\theta_1}{\partial y}\right),$$ (8.72)

$$\theta_1(0,y) = 0,$$ (8.73a)

$$\left(\frac{\partial\theta_1}{\partial y}\right)_{y=0} = 0, \quad \theta_1(x,L) = \frac{T_w - T_m}{(\Delta T)_0}.$$ (8.73b,c)

In regions away from the inlet, only the first term in the series in Equation 8.70 needs to be considered. Hence, the asymptotic solution, deleting the subscript 1, becomes

$$\frac{T - T_m}{(\Delta T)_0} = \theta_1(x,y) + ce^{-\alpha_n x}\sqrt{P^2(y) + Q^2(y)}\sin 2\pi(\beta t - \delta_n x + \varepsilon_n).$$ (8.74)

It is to be noted that solutions developed so far are also valid for laminar flow.

The form of Equation 8.74 suggests that the results can best be confirmed experimentally by the frequency response method, and the first values of the eigenvalues $\alpha_n$ and $\delta_n$ can be determined for various values of the inlet frequency for a wide range of the Reynolds number.

An experimental setup can be designed and used to study the decay of sinusoidal inlet conditions for turbulent forced convection in various channel geometries and to obtain experimentally the first eigenvalue and other parameters appearing in the general solutions by the frequency analysis.

At a fixed Reynolds number, the changes of amplitudes and phases of temperature waves along the channel can be measured. Phase data can be taken with respect to the thermocouple nearest to the inlet heater. An experiment can be carried out for the different frequency values of the sinusoidal variation of heat input to the inlet heater.

From the series of such measurements, the coefficients $\alpha$ and $\delta$ can be measured as a function of the Reynolds number, inlet frequency, and distance along the channel. The results of the general solution given in this section were confirmed experimentally by the frequency response method for a limited range of Reynolds number [42].

## Problems

**8.1**  Obtain the solution given by Equation 8.7.

**8.2**  Derive Equation 8.20.

**8.3**  Obtain the solution given by Equation 8.23.

**8.4**  Solve the following transient problem:

$$\frac{\partial T}{\partial t} + u\frac{\partial T}{\partial x} = \alpha\frac{1}{r}\frac{\partial}{\partial r}\left(r\frac{\partial T}{\partial r}\right),$$

with

$$T(x,r,0) = T_i; \quad T(0,r,t) = T_i,$$

$$\left(\frac{\partial T}{\partial r}\right)_{r=0} = 0; \quad \left(\frac{\partial T}{\partial r}\right)_{r=r_0} = \frac{q_w''}{k}.$$

Assume that $u$ = constant.

**8.5** Solve the transient problem defined by the energy equation (8.33) and the conditions (8.34a,b,c,d) by assuming that $u$ = constant.

**8.6** Resolve Problem 8.5 for the following boundary condition at $y = L$:

$$\left(\frac{\partial T}{\partial y}\right)_{y=L} = \frac{q_w''}{k}.$$

**8.7** Obtain the solution given by Equation 8.55.

**8.8** Obtain the solution given by Equation 8.57.

**8.9** Obtain the solution given by Equation 8.64.

## References

1. Dusinberre, G. M., Calculation of transient temperatures in pipes and heat exchangers by numerical methods, *Trans. ASME*, 76, 421–426, 1954.
2. Rizika, J. W., Thermal lags in flowing systems containing heat capacitors, *Trans. ASME*, 78, 411–420, 1954.
3. Deissler, R. G., Analysis of turbulent heat transfer, mass transfer, and friction in smooth tubes at high Prandtl and Schmidt numbers, NACA Report 1210, 1955.
4. Rizika, J. W., Thermal lags in flowing incompressible fluid systems containing heat capacitors, *Trans. ASME*, 78, 1407–1413, 1956.
5. Sparrow, E. M. and Siegel, R., Thermal entrance region of a circular tube under transient heating conditions, *Proceedings of the Third U.S. National Congress on Applied Mechanics*, Providence, RI, pp. 817–826, 1958.
6. Siegel, R. and Sparrow, E. M., Transient heat transfer for laminar forced convection in the thermal entrance region of flat ducts, *Trans. ASME, J. Heat Transfer*, 81C, 29–36, 1959.
7. Siegel, R., Transient heat transfer for laminar slug flow in ducts, *Trans. ASME, J. Appl. Mech.*, 81E, 140–142, 1959.
8. Siegel, R., Heat transfer for laminar flow in ducts with arbitrary time variation in wall temperature, *Trans. ASME, J. Appl. Mech.*, 82E, 241–249, 1960.
9. Perlmutter, M. and Siegel, R., Unsteady laminar flow in a duct with unsteady heat addition, *Trans. ASME, J. Heat Transfer*, 83, 432–440, 1961.
10. Perlmutter, M. and Siegel, R., Two-dimensional unsteady incompressible laminar duct flow with a step change in wall temperature, *Int. J. Heat Mass Transfer*, 3, 94–107, 1961.
11. Siegel, R. and Perlmutter, M., Laminar heat transfer in a channel with unsteady flow and wall heating varying with position and time, *Trans. ASME, J. Heat Transfer*, 85, 358–365, 1963.

12. Siegel, R., Forced convection in a channel with wall heat capacity and with wall heating variable with axial position and time, *Int. J. Heat Mass Transfer*, 6, 607–620, 1963.

13. Clark, J. A., Arpaci, V. S., and Treadwell, K. M., Dynamic response of heat exchangers having internal heat sources—Part I, *Trans. ASME*, 80, 612–624, 1958.

14. Arpaci, V. S. and Clark, J. A., Dynamic response of heat exchangers having internal heat sources—Part II, *Trans. ASME*, 80, 625–634, 1958.

15. Arpaci, V. S. and Clark, J. A., Dynamic response of heat exchangers having internal heat sources—Part III, *Trans. ASME, J. Heat Transfer*, 81C, 253–266, 1959.

16. Yang, J. W., Clark, J. A., and Arpaci, V. S., Dynamic response of heat exchangers having internal heat sources—Part IV, *Trans. ASME, J. Heat Transfer*, 83C, 321–388, 1961.

17. Kardas, A., On a problem in the theory of the unidirectional regenerators, *Int. J. Heat Mass Transfer*, 9, 567, 1966.

18. Namatame, K., Transient temperature response of an annular flow with step change in heat generating rod, *J. Nucl Sci. Technol.*, 6, 591–600, 1969.

19. Campo, A. and Yoshimura, T., Random heat transfer in flat channels with timewise variation of ambient temperature, *Int. J. Heat Mass Transfer*, 22, 5–12, 1979.

20. Lin, H. T. and Shih, Y. P., Unsteady thermal entrance heat transfer of power-law fluids in pipes and plate slits, *Int. J. Heat Mass Transfer*, 24, 1531–1539, 1981.

21. Sucec, J., An improved quasi-steady approach for transient conjugated forced convection problems, *Int. J. Heat Mass Transfer*, 24, 1711–1722, 1981.

22. Lin, T. F., Hawks, K. H., and Leidenfrost, W., Unsteady thermal entrance heat transfer in laminar pipe flows with a step change in ambient temperature, *Warme Stöffubertragung*, 17, 125–132, 1983.

23. Lin, T. F., Hawks, K. H., and Leidenfrost, W., Transient thermal entrance heat transfer in laminar flows with a step change in pumping pressure, *Warme Stöffubertragung*, 17, 201–209, 1983.

24. Lin, T. F., Hawks, K. H., and Leidenfrost, W., Transient conjugated heat transfer between a cooling coil and its surrounding enclosure, *Int. J. Heat Mass Transfer*, 16, 1661–1667, 1983.

25. Chen, S. C., Anand, N. K., and Tree, D. R., Analysis of transient laminar convective heat transfer inside a circular duct, *J. Heat Transfer*, 105, 922–924, 1983.

26. Cotta, R. M., Mikhailov, M. D., and Özisik, M. N., Transient conjugated forced convection in ducts with periodically varying inlet temperature, *Int. J. Heat Mass Transfer*, 30, 2073, 1987.

27. Sparrow, E. M. and De Farias, F. N., Unsteady heat transfer in ducts with time varying inlet temperature and participating walls, *Int. J. Heat Mass Transfer*, 11, 837–853, 1968.

28. Kakaç, S. and Yener, Y., Exact solution of the transient forced convection energy equation for timewise variation of inlet temperature, *Int. J. Heat Mass Transfer*, 16, 2205–2214, 1973.

29. Cotta, R. M., Özisik, M. N., and McRae, D. S., Transient heat transfer in channel flow with step change in inlet temperature, *Numer. Heat Transfer*, 9, 619, 1986.

30. Cotta, R. M. and Özisik, M. N., Laminar forced convection inside ducts with periodic variation of inlet temperature, *Int. J. Heat Mass Transfer*, 29, 1495, 1986.

31. Kakaç, S., Ding, Y., and Li, W., Transient fluid flow and heat transfer in ducts with a timewise variation of inlet temperature, *Proceedings of the 3rd International Symposium on Transport Phenomena in Thermal Control*, Taipei, Taiwan, Hemisphere Publishing Co., New York, 1989.

32. Kakaç, S., Ding, Y., and Li, W., Experimental investigation of transient laminar forced convection in ducts, *Proceedings of the International Conference on Experimental Heat Transfer, Fluid Mechanics and Thermodynamics*, September 4–9, 1988, Dubrovnik, Yugoslavia, Elsevier Science Publishing, Amsterdam, the Netherlands, 1989.

33. Kakaç, S., Li, W., and Cotta, R. M., Theoretical and experimental study of transient laminar forced convection in a duct with timewise variation of inlet temperature, *ASME Winter Annual Meeting*, San Francisco, CA, 1989.

34. Ding, Y., Experimental investigation of transient forced convection in ducts for a timewise varying inlet temperature, Master of Science thesis in Mechanical Engineering, University of Miami, Miami, FL, 1987.

35. Kakaç, S., Li, W., and Cotta, R. M., Unsteady laminar forced convection with periodic variation of inlet temperature, *Trans ASME, J. Heat Transfer*, 112, 913–920, 1990.
36. Özisik, M. N. and Muny, R. L., On the solution of linear diffusion problems with variable boundary condition parameters, *Trans. ASME, J. Heat Transfer*, 96C, 48, 1974.
37. Kakaç, S. and Yener, Y., Transient laminar forced convection in ducts, in *Low Reynolds Number Flow Heat Exchangers*, S. Kakaç, R. K. Shah, and A. E. Bergles (Eds.), Hemisphere Publishing Co., New York, 1983, pp. 205–227.
38. Yener, Y. and Kakaç, S., Unsteady forced convection in ducts, in *Handbook of Single-Phase Convective Heat Transfer*, S. Kakaç, R. K. Shah, and W. Aung (Eds.), John Wiley & Sons, New York, Chapter 11, 1987.
39. Li, W., Experimental and theoretical investigation of unsteady forced convection in ducts, Dissertation, Ph.D. in Mechanical Engineering, University of Miami, Miami, FL, May 1990.
40. Li, W. and Kakaç, S., Unsteady thermal entrance heat transfer in laminar flow with a periodic variation of inlet temperature, *Int. J. Heat Mass Transfer*, 34, 2581–2592, 1991.
41. Yener, Y. and Kakaç, S., *Heat Conduction*, 4th edn., Taylor & Francis Group, New York, 2008.
42. Kakaç, S. and Yener, Y., Frequency response analysis of transient turbulent forced convection for timewise variation of inlet temperature, in *Turbulent Forced Convection in Channels and Bundles*, S. Kakaç and D. B. Spalding (Eds.), Hemisphere Publishing Co., New York, 1979, Vol. 2, pp. 865–880.
43. Abbrecht, P. H. and Churchill, S.W., The thermal entrance region in fully developed turbulent flow, *AIChE J.*, 6(2), 268, 1960.
44. Sparrow, E. M. and Siegel, R., Unsteady turbulent heat transfer in tubes, *Trans. ASME, J. Heat Transfer*, 82C, 170–180, 1960.
45. Kakaç, S., Transient turbulent flow in ducts, *Warme Stoffübertragung*, 1, 169–176, 1968.
46. Kakaç, S., A general analytical solution to the equation of transient forced convection with fully developed flow, *Int. J. Heat Mass Transfer*, 18, 1449–1453, 1975.
47. Gartner, D., Instationarer warmeübergan bei tuibulenter ringspaltströmung, *Warme Stoffübertragung*, 9, 179–191, 1976.
48. Kawamura, H., Experimental and analytical study of transient heat transfer for turbulent flow in a circular tube, *Int. J. Heat Mass Transfer*, 20, 443–450, 1977.
49. Sparrow, E. M., Hallman, T. M., and Siegel, R., Turbulent heat transfer in the thermal entrance region of a pipe with uniform heat flux, *Appl. Sci. Res.*, 7A, 37–52, 1957.
50. Acker, M. T. and Fourcher, B., Analyse in regime thermique periodique du cauplage conduction-convection entre un fluide en eeoulement laminaiie et une paroi de stokage, *Int. J. Heat Mass Transfer*, 24, 1201–1210, 1981.
51. Sucec, J. and Sawant, A. M., Unsteady conjugated forced convection heat transfer in a parallel plate duct, *Int. J. Heat Mass Transfer*, 27, 45–101, 1984.
52. Kim, W. S. and Özisik, M. N., Turbulent forced convection inside a parallel-plate channel with periodic variation of inlet temperature, *Trans. ASME, J. Heat Transfer*, III, 883, 1989.

# 9

# Empirical Correlations for Single-Phase Forced Convection in Ducts

## Nomenclature

| | |
|---|---|
| $A$ | constant |
| $A_c$ | net free-flow cross-sectional area, m$^2$ |
| $C_p$ | specific heat at constant pressure, J/(kg K) |
| $D$ | a diameter of a circular annulus, m |
| $D_e$ | equivalent diameter for heat transfer = $4A_c/P_h$ m |
| $D_h$ | hydraulic diameter for pressure drop = $4A_c/P_w$ |
| $d$ | circular duct diameter, m |
| $f$ | Fanning friction factor = $2\tau_w/\rho U_m^2$ |
| $G$ | fluid mass velocity = $\rho U_m$, kg/(m$^2$ s) |
| $h$ | average heat-transfer coefficient, W/(m$^2$ K) |
| $h_x$ | local heat-transfer coefficient, W/(m$^2$ K) |
| $k$ | thermal conductivity of fluid, W/(m K) |
| $L$ | distance along the duct, m |
| $L_h$ | hydrodynamic entrance length, m |
| $L_t$ | thermal entrance length, m |
| $m$ | exponent, Equations 21.b and 22.b |
| $\dot{m}$ | mass flow rate, kg/s |
| $Nu$ | average Nusselt number = $hd/k$ |
| $n$ | exponent, Equations 21.a and 22.a |
| $Pe$ | Peclet number = $RePr$ |
| $Pr$ | Prandtl number = $c_p\mu/k = v/\alpha$ |
| $q''$ | heat flux, W/m$^2$ |
| $Re$ | Reynolds number = $\rho U_m d/\mu$, $\rho U_m D_h/\mu$ |
| $T$ | temperature, °C, K |
| $T_f$ | film temperature = $(T_w + T_b)/2$, °C, K |
| $u$ | velocity component in axial direction, m/s |
| $U_m$ | mean axial velocity, m/s |
| $x$ | Cartesian coordinate, axial distance, m |
| $y$ | Cartesian coordinate, distance normal to the surface, m |

**Greek Symbols**

| | |
|---|---|
| $\alpha$ | thermal diffusivity of fluid, m²/s |
| $\varepsilon_h$ | thermal eddy diffusivity, m²/s |
| $\varepsilon_m$ | momentum eddy diffusivity, m²/s |
| $\mu$ | dynamic viscosity of fluid, Pa s |
| $\nu$ | kinematic viscosity of fluid, m²/s |
| $\rho$ | density of fluid, kg/m³ |
| $\tau_w$ | shear stress at the wall, Pa |

**Subscripts**

| | |
|---|---|
| $a$ | arithmetic mean |
| $b$ | bulk fluid condition or properties evaluated at bulk mean temperature |
| $cp$ | constant property |
| $f$ | film fluid condition or properties evaluated at film temperature |
| $H$ | constant heat-flux boundary condition |
| $i$ | inlet condition or inner surface |
| $l$ | laminar |
| $o$ | outlet condition or outer surface |
| $T$ | constant temperature boundary condition |
| $t$ | turbulent |
| $w$ | wall condition or wetted |
| $x$ | local value at distance $x$ |
| $\infty$ | fully developed conditions ($x \to \infty$) |

## 9.1 Introduction

In the previous chapters, we discussed the type of convection problems that can be approached analytically. Unfortunately, it is not always possible to approach forced convection problems analytically, especially when the flow is turbulent. When the difference between the fluid bulk and the wall temperatures is high, the effect of property variations on the convective heat transfer has to be accounted for. The cases of turbulent flow, and even developing laminar flow, with property variations become much more complicated, and these are the important problems faced in the design of heat-exchanging devices. In such cases, the engineers are forced to perform experiments to obtain information for the design. Then by the use of dimensional analysis, a relation that relates important physical quantities, such as flow velocity and fluid properties in dimensionless groups such as Reynolds, Prandtl, and Nusselt numbers, is obtained. The precise functional relationship between these dimensionless groups is then determined by the use of experimental data.

In many two-phase flow heat exchangers such as boilers, steam generators, power condensers, and air conditioning evaporators and condensers, one side has single-phase fluid, while the other side has two-phase flow. Generally, the single-phase side represents higher thermal resistance, particularly with gas or oil flow. In this chapter, a comprehensive review is made of the available correlations for laminar and turbulent flow of single-phase Newtonian fluids through circular and noncircular ducts with and without the effect of property variations. A large number of experimental and analytical correlations are available for heat-transfer coefficient and flow friction factor for laminar and turbulent flows through ducts. In this chapter, recommended correlations for single-phase forced convection in ducts are given.

Laminar and turbulent forced convection correlations for single-phase fluids represent an important class of heat-transfer solutions for heat-exchanger applications. As discussed in the previous chapters, when a viscous fluid flows in a duct, a boundary layer will form along the duct. Gradually, the boundary layer fills the entire duct and the flow is then said to be fully developed. The distance at which the velocity becomes fully developed is called the hydrodynamic or velocity entrance length ($L_h$). Theoretically, the approach to the fully developed velocity profile is asymptotic, and it is, therefore, impossible to describe a definite location where the boundary layer completely fills the duct.

If the walls of the duct are heated or cooled, then a thermal boundary layer will also develop along the duct. At a certain point downstream, one can talk about fully developed temperature profile, where the thickness of thermal boundary layer is approximately equal to $d/2$. The distance at which the temperature profile becomes fully developed is called the thermal entrance length ($L_t$).

If the heating starts from the inlet of the duct, then both velocity and temperature profiles develop simultaneously. The associated heat-transfer problem is referred to as the combined hydrodynamic and thermal entry length problem or simultaneously developing region problem. Therefore, there are four types of duct flows with heating, namely, fully developed, hydrodynamically developing, thermally developing, and simultaneously developing, and the design correlations should be selected accordingly.

As discussed in the preceding chapters, the rate of development of velocity and temperature profiles in the combined entrance region depends on the fluid Prandtl number ($Pr = v/\alpha$). For high-Prandtl-number fluids, such as oils, even though both velocity and temperature profiles are uniform at the duct entrance, the velocity profile is established much more rapidly than the temperature profile. In contrast, for very-low-Prandtl-number fluids, such as the liquid metals, the temperature profile establishes much more rapidly than the velocity profile. However, for Prandtl numbers about unity, as for gases, both temperature and velocity profiles develop at a similar rate simultaneously along the duct, starting from uniform temperature and uniform velocity at the duct entrance.

For the limiting case of $Pr = \infty$, the velocity profile is developed before the temperature profile starts developing. For the other limiting case of $Pr = 0$, the velocity profile never develops and remains uniform while the temperature profile is developing. The idealized $Pr = \infty$ and 0 cases are good approximations for highly viscous fluids and liquid metals, respectively.

When fluids flow at very low velocities, the fluid particles move in definite paths called streamlines. This type of flow is called laminar flow. The shape of the fully developed velocity profile for laminar flow is parabolic in a circular duct or between parallel plates. There is no component of fluid velocity normal to the duct axis. Depending upon the roughness of the circular duct inlet and inside surfaces, fully developed laminar flow will be obtained for $Re \leq 2300$ within the duct length $L$ if it is longer than the hydrodynamic entry length $L_h$; however, if $L < L_h$, developing laminar flow would exist over the entire duct length. The hydrodynamic and thermal entrance lengths for laminar flow inside conduits have been given in [1,2]. The hydrodynamic entrance length $L_h$ for laminar flow inside ducts of various cross sections based on the definition discussed previously is presented in Table 9.1. Included in this table are the thermal entrance lengths for constant wall temperature and constant wall heat-flux boundary conditions for thermally

**TABLE 9.1**

Hydrodynamic Entrance Length ($L_h$) and Thermal Entrance Length ($L_t$) for Laminar Flow Inside Ducts[a]

| Geometry | $\dfrac{L_h/D_h}{Re}$ | $\dfrac{L_h/D_h}{Pe}$ Constant Wall Temperature | Constant Wall Heat Flux |
|---|---|---|---|
| circle, $D$ | 0.056 | 0.033 | 0.043 |
| parallel plates, $2b$ | 0.011 | 0.008 | 0.012 |
| rectangle $\dfrac{a}{b} = 0.25$ | 0.075 | 0.054 | 0.042 |
| 0.50 | 0.085 | 0.049 | 0.057 |
| 1.0 | 0.09 | 0.041 | 0.066 |

*Source:* Based on the results reported in Shah, R. K. and London, A. L., *Laminar Forced Convection in Ducts*, Academic Press, New York, 1978; Shah, R. K. and Bhatti, M. S., Laminar convective heat transfer in ducts, in *Handbook of Single-Phase Convective Heat Transfer*, S. Kakaç, R. K. Shah, and W. Aung (Eds.), John Wiley, New York, 1987, pp. 3.1–3.137.

[a] The thermal entry lengths are for the hydrodynamically developed, thermally developing flow conditions.

developing, hydrodynamically developed flow. In Table 9.1, the Reynolds number is based on the hydraulic diameter $D_h$.

As discussed in Chapter 6, if the velocity of the fluid is gradually increased, there will be a point where the laminar flow becomes unstable in the presence of small disturbances and the fluid no longer flows along parallel lines (streamlines), but by a series of eddies that result in a complete mixing of entire flow field (Figure 6.3). This type of flow is called turbulent flow. The Reynolds number at which the flow changes from laminar to turbulent is referred to as the critical (value of) Reynolds number. The critical Reynolds number in circular ducts is between 2100 and 2300. Although the value of critical Reynolds number depends on the duct cross-sectional geometry and surface roughness, for particular applications it can be assumed that the transition from laminar to turbulent flow in noncircular ducts will also take place at about $Re_{cr} = 2100$–$2300$ when the hydraulic diameter of the duct, which is defined as four times the cross-sectional (flow) area $A_c$ divided by the wetted perimeter $P$ of the duct, is used in calculating the Reynolds number.

At a Reynolds number $Re > 10^4$, the flow is completely turbulent. Between the lower and upper limits lies the transition zone from laminar to turbulent flow. Therefore, fully turbulent flow in a duct occurs at a Reynolds number $Re \geq 10^4$.

## 9.2 Dimensional Analysis of Forced Convection

The method of dimensional analysis is based on the fact that any relation between physical quantities must be dimensionally consistent.

Dimensional analysis is a mathematical tool that is of help in the planning and in the interpretation of experimental work. However, it cannot be used to deduce numerical values in a relationship between physical quantities. The basic principle in dimensional analysis is simply a theorem in mathematics named the $\pi$-theorem. A relationship or function of the form

$$f(x_1, x_2, \ldots, x_n) = 0 \tag{9.1}$$

is always reducible to a function of the form

$$\Psi(\pi_1, \pi_2, \ldots, \pi_{n-k}) = 0 \tag{9.2a}$$

or

$$\Psi(\pi_1, \pi_2, \ldots, \pi_i) = 0, \tag{9.2b}$$

where $k$ is the number of independent fundamental dimensions required to specify the $n$ quantities and each $\pi$ is a dimensionless product of the form

$$\pi = x_1^{a_1} x_2^{a_2} \cdots x_n^{a_n} \tag{9.3}$$

and will be dimensionless. The set of $\pi$ terms will include all independent dimensionless groupings of the variables. There are exceptions to the previous rule as noted by Buckingham [1] and Brigeman [2]. There are three possible cases:

1. $n < k$; no solution is possible.
2. $n > k$; this leads to $n - k$ dimensionless independent groups. This corresponds to Buckingham's statement of the $\pi$-theorem where $i = n - k$ is the maximum number of independent dimensionless groups that can be formed.
3. $n = k$; this corresponds to a unique dimensionless product. This last case is sometimes encountered in applications of the $\pi$-theorem when not all the necessary physical quantities are used in the original statement of Equation 9.1.

One, but not the only, procedure for obtaining $n - k$ dimensionless groups is to select $k$ of the $x$ variables and to combine them in turn with each of the remaining $(n - k)x$ variables. The selection of $k$ of the $x$ variables must together involve all the independent dimensions, but they must not form a dimensionless group by themselves. Further, the exponent of each of the $(n - k)x$ variables in each term is arbitrary, and we can select the simplest possible value, that is, unity.

In an application of the $\pi$-theorem, any set of $k$ independent variables ($x$'s) may be chosen; by independent it is meant that these $k$ variables will not by themselves form a dimensionless group. This is best shown by an example that follows.

As an application of dimensional analysis, consider the heat transfer in a fully developed turbulent flow at a moderate-velocity condition in a smooth tube with constant fluid properties. From the previous analysis, it will be reasonable to assume that the heat-transfer coefficient will depend on the diameter of the tube d, mean velocity of the fluid $U_m$, and the physical properties $k$, $c_p$, $\rho$, and $\mu$ of the fluid. Provided that the previous quantities are sufficient to define the heat-transfer coefficient, we may write

$$f(h,d,U_m,\rho,\mu,c_p,k) = 0. \tag{9.4}$$

There are four independent dimensions, that is, $M$ (mass), $L$ (length), $t$ (time), and $T$ (temperature).

Select any four independent variables such as $d$, $U_m$, $\rho$, and $h$. Note that we cannot choose $d$, $U_m$ $\mu$, and $\rho$, since they will form the dimensionless group, $d\rho U_m/\mu$. The dimensions of each of the variables can be arranged as follows:

$$[d] = L \qquad [\mu] = ML^{-1}t^{-1},$$

$$[U_m] = Lt^{-1} \qquad [c_p] = L^2 t^{-2} T^{-1},$$

$$[\rho] = ML^{-3} \qquad [k] = MLt^{-3}T^{-1},$$

$$[h] = Mt^{-3}T^{-1}.$$

The number of independent $\pi$'s is $7 - 4 = 3$. Form the dimensionless $\pi$'s by using the four independent variables and any other grouping of the remaining variables. The simplest procedure is to use each of the remaining variables at a time, that is,

$$\pi_1 = d^{a_1} U_m^{a_2} \rho^{a_3} h^{a_4} \mu^{d_1}, \tag{9.5}$$

$$\pi_2 = d^{b_1} U_m^{b_2} \rho^{b_3} h^{b_4} c_p^{e_1}, \tag{9.6}$$

$$\pi_3 = d^{c_1} U_m^{c_2} \rho^{c_3} h^{c_4} k^{f_1}. \tag{9.7}$$

Substituting dimensions for $n$, we obtain

$$[L]^{a_1} \left[\frac{L}{t}\right]^{a_2} \left[\frac{M}{L^3}\right]^{a_3} \left[\frac{M}{t^3 T}\right]^{a_4} \left[\frac{M}{Lt}\right]^{d_1} = 0. \tag{9.8}$$

From Equation 9.8, we get four equations in the exponents:

For $[M]: 0 + 0 + a_3(1) + a_4(1) + d_1(1) = 0.$ \tag{9.9}

For $[L]: a_1 + a_2 + a_3(-3) + 0 + d_1(-1) = 0.$ \tag{9.10}

$$\text{For } [t]: 0 + a_2(-1) + 0 + a_4(-3) + d_1(-1) = 0. \tag{9.11}$$

$$\text{For } [T]: 0 + 0 + 0 + a_4(-1) + 0 = 0. \tag{9.12}$$

The solution is $a_3 = -d_1$, $a_1 = -d_1$, $a_2 = -d_1$, $a_4 = 0$, or

$$\pi_1 = \left(\frac{\mu}{U_\infty \rho d}\right)^{d_1}.$$

Since $d_1$ is arbitrary, we select the simplest possible value, that is, unity. Thus,

$$\pi_1 = \frac{\mu}{U_m \rho d},$$

which is the reciprocal of the Reynolds number Re. Proceeding likewise for $\pi_2$ and $\pi_3$, we get the following:

$$\pi_2 = \frac{U_m \rho c_p}{h} = \frac{1}{St}, \text{ the reciprocal of the Stanton number.}$$

$$\pi_3 = \frac{k}{hd} = \frac{1}{Nu}, \text{ the reciprocal of the Nusselt number.}$$

The final result is

$$\Psi(Re, St, Nu) = 0. \tag{9.13}$$

Noting that the Prandtl number can be obtained from $Re$, $Nu$, and $St$, we can also write

$$\Psi(Nu, Re, Pr) = 0. \tag{9.14}$$

Or, more usually,

$$Nu = \Psi_2(Re, Pr). \tag{9.15}$$

It should be noted that dimensional analysis does not yield any information as to the nature of the function $\Psi$ nor does it permit deduction of numerical values for $Nu$ in terms of $Re$ and $Pr$.

A large number of measurements required to relate $h$ to the other six independent variables are greatly reduced. For example, if all possible combinations of five different values of each independent variable were chosen, according to Equation 9.4, the total number of measurements would be $5^6 = 15625$. On the other hand, in the experimental investigation one dimensionless group may be regarded as the dependent and the others as the independent variables, and if we chose five values of the independent variables, the number of measurements required for Equation 9.15 is reduced to 25.

If we are dealing with problems of heat transfer in which there are large temperature differences, it may be necessary to take into account the variations in the values of the physical properties of the fluid with temperature. We should then include characteristic temperatures, for instance, the wall temperature $T_w$, and the mean bulk temperature of the fluid $T_b$ or viscosities at these temperatures. This involves the introduction of another dimensionless ratio $(T_w/T_b)$ or $(\mu_w/\mu_b)$ so that instead of Equation 9.15 we get

$$Nu = \phi_1\left(Re, Pr, \frac{T_w}{T_b}\right) \quad \text{(for gases)} \tag{9.16}$$

or

$$Nu = \phi_2\left(Re, Pr, \frac{\mu_w}{\mu_b}\right) \quad \text{(for liquids).} \tag{9.17}$$

If we are concerned, for instance, with heat transfer at the entrance region as well, then Equation 9.17 becomes

$$Nu = \phi_3\left(Re, Pr, \frac{x}{d}, \frac{\mu_w}{\mu_b}\right). \tag{9.18}$$

## 9.3 Laminar Forced Convection

Laminar duct flow is encountered generally in compact heat exchangers, cryogenic cooling systems, heating or cooling of heavy (highly viscous) fluids such as oils, and in many other applications. Different investigators performed extensive experimental and theoretical studies with various fluids for numerous duct geometries and under different surface and entrance conditions. As a result, they formulated relations for the Nusselt number versus the Reynolds and Prandtl numbers for a wide range of these dimensionless groups. Shah and London [1] and Shah and Bhatti [2] have compiled the laminar flow solutions.

Laminar flow can be obtained for a specified mass velocity $G = \rho U_m$ for (1) small hydraulic diameter $D_h$ of the flow passage or (2) high fluid viscosity $\mu$. Flow passages with small hydraulic diameter are encountered in compact heat exchangers, since they result in large surface area per unit volume of the exchanger. The internal flow of oils and other liquids with high viscosity in noncompact heat exchangers is generally of laminar nature.

### 9.3.1 Hydrodynamically Developed and Thermally Developing Laminar Flow in Smooth Circular Ducts

The well-known Nusselt–Graetz problem for heat transfer to an incompressible fluid with constant properties flowing through a circular duct with constant wall temperature boundary condition and fully developed laminar velocity profile was solved numerically by several investigators [1,2]. The asymptotes of the mean Nusselt number for a circular duct of the length $L$ are

$$Nu_T = 1.61 \left( \frac{Pe_b d}{L} \right)^{1/3} \quad \text{for} \quad \frac{Pe_b d}{L} > 10^3 \tag{9.19}$$

or

$$Nu_T = 3.66 \quad \text{for} \quad \frac{Pe_b d}{L} < 10^2. \tag{9.20}$$

The superposition of two asymptotes for the mean Nusselt number derived by Schlünder [3] gives sufficiently good results for most of the practical cases:

$$Nu_T = \left[ 3.66^3 + 1.61^3 \left( \frac{Pe_b d}{L} \right) \right]^{1/3}. \tag{9.21}$$

An empirical correlation has also been developed by Hausen [4] for laminar flow in the thermal entrance region of a circular duct at constant wall temperature and is given as

$$Nu_T = 3.66 + \frac{0.19(Pe_b d/L)^{0.8}}{1 + 0.117(Pe_b d/L)^{0.467}}. \tag{9.22}$$

The results of Equations 9.21 and 9.22 are comparable with each other. These equations may be used for the laminar flow of gases and liquids in the range of $0.1 < Pe_b(d/L) < 10^4$. Axial conduction effects must be considered at $Pe_b(d/L) < 0.1$. All the physical properties are evaluated at the mean bulk temperature of fluid, defined as

$$T_b = \frac{(T_i + T_o)}{2}, \tag{9.23}$$

where $T_i$ and $T_o$ are the bulk temperatures of fluid at the inlet and exit of the duct, respectively.

The asymptotic mean Nusselt numbers in circular ducts with constant wall heat-flux boundary condition are [1] as follows:

$$Nu_H = 1.953 \left( \frac{Pe_b d}{L} \right)^{1/3} \quad \text{for} \quad \frac{Pe_b d}{L} > 10^2 \tag{9.24}$$

and

$$Nu_H = 4.36 \quad \text{for} \quad \frac{Pe_b d}{L} < 10. \tag{9.25}$$

The fluid properties are evaluated at the mean bulk temperature Tb as defined by Equation 9.23.

The results given by Equations 9.20 and 9.25 represent the dimensionless heat-transfer coefficients for laminar forced convection inside a circular duct in the hydrodynamically and thermally developed regions under constant wall temperature and constant wall heat-flux boundary conditions, respectively.

### 9.3.2 Simultaneously Developing Laminar Flow in Smooth Circular Ducts

When heat transfer starts as soon as the fluid enters a duct, the velocity and temperature profiles start developing simultaneously. The analysis of temperature distribution in the flow, and hence of the heat transfer between the fluid and the duct wall, for such situations is more complex because the velocity distribution varies in axial direction as well as normal to it. Heat-transfer problems involving simultaneously developing flow have been mostly solved by numerical methods for various duct cross sections. A comprehensive review of such solutions is given by Shah and Bhatti [2] and by Kakaç [5].

Shah and London [1,2] presented the numerical values of the mean Nusselt number for this region. In the case of a short duct length, *Nu* values are represented by the asymptotic equation of Pohlhausen [6] for simultaneously developing flow over a flat plate; for a circular duct, this equation becomes

$$Nu_T = 0.664 \left( \frac{Pe_b d}{L} \right)^{1/2} Pr_b^{-1/6}. \tag{9.26}$$

The range of validity is $0.5 < Pr_b < 500$ and $Pe_b\, d/L > 10^3$.

For most of the engineering applications with short circular ducts ($d/L > 0.1$), it is recommended that whichever of the Equations 9.21, 9.22, or 9.26 gives the highest Nusselt number be used.

### 9.3.3 Laminar Flow through Concentric Smooth Ducts

Correlations for concentric annular ducts are very important in heat-exchanger applications. The simplest form of a two-fluid heat exchanger is a double pipe; one fluid flows inside the inner tube, while the other flows through the annular passage (see Figure 9.1).

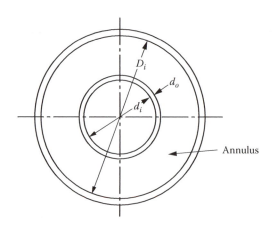

**FIGURE 9.1**
Concentric tube annulus.

Heat is usually transferred through the wall of the inner tube, while the outer wall of the annular duct is insulated. Heat-transfer coefficient in annular ducts depends on the ratio of the diameters $(D_i/d_0)$ because of the shape of the velocity profile.

The hydraulic (equivalent) diameter approach is the simplest method to calculate the heat transfer and the pressure drop in the annulus. In this approach, the hydraulic diameter of annulus $D_h$ is substituted instead of the tube diameter in internal flow correlations:

$$D_h = \frac{4 \times \text{Net free-flow area}}{\text{Wetted (or heat transfer) perimeter}}.$$ (9.27)

This approximation is acceptable for heat-transfer and pressure drop calculations. The validity of the hydraulic diameter approach has been substantiated by the results of experiments performed with finned annuli [7].

The total wetted perimeter of the annulus for pressure drop calculations is given by

$$P_w = \pi(D_i + d_o),$$ (9.28)

and the heat-transfer perimeter of the annulus can be calculated by

$$P_h = \pi d_0.$$ (9.29)

The only difference between $P_w$ and $P_h$ is $D_i$, which is the inner diameter of the shell (outer tube) of the annulus. This difference is due to the fluid friction on the inner surface of the shell; however, such is not the case for heat-transfer perimeter since the heat transfer takes place only through the wall of the inner tube. The net free-flow area of the annulus is given by

$$A_C = \pi \frac{D_i^2 - d_o^2}{4}.$$ (9.30)

Hydraulic diameter based on total wetted perimeter for pressure drop calculations is

$$D_h = \frac{4 A_C}{P_w},$$ (9.31)

and the hydraulic diameter based on the heat-transfer perimeter is given by Equation 9.32, which is hereafter named as the equivalent diameter.

$$D_e = \frac{4 A_C}{P_h}.$$ (9.32)

Reynolds number, Graetz number, and the ratio $d/L$ are to be calculated with $D_h$. The equivalent diameter $D_e$ is used to calculate the heat-transfer coefficient from the Nusselt number and in evaluating the Grashof number. Slightly higher heat-transfer coefficients are obtained while using $D_h$ instead of $D_e$ for heat-transfer calculations.

For the constant wall temperature boundary condition, Stephen [8] has developed a heat-transfer correlation based on Equation 9.32. The Nusselt number for hydrodynamically developed laminar flow in the thermal entrance region of an isothermal annulus, outer wall of which is insulated, may be calculated from the following correlation:

$$Nu_T = Nu_\infty + \left[ 1 + 0.14 \left( \frac{d_o}{D_i} \right)^{-1/2} \right] \frac{0.19(Pe_b D_h/L)^{0.8}}{1 + 0.117(Pe_b D_h/L)^{0.467}}, \qquad (9.33)$$

where $Nu_\infty$ is the Nusselt number for fully developed flow.

A detailed review of laminar convective heat transfer in ducts for various hydrodynamic and thermal boundary conditions is given in [2].

## 9.4 Effects of Variable Physical Properties

When the previously mentioned correlations are applied to practical heat-transfer problems with large temperature differences between the surface and the fluid, the constant-property assumption could cause significant errors, since the transport properties of most fluids vary with temperature, which influence the variation of velocity and temperature through the boundary layer or over the flow cross section of a duct.

For practical applications, a reliable and appropriate correlation based on the constant-property assumption can be modified and/or corrected so that it may be used when the variable-property effect becomes important.

Two methods of correcting constant-property correlations for the variable-property effect have been employed, namely, the reference temperature method and the property ratio method. In the former, a characteristic temperature is chosen at which the properties appearing in nondimensional groups are evaluated so that the constant-property results at that temperature may be used to consider the variable-property behavior; in the latter case, all properties are evaluated at the bulk temperature, and then all variable-property effects are lumped into a function of the ratio of one property evaluated at the wall (surface) temperature to that property evaluated at bulk temperature. Some correlations may involve a modification or combination of these two methods.

For liquids, the variation of viscosity is responsible for most of the property effects. Therefore, the variable-property Nusselt numbers and friction factors in the property ratio method for liquids are correlated by

$$\frac{Nu}{Nu_{cp}} = \left( \frac{\mu_b}{\mu_w} \right)^n, \qquad (9.34a)$$

$$\frac{f}{f_{cp}} = \left( \frac{\mu_b}{\mu_w} \right)^m, \qquad (9.34b)$$

where
   $\mu_b$ is the viscosity evaluated at the bulk mean temperature
   $\mu_w$ is the viscosity evaluated at the wall temperature
   the subscript $cp$ refers to the constant-property solution

The friction coefficient employed is the so-called Fanning friction factor based on the wall shear rather than the pressure drop.

For gases, the viscosity, thermal conductivity, and density vary with the absolute temperature. Therefore, in the property ratio method, temperature corrections of the following forms are found to be adequate in practical applications for the temperature-dependent-property effects in gases:

$$\frac{Nu}{Nu_{cp}} = \left(\frac{T_w}{T_b}\right)^n , \tag{9.35a}$$

$$\frac{f}{f_{cp}} = \left(\frac{T_w}{T_b}\right)^m , \tag{9.35b}$$

where $T_b$ and $T_w$ are the absolute bulk mean and wall temperatures, respectively.

It must be noted that the constant-property portion of the specific correlation is evaluated in terms of the parameters and conditions defined by its author(s).

Extensive theoretical and experimental investigations on convective heat transfer of fluids with variable properties have been reported in the literature to obtain the values of the exponents $n$ and $m$, which will be cited in the following sections of this chapter.

### 9.4.1 Laminar Flow of Liquids

Deissler [9] carried out a numerical analysis as described previously for laminar flow through a circular duct at constant heat-flux boundary condition for liquid viscosity variation with temperature given by

$$\frac{\mu}{\mu_w} = \left(\frac{T}{T_w}\right)^{-1.6} \tag{9.36}$$

and obtained $n = 0.14$ to be used with Equation 9.34a. This has been used widely to correlate experimental data for laminar flow for $Pr > 0.6$.

Deissler [9] also obtained $m = -0.58$ for heating and $m = -0.50$ for cooling of liquids to be used with Equation 9.34b.

Yang [10] obtained the solution for both constant wall heat-flux and constant wall temperature boundary conditions by assuming a viscosity dependence of a liquid on temperature as

$$\frac{\mu}{\mu_w} = \left[1 + A\left(\frac{T_w - T}{T_w - T_i}\right)\right]^{-1} , \tag{9.37}$$

where $A$ is a constant. His prediction for both constant wall heat-flux and constant wall temperature boundary conditions was correlated with $n = 0.11$ in Equation 9.34a, and he concluded that the effect of thermal boundary conditions is small and the influence on the friction coefficient is very substantial. He also found that the correction for variables properties is the same for developing and developed regions.

A simple empirical correlation has been proposed by Sieder and Tate [11] to predict the mean Nusselt number for laminar flow in a circular duct at constant wall temperature

$$Nu_T = 1.86 \left( \frac{Pe_b d}{L} \right)^{1/3} \left( \frac{\mu_b}{\mu_w} \right)^{0.14},$$

(9.38)

which is valid for smooth tubes, $0.48 < Pr_b < 16700$, and $0.0044 < (\mu_b/\mu_w) < 9.75$. This correlation has been recommended by Whitaker [12] for values of

$$\left( \frac{Pe_b d}{L} \right)^{1/3} \left( \frac{\mu_b}{\mu_w} \right)^{0.14} \geq 2.$$

(9.39)

All physical properties are evaluated at the fluid bulk mean temperature, except $\mu_w$, which is evaluated at the wall temperature.

It is not surprising that alternative correlations have been proposed for specific fluids. Oskay and Kakaç [13] performed experimental studies with mineral oil in laminar flow through a circular duct under constant wall heat-flux boundary condition in the range of $0.8 \times 10^3 < Re_b < 1.8 \times 10^3$ and $1 < (T_w/T_b) < 3$ and suggested that the viscosity ratio exponent for $Nu$ should be increased to 0.152 for mineral oil.

Kuznetsova [14] made experiments with transformer oil and fuel oil in the range of $400 < Re_b < 1900$ and $170 < Pr_b < 640$ and recommended that

$$Nu = 1.23 \left( \frac{Pe_b d}{L} \right)^{0.4} \left( \frac{\mu_b}{\mu_w} \right)^{1/6}.$$

(9.40)

Test [15] conducted an analytical and experimental study on heat transfer and fluid friction of laminar flow in a circular duct for liquids with temperature-dependent viscosity. The analytical approach is a numerical solution of continuity, momentum, and energy equations. The experimental approach involves the use of a hot-wire technique for determination of velocity profiles. He obtained the following correlation for the local Nusselt number:

$$Nu = 1.4 \left( \frac{Pe_b d}{L} \right)^{1/3} \left( \frac{\mu_b}{\mu_w} \right)^{n},$$

(9.41)

where

$$n = \begin{cases} 0.05 & \text{for heating liquids,} \\ 1/3 & \text{for cooling liquids.} \end{cases}$$

He also obtained the friction factor as

$$f = \frac{16}{Re} \cdot \frac{1}{0.89} \left( \frac{\mu_b}{\mu_w} \right)^{0.2}.$$

(9.42)

Equations 9.41 and 9.42 should not be applied to extremely long ducts.

## Example 9.1

Determine the total heat-transfer coefficient at 30 cm from the inlet of a heat exchanger where engine oil flows through the tubes that have a diameter of 0.5 in. Oil flows with a velocity of 0.5 m/s and at a bulk temperature of 30°C, while the local tube wall temperature is 60°C.

## Solution

Properties of the engine oil at $T_b = 30$°C from Appendix B are as follows:

$$\rho = 882.3 \text{ kg/m}^3, \quad c_p = 1922 \text{ J/(kg K)},$$

$$\mu = 0.416 \text{ N s/m}^2, \quad k = 0.144 \text{ W/(m K)},$$

$$Pr = 5550, \quad \mu_w = 0.074 \text{ N s/m}^2.$$

Heat-transfer coefficient may be obtained from the knowledge of the Reynolds number:

$$Re = \frac{\rho U_m d_i}{\mu} = \frac{882.3 \times 0.5 \times 0.0127}{0.416} = 13.47.$$

Since $Re < 2300$, the flow inside the tube is laminar. We can calculate the heat-transfer coefficient from the Sieder and Tate correlation, Equation 9.38:

$$Nu_T = 1.86(Re_b Pr_b)^{1/3} \left(\frac{d}{L}\right)^{1/3} \left(\frac{\mu_b}{\mu_w}\right)^{0.14},$$

which may be used as long as the following conditions are satisfied:

$$\left(\frac{\mu_b}{\mu_w}\right) = \left(\frac{0.416}{0.074}\right) = 5.62 < 9.75,$$

$$\left(\frac{Re_b Pr_b d}{L}\right)^{1/3} \left(\frac{\mu_b}{\mu_w}\right)^{0.14} = \left[\frac{13.8 \times 5550 \times 0.0127}{0.3}\right]^{1/3} \left(\frac{0.416}{0.074}\right)^{0.14} = 18.7 > 2.$$

Therefore, the previous correlation is applicable, and

$$Nu_T = 1.86 \times 18.7 = 34.8,$$

$$h = \frac{Nu_T k}{d_i} = \frac{34.8 \times 0.144}{0.0127} = 394.6 \text{ W/(m}^2 \text{ K)}.$$

The correlation given by Equation 9.24 that is applicable with constant heat-flux boundary condition can also be used since $Re_b Pr_b d/L = 5550 \times 13.47 \times 0.04233 = 3164 > 100$:

$$Nu_H = 1.953 \left(\frac{Pe_b \cdot d}{L}\right)^{1/3} = 1.953 \times (3164)^{1/3} = 28.67,$$

$$h = \frac{Nu_H k}{d_i} = \frac{28.67 \times 0.144}{0.0127} = 325 \text{ W/m}^2 \text{ K}.$$

The correlation (9.24) gives a more conservative answer.

### 9.4.2 Laminar Flow of Gases

The first reasonably complete solution for laminar heat transfer to a gas flowing in a tube with temperature-dependent properties was developed by Worsøe-Schmidt [16]. He solved the governing equations with a finite-difference technique for fully developed gas flow through a circular tube. Heating and cooling with a constant surface temperature and heating with constant heat flux were considered. In this solution, the radial velocity was included. He concluded that near the entrance, and also well downstream, the results can be satisfactorily correlated for heating $1 < (T_w/T_b) < 3$ by $n = 0$, $m = 1.00$ and for cooling $0.5 < (T_w/T_b) < 1$ by $n = 0$, $m = 0.81$.

Laminar forced convection and fluid flow in ducts have been studied extensively, and numerous results are available for circular and noncircular ducts under various boundary conditions. These results have been compiled by Shah and London [1] and Shah and Bhatti [2]. The laminar forced convection correlations discussed in the previous sections are summarized in Table 9.2. The constant-property correlations can be corrected for the variable physical properties by the use of Table 9.3, in which the exponents $m$ and $n$ are summarized. For fully developed laminar flow, $n = 0.14$ is generally recommended for heating of liquids.

---

## 9.5 Turbulent Forced Convection

Extensive experimental and theoretical efforts have been made to obtain the solutions for turbulent forced convection and flow friction problems in ducts because of their frequent occurrence and application in heat-transfer engineering. A complication of such solutions and correlations for circular and noncircular ducts has been summarized by Bhatti and Shah [17]. There are numerous correlations available in the literature for fully developed turbulent flow of single-phase Newtonian fluids in smooth, straight circular ducts with constant and temperature-dependent physical properties. The objective of this section is to highlight some of the existing correlations to be used in the design of heat-exchange equipment and to emphasize the conditions or limitations imposed on the applicability of these correlations.

### 9.5.1 Turbulent Flow in Circular Ducts with Constant Properties

Probably the most widely used relations in the past for fully developed turbulent flow of fluids with $Pr > 0.6$ in smooth tubes under moderate temperature and velocity conditions are recommended by Dittus–Boelter [3] and Colburn [4]:

$$Nu_b = 0.023 Re_b^{0.8} Pr_b^k \text{(Dittus}-\text{Boelter)}, \tag{9.43}$$

where

$$k = \begin{cases} 0.4 & \text{for heating} \\ 0.3 & \text{for cooling} \end{cases}$$

**TABLE 9.2**

Laminar Forced Convection Correlations in Smooth Straight Circular Ducts

| No. | References | Correlation | Limitations and Remarks |
|---|---|---|---|
| 1. | Nusselt–Graetz [1,2] | $Nu_T = 1.61(Pe_b d / L)^{1/3}$ <br><br> $Nu_T = 3.66$ | $Pe_b d/L > 10^3$, constant wall temperature. $Pe_b d/L < 10^2$, fully developed flow in a circular duct, constant wall temperature. |
| 2. | Schlünder [3] | $Nu_T = [(3.66)^3 + (1.61)^3 Pe_b d/L]^{1/3}$ | Superposition of two asymptotes given in No. 1 for the mean Nusselt number. $0.1 < Pe_b d/L < 10^4$. |
| 3. | Hausen [4] | $Nu_T = 3.66 + \dfrac{0.19(Pe_b d/L)^{0.8}}{1 + 0.117(Pe_b d/L)^{0.467}}$ | Thermal entrance region, constant wall temperature. $0.1 < Pe_b d/L < 10^4$. |
| 4. | Nusselt–Graetz [1,2] | $Nu_T = 1.953(Pe_b d/L)^{1/3}$ <br><br> $Nu_H = 4.36$ | $Pe_b d/L > 10^2$, constant heat flux. $Peb$ $d/L < 10$, fully developed flow in a circular duct, constant heat flux. |
| 5. | Pohlhausen [6] | $Nu_T = 0.664 \dfrac{1}{(Pr)^{1/6}} (Pe_b d/L)^{1/2}$ | $Pe_b d/L > 10^3$, $0.5 < Pr < 500$, simultaneously developing flow. |
| 6. | Stephan [8] | $Nu_T = Nu$ <br><br> $\quad + \phi(d_0/D_i) \dfrac{0.19(Pe D_h/L)^{0.8}}{1 + 0.117(Pe D_h/L)^{0.467}}$ | Circular annular duct, constant wall temperature, thermal entrance region. |
|  |  | $\phi(d_0/D_i) = 1 + 0.14(d_0/D_i)^{-1/2}$ | Outer wall is insulated, heat transfer through the inner wall. |
|  |  | $\phi(d_0/D_i) = 1 + 0.14(d_0/D_i)^{0.1}$ | Heat transfer through outer and inner wall. |
| 7. | Sieder and Tate [11] | $Nu_T = 1.86(Re_b Pr_b d/L)^{1/3}(\mu_b/\mu_w)^{0.14}$ | Thermal entrance region, constant wall temperature, $0.48 < Pr_b < 16{,}700$, $4.4 \times 10^{-3} < (\mu_b/\mu_w) < 9.75$, $(Re_b Pr_b d/L)^{1/3}$ $(\mu_b/\mu_w)^{0.14} > 2$. |
| 8. | Oskay and Kakaç [13] | $Nu_H = 1.86(Re_b Pr_b d/L)^{1/3}(\mu_b/\mu_w)^{0.152}$ | Thermal entrance region, constant wall heat flux, for oils $0.8 \times 10^3 < Re_b < 1.8 \times 10^3$, $1 < (T_w/T_b) < 3$. |
| 9. | Kuznetsova [14] | $Nu_H = 1.23(Re_b Pr_b d/L)^{0.4}(\mu_b/\mu_w)^{1/6}$ | Thermal entrance region, constant heat flux, $400 < Re_b < 1900$, $170 < Pr_b < 640$, for oils. |
| 10. | Test [15] | $Nu_b = 1.4(Re_b Pr_b d/L)^{1/3}(\mu_b/\mu_w)^n$ | Thermal entrance region, $n = 0.05$ for heating liquids, $n = 1/3$ for cooling liquids. |

*Source:* Kakaç, S. (Ed.), *Boilers, Evaporators and Condensers*, Wiley, New York, 1991.
*Note:* Unless otherwise stated, fluid properties are evaluated at the bulk mean fluid temperature, $T_b = (T_i + T_0)/2$.

and

$$Nu_b = 0.023 Re_b^{0.8} Pr_f^{1/3} \text{ (Colburn)}. \tag{9.44}$$

Extensive efforts have been made to obtain empirical correlations that either represent a best fit curve to the experimental data or have the constant in the theoretical equations adjusted to best fit the experimental data. An example of the latter is the correlation given by Petukhov and Popov [20]. Their theoretical calculations for the case of fully developed turbulent flow with constant properties in a circular tube with constant heat-flux

**TABLE 9.3**

Exponents *n* and *m* Associated with Equations 3.21 and 3.22 for Laminar Forced Convection through Circular Ducts, $Pr > 0.5$

| No. | References | Fluid | Condition | N | $m^a$ | Limitations |
|---|---|---|---|---|---|---|
| 1. | Deissler [9] | Liquid | Laminar, heating | 0.14 | −0.58 | Fully developed flow, $q''_w = $ constant, $Pr > 0.6$, $\mu/\mu_w = (T/T_w)^{-1.6}$. |
| | | Liquid | Laminar, cooling | 0.14 | −0.50 | |
| 2. | Yang [10] | Liquid | Laminar, heating | 0.11 | — | Developing and fully developed regions of a circular duct, $T_w = $ constant, $q''_w = $ constant. |
| 3. | Worsøe-Schmidt [16] | Gas | Laminar, heating | 0 | 1.00 | Developing and fully developed regions, $q''_w = $ constant, $T_w = $ constant, $1 < (T_w/T_b) < 3$. |
| | | Gas | Laminar, cooling | 0 | 0.81 | $T_w = $ constant, $0.5 < (T_w/T_b) < 1$. |

*Source:* Kakaç, S. (Ed.), *Boilers, Evaporators and Condensers*, Wiley, New York, 1991.

[a] Fanning friction factor *f* is defined as $f = 2\tau_w/(\rho u_m^2)$, and for hydrodynamically developed, isothermal laminar flow is $f = 16/Re$.

boundary condition yielded the following correlation, which is based on the three-layer turbulent boundary-layer model with constant adjusted to match the experimental data:

$$Nu_b = \frac{(f/2)Re_b Pr_b}{(1+13.6f)+(11.7+1.8Pr_b^{-1/3})(f/2)^{1/2}(Pr_b^{2/3}-1)}, \tag{9.45}$$

where

$$f = (3.64 \log Re_b - 3.28)^{-2}. \tag{9.46}$$

Equation 9.45 is applicable to fully developed turbulent flow in the range of $10^4 < Re_b < 5 \times 10^5$ and $0.5 < Pr_b < 2000$ with 1% error and in the range of $5 \times 10^5 < Re_b < 5 \times 10^6$ and $200 < Pr_b < 2000$ with 1%–2% error. Equation 9.45 is also applicable to rough tubes. A simpler correlation has also been given by Petukhov and Kirillov as reported in [21] as

$$Nu_b = \frac{(f/2)Re_b Pr_b}{1.07+12.7(f/2)^{1/2}(Pr_b^{2/3}-1)}. \tag{9.47}$$

Equation 9.47 predicts the results in the range of $10^4 < Re_b < 5 \times 10^6$ and $0.5 < Pr_b < 200$ with 5%–6% error and in the range of $0.5 < Pr_b < 2000$ with 10% error.

Webb [20] has examined a range of data for fully developed turbulent flow in smooth tubes; he concluded that the best relation developed by Petukhov and Popov, given previously, provides the best agreement with the measurements. Sleicher and Rouse [23] correlated analytical and experimental results for the range of $0.1 < Pr_b < 10^4$ and $10^4 < Re_b < 10^6$, obtaining

$$Nu_b = 5 + 0.015 Re_b^m Pr_b^n, \tag{9.48}$$

with

$$m = 0.88 - \frac{0.24}{4 + Pr_b},$$

$$n = \frac{1}{3} + 0.5\exp(-0.6Pr_b).$$

Equations 9.45, 9.47, and 9.48 are not applicable in the transition region. Gnielinski [24] further modified the Petukhov–Kirillov correlation by comparing it with experimental data so that the correlation covers a lower Reynolds number range. Gnielinski recommended the following correlation:

$$Nu_b = \frac{(f/2)(Re_b - 1000)Pr_b}{1 + 12.7(f/2)^{1/2}(Pr_b^{2/3} - 1)}, \tag{9.49}$$

where

$$f = (1.58 \ln Re_b - 3.28)^{-2}. \tag{9.50}$$

The effect of thermal boundary conditions is almost negligible in turbulent forced convection [26]; therefore, the empirical correlations given in Table 9.4 can be used for both constant wall temperature and constant wall heat-flux boundary conditions.

## 9.6 Turbulent Flow in Smooth Straight Noncircular Ducts

Heat-transfer and friction coefficients for turbulent flow in noncircular ducts are compiled in [17]. A common practice is to employ the hydraulic diameter in the circular duct correlations to predict $Nu$ and $f$ for the turbulent flow in noncircular ducts. For most of the noncircular smooth ducts, the accurate constant-property experimental friction factors are within ±10% of those predicted using the smooth circular duct correlation with hydraulic (equivalent) diameter $Dh$ instead of circular duct diameter $d$. The constant-property experimental Nusselt numbers are also within ±10% to ±15% except for some sharp-cornered and narrow channels. This order of accuracy is adequate for the overall heat-transfer coefficient and the pressure drop calculations in most of the practical design problems.

Many attempts have been reported in the literature to arrive at a universal characteristic dimension for internal turbulent flows that would correlate the constant-property friction factors and Nusselt numbers for all noncircular ducts [30–32]. It must be emphasized that any improvement made by these attempts is only a few percent, and therefore, the circular duct correlations may be adequate for many engineering applications.

Correlations given in Table 9.4 do not account for entrance effects occurring in short ducts. Gnielinski [3] recommends the entrance correlation factor derived by Hausen [25] to obtain the Nusselt number for short ducts from the following correlation:

**TABLE 9.4**

Correlations for Fully Developed Turbulent Forced Convection through a Circular Duct with Constant Properties

| No. | References | Correlation[a] | Remarks and Limitations |
|-----|-----------|----------------|-------------------------|
| 1. | Prandtl [25,26] | $Nu = \dfrac{(f/2)Re_b Pr_b}{1 + 8.7(f/2)^{1/2}(Pr_b - 1)}$ | Based on three-layer turbulent boundary-layer model. $Pr > 0.5$. |
| 2. | McAdams [27] | $Nu = 0.021\,Re_b^{0.8} Pr_b^{0.4}$ | Based on data for common gases; recommended for Prandtl numbers = 0.7. |
| 3. | Petukhov and Kirillov [21] | $Nu = \dfrac{(f/2)Re_b Pr_b}{1.07 + 12.7(f/2)^{1/2}(Pr_b^{2/3} - 1)}$ | Based on three-layer model with constants adjusted to match experimental data. $0.5 < Pr_b < 2000$, $10^4 < Re_b < 5 \times 10^6$. |
| 4. | Webb [22] | $Nu = \dfrac{(f/2)Re_b Pr_b}{1.07 + 9(f/2)^{1/2}(Pr_b - 1)Pr_b^{-1/4}}$ $f = (1.58\ln Re_b - 3.28)^{-2}$ | Theoretically based. Webb found No. 3 better at high $Pr$ and this one the same at other $Pr$. |
| 5. | Sleicher and Rouse [23] | $Nu = 5 + 0.015Re_b^m Pr_b^n$ $m = 0.88 - 0.24/(4 + Pr_b)$ $n = 1/3 + 0.5\exp(-0.6Pr_b)$ $Nu = 5 + 0.012Re_b^{0.83}(Pr_b + 0.29)$ | Based on numerical results obtained for $0.1 < Pr_b < 10^4$, $10^4 < Re_b < 10^6$. Within 10% of No. 6 for $> 10^4$. Simplified correlation for gases, $0.6 < Pr_b < 0.9$. |
| 6. | Gnielinski [24] | $Nu_b = \dfrac{(f/2)(Re_b - 1000)Pr_b}{1 + 12.7(f/2)^{1/2}(Pr_b^{2/3} - 1)}$ $f = (1.58\ln Re_b - 3.28)^{-2}$ $Nu_b = 0.0214(Re_b^{0.8} - 100)Pr_b^{0.4}$ $Nu_b = 0.012(Re_b^{0.87} - 280)Pr_b^{0.4}$ | Modification of No. 3 to fit experimental data at low $Re(2300 < Re_b < 10^4)$. Valid for $2300 < Re_b < 5 \times 10^6$ and $0.5 < Pr_b < 2000$. Simplified correlation for $0.5 < Pr < 1.5$. Agrees with No. 4 within −6% and +4%. Simplified correlation for $1.5 < Pr < 500$. Agrees with No. 4 within −10% and +0% for $3 \times 10^3 < Re_b < 10^6$. |
| 7. | Kays and Crawford [25] | $Nu_b = 0.022Re_b^{0.8} Pr_b^{0.5}$ | Modified Dittus–Boelter correlation for gases ($Pr = 0.5$–1.0). Agrees with No. 6 within 0%–4% for $Re_b \geq 5000$. |

*Source:*  Kakaç, S. (Ed.), *Boilers, Evaporators and Condensers*, Wiley, New York, 1991.
[a]  Properties are evaluated at bulk temperatures.

$$Nu_L = Nu_\infty\left[1 + \left(\frac{d}{L}\right)^{2/3}\right], \tag{9.51}$$

where $Nu_\infty$ represents the fully developed Nusselt numbers calculated from correlations given in Table 9.4. It should be noted that the entrance length depends on Reynolds and Prandtl numbers and thermal boundary condition. Thus, Equation 9.51 should be used cautiously.

**Example 9.2**

A 5000 kg/h of water will be heated from 20°C to 35°C by hot water at 140°C. A 15°C hot water temperature drop is allowed. A number of 15 ft (4.5 m) hairpins of 3 in. (ID = 3.068 in., OD = 3.5 in.) by 2 in. (ID = 2.067 in., OD = 2.375 in.) double-pipe heat exchanger with annuli and pipes each connected in series will be used. Hot water flows through the inner tube. Calculate the following:

a. Heat-transfer coefficient in the inner tube
b. Heat-transfer coefficient inside the annulus (assuming the outside of the annulus is insulated against heat loss)

**Solution**
We first calculated the Reynolds number to determine if the flow is laminar or turbulent and then select the proper correlation to calculate the heat-transfer coefficient.

a. *Heat-transfer coefficient in the inner tube.* The properties of hot water at $T_b = 132.5°C$ from Appendix B are as follows:

$$\rho = 932.4 \text{ kg/m}^3, \qquad c_p = 4268.1 \text{ J/(kg K)},$$

$$k = 0.688 \text{ W/(m}^2 \text{ K)}, \quad \mu = 0.208 \times 10^{-3} \text{ N s/m}^2,$$

$$Pr = 1.29.$$

We now make an energy balance to calculate the hot water mass flow rate:

$$(mc_p)_h \Delta T_h = (mc_p)_c \Delta T_c,$$

$$m_h = \frac{m_c c_{pc}}{c_{ph}} = \frac{(5000/3600)(4179)}{4268.1} = 1.360 \text{ kg/s}.$$

where $C_{pc} = 4179$ J/(kg K) is at $T_b = 27.5°C$. Reynolds number of the heated water flow can be calculated as

$$Re = \frac{\rho U_m d_i}{\mu} = \frac{4\dot{m}}{\pi d_i \mu} = \frac{4 \times 1.389}{\pi \times 0.0525 \times 0.208 \times 10^{-3}} = 161,940,$$

so the flow is turbulent. We can therefore select a correlation from Table 9.4. The Petukhov–Kirillov correlation, Equation 9.47, is used here:

$$Nu_b = \frac{(f/2)Re_b Pr_b}{1.07 + 12.7(f/2)^{0.5}(Pr_m^{2/3} - 1)},$$

where

$$f = (1.58 \ln Re_b - 3.28)^{-2}$$

$$f = [1.58 \ln(161940) - 3.28]^{-2} = 0.00407,$$

$$Nu_b = \frac{\frac{1}{2}(4.0 \times 10^{-3})(161940 \times 1.29)}{1.07 + 12.7(4.07 \times 10^{-3}/2)^{0.5}(1.29^{2/3} - 1)} = 361.4,$$

$$h_i = \frac{Nu_b k}{d_i} = \frac{361.4 \times 0.688}{2.067 \times 2.54 \times 10^{-2}} = 4736 \text{ W/(m}^2 \text{ K)}.$$

The effect of property variations can be found from Equation 9.34 with $n = 0.25$ for cooling of a liquid in turbulent flow (Table 9.4).

b. *Heat-transfer coefficient in the annulus.* The properties of cold water at $T_b = 27.5°C$ from Appendix B are as follows:

$$\rho = 996.8 \text{ kg/m}^3, \quad c_p = 4179 \text{ J/(kg K)},$$

$$k = 0.614 \text{ W/(m K)}, \quad \mu = 0.846 \times 10^{-3} \text{ N s/m}^2,$$

$$Pr = 5.77.$$

Hydraulic diameter of annulus from Equation 9.31 is

$$D_i - d_o = (3.068 - 2.375) \times 2.54 \times 10^{-2} = 0.0176 \text{ m},$$

$$Re = \frac{4D_h m_c}{\pi\mu(D_i^2 - d_o^2)} = \frac{4(0.0176)(5000/3600)}{\pi \times (846 \times 10^{-6})(0.002432)} = 15117.53.$$

Therefore, the flow inside annulus is turbulent. One of the correlations can be selected from tables. Gnielinski correlation is used here. It should be noted that for the annulus, Nusselt number should be based on hydraulic diameter (or equivalent diameter) calculated from Equation 9.32:

$$D_e = \frac{4A_c}{P_h} = \frac{4\left[\pi/4(D_i^2 - d_o^2)\right]}{\pi d_o}$$

$$D_e = \frac{\left[(0.0779)^2 - (0.0603)^2\right]}{0.0603} = 0.0403 \text{ m},$$

$$Nu_b = \frac{(f/2)(Re_b - 1000)Pr_b}{1.07 + 12.7(f/2)^{1/2}(Pr_b^{2/3} - 1)},$$

$$f = \left[1.58\ln Re_b - 3.28\right]^{-2}$$

$$f = \left[1.58\ln(15135.13) - 3.28\right]^{-2}$$

$$= 0.00703,$$

$$Nu_b = \frac{(0.00703/2)(15117.53 - 1000)(5.77)}{1.07 + 12.7(0.00703/2)^{1/2}(5.77^{2/3} - 1)} = 244.9,$$

$$h_0 = \frac{Nu_b k}{D_e} = \frac{244.9 \times 0.614}{0.0403} = 3732 \text{ W/(m}^2 \text{ K)}.$$

## 9.7 Effects of Variable Physical Properties in Turbulent Forced Convection

When there is a large difference between the duct wall and fluid bulk temperatures, heating and cooling influence the heat transfer and the fluid friction in turbulent duct flow due to the distortion of turbulent transport mechanisms, in addition to the variation of fluid properties with temperatures as in laminar flow.

### 9.7.1 Turbulent Liquid Flow in Ducts

Petukhov [21] reviewed the status of heat transfer and wall friction in fully developed turbulent pipe flow both constant and variable physical properties.

To choose the correct value of n in Equation 9.34a, the heat-transfer experimental data corresponding to heating and cooling for several liquids over a wide range of values $(\mu_w/\mu_b)$ were collected by Petukhov [21]. He found that the data are well correlated by

$$\frac{\mu_w}{\mu_b} < 1, \quad n = 0.11 \text{ for heating liquids}, \tag{9.52}$$

$$\frac{\mu_w}{\mu_b} < 1, \quad n = 0.25 \text{ for cooling liquids}, \tag{9.53}$$

which are applicable for fully developed turbulent flow in the range of $10^4 < Re_b < 5 \times 10^6$, $2 < Pr_b < 140$, and $0.08 < (\mu_w/\mu_b) < 40$. The value of $Nu_{cp}$ in Equation 9.34a is calculated from Equation 9.45 or 9.47.

The value of $Nu_{cp}$ can also be calculated from the correlations listed in Table 9.4.

Petukhov [21] collected data from various investigators for variable viscosity influence on friction in water for both heating and cooling and suggested the following correlations for the friction factor:

$$\frac{\mu_w}{\mu_b} < 1, \quad \frac{f}{f_{cp}} = \frac{1}{6}\left(7 - \frac{\mu_b}{\mu_w}\right) \text{ for heating liquids}, \tag{9.54}$$

$$\frac{\mu_w}{\mu_b} < 1, \quad \frac{f}{f_{cp}} = \left(\frac{\mu_b}{\mu_w}\right)^{0.24} \text{ for cooling liquids}. \tag{9.55}$$

The friction factor for an isothermal (constant property) flow, $f_{cp}$, can be calculated by the use of Table 9.5 or directly from Equation 9.46 for the range of $0.35 < (\mu_w/\mu_b) < 2$, $10^4 < Re_b < 23 \times 10^4$, and $1.3 < Pr_b < 10$.

### 9.7.2 Turbulent Gas Flow in Ducts

Heat-transfer and friction coefficients for turbulent fully developed gas flow in a circular duct were obtained theoretically by Petukhov and Popov [20] by assuming physical properties $\rho$, $c_p$, $k$, and $\mu$ as given functions of temperature. This analysis is valid only for low subsonic velocities, since the variations of density with pressure and heat dissipation in the flow were neglected. The eddy diffusivity of momentum was extended to the case of variable properties. The turbulent Prandtl number was taken to be unity (i.e., $\varepsilon_h = \varepsilon_m$). The analyses were carried out for hydrogen and air for the following range of parameters: $0.37 < (T_w/T_b) < 3.1$ and $10^4 < Re_b < 4.3 \times 10^6$ for air and $0.37 < (T_w/T_b) < 3.7$ and $10^4 < Re_b < 5.8 \times 10^6$ for hydrogen. The analytical results were

**TABLE 9.5**

Turbulent Flow Isothermal Fanning-Friction-Factor Correlations in Smooth Circular Ducts

| No. | References[a] | Correlation[b] | Remarks and Limitations |
|-----|-----------|-------------|------------------------|
| 1. | Blasius | $f = 2\tau_w / \rho u_m^2 = 0.0791 Re^{-1/4}$ | This approximate explicit equation agrees with No. 3 within $\pm 2.5\%$. $4 \times 10^3 < Re < 10^5$. |
| 2. | Drew, Koo, and McAdams | $f = 0.00140 + 0.125 Re^{-0.32}$ | This correlation agrees with No. 3 within $-0.5\%$ and $+3\%$. $4 \times 10^3 < Re < 5 \times 10^6$. |
| 3. | von Karman and Nikuradse | $1/\sqrt{f} = 1.737 \ln(Re\sqrt{f}) - 0.4$<br><br>or,<br><br>$1/\sqrt{f} = 4\log(Re\sqrt{f}) - 0.4$<br><br>approximated as<br><br>$f = (3.64\log Re - 3.28)^{-2}$ | von Karman's theoretical equation with the constants adjusted to best fit Nikuradse's experimental data. Also referred to as the Prandtl correlation. Should be valid for very high values of $Re$. $4 \times 10^3 < Re < 3 \times 10^6$. |
| | | $f = 0.046 Re^{-0.2}$ | This approximate explicit equation agrees with the previous one within $-0.4\%$ and $+2.2\%$ for $3 \times 10^4 < Re < 10^6$. |
| 4. | Flonenko | $f = 1/(1.58\ln Re - 3.28)^2$ | Agrees with No. 3 within $\pm 0.5\%$ for $3 \times 10^4 < Re < 10^7$ and within $\pm 1.8\%$ at $Re = 10^4$. $10^4 < Re < 5 \times 10^5$. |
| 5. | Techo, Tickner, and James | $1/f = \left(1.7372 \ln \dfrac{Re}{1.964 \ln Re - 3.8215}\right)^2$ | An explicit form of No. 3; agrees with it within $\pm 0.1\%$, $10^4 < Re < 2.5 \times 10^8$ |

*Source:* Kakaç, S. (Ed.), *Boilers, Evaporators and Condensers*, Wiley, New York, 1991.
[a] Cited in References 17,25,26,28.
[b] Properties are evaluated at bulk temperatures.

correlated by Equation 9.35a, where $Nu_{cp}$ is given by Equation 9.45 or 9.47, and the following values for $n$ are obtained:

$$\frac{T_w}{T_b} < 1, \quad n = -0.36 \text{ for cooling gases,} \tag{9.56}$$

$$\frac{T_w}{T_b} > 1, \quad n = \left[0.3\log\left(\frac{T_w}{T_b}\right) + 0.36\right] \text{ for heating gases.} \tag{9.57}$$

With these values for $n$, Equation 9.35a describes the solution for air and hydrogen with an accuracy of $\pm 4\%$. For simplicity, one can take $n$ to be constant for heating as $n = -0.47$; then, Equation 9.35a describes the solution for air and hydrogen within $+6\%$. These results have also been confirmed experimentally and they can be used for practical calculations when $1 < (T_w/T_b) < 4$. Exponents $n$ and $m$ associated with Equations 9.34 and 9.35 for turbulent forced convection in duct are presented in Table 9.6.

A great number of experimental studies are available in the literature for the heat transfer between the tube wall and the gas flow with large temperature differences and temperature-dependent physical properties. The majority of the work deals with gas heating at constant wall temperature in a circular duct; experimental studies on gas cooling are limited.

**TABLE 9.6**

Exponents *n* and *m* Associated with Equations 3.24 and 3.25 for Turbulent Forced Convection through Circular Ducts

| No. | References | Fluid | Condition | *n* | *m* | Limitations |
|-----|-----------|-------|-----------|-----|-----|-------------|
| 1. | Petukhov [21] | Liquid | Turbulent heating | 0.11 | — | $10^4 < Re_b < 1.25 \times 10^5, 2 < Pr_b < 140, 0.08 < \mu_w/\mu_b < 1$ |
| | | Liquid | Turbulent cooling | 0.25 | — | $1 < \mu_w/\mu_b < 40$ |
| | | Liquid | Turbulent heating | — | Equation 9.39 or −0.25 | $10^4 < Re_b < 23 \times 10^4, 1.3 < Pr_b < 10$ $0.35 < \mu_w/\mu_b < 1$ |
| | | Liquid | Turbulent cooling | — | −0.24 | $1 < \mu_w/\mu_b < 2$ |
| 2. | Petukhov and Popov [20] | Gas | Turbulent heating | −0.47 | — | $10^4 < Re_{b'} < 4.3 \times 10^6, 1 < T_w/T_b < 3.1$ |
| | | Gas | Turbulent cooling | −0.36 | — | $0.37 < T_w/T_b < 1$ |
| | | Gas | Turbulent heating | — | −0.52 | $14 \times 10^3 < Re^* < 10^6, 1 < T_w/T_b < 3.7$ |
| | | Gas | Turbulent cooling | — | −0.38 | $0.37 < T_w/T_b < 1$ |
| 3. | Perkings and Worsøe-Schmidt [33] | Gas | Turbulent heating | — | −0.264 | $1 \le T_w/T_b \le 4$ |
| 4. | McElligot et al. [34] | Gas | Turbulent heating | — | −0.1 | $1 < T_w/T_b < 2.4$ |

The results of heat-transfer measurements at large temperature differences between the wall and the gas flow are usually presented as

$$Nu_b = CRe_b^{0.8} Pr_b^{0.4} \left( \frac{T_w}{T_b} \right)^n. \tag{9.58}$$

For fully developed temperature and velocity profiles (i.e., $L/d < 60$), *C* becomes constant and *n* becomes independent of $L/d$.

A number of heat-transfer correlations have been developed for variable-property fully developed turbulent liquid and gas flow in a circular duct, some of which are also summarized in Tables 9.7 and 9.8.

### Example 9.3

Air at 40°C flows through a heated pipe section with a velocity of 6 m/s. The length and diameter of the pipe are 300 and 2.54 cm, respectively. The average pipe wall temperature is 300°C. Determine the average heat-transfer coefficient.

### Solution
Since the wall temperature is so much higher than the initial air temperature, variable-property flow must be considered.

*Properties.* From Appendix B, for air at ($T_b = 40$°C)

$$\rho = 1.128 \text{ kg/m}^3, \qquad c_p = 1005.3 \text{ J/(kg K)},$$

$$k = 0.0267 \text{ W/(m K)}, \quad \mu = 1.912 \times 10^{-5} \text{ N s/m}^2,$$

$$Pr = 0.719.$$

**TABLE 9.7**

Turbulent Forced Convection in Circular Ducts for Liquids with Variable Properties

| No. | References | Correlation | Comments and Limitations |
|---|---|---|---|
| 1. | Colburn [19] | $St_b Pr_f^{2/3} = 0.023 Re_f^{-0.2}$ | $L/d > 60$, $Pr_b > 0.6$, $T_f = (T_b + T_w)/2$; inadequate for large $(T_w - T_b)$. |
| 2. | Sieder and Tate [11] | $Nu_b = 0.023 Re_b^{0.8} Pr_b^{1/3} \left( \dfrac{\mu_b}{\mu_w} \right)^{0.14}$ | $L/d > 60$, $Pr_b > 0.6$, for moderate $(T_w - T_b)$. |
| 3. | Petukhov and Kirillov [21] | $Nu_b = \dfrac{(f/8) Re_b Pr_b}{1.07 + 12.7\sqrt{f/8}(Pr_b^{2/3} - 1)} \left( \dfrac{\mu_b}{\mu_w} \right)^{n}$ | $L/d > 60$, $0.08 < \mu_w/\mu_b < 40$, $10^4 < Re_b < 5 \times 10^6$, $2 < Pr_b < 140$, $f = (1.82 \log Re_b - 1.64)^{-2}$, $n = 0.11$ (heating), $n = 0.25$ (cooling). |
| 4. | Hufschmidt et al. [35] | $Nu_b = \dfrac{(f/8) Re_b Pr_b}{1.07 + 12.7\sqrt{f/8}(Pr_b^{2/3} - 1)} \left( \dfrac{Pr_b}{Pr_w} \right)^{0.11}$ | Water, $2 \times 10^4 < Re_b < 6.4 \times 10^5$, $2 < Pr_b < 5.5$, $f = (1.82 \log Re_b - 1.64)^{-2}$, $0.1 < Pr_b/Pr_w < 10$. |
| 5. | Yakovlev [36] | $Nu_b = 0.0277 Re_b^{0.8} Pr_b^{0.36} \left( \dfrac{Pr_b}{Pr_w} \right)^{0.11}$ | Fully developed conditions. The use of Prandtl group was first suggested by the author in 1960. |
| 6. | Oskay and Kakaç [13] | $Nu_b = 0.023 Re_b^{0.8} Pr_b^{0.4} \left( \dfrac{\mu_b}{\mu_w} \right)^{0.262}$ | Water, $L/d > 10$, $1.2 \times 10^4 < Re_b < 4 \times 10^4$. |
| | | $Nu_b = 0.023 Re_b^{0.8} Pr_b^{0.4} \left( \dfrac{\mu_b}{\mu_w} \right)^{0.487}$ | 30% glycerin–water mixture $L/d > 10$, $0.89 \times 10^4 < Re_b < 2.0 \times 10^4$. |
| 7. | Hausen [37] | $Nu_b = 0.0235(Re_b^{0.8} - 230)(1.8 Pr_b^{0.3} - 0.8)$ $\times \left[ 1 + \left( \dfrac{d}{L} \right)^{2/3} \right] \left( \dfrac{\mu}{\mu_w} \right)^{0.14}$ | Altered form of equation presented in 1959 [4]. |
| 8. | Sleicher and Rouse [23] | $Nu_b = 5 + 0.015 Re_f^m Pr_w^n$ $m = 0.88 - 0.24/(4 + Pr_w)$ $n = \dfrac{1}{3} + 0.5e^{-0.6 Pr_w}$ | $L/d > 60$, $0.1 < Pr_b < 10^5$, $10^4 < Re_b < 10^6$. |
| | | $Nu_b = 0.015 Re_f^{0.88} Pr_w^{1/3}$ | $Pr_b > 50$. |
| | | $Nu_b = 4.8 + 0.015 Re_f^{0.85} Pr_w^{0.93}$ | $Pr_b < 0.1$, uniform wall temperature. |
| | | $Nu_b = 6.3 + 0.0167 Re_f^{0.85} Pr_w^{0.93}$ | $Pr_b < 0.1$, uniform wall heat flux. |

*Source:* Kakaç, S. (Ed.), *Boilers, Evaporators and Condensers*, Wiley, New York, 1991.

The inside heat-transfer coefficient can be obtained from the knowledge of flow regime, that is, the Reynolds number:

$$Re_b = \frac{\rho U_m d_i}{\mu} = \frac{1.128 \times 6 \times .0254}{1.912 \times 10^{-5}} = 8991.$$

Hence, the flow in the tube is turbulent. On the other hand, $L/d = 3/0.0254 = 118 > 60$; fully developed conditions are assumed. Since $Pr > 0.6$, we can use the

**TABLE 9.8**

Turbulent Forced Convection in Circular Ducts for Gases with Variable Properties

| No. | References | Correlation | Gas | Comments and Limitations |
|-----|-----------|-------------|-----|--------------------------|
| 1. | Humble et al. [38] | $Nu_b = 0.023 Re_b^{0.8} Pr_b^{0.4} \left( \dfrac{T_w}{T_b} \right)^n$ $T_w/T_b < 1 \quad n = 0 \text{ (cooling)}$ $T_w/T_b > 1 \quad n = -0.55 \text{ (heating)}$ | Air | $30 < L/d < 120, 7 \times 10^3 < Re_b < 3 \times 10^5,$ $0.46 < T_w/T_b < 3.5$ |
| 2. | Bialokoz and Saunders [21] | $Nu_b = 0.022 Re_b^{0.8} Pr_b^{0.4} \left( \dfrac{T_w}{T_b} \right)^{-0.5}$ | Air | $29 < L/d < 72, 1.24 \times 10^5 < Re_b < 4.35 \times 10^5, 1.1 < T_w/T_b < 1.73$ |
| 3. | Barnes and Jackson [39] | $Nu_b = 0.023 \, Re_b^{0.8} Pr_b^{0.4} \left( \dfrac{T_w}{T_b} \right)^n$ $n = -0.4 \text{ for air}$ $n = -0.185 \text{ for helium}$ $n = -0.27 \text{ for carbon dioxide}$ | Air, helium, carbon dioxide | $1.2 < T_w/T_b < 2.2, 4 \times 10^3 < Re_b < 6 \times 10^4,$ $L/d > 60$ |
| 4. | McEligot et al. [34] | $Nu_b = 0.021 Re_b^{0.8} Pr_b^{0.4} \left( \dfrac{T_w}{T_b} \right)^{-0.5}$ $Nu_b = 0.021 Re_b^{0.8} Pr_b^{0.4} \left( \dfrac{T_w}{T_b} \right)^{-0.5}$ $\times \left[ 1 + (L/d)^{-0.7} \right]$ | Air, helium, nitrogen | $L/d > 30, 1 < T_w/T_b < 2.5, 1.5 \times 10^4 < Re_{ib} < 2.33 \times 10^5, L/d > 5,$ local values |
| 5. | Perkins and Worsoe-Schmidt [33] | $Nu_b = 0.024 Re_{bb}^{0.8} Pr_b^{0.4} \left( \dfrac{T_w}{T_b} \right)^{-0.7}$ $Nu_w = 0.023 Re_w^{0.8} Pr_w^{0.4}$ $Nu_b = 0.024 Re_b^{0.8} Pr_b^{0.4} \left( \dfrac{T_w}{T_b} \right)^{-0.7}$ $\times \left[ 1 + \left( \dfrac{L}{d} \right)^{-0.7} \left( \dfrac{T_w}{T_b} \right)^{0.7} \right]$ | Nitrogen | $L/d > 40, 1.24 < T_w/T_b < 7.54, 18.3 \times 10^3 < Re_{ib} < 2.8 \times 10^5.$ Properties evaluated at wall temperature, $L/d > 24 \, 1.2 \le L/d \le 144$ |
| 6. | Petukov et al. [21] | $Nu_b = 0.021 Re_b^{0.8} Pr_b^{0.4} \left( \dfrac{T_w}{T_b} \right)^n$ $n = -\left( 0.9 \log \dfrac{T_w}{T_b} + 0.205 \right)$ | Nitrogen | $80 < L/d < 100, 13 \times 10^3 < Re_b < 3 \times 10^5 \, 1 < T_w/T_b < 6$ |
| 7. | Sleicher and Rouse [20] | $Nu_b = 5 + 0.012 Re_f^{0.83} (Pr_w + 0.29)$ | | For gases, $0.6 < Pr_b < 0.9$ |

*(continued)*

**TABLE 9.8 (continued)**

Turbulent Forced Convection in Circular Ducts for Gases with Variable Properties

| No. | References | Correlation | Gas | Comments and Limitations |
|-----|-----------|-------------|-----|--------------------------|
| 8. | Gnielinski [3] | $Nu_b = 0.0214(Re_b^{0.8} - 100)Pr_b^{0.4}\left(\dfrac{T_b}{T_w}\right)^{0.45}$ $\times\left[1+\left(\dfrac{d}{L}\right)^{2/3}\right]$ $Nu_b = 0.012(Re_b^{0.87} - 280)Pr^{0.4}\left(\dfrac{T_b}{T_w}\right)^{0.4}$ $\times\left[1+\left(\dfrac{d}{L}\right)^{2/3}\right]$ | Air, helium, carbon dioxide | $0.5 < Pr_b < 1.5$, for heating carbon dioxide of gases. The author collected the data from the literature. Second for $1.5 < Pr_b < 500$ |
| 9. | Dalle-Donne and Bowditch [40] | $Nu_b = 0.022Re_b^{0.8}Pr_b^{0.4}\left(\dfrac{T_w}{T_b}\right)^{-[0.29+0.0019L/d]}$ | Air, helium | $10^4 < Re_b < 10^5$, $18 < L/d < 316$ |

*Source:* Kakaç, S. (Ed.), *Boilers, Evaporators and Condensers*, Wiley, New York, 1991.

correlations given in Table 9.4. Hence, Gnielinski's correlation, Equation 9.49, with constant properties

$$Nu_b = \frac{(f/2)(Re_b - 1000)Pr_b}{1 + 12.7 \times (f/2)^{1/2}(Pr^{2/3} - 1)}$$

may be used to determine the Nusselt number:

$$f = (1.58\ln Re_b - 3.28)^{-2}$$

$$f = [1.58\ln(8991) - 3.28]^{-2} = 0.00811,$$

$$Nu_b = \frac{hd}{k} = \frac{(0.00811/2)(8991 - 1000)(0.719)}{1 + 12.7(0.00811/2)^{0.5}(0.719^{2/3} - 1)},$$

$$h = Nu_b \times \frac{k}{d} = \frac{27.72 \times 0.0267}{0.0254} = 29.14 \text{ W}/(\text{m}^2 \text{ K}).$$

The heat-transfer coefficient with variable properties can be calculated from Equation 9.35a, where $n$ is given in Table 9.6 as $n = -0.47$:

$$Nu_b = Nu_{cp}\left(\frac{T_w}{T_b}\right)^{-0.47}$$

$$Nu_b = 27.712\left(\frac{573}{313}\right)^{-0.47} = 20.856.$$

Then

$$h = \frac{Nu_b k}{d} = \frac{(20.856 \times 0.0267)}{0.0254} = 22 \text{ W}/(\text{m}^2 \text{ K}).$$

As seen, in the case of a gas with temperature-dependent properties, heating a gas decreases the heat-transfer coefficient.

## 9.8 Liquid Metal Heat Transfer

In recent years, liquid metals have been used as reactor coolants. The usefulness of liquid metals as reactor coolant is partly due to their high thermal conductivity that results in high heat-transfer rates. As a result, they are especially applicable to fast nuclear reactors where large energy quantities must be removed from a relatively small space. The Prandtl numbers of liquid metals vary from 0.003 to 0.06.

The liquid metals, such as sodium, mercury, lead, sodium–potassium, and lead–bismuth alloys, have low melting points and can be used over a wide range of temperature; but they pose some difficulties in handling and pumping.

In laminar flow, the measure of the relative magnitudes of heat transfer by convective transport and by molecular diffusion is given by the Peclet number. When the Peclet number is large, convective transport is large relative to molecular diffusion and it can be expected that axial heat conduction will be negligible. Nonmetallic fluids in laminar flow will rarely have Peclet numbers less than 100. With liquid metals, however, Peclet numbers less than unity are not unreasonable and neglecting the axial heat conduction may not be justified. There has been little technological interest in the forced convection laminar flow of liquid metals; consequently research in this area has been relatively small. From a practical point of view, the heat-transfer coefficient for laminar flow of liquid metals can be calculated using the appropriate analytical expressions given in Chapter 6.

For any fluid flowing in a long tube, the Nusselt number rapidly approaches a constant value of 4.36 for both constant wall heat flux and when the wall temperature increase is linear and 3.65 when the wall temperature is uniform.

Since liquid metals have relatively small viscosities, their use as coolant or heat-exchange fluid almost always involves turbulent flow. Further, both from experimental data and from fundamental principles, analysis has been content to ignore dependence of physical properties on temperature. With turbulent flow in ducts, ordinary molecular conduction of heat dominates the mechanism of heat transfer in regions adjacent to the duct wall.

At extremely small distances from the duct walls and throughout the remainder of the duct cross section, heat transfer by turbulent diffusion becomes important. With liquid metals, because of their relatively large thermal conductivities, molecular conduction is usually significant throughout the duct cross section. As a result, temperature gradients are not localized as with nonmetallic fluids.

Extensive experimental data on liquid metals are given in [41], and the heat-transfer characteristics are summarized in [42].

First theoretical analysis of liquid metals was given by Martinelli in 1947. The results of this analysis have been outlined in Chapter 7.

By assuming that the eddy diffusivities of heat and momentum are identical at each point, Lyon [43] showed that for Prandtl numbers less than 0.1, the results of Martinelli, Equation 7.117, can be approximated by the following equation for liquid metals in turbulent flow through long round tubes with constant heat input along the tube wall:

$$Nu_b = 7 + 0.025(RePr)_b^{0.8}. \tag{9.59}$$

In Equation 9.59, the last term shows the relative importance of eddy conductivity and the constant 7 shows the relative importance of ordinary thermal conduction. For $Pe < 10^4$, the ordinary conductivity has an important influence on Nusselt number because of the conduction in the turbulent core. In practical liquid metal reactor applications, the range of Peclet number with turbulent flow will be between about 20 and 10,000.

The idea of a fictitious fluid film that incorporates all of the thermal resistance is of little use in predicting liquid metal heat transfer. In liquid metal heat transfer, important thermal resistance may extend into the turbulent core.

The concept of an effective diameter, which is used for noncircular passages with fluids of $Pr > 0.7$, cannot be used directly in relating heat transfer in differently shaped channels for low Prandtl numbers, such as liquid metals.

It was found theoretically by Seban and Shimazaki [44] that the ordinary conduction contribution with constant wall temperature boundary condition is about 5, compared with 7 when constant heat flows through the wall. Thus, for constant wall temperature in a long tube, the heat transfer to liquid metals can be calculated by the following relation:

$$Nu_b = 5 + 0.025(RePr)_b^{0.8}, \tag{9.60}$$

where all properties are evaluated at the bulk temperature. Equation 9.60 is valid for $Pe > 100$ and $L/d > 60$.

Skupinski et al. [45] presented the experimental results for a Na–K mixture (44% Na/56% K) flow through a long horizontal circular pipe under constant heat-flux boundary condition, for the range of Reynolds numbers from $3.6 \times 10^3$ to $9.05 \times 10^5$ and Peclet numbers from $5.8 \times 10$ to $1.31 \times 10^4$ in the form of

$$Nu_b = 4.82 + 0.0185(RePr)_b^{0.827}. \tag{9.61}$$

Other approximate theoretical equations of interest for turbulent flows are as follows:

$$Nu_b = 5.8 + 0.02(RePr)_b^{0.8}, \tag{9.62}$$

for a channel between two plates with heat passing through only one side [46], and

$$Nu_b = \frac{h(D_o - D_i)}{k} = 0.75\left(\frac{D_o}{D_i}\right)^{1/3}\left[7 + 0.025(RePr)_b^{0.8}\right], \tag{9.63}$$

for an annulus [41] with constant heat flux at $D_o$, where $D_o$ is the outer diameter and $D_i$ is the inner diameter of the annulus.

Unfortunately, the theoretical expressions do not always predict the exact values that are found in experiments. A very complete summary of the experimental liquid metal heat-transfer information in long and short circular tubes and in annuli has been published by Lubarsky and Kaufmann [42].

For design purposes, Lubarsky and Kaufman recommended the following empirical equation:

$$Nu_b = 0.625(RePr)_b^{0.4}, \tag{9.64}$$

which correlates most of the constant-heat-flux data in the fully developed turbulent flow of liquid metals in smooth tubes.

**Example 9.4**

Liquid Na–K alloy (56% Na, 44% K) flows with a velocity of 6 m/s through a long 1 in. diameter electrically heated stainless steel tube and is to be heated from 90°C to 650°C as it passes through the tube. If the tube wall temperature is not to exceed 700°C, calculate the heat transfer per unit length of the tube.

**Solution**

The properties of Na–K alloy at the average bulk temperature 370°C, from Appendix B, are as follows:

$$\mu = 2.356 \times 10^{-4} \text{ kg/m s}, \quad c_p = 1055 \text{ J/(kg K)},$$

$$k = 27.56 \text{ W/(m K)}.$$

The Reynolds number is

$$Re = \frac{\rho U_m d}{\mu} = \frac{820 \times 6 \times 2.54 \times 10^{-2}}{2.356 \times 10^{-4}} = 5.305 \times 10^5.$$

Therefore, the flow is turbulent. From the properties we have

$$Pr = \frac{c_p \mu}{k} = \frac{1055 \times 2.356 \times 10^{-4}}{27.56} = 0.009.$$

We may use Equation 9.64 to calculate the heat-transfer coefficient since a constant heat flux is maintained on the tube wall:

$$Nu = \frac{hd}{k} = 0.625 (Re Pr)^{0.4}$$

$$= 0.625 \times (5.305 \times 10^5 \times 0.009)^{0.4} = 18.526,$$

which gives

$$h = Nu \frac{k}{d} = 18.526 \times \frac{27.56}{1 \times 2.54 \times 10^{-2}} = 2.01 \times 10^4 \text{ W/(m}^2 \text{ K)}.$$

Heat transfer per unit length of the tube can now be calculated from Newton's law of cooling:

$$q = hA(T_w - T_m)$$

$$= 2.01 \times 10^4 \times \pi \times 2.54 \times 10^{-2} \times (700 - 370)$$

$$= 5.29 \times 10^5 \text{ W/m}.$$

Equation 9.64 predicts values of about one-half those predicted by Equation 9.59 at $Pe = 100$ and values about 30% below by Equation 9.59 at $Pe = 10^3$.

There is a discrepancy between the theoretical prediction and the experimental results. In the experiments at low values of $Pe$, because of lower velocity, some natural convection may have become important, or as suggested by Donald and Quittenton [47], dissolved or entrained gas may be liberated and accumulated at poorly wetted regions of the wall. Such a condition might explain the drastic drop in experimental $Nu$ below $Pe = 100$.

Isakoff and Drew [48] experimentally studied the heat transfer in liquid metals. They have actually measured the temperature and velocity profiles of a liquid metal stream inside a heat exchanger. Their overall data agree with Equation 9.59 reasonably well.

The theoretical development is not quite satisfactory in predicting liquid metal heat transfer. Progress in theoretical development requires much more experimental work in liquid metal heat transfer.

Empirical correlations for liquid metal heat transfer in fully developed turbulent flow in various geometries derived from both experimental and theoretical results are summarized in Table 9.9.

For liquid metals, the effect of the entrance region upon the heat-transfer coefficient is not very pronounced and disappears in about 10-diameter downstream. Long-tube equations can be corrected [53] to give the short-tube local $Nu$ by multiplying the right-hand side by the correction factor $[1.72(d/L)^{0.16}]$.

Comprehensive information and correlations for convective heat transfer and friction factor in noncircular curved ducts and coils, in cross flow arrangements, over rod bundles, in various fittings, and for liquid metals are given in [54].

**TABLE 9.9**

Correlation of Turbulent Flow Heat Transfer for Liquid Metals

| Geometry | Correlations | Limitations |
|---|---|---|
| Long tube [43] | $Nu_b = 7 + 0.0025 Pe_b^{0.8}$ | $Pe > 100, L/d > 60, q_w'' = \text{constant}$ |
| Long tube [44] | $Nu_b = 5 + 0.0025 Pe_b^{0.8}$ | $Pe > 100, L/d > 60, T_w = \text{constant}$ |
| Long tube [45] | $Nu_b = 4.82 + 0.0185 Pe_b^{0.827}$ | $5.8 \times 10 < Pe < 1.31 \times 10^4,$ |
|  |  | $q_w'' = \text{constant}$ |
| Long tube [42] | $Nu_b = 0.625 Pe_b^{0.4}$ | $200 < Pe < 21 \times 10^4, L/d > 60,$ |
|  |  | $q_w'' = \text{constant}$ |
| Long tube [49] | $Nu_b = 7.0 + 0.05 Pr_b^{0.25} Pe_b^{0.77}$ | $Pr < 0.1; Pe < 15,000, q_w'' = \text{constant}$ |
| Long tube [50] | $Nu_b = 4.8 + 0.014 Pe_b^{0.8}$ | $100 < Pe < 1400, q_w'' = \text{constant}$ |
| Parallel plates with heat flow through one side or annuli with $D_o/D_i < 1.4$ [46] | $Nu_b = 5.8 + 0.020 Pe_b^{0.8}$ | $Pe > 50, q_w'' = \text{constant}, Na, K, \text{and } Na - K,$ |
| Narrow concentric annulus-wide rectangular channels $D_o/D_i \cong 1$ [51] | $Nu_b = 7.0 + 0.025 Pe_b^{0.8}$ | Constant $q_w'''$ through one wall |
| Concentric annulus $D_o/D_i > 1$ [51] | $Nu_b = 0.75(D_o/D_i)^{0.3}(7 + 0.025 Pe_b^{0.8})$ | Constant $q_w''$ at $D_o$ |
| Narrow concentric annulus-wide rectangular channels $D_o/D_i \cong 1$ [52] | $Nu_b = 10.5 + 0.036 Pe_b^{0.8}$ | Constant $q_w''$ through both walls |

## 9.9 Summary

Important and reliable correlations for Newtonian fluids in single-phase laminar and turbulent flows through ducts have been summarized, and they can be used in the design of heat-transfer equipment [55].

Table 9.2 covers the recommended specific correlations for laminar forced convection through ducts with constant and variable fluid properties. Table 9.3 provides exponents $m$ and $n$ associated with Equations 9.34 and 9.35 for laminar forced convection in circular ducts. By the use of this table, the effect of variable properties in laminar flow is incorporated by the property ratio method.

Turbulent forced convection correlations for fully developed flow through a circular duct with constant properties are summarized in Table 9.4. Gnielinski, Petukhov and Kirilov, Webb, Sleicher, and Rause correlations are recommended for constant-property Nusselt number evaluation for gases and liquids, and the entrance correlation factor is given by Equation 9.51. Recommended turbulent flow isothermal Fanning friction factor correlations for smooth circular ducts are listed in Table 9.5. Correlations given in Tables 9.4, 9.5, 9.7, and 9.8 can also be utilized for turbulent flow in smooth straight noncircular ducts for engineering application by the use of hydraulic diameter concept for heat-transfer and pressure drop calculations as discussed in section 9.33. Except for sharp-cornered and/or very irregular duct cross sections, the fully developed turbulent Nusselt number and friction factor vary from their actual values within ±15% and ±10%, respectively, when the hydraulic diameter is used in circular duct correlations.

When there is a large difference between the wall and fluid bulk temperatures, the influence of variable fluid properties on turbulent forced convection and pressure drop in circular ducts is taken into account by using the exponents $m$ and $n$ given in Table 9.6 with Equations 9.34 and 9.35. The correlations for turbulent flow of liquids and gases with variable properties in circular ducts are also summarized in Tables 9.7 and 9.8. Correlations for turbulent flow heat transfer for liquid metals are listed in Table 9.9.

## Problems

**9.1** Convert the following equation to the form of Equation 9.44:

$$\frac{h}{U_m \rho c_p}\left(\frac{\mu c_p}{k}\right)^{2/3} = 0.023\left(\frac{U_m d}{v}\right)^{-0.2}.$$

**9.2** In fully developed turbulent flow through a smooth pipe, the friction coefficient is correlated by $f = 0.046\,Re^{-0.2}$, where $f$ is defined as $\Delta p = 4f(L/d)\rho(U_m^2/2)$ in the range of Reynolds numbers $3 \times 10^4 < Re < 10^6$. Obtain a correlation for the Nusselt number under the previous condition if $Pr > 1$.

**9.3** A fluid flows steadily with a velocity of 6 m/s through a commercial iron rectangular duct whose sides are 1 in. × 2 in. The duct is 6 m long. The average temperature of the fluid is 60°C. The fluid completely fills the duct. Calculate the surface heat-transfer coefficient if the fluid is

a. Water

b. Air at atmospheric pressure

c. Engine oil ($\rho$ = 864 kg/m³, $c_p$ = 2047 J/(kg K), $v$ = 0.0839 × 10⁻³ m²/s, $k$ = 0.140 W/(m K), Pr = 1050)

**9.4**  Air at 1.5 atm and 40°C flows through a 10 m long rectangular duct 40 cm × 25 cm. The duct surface temperature is maintained at 120°C and the average air temperature at exit is 80°C. Calculate the total heat transfer.

**9.5**  Calculate the heat-transfer coefficient for water flowing through a 2 cm diameter tube with a velocity of 1 m/s. The average temperature of water is 60°C and the surface temperature (a) is slightly above this temperature and (b) is 120°C.

**9.6**  Air at 20°C and 1 bar is to be heated to 140°C at a rate of 0.4 m³/min, while flowing through a 2.5 cm diameter tube that is maintained at a temperature of 420°C. Estimate the length of tube required.

**9.7**  An oil with $k$ = 0.120 W/(m K), $c_p$ = 2000 J/(kg K), $\rho$ = 895 kg/m³, and $\mu$ = 0.0041 kg/m s flows through a 2 cm diameter tube that is 2 m long. The oil is cooled from 60°C to 30°C. The mean flow velocity is 0.4 m/s, and the tube wall temperature is maintained at 24°C ($\mu_w$ = 0.021 kg/m s). Calculate the heat-transfer rate.

**9.8**  Air at atmospheric pressure and 0°C enters a long rectangular duct of 1 cm × 2 cm cross section with a velocity of 50 m/s. There is a heat flux of 1 W/cm² through all surfaces. Calculate the surface temperature of the duct at a point of 200 cm from the entrance.

**9.9**  In a pressurized water reactor, the water flows in channels between flat plates of 0.10 in. thick with 0.10 in. gap between the plates. The length of the fuel elements is 150 cm. The water inlet temperature is 293°C, the outlet temperature is 321°C, and the water velocity is 12 m/s. Calculate the heat-transfer coefficient.

**9.10**  A shell-and-tube-type condenser is to be made with 3/4 in. outer diameter (0.654 in. inner diameter) brass tubes, and the length of the tubes between tube plates is 3 m. Under the worst conditions, cooling water is available at 21°C and the outlet temperature is to be 31°C. Water velocity inside the tubes is to be approximately 2 m/s. Vapor side film coefficient can be taken as 10,000 W/(m² K). Calculate the overall heat-transfer coefficient for this heat exchanger.

**9.11**  Water at 1.15 bar and 30°C is heated as it flows through a 1 in. inside diameter tube at a velocity of 3 m/s. The pipe surface temperature is kept constant by condensing steam outside the tube. If water outlet temperature is 80°C, calculate the surface temperature of the tube by assuming the inner surface of the tube to be

a. Smooth

b. A rough surface with a relative roughness of 0.001

What conclusion can be drawn from the previous two cases?

**9.12**  Water at the rate 0.5 kg/min is heated from 10°C to 30°C by passing it through a 2.5 cm inside diameter steel tube. The tube wall temperature is maintained at 90°C by condensing steam. Calculate the length of the tube.

**9.13**  Liquid bismuth flows through an annulus (2.0 in. inside diameter, 2.5 in. outside diameter) at a velocity of 4 m/s. The bulk temperature is 300°C. Calculate the average heat-transfer coefficient.

**9.14** Mercury at an inlet bulk temperature of 150°C flows through a 1.25 cm inside diameter tube at a flow rate of 5000 kg/h. Heat is generated uniformly within the tube. Determine the length of tube required to raise the bulk temperature of the mercury to 250°C.

**9.15** Determine the heat-transfer coefficient for liquid sodium flowing through an annulus (2 in. ID, 2.5 in. OD) at a velocity of 4 m/s. The wall temperature of inner surface is 400°C and the sodium is at 300°C. The outer surface is insulated.

# References

1. Shah, R. K. and London, A. L., *Laminar Forced Convection in Ducts*, Academic Press, New York, 1978.
2. Shah, R. K. and Bhatti, M. S., Laminar convective heat transfer in ducts, in *Handbook of Single-Phase Convective Heat Transfer*, S. Kakaç, R. K. Shah, and W. Aung (Eds.), John Wiley, New York, pp. 3.1–3.137, 1987.
3. Schunder, E. U. (Ed.), *Heat Exchanger Design Handbook*, Hemisphere, New York, pp. 2.5.1–2.5.13, 1983.
4. Hausen, H., Neue Gleichungen fur die Warmeubeitragung bei freier oder erzwungener Stromung, *Allg. Waermetech.*, 9, 75–79, 1959.
5. Kakaç, S., Laminar forced convection in the combined entrance region of ducts, in *Natural Convection: Fundamentals and Applications*, S. Kakaç, W. Aung, and R. Viskanta (Eds.), Hemisphere, New York, pp. 165–204, 1985.
6. Pohlhausen, E., Der Warmeaustausch Zwischen festen Korperm und Flussigkeiten mit Kleiner Reibung and Kleiner Wa-meleitung, *Z. Angew. Math. Mech.*, 1, 115–121, 1921.
7. Delorenzo, B. and Anderson, E. D., Heat transfer and pressure drop of liquids in double pipe fintube exchangers, *Trans. ASME*, 67, 697, 1945.
8. Stephen, K., Warmeubergang und Druckabfall beinichtausgebildeter Laminar Stormung in Rohren and ebenen Spalten, *Chem. Ing. Tech.*, 31, 773–778, 1959.
9. Deissler, R. G., analytical investigation of fully developed laminar flow in tubes with heat transfer with fluid properties variable along the radius, NACA TN 2410, 1951.
10. Yang, K. T., Laminar forced convection of liquids in tubes with variable viscosity, *J. Heat Transfer*, 84, 353–362, 1962.
11. Sieder, E. N. and Tate, G. E., Heat transfer and pressure drop of liquids in tubes, *Ind. Eng. Chem.*, 28, 1429–1453, 1936.
12. Whitaker, S., Forced convection heat-transfer correlations for flow in pipes, past flat plates, single cylinders, single spheres, and flow in packed beds and tube bundles, *AIChE J.*, 18, 361–371, 1972.
13. Oskay, R. and Kakaç, S., Effect of viscosity variations on turbulent and laminar forced convection in pipes, *METU J. Pure Appl. Sci.*, 6, 211–230, 1973.
14. Kuznetsova, V. V., Convective heat transfer with flow of a viscous liquids in a horizontal tube (in Russian), *Teploenergetika*, 19(5), 84, 1972.
15. Test, F. L., Laminar flow heat transfer and fluid flow for liquids with a temperature dependent viscosity, *J. Heat Transfer*, 90, 385–393, 1968.
16. Worsøe-Schmidt, P. M., Heat transfer and friction for laminar flow of helium and carbon dioxide in a circular tube at high heating rate, *Int. J. Heat-Mass Transfer*, 9, 1291–1295, 1966.
17. Bhatti, M. S. and Shah, R. K., Turbulent forced convection in ducts, in *Handbook of Single-Phase Convective Heat Transfer*, S. Kakaç, R. K. Shah, and W. Aung (Eds.), John Wiley, New York, pp. 4.1–4.166, 1987.

18. Dittus, F. W. and Boelter, L. M. K., Heat transfer in automobile radiators of the tubular type, *Univ. Calif. (Berkeley) Pub. Eng.*, 2, 443, 1930.

19. Colbum, A. P., A method of correcting forced convection heat transfer data and comparison with fluid friction, *Trans. AIChE*, 29, 174–210, 1933.

20. Petukhov, B. S. and Popov, V. N. N., Theoretical calculation of heat exchange and frictional resistance in turbulent flow in tubes of incompressible fluid with variable physical properties, *High Temperature*, 1, 69–83, 1963.

21. Petukhov, B. S., Heat transfer and friction in turbulent pipe flow with variable physical properties, in *Advances in Heat Transfer*, J. P. Hartnett and T. V. Irvine (Eds.), Academic Press, New York, Vol. 6, pp. 504–564, 1970.

22. Webb, R. I., A critical evaluation of analytical solutions and Reynolds analogy equations for heat and mass transfer in smooth tubes, *Warme Stqffubertragung*, 4, 197–204, 1971.

23. Sleicher, C. A. and Rause, M. W., A convenient correlation for heat transfer to constant and variable property fluids in turbulent pipe flow, *Int. J. Heat Mass Transfer*, 18, 667–683, 1975.

24. Gnielinski, V., New equations for heat and mass transfer in turbulent pipe and channel flow, *Int. Chem. Eng.*, 16, 359–368, 1976.

25. Kays, W. M. and Crawford, M. E., *Convective Heat and Mass Transfer*, 2nd edn., McGraw-Hill, New York, 1981.

26. Kakaç, S. (Ed.), *Boilers, Evaporators and Condensers*, Wiley, New York, 1991.

27. McAdams, W. H., *Heat Transmission*, 3rd edn., McGraw-Hill, New York, 1954.

28. Kakaç, S., The effects of temperature-dependent fluid properties on convective heat transfer, in *Handbook of Single-Phase Convective Heat Transfer*, S. Kakaç, R. K. Shah, and W. Aung (Eds.), John Wiley, New York, pp. 18.1–18.56, 1987.

29. Hausen, H., Darstellung des Warneuberganges un Rohren durch verallgeineinerte Potenzbeziebungen, *Z Ver. Dtsch. Ing. Beiheft Verfahrenstech.*, 4, 91–134, 1943.

30. Rehme, K., A simple method of predicting friction factors of turbulent flow in noncircular channels, *Int. J. Heat Mass Transfer*, 16, 933–950, 1973.

31. Malak, J., Hejna, J., and Schmid, J., Pressure losses and heat transfer in noncircular channels with hydraulically smooth walls, *Int. J. Heat Mass Transfer*, 18, 139–149, 1975.

32. Brundrett, E., Modified hydraulic diameter, turbulent forced convection in channels and bundles, in *Two-Phase Flow Heat Exchangers: Thermal-Hydraulic Fundamentals and Design*, S. Kakaç and D. B. Spalding (Eds.), Hemisphere, New York, Vol. 1, pp. 361–367, 1979.

33. Perkis, H. C. and Worsøe-Schmidt, P., Turbulent heat and momentum transfer for gases in a circular tube at wall to bulk temperature ratios to seven, *Int. J. Heat Mass Transfer*, 8, 1011–1031, 1965.

34. McElligot, D. M., Magee, P. M., and Leppert, G., Effect of large temperature gradients on convective heat transfer, the downstream region, *J. Heat Transfer*, 87, 67–76, 1965.

35. Hufschmidt, W., Burck, E., and Riebold, W., Die Bestimmung Orlicher und Warmeubergangs-Zahlen in Rohlen bei Hohen Warmestromdichten, *Int. J. Heat Mass Transfer*, 9, 539–565, 1966.

36. Rogers, D. G., Forced convection heat transfer in single phase flow of a Newtonian fluid in a circular pipe, CSIR Report CENG 322, Pretoria, South Africa, 1980.

37. Hausen, H., Extended equation for heat transfer in tubes at turbulent flow, *Warme Stoffubertragung*, 7, 222–225, 1974.

38. Humble, L. V., Lowdermilk, W. H., and Desmon, L. G., Measurement of average heat transfer and friction coefficient for subsonic flow of air in smooth tubes at high surface and fluid temperature, NACA Report 1020, 1951.

39. Barnes, J. F. and Jakson, J. D., Heat transfer to air, carbon dioxide and helium flowing through smooth circular tubes under conditions of large surface/gas temperature ratio, *J. Mech. Eng. Sch.*, 3(4), 303–314, 1961.

40. Dalle-Donne, M. and Bowditch, P. W., Experimental local heat transfer and friction coefficient for subsonic laminar transitional and turbulent flow of air or helium in a tube at high temperatures, Dragon Project Rept. 184, Winfirth, Dorchester, Dorset, U.K., 1963.

41. Lyon, R. D., *Liquids Metals Handbook*, 3rd edn., U.S. Atomic Energy Commission, Washington, DC, 1952.

42. Luborsky, B. and Kaufman, S. J., Review of experimental investigations of liquid metal heat transfer, NACA Technical Notes, 3336, 1955.
43. Lyon, R. N., Liquid metal heat transfer coefficient, *Chem. Eng. Progr.*, 47, 75–79, 1951.
44. Seban, R. A. and Shimazaki, T. T., Heat transfer to a fluid flowing turbulently in a smooth pipe with walls at constant temperature, *Trans. ASME*, 73, 803, 1951.
45. Skupinshi, E., Tortel, J., and Vautrey, L., Determination des coefficients de convection d'un alliage sodium-potassium dans un tube circulaire, *Int. J. Heat Mass Transfer*, 8, 937, 1965.
46. Seban, R. A., Heat transfer to a fluid flowing turbulently between parallel walls with asymmetric wall temperatures, *Trans. ASME*, 72, 789, 1950.
47. McDonald, W. C. and Quittenton, R. C., A critical analysis of metal "wetting" and gas entertainment in heat transfer to molten metals, Preprint No. 8, *Am. Inst. Chem. Eng.*, 1953.
48. Isakoff, S. E. and Drew, T. B., Heat and momentum transfer in turbulent flow of mercury, Institute of Mechanical Engineers and ASME, Proceedings, *General Discussion on Heat Transfer*, pp. 405–409, 1951.
49. Azer, N. Z. and Chao, B. T., A mechanism of turbulent heat transfer in liquid metals, *Int. J. Heat Mass Transfer*, 1, 121–138, 1960.
50. Mikhaev, M. A. (Ed.), *Problems of Heat Transfer*, Academy of Sciences, Moscow, Russia, USAEC Report AEC-TR-4511, 1962.
51. Khbkhpaheva, E. M. and Illini, Y. V., Heat transfer to a sodium-potassium alloy in an annulus, *At. Energy Suppl.*, 9, 494–496, 1960.
52. Hogereton, J. F. and Grass, R. C. (Eds.), *Reactor Handbook*, 1st edn., Vol. 2, Engineering, USAEC Report AECD-3646, 1955.
53. Poppendiek, H. F. and Palmer, L. D., *Forced Convection Heat Transfer in Thermal Entrance Regions*, Part II, ORNL-914, Oakridge, TN, 1952.
54. Kakaç, S., Shah, R. K., and Aung, W. (Eds.), *Handbook of Single-Phase Convective Heat Transfer*, John Wiley, New York, 1987.
55. Kakaç, S. Liu, H., and Pramuanjaroenkij, A., *Heat Exchangers: Selection, Rating, and Thermal Design*, CRC Press, Boca Raton, FL, 2012.

# 10

## Heat Transfer in Natural Convection

## Nomenclature

| | |
|---|---|
| $c_p$ | specific heat at constant pressure, J/(kgm) |
| $D$ | diameter of cylinder or sphere, m |
| $d$ | diameter of circular duct, m |
| $f$ | dimensionless stream function, defined by Equation 10.23 |
| $Gr$ | Grashof number $= g\beta\Delta T L^3/v^2$ or $g\beta\Delta T d^3/v^2$ |
| $Gr_x$ | local Grashof number $= g\beta\Delta T x^3/v^2$ |
| $Gr^*$ | heat-flux Grashof number $= g\beta L^4 q_w''/kv^2$ |
| $Gz$ | Graetz number $= (RePr)d/L$ |
| $g$ | gravitational acceleration, m/s$^2$ |
| $h$ | average heat-transfer coefficient, W/(m$^2$ K), height of enclosure, m |
| $h_x$ | local heat-transfer coefficient, W/(m$^2$ K) |
| $h^*$ | dimensionless height of enclosure $= H/L$ |
| $k$ | thermal conductivity, W/(m K) |
| $L$ | characteristic length, height of vertical plate, m |
| $m, n$ | exponents in exponential and power-law distributions |
| $Nu$ | average Nusselt number $= hL/k$ |
| $Nu_x$ | local Nusselt number $= h_x x/k$ |
| $P$ | pressure, Pa |
| $p_o$ | hydrostatic pressure, Pa |
| $Pr$ | Prandtl number $= c_p\mu/k$ |
| $q''$ | heat flux, W/m$^2$ |
| $q_w''$ | wall heat flux, W/m$^2$ |
| $Ra$ | Rayleigh number $= GrPr$ |
| $Ra_x$ | local Rayleigh number $= Gr_x Pr$ |
| $Re$ | Reynolds number $= \rho u d/\mu$ |
| $r_i, r_o$ | inner and outer radii of a concentric annulus |
| $St$ | Stanton number $= Nu/RePr$ |
| $t$ | time, s |
| $T$ | local temperature, °C, K |
| $T_w$ | wall temperature, °C, K |
| $T_\infty$ | ambient temperature, °C, K |
| $\Delta T$ | temperature difference $= T_w - T_\infty$, °C, K |
| $u$ | velocity component in $x$-direction, m/s |

| | |
|---|---|
| $v$ | velocity component in $r$- or $y$-direction, m/s |
| $x$ | axial coordinate in Cartesian and radial coordinates, m |
| $y, z$ | transverse coordinate, m |

## Greek Symbols

| | |
|---|---|
| $\alpha$ | thermal diffusivity, $m^2/s$ |
| $\beta$ | coefficient of thermal expansion $= -\dfrac{1}{\rho}\left(\dfrac{\partial\rho}{\partial T}\right)_p$, 1/K |
| $\gamma$ | inclination with the vertical |
| $\delta$ | velocity boundary-layer thickness, m |
| $\delta_T$ | thermal boundary-layer thickness, m |
| $\eta$ | similarity variable |
| $\theta$ | dimensionless temperature $= (T - T_\infty)/(T_w - T_\infty)$ |
| $\mu$ | dynamic viscosity, Pa s |
| $\nu$ | kinematic viscosity, $m^2/s$ |
| $\rho$ | density of fluid, $kg/m^3$ |
| $\tau$ | shear stress, Pa |
| $\psi$ | stream function, $m^2/s$ |

## Subscripts

| | |
|---|---|
| $b$ | bulk value |
| $c$ | cold wall, critical, center line |
| $F$ | pure forced convection |
| $f$ | evaluated at fluid film temperature $(T_w + T_b)/2$ |
| $h$ | hot wall |
| $i$ | value at the inlet or value on the inner wall |
| $m$ | average value |
| $N$ | pure natural convection |
| $o$ | value on the outer wall |
| $w$ | value at the wall or properties evaluated at the wall temperature |
| $x$ | local value |
| $\infty$ | ambient |

## 10.1 Introduction

Fluid motions due entirely to the action of a body force field such as gravitational field are usually called natural flows, in contrast to forced-convection flows brought about by external agents, and the corresponding heat-transfer process is termed *natural convection or free convection*. The movement of the fluid in natural convection, whether it is a gas or a liquid, results from the buoyancy forces imposed on the liquid when its density in the proximity of a heat-transfer surface is decreased as a result of heating or increased as a result of cooling. The density difference resulting from concentration difference also gives rise to buoyancy forces due to which the flow is generated. The presence of a buoyancy force is a requirement for the existence of a natural-convection flow. Ordinarily, the buoyancy arises from density differences that are the consequences of temperature or concentration gradients within the fluid. The buoyant flow arising from heat rejection to the atmosphere, heating of rooms, fires, and many other such heat-transfer processes, both natural and artificial, are examples of natural convection.

## 10.2 Basic Equations of Laminar Boundary Layer

Equations expressing the conservation of mass, momentum, and energy for a viscous and heat conducting fluid subject to a body force together with an equation of state govern the flow and associated temperature distribution in natural or free convection.
In the discussion that follows, we shall introduce the following assumptions:

1. Fluid is incompressible in the sense that the density does not change appreciably with pressure. The density is therefore considered to be a function of temperature only, that is, $\rho = \rho(T)$.

2. Fluid properties (specific heat, thermal conductivity, viscosity) are constant.

3. Viscous dissipation is negligible.

To analyze heat transfer in natural or free convection, we must first obtain the governing equations of motion. For this purpose, consider a vertical heated flat plate placed in an extensive quiescent medium at uniform temperature; we choose the $x$-coordinate along the plate positive upward, $y$-coordinate perpendicular to the plate as shown in Figure 10.1.

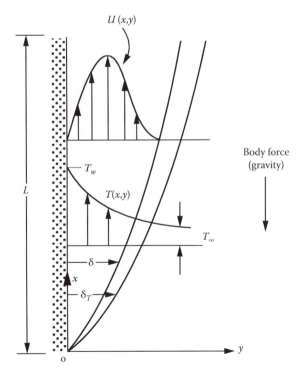

**FIGURE 10.1**
Temperature and velocity profiles in free convection on a heated flat plate ($T_w > T_\infty$).

The body force that must be considered in the derivation is the weight of the element of the fluid. Let us assume that the boundary layer that develops over the plate is steady and 2D, and the flow in the boundary layer is laminar. As in Section 2.2, the law of conservation of mass applied to an elemental control volume $\Delta x \Delta y \Delta z$ gives

$$\nabla \cdot (\rho V) = 0, \tag{10.1}$$

where $V = \hat{i}u + \hat{j}v$. If we equate the sum of the external forces on an element of fluid of mass $\rho \Delta x \Delta y \Delta z$ in the $x$- and $y$-directions to the acceleration of the particle in these directions (Newton's second law of motion) as in Section 2.3, or if we apply the law of conservation of linear momentum to an elemental control volume $\Delta x \Delta y \Delta z$, we get

$$u \frac{\partial u}{\partial x} + v \frac{\partial u}{\partial y} = -\frac{1}{\rho} \frac{\partial p}{\partial x} + \nu \nabla^2 u + \frac{1}{3} \nu \frac{\partial}{\partial x} (\nabla \cdot V) - g, \tag{10.2}$$

$$u \frac{\partial v}{\partial x} + v \frac{\partial v}{\partial y} = -\frac{1}{\rho} \frac{\partial p}{\partial y} + \nu \nabla^2 v + \frac{1}{3} \nu \frac{\partial}{\partial y} (\nabla \cdot V) - g, \tag{10.3}$$

where $g$ is the gravity per unit mass of the fluid. The Navier–Stokes equation in the $y$-direction (10.3) reduces by the Prandtl order of magnitude analysis to

$$\frac{\partial p}{\partial y} = 0. \tag{10.4}$$

Since $p \neq p(y)$, pressure distribution can be calculated at the outside of the boundary layer where $T = T_\infty$ and $u = v = 0$ everywhere. Thus, the evaluation of Equation 10.2 outside the boundary layer gives

$$-\frac{dp}{dx} = \rho_\infty g, \tag{10.5}$$

where $\rho_\infty$ is the fluid density outside the boundary layer. Substitution of Equation 10.5 into Equation 10.2 yields

$$\rho \left( u \frac{\partial u}{\partial x} + v \frac{\partial u}{\partial y} \right) = \mu \nabla^2 u + \frac{1}{3} \mu \frac{\partial}{\partial x} (\nabla \cdot V) - g(\rho_\infty - \rho). \tag{10.6}$$

The thermal expansion coefficient of the fluid is defined as

$$\beta = -\frac{1}{\rho} \left( \frac{\partial \rho}{\partial T} \right)_p. \tag{10.7}$$

Since $\rho = \rho(T)$, the derivative term in Equation 10.7 can be approximated to obtain

$$\rho \cong \rho_\infty \left[ 1 - \beta(T - T_\infty) \right]. \tag{10.8}$$

Substituting Equation 10.8 into Equation 10.6, we get

$$\rho\left(u\frac{\partial u}{\partial x}+v\frac{\partial u}{\partial y}\right)=\mu\nabla^2 u+\frac{1}{3}\mu\frac{\partial}{\partial x}(\nabla\cdot\mathbf{V})+g\rho_\infty\beta(T-T_\infty). \tag{10.9}$$

It is interesting to note that the term $g\rho_\infty\beta(T-T_\infty)$ in Equation 10.9 is not a body force but is a buoyancy force.

If it is now assumed that the density variations are negligible in the continuity equation (10.1) and in the x-momentum equation (10.9) except in the buoyancy term (*Boussinesq approximation*), we can write

$$\frac{\partial u}{\partial x}+\frac{\partial v}{\partial y}=0, \tag{10.10}$$

$$u\frac{\partial u}{\partial x}+v\frac{\partial u}{\partial y}=\nu\nabla^2 u+g\beta(T-T_\infty), \tag{10.11}$$

with the following boundary conditions:

$$\text{at } y=0: \quad u=v=0 \tag{10.12a,b}$$

$$\text{as } y\to\infty: \quad u=0 \tag{10.12c}$$

The momentum equation in the x-direction (10.11) also reduces by the Prandtl order of magnitude analysis to (see Chapter 3)

$$u\frac{\partial u}{\partial x}+v\frac{\partial u}{\partial y}=\nu\frac{\partial^2 u}{\partial y^2}+g\beta(T-T_\infty), \tag{10.13}$$

which is the Boussinesq approximated momentum equation. Equation 10.13 shows the coupling between the temperature field and the velocity field.

Energy equation in natural convection is the same for a forced-convection system at low velocity. Therefore, assuming that viscous dissipation is negligible, we have

$$\rho c_p\left(u\frac{\partial T}{\partial x}+v\frac{\partial T}{\partial y}\right)=k\nabla^2 T, \tag{10.14}$$

where we have assumed that the fluid density is constant (Boussinesq approximation) in addition to thermal conductivity and specific heat.

The energy equation (10.14) reduces by the Prandtl order of magnitude analysis to

$$u\frac{\partial T}{\partial x}+v\frac{\partial T}{\partial y}=\alpha\frac{\partial^2 T}{\partial y^2} \tag{10.15}$$

and the boundary conditions are (Figure 10.1)

$$\text{at } y = 0: \quad T = T_w \quad \text{or} \quad -k\frac{\partial T}{\partial y} = q_w'', \tag{10.16a}$$

$$\text{as } y \to \infty: \quad T = T_\infty. \tag{10.16b}$$

Equations 10.10, 10.13, and 10.15 are the well-known Boussinesq equations for natural convection on a vertical plate, and they have to be solved simultaneously under the boundary conditions (10.12) and (10.16) to find the velocity and temperature distributions in the steady laminar boundary layer that develops over a vertical heated plate. It is to be noted that the velocity and temperature fields are interdependent; the natural flow motion that arises is not known at the onset and has to be determined from a consideration of the heat and mass transfer processes coupled with fluid flow mechanisms. This makes natural-convection problems more difficult.

The Boussinesq approximation is very extensively used for a wide range of problems in natural convection. An important condition for the validity of this approximation is that $\beta(T - T_\infty) \ll 1$ [1]. Therefore, the approximation is valid for small temperature differences.

## 10.3 Pohlhausen Solution for Laminar Boundary Layer over a Constant Temperature Vertical Flat Plate

Natural convection from a heated vertical plate has been of interest to many investigators. The governing equations to be solved are Equations 10.13 and 10.15. Although the problem is considerably simplified, the complexities due to these coupled partial differential equations remain. An important method for finding the boundary-layer flow over a heated vertical plate is the similarity variable method discussed in Chapter 4.

Pohlhausen [2] demonstrated that if a stream function is introduced by defining

$$u = \frac{\partial \psi}{\partial y} \quad \text{and} \quad v = -\frac{\partial \psi}{\partial x} \tag{10.17a,b}$$

to satisfy the continuity equation, then the resulting differential equation for $\psi(x, y)$ can be reduced to an ordinary differential equation by the similarity transformation as discussed in Chapter 4.

Introducing the stream function into Equations 10.13 and 10.15, we get

$$\frac{\partial \psi}{\partial y}\frac{\partial^2 \psi}{\partial x \partial y} - \frac{\partial \psi}{\partial x}\frac{\partial^2 \psi}{\partial y^2} = v\frac{\partial^3 \psi}{\partial y^3} + \beta g\Theta(T_w - T_\infty) \tag{10.18}$$

and

$$\frac{\partial \psi}{\partial y}\frac{\partial \Theta}{\partial x} - \frac{\partial \psi}{\partial x}\frac{\partial \Theta}{\partial y} = \alpha\frac{\partial^2 \Theta}{\partial y^2}, \tag{10.19}$$

where

$$\Theta = \frac{T - T_\infty}{T_w - T_\infty}.$$

The boundary conditions become

$$\frac{\partial \psi(x,0)}{\partial x} = \frac{\partial \psi(x,0)}{\partial y} = 0, \tag{10.20a,b}$$

$$\Theta(x,\ 0)\ =\ 1, \tag{10.20c}$$

$$\frac{\partial \psi(x,\infty)}{\partial y} = \Theta(x,\infty) = 0. \tag{10.20d,e}$$

It should be noted that the term $g\beta(T_w - T_\infty)$ in Equation 10.18 is constant if the wall temperature is constant.

Pohlhausen obtained a solution for laminar boundary-layer flow of air along a heated vertical plate by applying the following similarity variable:

$$\eta = Cyx^{-1/4}, \tag{10.21}$$

where

$$C = \left[ \frac{g\beta(T_w - T_\infty)}{4v^2} \right]^{1/4}. \tag{10.22}$$

General solution of these equations proceeds as before for forced convection over a flat plate (Chapter 4):

$$\psi(x,y) = 4vCx^{3/4}f(\eta), \tag{10.23}$$

$$\Theta(x,y)\ = \Theta(\eta, Pr). \tag{10.24}$$

Equations 10.18 and 10.19 are transformed into

$$\frac{d^3f}{d\eta^3} + 3f\frac{d^2f}{d\eta^2} - 2\left(\frac{df}{d\eta}\right)^2 + \Theta = 0, \tag{10.25}$$

$$\frac{d^2\Theta}{d\eta^2} + 3prf\frac{d\Theta}{d\eta} = 0. \tag{10.26}$$

The boundary conditions for the uniform wall temperature (UWT) case are as follows:

$$f'(0) = f(0) = 0, \quad \Theta(0) = 1, \tag{10.27a,b,c}$$

$$f'(\infty) = 0, \quad \Theta(\infty) = 0, \tag{10.27d,e}$$

where the primes denote differentiations with respect to the similarity variable $\eta$.

These equations were first solved in this form by Pohlhausen [2] and Schmidt and Beckmann [3] during the period 1930–1932 for only one Prandtl number of 0.733. The advance of the computing facilities permitted Ostrach [4] to solve these equations numerically to obtain the solution for the Prandtl number range of 0.01–1000. Figures 10.2 and 10.3 show the velocity and temperature distribution thus obtained, respectively, in the free-convection laminar boundary layer on a heated vertical plate for various values of the Prandtl number. From these solutions, the values of $\theta'(0)$ and $f''(0)$ are given in Table 10.1. Comparison with the experimental work of Schmidt and Beckmann shows that the velocity and temperature profiles are in good agreement for air. The maximum values of the dimensionless velocity distributions can be seen to occur at larger values of the parameter $\eta$ as the Prandtl number decreases, and the velocities decrease with increasing the Prandtl number. For $Pr \gg 1$, the velocity boundary layer is much thicker than the thermal boundary layer.

The local heat flux from the surface to the fluid at any $x$-location can readily be calculated by using Fourier's law of heat conduction, that is,

$$q_w'' = -k\left(\frac{\partial T}{\partial y}\right)_{y=0} = -k(T_w - T_\infty)Cx^{-1/4}\left(\frac{d\Theta}{d\eta}\right)_{\eta=0}. \tag{10.28}$$

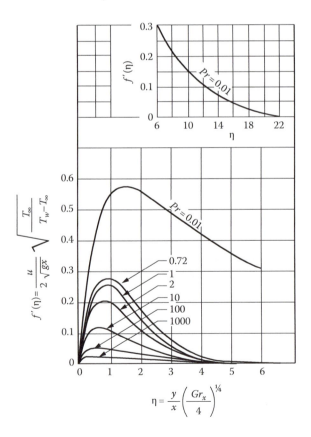

**FIGURE 10.2**
Dimensionless velocity distribution for laminar free convection on a vertical flat plate. (From Ostrach, S., An analysis of laminar free-convection flow and heat transfer about a flat plate parallel to the direction of the generating body force, NACA Report 1111, 1953.)

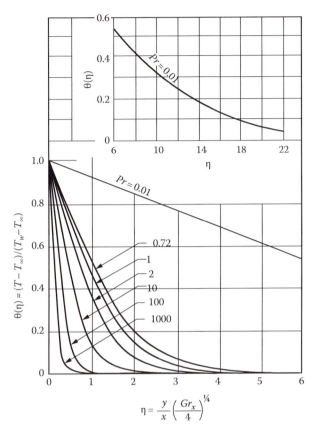

**FIGURE 10.3**

Dimensionless temperature distribution for laminar free convection on a vertical flat plate. (From Ostrach, S., An analysis of laminar free-convection flow and heat transfer about a flat plate parallel to the direction of the generating body force, NACA Report 1111, 1953.)

**TABLE 10.1**

Computed Values of $\Theta'(0)$ and $f''(0)$ for Various Values of Prandtl Number

| Pr | $\Theta'(0)$ | $f''(0)$ |
|---|---|---|
| 0.01 | 0.080592 | 0.9862 |
| 0.72 | 0.50463 | 0.6760 |
| 0.733 | 0.50789 | 0.6741 |
| 1.0 | 0.56714 | 0.6421 |
| 2.0 | 0.716483 | 0.5713 |
| 10.0 | 1.168 | 0.4192 |
| 100 | 2.1914 | 0.2517 |
| 1000 | 3.97 | 0.1450 |

*Source:* Ostrach, S., An analysis of laminar free-convection flow and heat transfer about a flat plate parallel to the direction of the generating body force, NACA Report 1111, 1953.

The derivative $(d\Theta/d\eta)_{\eta=0}$, normally abbreviated as $\Theta'(0)$, is found from the solutions of Equations 10.25 and 10.26 for various values of the Prandtl number.

A dimensionless representation of the results is achieved by the use of the local heat-transfer coefficient and local Nusselt number, which are written in the usual way as

$$h_x = \frac{q_w''}{T_w - T_\infty}, \quad Nu_x = \frac{h_x x}{k}.$$

Using these definitions of hx and Nux, from Equation 10.28, we have

$$Nu_x = -\frac{\Theta'(0)Gr_x^{1/4}}{\sqrt{2}}, \tag{10.29}$$

where $Gr_x$ is the so-called local Grashof *number* defined as

$$Gr_x = \frac{g\beta |(T_w - T_\infty)| x^3}{v^2}, \tag{10.30}$$

which is applicable to both $T_w > T_\infty$ and $T_w < T_\infty$. For a perfect gas, the *Grashof number* becomes

$$Gr_x = \frac{g |(T_w - T_\infty)| x^3}{T_\infty v^2}. \tag{10.31}$$

The Grashof number is a dimensionless group that represents the ratio of the buoyancy forces to the viscous forces in free convection. A large value of the Grashof number, therefore, indicates small effects in the momentum equation, similar to the physical significance of the Reynolds number in forced flow.

Ostrach's calculations are approximated for a vertical surface by the expression

$$\frac{Nu_x}{\sqrt[4]{Gr_x/4}} = \frac{0.676 Pr^{1/2}}{(0.861 + Pr)^{1/4}}. \tag{10.32}$$

Since $h \sim x^{1/4}$, the average heat-transfer coefficient from 0 to $L$ is given by $h = 4h_L/3$. The average Nusselt number $Nu$, therefore, becomes

$$\frac{Nu}{\sqrt[4]{Gr_L/4}} = \frac{0.902 Pr^{1/2}}{(0.861 + Pr)^{1/4}}. \tag{10.33}$$

A semiempirical equation as given in Reference 5 relating the average (over the length $L$) Nusselt number to the Prandtl and Grashof numbers that has been used in the heat-transfer calculations up to the recent years is

$$Nu = 0.548(Gr_L Pr)^{1/4}, \quad 10 \le Gr_L \le 10^9. \tag{10.34}$$

The constant 0.548 pertains specifically to air; for oils it should be 0.555 [4] and for mercury approximately 0.33 [6]. When the results of Equation 10.34 are compared with those of Equation 10.33, it is seen that these agree only for a restricted range of the Prandtl numbers.

Le Fevre [7] solved the asymptotic forms of Equations 10.25 and 10.26 for vanishing and infinite values of the Prandtl number. On the basis of these solutions, he developed the following correlation for the mean Nusselt that agrees very well with the exact results of Ostrach [4]:

$$Nu = \left( \frac{Gr_L Pr^2}{2.43478 + 4.884\sqrt{Pr} + 4.95283Pr} \right)^{1/4}. \tag{10.35}$$

The product $Ra = PrGr$ is called *Rayleigh number*, that is,

$$Ra_L = Gr_L Pr = \frac{\beta g (T_w - T_\infty)L^3}{v\alpha}. \tag{10.36}$$

The earlier results all refer to laminar flows. It is found experimentally that transition from laminar to turbulent flow in free convection will ordinarily occur when $Ra \sim 10^9$.

### Example 10.1

A vertical wall of an oven is 60 cm long and is covered with a metal sheet maintained at 170°C. Air temperature inside the oven is 90°C and the pressure is atmospheric. Calculate the local heat-transfer coefficient at the end of the wall and the average heat-transfer coefficient over the wall.

**Solution**
The film temperature is

$$T_f = \frac{170 + 90}{2} = 130°C.$$

Properties of air at 130°C from Appendix B are as follows:

$$k = 0.0336\,W/(m\,K),$$

$$v = 2.639 \times 10^{-5}\,m^2/s,$$

$$\beta = 2.49 \times 10^{-3}\,1/K,$$

$$Pr = 0.697.$$

The Grashof number at the end of the wall is

$$Gr_L = \frac{g\beta(T_w - T_\infty)L^3}{v^2} = \frac{9.81 \times 2.49 \times 10^{-3} \times (170 - 90) \times (0.6)^3}{(2.639 \times 10^{-5})^2}$$

$$= 6.055 \times 10^8.$$

Hence, the flow is laminar. The local heat-transfer coefficient at the end of the plate can be calculated from Equation 10.32, that is,

$$Nu_L = \frac{0.676 Pr^{1/2}}{(0.861 + Pr)^{1/4}} \left( \frac{Gr_L}{4} \right)^{1/4} = \frac{0.676 \times (0.697)^{1/2}}{(0.861 + 0.697)^{1/4}} \left( \frac{6.055 \times 10^8}{4} \right)^{1/4}$$

$$= 56.031.$$

Thus,

$$h_L = \frac{Nu_L k}{L} = \frac{56.031 \times 0.0336}{0.6} = 3.14 \text{ W}/(\text{m}^2\text{ K})$$

and the average heat-transfer coefficient is then calculated as

$$h = \frac{4}{3} h_L = \frac{4}{3} \times 3.14 = 4.15 \text{ W}/(\text{m}^2\text{ K}).$$

One of the first practical solutions of the steady-state free-convection equations was given in 1881 by Lorenz [8] who considered the heat transfer from a single vertical plate at a constant temperature in air with gravity being the only force acting on the gas. Considering a flow that was streamlined with constant gas properties independent of temperature and by integrating the heat balance equation for a cross section of fluid, Lorenz obtained

$$h = 0.548 \left[ \frac{g\rho^2 c_p k^3 (T_w - T_\infty)}{\mu L T_\infty} \right]^{1/4}. \tag{10.37}$$

It is of particular importance to note that the rate of heat transfer from surface to air is proportional to $(T_w - T_\infty)^{5/4}$, a result that has several times been experimentally verified.

## 10.4 Exact Solution of Boundary-Layer Equations for Uniform Heat Flux

An exact solution of the laminar boundary-layer equations using similarity analysis from a vertical plate having uniform surface heat flux has been obtained by Sparrow and Gregg [9]. Heat-transfer parameters have been calculated for the Prandtl numbers in the range 0.1–100.

Equations 10.13 and 10.15 can be transformed to ordinary differential equations by introducing the following independent and dependent variables:

$$\eta = C_1 y x^{-15}, \tag{10.38a}$$

$$\Psi = C_2 x^{4/5} f(\eta), \tag{10.38b}$$

$$\Theta(\eta) = \frac{C_1 (T_\infty - T)}{x^{1/5} \left( q_w''/k \right)}, \tag{10.38c}$$

where

$$C_1 = \left( \frac{g\beta q_w''}{5k v^2} \right)^{1/5}, \tag{10.39a}$$

$$C_2 = \left( \frac{5^4 g\beta q_w'' v^3}{k} \right)^{1/5}, \tag{10.39b}$$

and $q''_w$ is the heat flux on the wall. Expression for the velocity components can be derived using Equations 10.17 and 10.38:

$$u = C_1 C_2 x^{3/5} f'(\eta), \tag{10.40a}$$

$$v = \frac{C_2}{5x^{1/5}} \left[ \eta f'(\eta) - 4f(\eta) \right]. \tag{10.40b}$$

After various terms in Equations 10.13 and 10.15 are evaluated, we get

$$\frac{d^3 f}{d\eta^3} - 3\left( \frac{df}{d\eta} \right)^2 + 4f \frac{d^2 f}{d\eta^2} - \Theta = 0, \tag{10.41}$$

$$\frac{d^3 f}{d\eta^3} - Pt\left( 4f \frac{df}{d\eta} - \Theta \frac{df}{d\eta} \right) = 0. \tag{10.42}$$

Boundary conditions are

$$f(0) = f'(0) = 0, \quad \Theta'(0) = 1, \tag{10.43a,b,c}$$

$$f'(\infty) = 0, \quad \Theta(\infty) = 0. \tag{10.43d,e}$$

The solutions to the differential Equations 10.41 and 10.42 subject to the boundary conditions (10.43) have been obtained numerically.

When Equation 10.38c is evaluated at the surface ($\eta = 0$), we obtain the following expression for the surface temperature:

$$T_w - T_\infty = -5^{1/5}\Theta(0)\frac{q''_w x}{k}\left( \frac{\beta g q''_w x^4}{v^2 k} \right)^{-1/5} = -\Theta(0)\left( \frac{k^4 g \beta}{5 v^2 q''^4_w x} \right)^{-1/5}. \tag{10.44}$$

In terms of modified Grashof number $Gr^*$, Equation 10.44 becomes

$$\frac{T_w - T_\infty}{\left( q''_w x/k \right)} Gr^{*1/5}_x = -5^{1/5}\Theta(0) \tag{10.45}$$

or

$$\frac{Nu_x}{Gr^{*1/5}_x} = -\frac{1}{5^{1/5}\Theta(0)}, \tag{10.46}$$

where

$$Nu_x = \frac{hx}{k} = \frac{q''_w x}{k(T_w - T_\infty)} \tag{10.47a}$$

$$Gr^*_x = \frac{\beta g q''_w x^4}{v^2 k} \tag{10.47b}$$

**TABLE 10.2**

$\Theta(0)$ and $f''(0)$ Values

| Pr | $\Theta(0)$ | $f''(0)$ |
|---|---|---|
| 0.1 | −2.7507 | 1.6434 |
| 1 | −1.3574 | 0.72196 |
| 10 | −0.76746 | 0.30639 |
| 100 | −0.46566 | 0.12620 |

*Source:* Sparrow, E. M. and Gregg, J. L., *Trans. ASME*, 78, 435, 1956.

The values of $\Theta(0)$ and $f''(0)$ associated with the solution of the differential equations (10.41) and (10.42) subject to the boundary conditions (10.43) are listed in Table 10.2.

Since there is not any characteristic temperature difference in the problem, the choice of a temperature difference in defining an average heat-transfer coefficient is purely arbitrary. In the literature, the average Nusselt numbers have been defined usually using the average plate temperature minus ambient temperature, which is

$$\overline{T_w - T_\infty} = \frac{1}{L} \int_0^L (T_w - T_\infty) dx = \frac{5}{6}(T_w - T_\infty)_L. \tag{10.48}$$

The average Nusselt number based on this temperature difference is

$$Nu = \frac{hL}{k} = \frac{6}{5^{6/5}\Theta(0)} Gr_L^{*1/5}. \tag{10.49}$$

But, since

$$Gr_L^* = \left(\frac{g\beta q_w'' L^4}{Kv^2}\right) = \left[\frac{g\beta \overline{(T_w - T_\infty)}L^3}{v^2}\right]\frac{hL}{k}, \tag{10.50}$$

then

$$\frac{Nu}{Gr_L^{1/4}} = \left[-\frac{6}{5^{6/5}\Theta(0)}\right]^{5/4}, \tag{10.51}$$

where

$$Gr_L = \frac{\beta g(T_w - T_\infty)L^3}{v^2}. \tag{10.52}$$

The values of $Nu/Gr_L^{1/4}$ for a flat plate having uniform surface temperature have also been calculated from the results of Ostrach [4] and are tabulated in Table 10.3 with the values from Equation 10.51.

Average Nusselt numbers have also been defined by using the temperature difference halfway along the plate [6].

**TABLE 10.3**

Comparison of Average Nusselt Numbers for
Uniform Wall Heat Flux and Constant Wall
Temperature

| Pr | $Nu/(Gr_L)^{1/4}$ (Equation 10.51) | $Nu/(Gr_L)^{1/4}$ [4] |
|---|---|---|
| 0.1 | 0.237 | 0.219 |
| 1.0 | 0.573 | 0.535 |
| 10.0 | 1.17 | 1.10 |
| 100.0 | 2.18 | 2.07 |

*Source:* Ostrach, S., An analysis of laminar free-convection
flow and heat transfer about a flat plate parallel to
the direction of the generating body force, NACA
Report 1111, 1953.

## Example 10.2

A vertical wall of an oven is 60 cm long and it is covered with a metal plate that is heated
electrically. The wall heat flux is 15 W/m² on the plate. The inside air temperature is
90°C and pressure is atmospheric; $Pr = 1$ is assumed. Calculate

a. The local heat-transfer coefficient at the end of the plate
b. Average heat-transfer coefficient over the plate

## Solution

a. Since the film temperature is now known, we shall evaluate the properties at
90°C. From Appendix B, we have

$$k = 0.0307 \ W/(m\,K), \quad \beta = 2.78 \times 10^{-3} \, 1/K, \quad v = 2.2 \times 10^{-5} \, m^2/s.$$

From Equation 10.44, we get

$$(T_w - T_\infty)_L = -5^{1/5}(-1.3574)\frac{15 \times 0.6}{0.0307}\left[\frac{2.78 \times 10^{-3} \times 9.8 \times 15(0.6)^4}{(2.2 \times 10^{-5})^2 \times 0.0307}\right]^{-1/5}$$

$$= 7.9°C,$$

and therefore Equation 10.48 yields

$$\overline{(T_w - T_\infty)} = \frac{5}{6}(T_w - T_\infty)_L = \frac{5}{6} \times 7.9 = 6.58°C.$$

The local heat-transfer coefficient can be calculated from Equation 10.46 as

$$Gr_x^* = \frac{\beta g q_w'' x^4}{v^2 k} = \frac{2.78 \times 10^{-3} \times 9.8 \times 15 \times (0.64)^4}{(2.2 \times 10^{-5})^2 \times 0.0307} = 16.2 \times 10^8,$$

$$Nu_x = \frac{Gr^{*1/5}}{5^{1/5}\Theta(0)} = \frac{(16.2 \times 10^8)^{1/5}}{5^{1/5} \times (1.3574)} = 37.1,$$

$$h_x = \frac{Nu_x k}{x} = \frac{37.1 \times 0.0307}{0.60} = 1.898 \text{ W}/(\text{m}^2 \text{ K}),$$

and the wall temperature is

$$T_w = \frac{q''_w}{h_x} + T_\infty = \frac{15}{1.898} + 90 \cong 98^\circ\text{C}.$$

Since $T_f = 94^\circ\text{C}$, the properties should be reevaluated. From Appendix B at $94^\circ\text{C}$, we have

$$k = 0.0310 \text{ W}/(\text{m K}), \quad \beta = 2.72 \times 10^{-3} \, 1/\text{K}, \quad v = 2.242 \times 10^{-5} \text{ m}^2/\text{s}.$$

If the earlier calculation is repeated with these properties, we obtain

$$h_x = 2.21 \text{ W}/(\text{m}^2 \text{ K})$$

and

$$T_w = \frac{q''_w}{h_x} + T_\infty = \frac{15}{2.21} + 90 = 96.77^\circ\text{C},$$

which is very close to the wall temperature.
b. Since

$$\overline{T_w - T_\infty} = \frac{5}{6}(T_w - T_\infty)_L = 5.64^\circ\text{C},$$

from Equation 10.52, we get

$$Gr_L = \frac{g\beta \overline{(T_w - T_\infty)}L^3}{v^2} = \frac{9.8 \times 2.72 \times 10^{-3} \times 5.6470 \times (0.6)^3}{(2.242 \times 10^{-5})}$$

$$= 6.468 \times 10^7.$$

From Table 10.3, we have (or Equation 10.51)

$$Nu = 0.573 Gr_L^{1/4} = 51.387,$$

which yields

$$h = 2.655 \text{ W}/(\text{m}^2 \text{ K}).$$

Ostrach's solution, on the other hand, gives

$$Nu = 0.535 Gr_L^{1/4} = 47.98,$$

which yields

$$h = 2.479 \text{ W}/(\text{m}^2 \text{ K}).$$

## 10.5 Inclined and Horizontal Surfaces

In many applications of natural convection, the heated surface may be curved or inclined with respect to the direction of the gravity field. Boundary-layer equations, similar to those for a vertical surface, may be derived for such flows. It can be shown that if $x$ is taken along the surface and $y$ normal to it, the continuity and energy equations (10.10) and (10.15) remain unchanged and the $x$-momentum equation (10.13) becomes (Figure 10.4)

$$u\frac{\partial u}{\partial x}+v\frac{\partial u}{\partial y}=v\frac{\partial^2 u}{\partial y^2}+\beta g(T-T_\infty)\cos\gamma. \tag{10.53}$$

This problem is identical to the natural-convection problem over a vertical flat plate except that $g$ is replaced by $g\cos\gamma$. This implies that the heat-transfer coefficient for natural convection on an inclined surface can be predicted by the results of a vertical surface if $g$ is replaced by $g\cos\gamma$ in all relationships derived in preceding sections. Therefore, the Grashof number is adjusted to accommodate the effect of the inclination, that is, $Gr_x\cos\gamma$ is used for $Gr_x$ for the prediction of the heat-transfer coefficient.

Natural convection on inclined surfaces has been studied by several investigators [10–13]. The earlier procedure for natural forced convection over an inclined surface was first suggested theoretically by Rich [10]. Vliet [11] studied natural convection on constant heat-flux inclined surfaces in air and in water. His data verify the validity of the earlier procedure up to inclination angles as large as 60°. Fujii and Imura [12] conducted detailed experiments for natural convection from an inclined surface with the heated surface facing downward and the heated surface facing upward subjected to approximately uniform wall heat flux to water (Figure 10.4a and b). They also discuss the transition to turbulent flow for the inclined surface facing upward. Pern and Gebhart [13] have considered flow over surfaces slightly inclined from the horizontal.

Consider now a horizontal flat plate as shown in Figure 10.5. The continuity, momentum, and energy equations under the conditions stated in Section 10.2 are given by the following:

$$\frac{\partial u}{\partial x}+\frac{\partial v}{\partial y}=0, \tag{10.54}$$

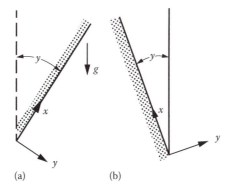

(a)                    (b)

**FIGURE 10.4**
Inclined flat surfaces (a) with the heated surface facing downward and (b) with the heated surface facing upward.

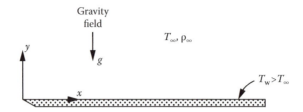

**FIGURE 10.5**
Geometry for the natural-convection boundary layer over a horizontal flat plate.

$$u\frac{\partial u}{\partial x} + v\frac{\partial u}{\partial v} = v\frac{\partial^2 u}{\partial y^2} - \frac{1}{\rho}\frac{\partial p}{\partial x},$$ (10.55)

$$\rho\beta(T - T_\infty) = \frac{\partial p}{\partial y},$$ (10.56)

$$u\frac{\partial T}{\partial x} + v\frac{\partial T}{\partial y} = \alpha\frac{\partial^2 T}{\partial y^2}.$$ (10.57)

Pern and Gebhart [13] introduced the following similarity variables

$$\eta = \frac{y}{x}\left(\frac{Gr_x}{5}\right)^{1/5}$$ (10.58)

and

$$\psi = 5vf(\eta)\left(\frac{Gr_x}{5}\right)^{1/5}.$$ (10.59)

They solved the resulting ordinary differential equations numerically to obtain velocity and temperature profiles for natural flow over the heated horizontal surface facing upward, the results of which are presented in Figure 10.6.

Pern and Gebhart [13] obtained the local Nusselt numbers for horizontal surfaces for both the UWT and uniform wall heat-flux boundary conditions. The correlations obtained for the Prandtl numbers over the range 0.1–100 are

$$Nu_x = \frac{hx}{k} = 0.394Gr_x^{1/5}Pr^{1/4}$$ (10.60)

for a uniform surface temperature and

$$Nu_x = 0.5013Gr_x^{1/5}Pr^{1/4}$$ (10.61)

for a uniform wall heat flux.

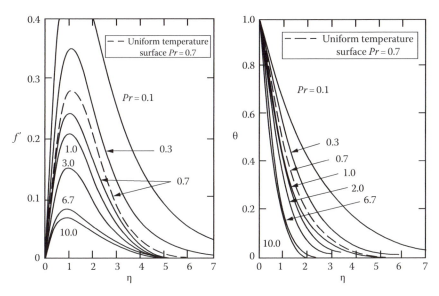

**FIGURE 10.6**
Velocity and temperature profiles for a natural convection over a horizontal surface with UHF. (From Pern, L. and Gebhart, B., *Int. J. Heat Mass Transfer*, 16, 1131, 1972.)

## 10.6 Property Variation in Free Convection

In situations where there are large temperature differences, the constant-property analysis leads to erroneous results. Therefore, when there are large temperature differences, variations in the fluid properties have to be taken into consideration, and for such an analysis the continuity, momentum, and energy equations for steady laminar boundary layer on a vertical plate as illustrated in Figure 10.7, with the Prandtl order of magnitude analysis, become

$$\frac{\partial}{\partial x}(\rho u) + \frac{\partial}{\partial y}(\rho v) = 0, \tag{10.62}$$

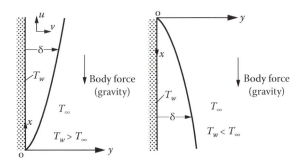

**FIGURE 10.7**
Free convection from a hot and a cold vertical plate.

$$\rho\left(u\frac{\partial u}{\partial x}+v\frac{\partial u}{\partial y}\right)=g(\rho_\infty-\rho)+\frac{\partial}{\partial y}\left(\mu\frac{\partial u}{\partial y}\right),\tag{10.63}$$

$$\rho c_p\left(u\frac{\partial T}{\partial x}+v\frac{\partial T}{\partial y}\right)=\frac{\partial}{\partial y}\left(k\frac{\partial T}{\partial y}\right),\tag{10.64}$$

where viscous dissipation and work of compression have been neglected.

The boundary conditions appropriate to the problem are

$$y=0:\quad u=v=0,\quad T=T_w,\tag{10.65a,b,c}$$

$$y\to\infty:\quad u=0,\quad T=T_\infty,\tag{10.65d,e}$$

where $T_w$ and $T_\infty$ are prescribed constant temperatures.

The problem of free convection with variable fluid properties about an isothermal vertical plate was analyzed by Sparrow and Gregg [14]. They introduced a stream function $\psi$ defined by the relations

$$\frac{\rho}{\rho_w}u=\frac{\partial\psi}{\partial y},\quad\frac{\rho}{\rho_w}v=-\frac{\partial\psi}{\partial x},\tag{10.66a,b}$$

where the fluid density at the wall, $\rho_w$, is regarded as a constant. Reduction of the momentum equation (10.63) and the energy equation (10.64) to ordinary differential equations can be done following the same procedure as outlined in Section 10.3. A new independent variable $\eta$, called the similarity variable, is defined by

$$\eta=Cx^{-1/4}\int_0^y\frac{\rho}{\rho_w}\,dy,\quad C=\frac{g(\rho_\infty-\rho)/\rho_w}{4v_w^2}.\tag{10.67a,b}$$

New dependent variables $f$ and $\Theta$ are given by

$$f(\eta)=\left(\frac{\psi}{x^{3/4}}\right)\left(\frac{1}{4v_mC}\right),\quad\Theta(\eta)=\frac{T-T_\infty}{T_w-T_\infty}.\tag{10.68a,b}$$

The function $\Theta$ is a dimensionless temperature and $f$ is related to the velocities in the following way:

$$u=4v_wC^2x^{1/2}f',\quad v=\left(\frac{\rho_w}{\rho}\right)\left(\frac{v_mC}{x^{1/4}}\right)(\eta f'-3f),\tag{10.69a}$$

where the primes denote differentiation with respect to $\eta$. The momentum equation (10.63) and the energy equation (10.64) can now be reduced to the following two ordinary differential equations [14]:

$$\frac{d}{d\eta}\left(\frac{\rho\mu}{\rho_w\mu_w}f''\right)+3ff''-2(f')^2+\frac{(\rho_\infty/\rho)-1}{(\rho_\infty/\rho_w)-1}=0,\tag{10.70}$$

$$\frac{d}{d\eta}\left(\frac{\rho k}{\rho_w k_w}\Theta''\right)+3Pr_w\left(\frac{c_p}{c_{pw}}\right)f\Theta'=0,$$ (10.71)

with the following boundary conditions:

$$f(0)=f'(0)=0,\quad\Theta(0)=1,$$ (10.72a,b,c)

$$f'(\infty)=\Theta(\infty)=0.$$ (10.72d,e)

Sparrow and Gregg [14] made an analysis of laminar free convection with variable properties on an isothermal vertical flat plate for gases and mercury. They presented a simple shorthand procedure for calculating free convective heat transfer under variable-property conditions by solving Equation 10.71 on a digital computer. It involves the use of the results that have been derived for constant-property fluids with a reference temperature.

According to Sparrow's findings, for gases, the constant-property heat-transfer results are generalized to the variable-property situation by replacing $\beta$ by $1/T_\infty$ and evaluating the other properties at [15]

$$T_r=T_w-0.38(T_w-T_\infty).$$ (10.73)

By the use of reference temperature according to Sparrow, the constant-property heat-transfer results coincide with the results of variable-property heat-transfer results. The error in the heat transfer predicted from the constant-property results by using this reference temperature is at most 0.6% over the entire range of $1/4 < T_w/T_\infty < 4$.

It can be shown that, from the mathematical point of view, the constant-property problem is identical to that of the special variable-property fluid with $p = \rho RT$, $\rho k = $ constant, $\rho\mu = $ constant, and $c_p = $ constant. It is interesting to note that a similar finding also applies to forced convection.

## 10.7 Approximate Solution of Laminar Free Convection on a Vertical Plate: von Karman–Pohlhausen Integral Method

Integral forms of the momentum and energy equations can be obtained either by integrating the corresponding differential equations or by applying the law of conservation of momentum and the first law of thermodynamics to an element of the boundary layer as shown in Figure 10.8. For the free-convection system, the integral momentum and energy equations become

$$\frac{d}{dx}\left(\int_0^\delta \rho u^2 dy\right)=\int_0^\delta \rho g\beta(T-T_\infty)dy-\mu\left(\frac{\partial u}{\partial y}\right)_w,$$ (10.74)

$$\frac{d}{dx}\left[\int_0^\delta u(T-T_\infty)dy\right]=-\alpha\left(\frac{dT}{dy}\right)_w.$$ (10.75)

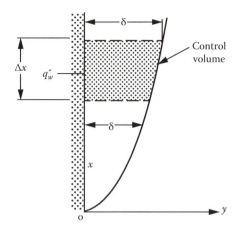

**FIGURE 10.8**
Control volume for the derivation of integral momentum and energy equations.

It is assumed that a common boundary-layer thickness δ can be used for both the velocity and thermal boundary layer. This assumption has its justification in the fact that the results of calculations performed with it are in good agreement with those from exact solutions of the boundary-layer differential equations for the cases of UWT and uniform heat flux (UHF).

To obtain the solutions to these equations, both the velocity and temperature distributions must be assumed as in the analyses of forced convection in Chapter 5.

### 10.7.1 Constant Wall Temperature Boundary Condition

The following conditions apply for the velocity and temperature distributions:

For velocity:

$$u(x,0) = 0, \quad v\frac{\partial^2 u(x,0)}{\partial y^2} = -g\beta(T_w - T_\infty), \tag{10.76a,b}$$

$$u(x,\delta) = 0, \quad \frac{\partial u(x,\delta)}{\partial y} = 0. \tag{10.76c,d}$$

For temperature:

$$T(x,0) = T_w = \text{constant}, \tag{10.77a}$$

$$T(x,\delta) = T_\infty, \quad \frac{\partial T(x,\delta)}{\partial y} = 0. \tag{10.77b,c}$$

The condition (10.76b) is obtained from Equation 10.13 at $y = 0$.

A cubic polynomial form can be chosen for the velocity profile, that is,

$$\frac{u}{u_x} = A + B\left(\frac{y}{\delta}\right) + C\left(\frac{y}{\delta}\right)^2 + D\left(\frac{y}{\delta}\right)^3, \tag{10.78}$$

where $u_x$ is a function of $x$ that accounts for the increase in local velocity with $x$. Constants in Equation 10.78 are determined by the use of the conditions (10.76). The final velocity profile to be assumed becomes

$$u = u_x \frac{y}{\delta}\left(1 - \frac{y}{\delta}\right)^2. \tag{10.79}$$

The temperature profile can be approximated by the parabolic curve:

$$T - T_\infty = (T_w - T_\infty)\left(1 - \frac{y}{\delta}\right)^2. \tag{10.80}$$

The integral equations, Equations 10.74 and 10.75, can be evaluated by substituting the velocity and temperature distributions from Equations 10.79 and 10.80. The results are

$$\frac{1}{105}\frac{d}{dx}\left(u_x^2\delta\right) = \frac{1}{3}g\beta(T_w - T_\infty)\delta - \frac{vu_x}{\delta}, \tag{10.81}$$

$$\frac{1}{30}\frac{d}{dx}(u_x\delta) = \frac{2\alpha}{\delta}. \tag{10.82}$$

For the solution of Equations 10.81 and 10.82, additional assumptions are needed. Therefore, we assume that $u_x$ and $\delta$ vary as simple power functions of $x$:

$$u_x = C_1 x^m, \quad \delta = C_2 x^n \tag{10.83a,b}$$

When these are inserted into Equations 10.81 and 10.82, we obtain

$$\left(\frac{2m+n}{105}\right)C_1^2 C_2 x^{2m+n-1} = g\beta(T_w - T_\infty)\frac{C_2}{3}x^n - \frac{C_1}{C_2}vx^{m-n}, \tag{10.84}$$

$$\left(\frac{m+n}{30}\right)C_1 C_2 x^{m+n-1} = \frac{2\alpha}{C_2}x^{-n}. \tag{10.85}$$

From dimensional reasoning, the value of the exponent of $x$ must be the same on both sides of each equation, that is,

$$2m+n-1 = n = m-n \quad \text{and} \quad m+n-1 = -n,$$

which are satisfied by $m = 1/2$ and $n = 1/4$. Substitution of these values back into Equations 10.84 and 10.85 gives

$$C_1 = 5.17v\left(\frac{20}{21} + Pr\right)^{-1/2}\left[\frac{\beta g(T_w - T_\infty)}{v^2}\right]^{1/2}, \tag{10.86}$$

$$C_2 = 3.93\left(\frac{20}{21} + Pr\right)^{-1/4}\left[\frac{\beta g(T_w - T_\infty)}{v^2}\right]^{1/4}Pr^{-1/2}. \tag{10.87}$$

When the values of $C_1$, $C_2$, $m$, and $n$ are substituted into Equations 10.83a and b, the resultant equations for $u_x$ and $\delta$ become

$$\frac{u_x}{x} = 5.17v(0.952 + Pr)^{-1/2}Gr_x^{1/2}, \tag{10.88}$$

$$\frac{\delta}{x} = 3.93Pr^{-1/2}(0.952 + Pr)^{-1/4}Gr_x^{1/4}. \tag{10.89}$$

The heat-transfer coefficient may be evaluated from

$$q''_w(x) = -kA\left(\frac{\partial T}{\partial y}\right)_w = h_x A(T_w - T_\infty). \tag{10.90}$$

Introducing the temperature distribution (10.80), we obtain

$$h_x = \frac{2k}{\delta}, \quad Nu_x = 2\frac{x}{\delta} \tag{10.91a,b}$$

so that

$$h_x = 0.508Pr^{1/2}(0.952 + Pr)^{-1/4}\left[\frac{\beta g(T_w - T_\infty)}{v^2}\right]^{1/4}kx^{-1/4} \tag{10.92}$$

and

$$Nu_x = 0.508Pr^{1/2}(0.952 + Pr)^{-1/4}(Gr_x)^{-1/4}. \tag{10.93}$$

Equation 10.93 agrees well with Equation 10.32. Equation 10.93 can also be expressed as

$$Nu_x = 0.508Ra_x^{1/4}\left(\frac{Pr}{0.952 + Pr}\right)^{1/4}, \tag{10.94}$$

where $Ra_x$ is the local Rayleigh number that is defined as

$$Ra_x = \frac{g\beta(T_x - T_\infty)x^3}{v\alpha}. \tag{10.95}$$

The average heat-transfer coefficient, $\bar{h}$ over $x = 0$ to $L$, is 4/3 times the heat-transfer coefficient evaluated at $x = L$, that is,

$$Nu = 0.68Ra_L^{1/4}\left(\frac{Pr}{0.952 + Pr}\right)^{1/4}. \tag{10.96}$$

As it can be seen from Equation 10.89, as the Prandtl number decreases, the boundary-layer thickness increases; therefore, at low Rayleigh numbers (i.e., $Pr \to 0$), Equation 10.93 does not give accurate results (why?).

Churchill and Chu [16] proposed the following correlation for free-convection heat transfer on a vertical flat plate with constant wall temperature:

$$Nu = 0.68 + \frac{0.67 Ra_L^{1/4}}{\left[1 + (0.492/Pr)^{9/16}\right]^{4/9}}, \quad \text{for } 10^{-1} < Ra_L < 10^9. \tag{10.97}$$

The comparison of the results of the integral method with exact solutions and experimental results indicates that the prediction of the heat-transfer coefficient with the integral method is quite satisfactory.

### 10.7.2 Nonuniform Wall Heat Flux or Nonuniform Wall Temperature Boundary Condition

Problem of laminar free convection on a vertical plate with prescribed nonuniform wall temperature was solved by Sparrow [15]. The flow was taken to be of the boundary-layer type, and the problem was formulated by the von Karman–Pohlhausen method. The solution of the resulting equations is achieved by series expansion. The first term in the series corresponds to the result for uniform thermal conditions on the wall. The succeeding terms give the influence of the nonuniform thermal conditions. The first five terms of the series have been calculated.

In a large number of technical applications, the thermal conditions on the surface are nonuniform. These nonuniformities in thermal conditions may be grouped into two categories:

1. The heat flux may be prescribed to vary over the surface. It is then of interest to calculate the resulting variation of the surface temperature.
2. The variation of the temperature on the surface may be prescribed. It is then of interest to calculate either the local rate heat transfer at various locations on the surface, or the overall rate of heat transfer from the surface, or both.

The analysis of prescribed nonuniform wall heat flux proceeds with the use of the von Karman–Pohlhausen method, according to which the velocity and temperature distributions in the boundary layer are written as polynomials in $y$ whose coefficients are functions of $x$. The following conditions apply for temperature and velocity distributions:

$$\frac{\partial T(x,0)}{\partial y} = \frac{q_w''(x)}{k}, \quad u(x,0) = 0, \tag{10.98a,b}$$

$$T(x,\delta) = T_\infty, \quad u(x,\delta) = \frac{\partial u(x,\delta)}{\partial y} = \frac{\partial T(x,\delta)}{\partial y} = 0. \tag{10.98c,d,e,f}$$

The following polynomials are chosen to satisfy conditions (10.98):

$$T - T_\infty = \frac{q_w''\delta}{2k}\left(1 - \frac{y}{\delta}\right)^2, \tag{10.99}$$

$$\frac{u}{u_x} = \frac{y}{\delta}\left(1 - \frac{y}{\delta}\right)^2. \tag{10.100}$$

The polynomials representing the velocity and temperature distributions are introduced into Equations 10.74 and 10.75, and after the integration is carried out, there result a pair of first-order ordinary differential equations for $u_x$ and $\delta$. In dimensionless form, these equations are

$$\frac{1}{105}\frac{d}{dX}(\Omega^2\Delta) = \frac{\Delta^2}{6}\frac{q_w''}{q_w''} - \frac{\Omega}{\Delta}, \tag{10.101}$$

$$\frac{1}{30}\frac{d}{dX}\left(\Omega\Delta^2\frac{q_w''}{q_w''}\right) = \frac{2}{Pr}\frac{q_w''}{q_0''}, \tag{10.102}$$

where $q_w'' = q_w''$ at $x = 0$ and

$$\Delta = \left(\frac{g\beta q_0''}{k v^2}\right)^{1/4}\delta, \quad \Omega = u_x\left(\frac{g\beta q_0'' v^2}{k}\right)^{-1/4}, \tag{10.103a,b}$$

$$X = \left(\frac{g\beta q_0''}{k v^2}\right)^{1/4}x. \tag{10.104}$$

If the variation of $q_w''/q_0''$ is prescribed, solutions of Equations 10.101 and 10.102 are obtained. When $q_w''/q_0''$ for all values of $x$ (constant heat-flux case), solutions of Equations 10.101 and 10.102 become

$$\Omega_0 = (6000)^{1/5}Pr^{-1/5}\left(\frac{4}{5} + Pr\right)^{-2/5}X^{3/5}, \tag{10.105}$$

$$\Delta_0 = (360)^{1/5}\left(\frac{4}{5} + Pr\right)^{1/5}Pr^{-2/5}X^{1/5}. \tag{10.106}$$

Equation 10.106 can be rewritten as

$$\frac{\Delta_0}{X} = \frac{\delta}{x} = (360)^{1/5}\left[\frac{(4/5) + Pr}{Pr^2 Gr_x^*}\right]^{1/5}, \tag{10.107}$$

where $Gr_x^*$ is a modified Grashof number based on $x$ and defined by

$$Gr_x^* = \frac{g\beta q_0'' x^4}{k v^2}. \tag{10.108}$$

The surface temperature from Equation 10.99 is

$$T_w - T_\infty = \frac{q_0''\delta}{2k} \tag{10.109a}$$

or

$$T_w - T_\infty = 1.622 \frac{q_0'' x}{k} \left( \frac{0.8 + Pr}{Pr^2 Gr_x^*} \right)^{1/5}.$$ (10.109b)

Rearrangement of Equation 10.109 gives the local Nusselt number for the case of uniform wall heat flux as derived by the von Karman–Pohlhausen method:

$$Nu_x = 2 \left[ \frac{Pr^2 Gr_x^*}{360(0.8 + Pr)} \right]^{1/5}.$$ (10.110)

This expression is compared to that found from the exact solution, Equation 10.46, in Figure 10.9 and the agreement is again good everywhere except at the lower Prandtl numbers.

The fact that such a good agreement is obtained on the basis of such simple assumed quadratic profiles as given by Equations 10.79, 10.80, and 10.99 is perhaps somewhat surprising in view of the rather sophisticated profiles assumed in the application of the von Karman–Pohlhausen method for forced-convection problems. However, in forced convection, it is necessary to approximate not only the heat-transfer characteristics but also flow characteristics (as skin friction). The earlier procedure does not give good agreement with the flow parameters for free convection. Braun and Heighway [17] have modified the von Karman–Pohlhausen method, not only to give better agreement with flow characteristics for free convection but also to improve the accuracy of the heat-transfer results for very low and high Prandtl numbers.

For the nonuniform wall temperature, the temperature profile in the boundary layer is approximated by the following polynomial:

$$T - T_\infty = (T_w - T_\infty)\left(1 - \frac{y}{\delta}\right)^2 = \Theta(x)\left(1 - \frac{y}{\delta}\right)^2.$$ (10.111)

Solution to this problem is also given in Reference 17.

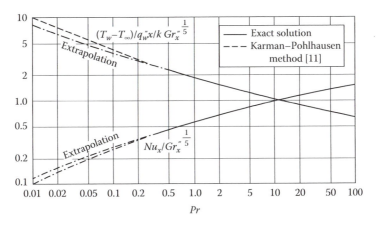

**FIGURE 10.9**
Variation of $Nu_x(Gr_x^*)^{1/5}$ with Prandtl numbers.

## 10.8 Turbulent Heat Transfer on a Vertical Plate

Most natural-convection flows of interest, in nature and in technology, are turbulent flows. Because of the importance of turbulent natural-convection flows, a considerable amount of effort, experimental and analytical, has been directed at understanding and determining the transport mechanisms and the rates of energy transfer. The understanding of turbulent natural convection is derived from the extensive work done in turbulent forced convection. Various regimes of boundary-layer growth from laminar to turbulent flow are shown in Figure 10.10. From experimental observations, it is found that when $Gr_L > 10^9$, flow over the plate becomes turbulent.

The turbulent boundary-layer continuity, momentum, and energy equations can be written in terms of mean flow parameters by replacing $v$ and $\alpha$ by $(v + \varepsilon_m)$ and $(\alpha + \varepsilon_h)$ in Equations 10.13 and 10.15, respectively:

$$\frac{\partial u}{\partial x} + \frac{\partial v}{\partial y} = 0, \tag{10.112}$$

$$u\frac{\partial u}{\partial x} + v\frac{\partial u}{\partial y} = g\beta(T - T_\infty) + \frac{\partial}{\partial y}\left[(v + \varepsilon_m)\frac{\partial u}{\partial y}\right], \tag{10.113}$$

$$u\frac{\partial T}{\partial x} + v\frac{\partial T}{\partial y} = \frac{\partial}{\partial y}\left[(\alpha + \varepsilon_h)\frac{\partial T}{\partial y}\right]. \tag{10.114}$$

Turbulence models outlined in Chapter 7 can be employed for solving the earlier governing equations.

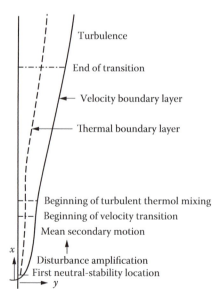

**FIGURE 10.10**
Growth of the boundary layer and the sequence of events during transition in water, $Pr = 6.7$. (From Jaluria, Y., Kakaç, S., Shah, R. K., and Aung, W. (Eds.): *Handbook of Single-Phase Convective Heat Transfer*. 1987. Copyright Wiley-VCH Verlag GmbH & Co. KGaA.)

The momentum and energy equations in integral forms can also be used to calculate free convection on the turbulent region [19]. Velocity and temperature distributions in such a turbulent boundary layer can well be approximated by the following equations:

$$u = u_x \left(\frac{y}{\delta}\right)^{1/7} \left(1 - \frac{y}{\delta}\right)^4,$$

(10.115)

$$\Theta = \frac{T - T_\infty}{T_w - T_\infty} = \left[1 - \left(\frac{y}{\delta}\right)^{1/7}\right].$$

(10.116)

But we can no longer calculate the shear at the wall from the expression $\mu(\partial u/\partial y)_w$ and heat flux from the expression $-k(\partial T/\partial y)_w$ by the use of Equations 10.115 and 10.116, respectively. It is true that the stress must ultimately be transmitted to the wall through the mechanism of viscosity in the laminar sublayer, but we cannot expect to represent the thickness or the velocity distribution in the laminar sublayer with complete accuracy. Therefore, the velocity and temperature profiles chosen are valid in the turbulent zone outside the laminar sublayer, and we make use of the following empirical expression for the wall shear stress that is obtained from measurements on turbulent flow in circular ducts:

$$\tau_w = 0.0225 \rho u_x^2 \left(\frac{v}{u_x \delta}\right)^{1/4}.$$

(10.117)

This equation also proved to be valid for flat plates. In free convection, $U_\infty$ is replaced with $u_x$.

The last terms in Equations 10.74 and 10.75 should be replaced by $\tau_w$ and $q_w''/\rho c_p$ for the turbulent case. Inserting the velocity and temperature distributions together with Equation 10.117 into the integral momentum equation, we get

$$0.0523 \frac{d}{dx}\left(\delta u_x^2\right) = 0.1255(T_w - T_\infty)g\beta\delta - 0.0225 u_x^2 \left(\frac{v}{u_x \delta}\right)^{1/4}.$$

(10.118)

Wall heat flux can be obtained from Colburn analogy:

$$\frac{q_w''}{\rho c_p (T_w - T_\infty)} Pr^{2/3} = \frac{\tau_w}{\delta u_x}.$$

(10.119)

Then the integral energy Equation 10.75 becomes

$$0.366 \frac{d}{dx}(\delta u_x) = 0.0228 u_x \rho c_p \left(\frac{v}{u_x \delta}\right)^{1/4} (Pr)^{-2/3}.$$

(10.120)

The solutions of Equations 10.118 and 10.120 proceed exactly in the same way as for laminar boundary layer. Assume

$$u_x = Ax^m,$$

(10.121a)

$$\delta = Bx^n.$$

(10.121b)

Introducing these expressions into Equations 10.118 and 10.120, we get

$$0.0523A^2B(2m+n)x^{2m+n-1} = 0.125(T_w - T_\infty)g\beta x^n - 0.0225A^2 x^{2m}\left(\frac{v}{ABx^{m+n}}\right)^{1/4}, \quad (10.122)$$

$$0.0366AB(m+n)x^{m+n-1} = 0.0228\rho c_p A x^m \left(\frac{v}{ABx^{m+n}}\right)^{1/4} Pr^{-2/3}. \quad (10.123)$$

These equations must be valid for any value of $x$; the exponents must have the same value for every summand:

$$2m+n-1 = 2m - \frac{(m+n)}{4},$$

$$m+n-1 = m - \frac{(m+n)}{4}.$$

These give

$$m = \frac{1}{2}, \quad n = \frac{7}{10}.$$

Then $A$ and $B$, and in turn $u_x$ and $\delta$ can be calculated. The results are the following equations:

$$u_x = 1.185\frac{v}{x}(Gr_x)^{1/2}\left[1+0.494Pr^{2/3}\right]^{-1/2}, \quad (10.124)$$

$$\delta = 0.565x(Gr_x)^{-1/10}(Pr)^{-8/15}\left[1+0.494Pr^{2/3}\right]^{-1/10}. \quad (10.125)$$

Inserting these expressions into Equation 10.100, $\tau_w$ is obtained as a function of $x$, and then Colburn analogy (Equation 10.119) yields the Nusselt number, resulting in

$$Nu_x = \frac{h_x x}{k} = 0.0295(Gr_x)^{2/5}(Pr)^{7/5}(1+0.494Pr^{2/3})^{-2/5} \quad (10.126a)$$

or

$$Nu_x = 0.0295(Ra_x)^{2/5}\frac{Pr^{1/15}}{(1+0.494Pr^{2/3})^{2/5}}. \quad (10.126b)$$

The average Nusselt number can be calculated from Equation 10.126 as $Nu = 5Nu_L/6$. Equation 10.126 agrees well with experimental measurements. Experimental work on turbulent natural convection is limited. From experimental observations, it is found that for values of the product $GrPr$ greater than $10^9$ the motion is generally turbulent, and it is found that the Nusselt number is then proportional to $(GrPr)^{1/3}$ [20]. The following working formulas have been suggested by Saunders [6] for vertical surfaces:

$$\text{For gases:} \quad Nu = 0.12Gr^{1/3}Pr^{1/3}. \tag{10.127a}$$

$$\text{For liquids:} \quad Nu = 0.17Gr^{1/3}Pr^{1/3}. \tag{10.127b}$$

Bayley [21] recommends the following correlation for vertical surfaces:

$$Nu = 0.1Ra^{1/3}, \quad 2\times10^9 \leq Ra \leq 10^{12} \tag{10.128}$$

and

$$Nu = 0.183Ra^{0.31}, \quad 2\times10^9 \leq Ra \leq 10^{15}, \tag{10.129}$$

which agrees well with experimental measurements [22]. Vliet and Liu [23] and Vliet [11] performed experimental studies with vertical and inclined surfaces under uniform wall heat-flux boundary conditions using both water and air. They correlated the turbulent heat-transfer data as

$$Nu_x = 0.568\left(Gr_x^* Pr\right)^{0.22}, \quad 2\times10^{13} < Gr_x^* Pr < 10^6, \tag{10.130}$$

where

$$Gr_x^* = \frac{g\beta q_w'' x^4}{kv^2}. \tag{10.131}$$

Vliet [11] found the exponent to vary from 0.22 for vertical surfaces to 0.25 for horizontal. Vliet and Rose [24] studied turbulent natural convection on upward and downward facing inclined surfaces with constant heat-flux boundary conditions and found their data to correlate well with the following relationship:

$$Nu_x = 0.17\left(Gr_x^* Pr\right)^{0.25}. \tag{10.132}$$

In the Grashof number $Gr_x^*$, $g$ is replaced by $g\cos\gamma$, where $\gamma$ is the angle at which the surface is inclined with the vertical.

Churchill and Chu [16] proposed the following correlation for both laminar and turbulent free convection over a vertical flat plate with constant wall temperature:

$$Nu_m^{1/2} = 0.825 + \frac{0.387Ra_L^{1/6}}{\left[1+(0.492/Pr)^{9/16}\right]^{8/27}}, \quad 10^{-1} < Ra_L < 10^{12}. \tag{10.133}$$

## 10.9 Dimensional Analysis in Natural Convection

The procedure outlined in Chapter 9 will now be applied to obtain the dimensionless groups in natural convection. The presence of a buoyancy force is a requirement for the existence of a natural-convection flow. The buoyancy arises from density differences, which are consequences of temperature gradients within the fluid. If $\rho_\infty$ is the density of undisturbed fluid with the corresponding temperature $T_\infty$, then the buoyancy force per unit volume of an element of fluid at temperature $T$ and density $\rho$ is

$$(\rho_\infty - \rho)g \qquad (10.134)$$

and $\rho_\infty$ is related to $\rho$ by

$$\rho_\infty = \rho(1 + \beta\Delta T), \qquad (10.135)$$

where
  $\beta$ is the thermal expansion coefficient
  $\Delta T = T - T_\infty$

Thus, the buoyancy force per unit mass is

$$\frac{[\rho(1 + \beta\Delta T) - \rho]g}{\rho} = g\beta\Delta T. \qquad (10.136)$$

Therefore, heat flux at the wall in natural convection will depend upon the fluid properties $\rho$, $\mu$, $c_p$, and $k$, linear dimension $L$, the overall temperature difference $\Delta T$, and the buoyancy force $\beta\Delta T$. Since $\Delta T$ has already been listed, then it is only necessary to add $\beta g$, that is,

$$q''_w = f(\Delta T, \beta g, \rho, \mu, c_p, k, L). \qquad (10.137)$$

There are eight physical quantities and five fundamental dimensional quantities that are mass $M$, length $L$, time $t$, temperature $T$, and heat $H$. Although heat can be expressed as $ML^2/t^2$, it can be regarded as independent provided there is no transfer of energy from one form to another. Therefore, according to the $\pi$-theorem, three $\pi$-terms are expected. Using the method of Chapter 9, let us take the primary quantities as follows:

| Quantity: | $\Delta T$ | $k$ | $\rho$ | $v$ | $L$, |
|-----------|------------|-----|--------|-----|------|
| Dimension: | $T$ | $\dfrac{ML}{t^3T}$ | $\dfrac{M}{L^3}$ | $\dfrac{M}{Lt}$ | $L$. |

Note that $q''_w, \beta g, c_p$ will each appear in a separate $\pi$-term. Then for $q''_w$, $\pi_1 = q''_w L/\Delta Tk = hL/k$, the Nusselt number; for $\beta g$, $\pi_2 = \beta g\Delta TL^3/v^2$, the Grashof number; and for $c_p$, $\pi = \mu c_p/k$, the Prandtl number. The dimensionless relation for free convection from a vertical plate may now be expressed as

$$Nu = \phi(Gr, Pr). \qquad (10.138)$$

**TABLE 10.4**

Summary of the Typical Natural-Convection Correlations for External Flows

| Geometry | Recommended Correlation | Conditions |
|---|---|---|
| 1. Vertical flat surface | $Nu_T = \left\{ 0.825 + \dfrac{0.387 Ra^{1/6}}{\left[ 1 + (0.492/Pr)^{9/16} \right]^{8/27}} \right\}$ | $10^{-1} < Ra < 10^{12}$ isothermal [16] |
| 2. Vertical flat surface | $Nu_T = 0.68 + \dfrac{0.67 Ra^{1/4}}{\left[ 1 + (0.492/Pr)^{9/16} \right]^{4/9}}$ | $0 < Ra < 10^7$ isothermal [16] |
| 3. Inclined flat surface | Earlier equations with $g$ replaced by $g \cos \gamma$ | $\gamma \le 60°$ |
| 4. Inclined surface | $Nu_H = 0.14[(GrPr)^{1/3} - (Gr_L Pr)^{1/3}]$ $+ 0.56(Gr_{cr} Pr \cos \gamma)^{1/4}$ | $10^5 < Gr_L Pr \cos \gamma < 10^{11}$, $15° < \gamma < 75°$ UHF, facing downward, $Gr_{cr} =$ $5 \times 10^9$ $2 \times 10^9$, $10^8$, and $10^6$ for $\gamma = 15°$, $30°$, $60°$, and $70°$, respectively [12] |
| 5. Horizontal flat surface | a. Heated, facing upward $\quad \bar{Nu} = 0.54 Ra^{1/4}$ b. Heated, facing downward $\quad \bar{Nu} = 0.15 Ra^{1/3}$ $\bar{Nu} = 0.27 Ra^{1/4}$ | $10^5 < Ra < 10^7$, isothermal [5] $10^7 < Ra < 10^{10}$, isothermal [5] $3 \times 10^5 < Ra < 10^{10}$, isothermal [5] |
| 6. Horizontal cylinders | $Nu^{1/2} = 0.60 + \dfrac{0.387 Ra_d^{1/6}}{[1 + (0.559/Pr)^{9/16}]^{8/27}}$ | $10^{-5} < Ra_d < 10^{12}$, isothermal [10] |
| 7. Sphere | $Nu = 2 + 0.43 Ra_d^{1/4}$ | $Pr \cong 1$, $1 < Ra_d < 10^5$ [25] |

*Source:* Jaluria, Y., Kakaç, S., Shah, R. K., and Aung, W. (Eds.): *Handbook of Single-Phase Convective Heat Transfer*. 1987. Copyright Wiley-VCH Verlag GmbH & Co. KGaA.

*Note:* All fluid properties are to be evaluated at the film temperature, $T_f = (T_w + T_\infty)/2$, unless mentioned otherwise. $Nu$ and $Ra$ are based on height $L$ for vertical plate and for inclined and horizontal surfaces on diameter $(d)$.

If a functional relationship is assumed to be a product of powers, we get

$$Nu = C(GrPr)^m. \tag{10.139}$$

In practical applications, many problems of heat transfer and flow process are so complicated that analytical methods cannot easily be handled and the experimental analysis must be performed to obtain heat-transfer correlations for design purposes.

Over the years, a considerable amount of heat-transfer information for various flow configuration and boundary conditions has been gathered, and correlations have been recommended. These correlations can be found in text books such as References 26,27. Some of the typical correlations available in the literature for external flows are presented in Table 10.4.

## 10.10 Interferometric Studies

The foregoing analyses of free-convection heat transfer are the cases that can be solved mathematically. But especially in turbulent convection, it is difficult to predict the temperature and velocity profiles analytically. Hence, interferometric methods are used to study the temperature field. The laser anemometer shows high promise in free-convection measurements as it does not disturb the flow field.

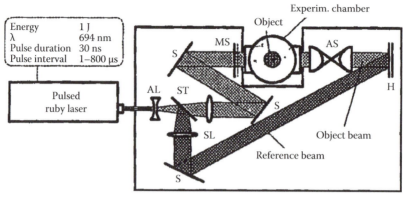

**FIGURE 10.11**
Schematic representation of holographic interferometer. *Note:* AL, divergent lens; AS, imaging system; H, holoplate; MS, ground glass; S, mirror; SL, convergent lens; ST, beam splitter. (From Chavez, A., Holografische Untersuchung an Einspritzstrahlen-Fluiddynamik und Wärmeübergang durch Kondensation, Dissertation, Technische Universität München, Germany, 1991.)

Modern interferometers make use of holographic techniques to overcome classical problems related to the optical quality of the interferometer components and adjustments of the light beam paths while the experiment is in progress. These modern interferometers are called "holographic interferometers." Their design is based upon the Mach–Zehnder interferometer, in which the second beam splitter is substituted by a holographic plate as shown in Figure 10.11. As a first step, the object and the reference beams meet at the plane of the holographic plate, building an interferogram as reference state for the experiment. In the second step, after photographic development and repositioning of the holographic plate, the recorded interferogram is reconstructed by illuminating the plate with the reference beam. More interesting is the case in which the two beams, object and reference beams, meet again at the plane of the repositioned holographic plate. In this case, the interferogram is a real-life representation of the difference between the actual state of the experiment and its reference state. The produced interferograms can be recorded by using kinematography or video tape. In this way, all possible disturbances arising due to imperfection in the optical components are automatically compensated for. This last method is called "real-time method of holographic interferometry" and is very suitable for free-convection heat-transfer studies.

An interferogram indicates lines of constant refractive index. For a gas in free convection at low pressures, the lines of constant refractive index are equivalent to lines of constant temperature, analogous to the potential flow theory, and can be interpreted as potential lines. The corresponding streamlines describe the heat flow field. In this way, the interferogram gives a clear indication of the development of the layer along and around surfaces. Examples of interferometric studies can be found in References 28–34. Figure 10.12 shows an interferogram of the flow field through an array of heated cylinders. The interference fringes can be interpreted as lines of constant temperature. Once the temperature field is obtained, the heat transfer from the surfaces in free convection can be calculated by using the temperature gradient at the surfaces.

Figure 10.13 presents isotherms in natural convection between concentric cylinders. An interferogram shows the temperature distribution; the inner tube is heated at constant

**FIGURE 10.12**
An interferogram (isotherms) of the flow field around an array of heated cylinders. (From Chavez, A., *Holografische Untersuchung an Einspritzstrahlen-Fluiddynamik und Warmeubergang durch Kondensation*, Dissertation, Technische Universitat Munchen, Germany, 1991; Chavez, A. and Mayinger, F., Evaluation of pulsed laser holograms of spray droplets using digital image processing, *Proceedings of the 2nd International Congress on Particle Sizing*, Tempe, AZ, E. Dan Hirleman (Ed.), pp. 462–471, 1990.)

heat flux and the outer tube is at a constant temperature. More details of the interferometric studies can be found in References 28,31.

## 10.11 Natural Convection in Enclosed Spaces

Enclosures are defined as finite spaces bounded by walls and filled with fluid media. Natural convection in such enclosures occurs as a result of buoyancy caused by a body force field with density variations within the fluid.

Natural convection in enclosures is important in many engineering applications such as in furnace design, in the operation and design of solar collectors, in calculating heat losses through double windows, in energy storage systems, in the cooling of electronic equipment and devices, in fire protection in buildings and other confined spaces, and in the production of high-purity crystals.

Natural-convection flows in enclosures is also known as buoyancy-driven enclosure flows. In writing this section, the authors utilized two important references on the natural convection in enclosed spaces [35,36].

Several examples of enclosed spaces are shown in Figure 10.14.

2D rectangular enclosures shown in Figure 10.14a and b have two characteristic dimensions that are the height $H$ and the width $W$. The ratio of height to width $H/W$ is defined as the aspect ratio. When the aspect ratio, $H/W < 1$, the enclosure is known as shallow enclosure or shallow cavity. If the aspect ratio $H/W > 1$, it is called a vertical slot (enclosure).

### 10.11.1 Governing Equations for Enclosure Flows

The governing equations for natural convection in enclosed spaces are for the conservation of mass, momentum, and energy. Under the conditions of incompressible Boussinesq

**FIGURE 10.13**
Isotherms in natural convection between concentric cylinders, $Gr$ = 122,000, $T$ = 13°C. (From Hauf, W. and Grigull, U., Optical methods in heat transfer, in *Advances in Heat Transfer*, Vol. 6, Academic Press, New York, 1970.)

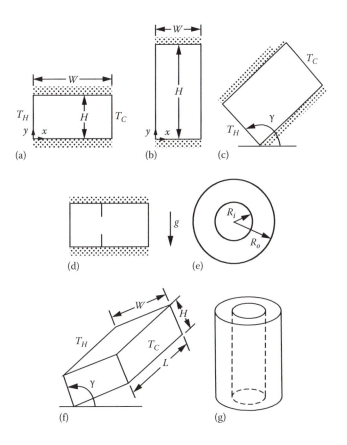

**FIGURE 10.14**
Some examples of 2D and 3D enclosures (Adapted from Yang, K. T., Kakaç, S., Shah, R. K., and W. Aung (Eds.): *Handbook of Single-Phase Convective Heat Transfer.* 1987. Copyright Wiley-VCH Verlag GmbH & Co. KGaA. With permission.): (a) 2D rectangular horizontal enclosure, (b) 2D rectangular vertical enclosure, (c) an enclosure tilted relative to the gravity field, (d) a rectangular enclosure with a pair of vertical partitions, (e) an annular enclosure bounded by two concentric cylinders, (f) 3D enclosure—a paralleled-piped enclosure, and (g) a truncated annular enclosure.

fluid with the assumption of constant thermophysical properties, the governing equations for laminar flows can be written as

$$\nabla \cdot V = 0, \tag{10.140}$$

$$\rho \frac{D\mathbf{V}}{Dt} = \mu \nabla^2 \mathbf{V} - \nabla(p - p_0) - \rho \beta g (T - T_0), \tag{10.141}$$

$$\rho c_P \frac{DT}{Dt} = k \nabla^2 T + S, \tag{10.142}$$

where
   $p_0$ and $T_0$ both refer to the hydrostatic conditions
   $S$ is the source term

It should be noted that the energy equation (10.142) does not contain the dissipation and pressure work terms, since both can be neglected for natural-convection phenomena [36].

The governing equations must be solved simultaneously for the five unknowns $u$, $v$, $w$, $p$, and $T$. For the solution of Equations 10.140 through 10.142, hydrostatic pressure field, the boundary, and initial conditions for a given enclosure must be specified.

There are also open enclosures when one of the walls is missing. They are also called open cavities. If the opening is small such as a window of a room, it is referred to as a vent. The boundary conditions at the openings must also be specified [35,36].

Solutions for enclosure natural-convection problems are generally obtained by numerical methods because of the complexity of the governing equations that must be solved simultaneously. There are no exact solutions available for natural convection in enclosed spaces except for some limiting cases. For purposes of illustration, in the following section, a solution to a vertical slot with isothermal walls is given.

### 10.11.2 Laminar Natural Convection in a Vertical Slot with Isothermal Walls

A viscous fluid confined in a vertical slot between two parallel plates moves upward adjacent to the hot wall and downward adjacent to the cold wall. If the ends of the slot are closed, then a unicellular flow pattern in the slot is observed provided that the difference in the temperature of the wall is small. If the wall temperature difference exceeds a critical value, then the unicellular motion may become unstable and a secondary multicellular motion appears in the slot.

When the *aspect ratio*, defined as the ratio of the height of the slot to the distance between the plates, is large (>100), the temperature profile in the slot away from the ends is observed to be linear and heat is transferred across the slot primarily by conduction. This solution is called the *conduction regime*. In this regime, a cubic velocity distribution is also observed in the slot.

When the aspect ratio is small (10–100) depending on the Rayleigh number, different regimes are observed: if the Rayleigh number is small (<$10^3$), the regime is a conduction regime [37]. At higher Rayleigh numbers, however, the existence of a new regime consisting of a thin boundary layer around an isothermal core is observed, with convection being the predominant mode of heat transfer. This new regime is called the *convection regime* or the *boundary-layer regime*. In the boundary-layer regime, a vertical temperature gradient is also observed in the core. There is also a *transition* or *asymptotic regime* between the conduction and convection regimes. The vertical temperature gradient in the core starts to develop in the transition regime and reaches a constant value in the boundary-layer regime. Figure 10.15 shows the distributions of the temperature and velocity in these three regimes [38]. In this figure, the variation of the Nusselt number with the Rayleigh number is also indicated. The flow in the slot in these regimes is observed to be laminar. When the Rayleigh number is further increased (>$10^7$), the flow becomes turbulent in the slot. Before the flow becomes turbulent, however, depending on the disturbances, the unicellular flow may decompose into a multicellular pattern, that is, the flow may become unstable [39].

The first analytical investigation of the natural convection in a vertical slot of finite height was given by Batchelor [40]. He concluded that at low Rayleigh numbers heat is transferred across the slot by conduction and suggested for the first time the existence of an isothermal core at higher Rayleigh numbers of the boundary-layer regime. Interferometric temperature measurements, performed with air by Eckert and Carlson [30], confirmed the existence of two such flow regimes and the vertical temperature gradient. The same behavior was also observed in high Prandtl number fluids by Elder [41].

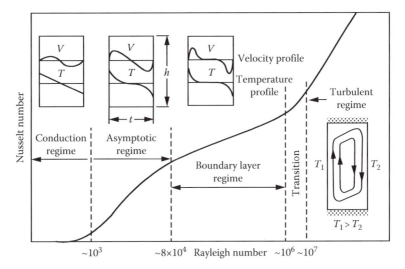

**FIGURE 10.15**
Natural-convection regimes in a vertical slot. (From MacGregor, R. K. and Emery, A. P., *J. Heat Transfer*, 91, 391, 1969.)

### 10.11.2.1 Mathematical Formulation

Consider the steady 2D natural convection of a viscous fluid in the slot shown in Figure 10.16. For this vertical enclosure, governing equations (10.140) and (10.141) become

$$\frac{\partial u}{\partial x} + \frac{\partial v}{\partial y} = 0, \tag{10.143a}$$

$$u\frac{\partial u}{\partial x} + v\frac{\partial v}{\partial y} = -\frac{1}{\rho}\frac{\partial p'}{\partial x} + g\beta T' + v\nabla^2 u, \tag{10.143b}$$

$$u\frac{\partial v}{\partial x} + v\frac{\partial v}{\partial y} = -\frac{1}{\rho}\frac{\partial p'}{\partial y} + v\nabla^2 v, \tag{10.143c}$$

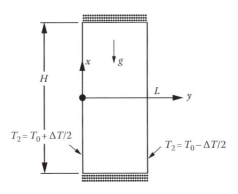

**FIGURE 10.16**
Vertical slot with isothermal walls.

$$u\frac{\partial T'}{\partial x}+v\frac{\partial T'}{\partial y}=\alpha\nabla^2 T',\tag{10.144}$$

where
  The fluids properties $v$, $\beta$, and $\alpha$ are assumed to be constant
  $p'$ and $T'$ are the deviations of pressure and temperature from their values at the hydrostatic condition at $P_0$ and $T_0$, that is,

$$P = P_0 + P',$$

$$T = T_0 + T'.$$

Further, $u$ and $v$ are the velocity components in the $x$- and $y$-directions.
  The boundary conditions are

$$\text{At }y = 0: \quad u = v = 0, \quad T = \frac{T_0 + \Delta T}{2}$$

$$\text{At }y = L: \quad u = v = 0, \quad T = \frac{T_0 - \Delta T}{2}.$$

$$\text{At }x = \mp\frac{H}{2}: \quad u = v = 0, \quad \frac{\partial T}{\partial y} = 0$$

For large aspect ratios $h^* = H/L$, the motion may be assumed to be parallel away from the ends, that is, $u = u(y)$ and $v = 0$. Hence, Equation 10.143 reduces to

$$-\frac{1}{\rho}\frac{dp'}{dx}+g\beta T'+v\frac{d^2 u}{dy^2}=0,\tag{10.145a}$$

$$u\frac{\partial T'}{\partial x}=\alpha\left(\frac{\partial^2 T'}{\partial x^2}+\frac{\partial^2 T'}{\partial y^2}\right).\tag{10.145b}$$

Equations 10.145a and b may be written in the following dimensionless forms:

$$\frac{d^2 U}{dY^2}=-\phi(X,Y)+\frac{1}{Grh^*}\frac{d\bar{p}}{dX},\tag{10.145c}$$

$$U\frac{\partial\phi}{\partial X}=\frac{h^*}{Ra}\left(\frac{1}{h^{*2}}\frac{\partial^2\phi}{\partial X^2}+\frac{\partial^2\phi}{\partial Y^2}\right),\tag{10.145d}$$

with the boundary conditions

$$U(0) = U(1) = 0,\tag{10.146a}$$

$$\phi(X,0)=\frac{1}{2},\quad\phi(X,1)=-\frac{1}{2},\tag{10.146b}$$

where we have introduced the following dimensionless quantities:

$$U = \frac{u}{\bar{u}}, \quad \bar{u} = Gr\frac{v}{L}, \quad Gr = \frac{g\beta\Delta TL^3}{v^2},$$

$$\bar{p} = \frac{p'}{\rho(v/L)^2},$$

$$\phi = \frac{T - T_0}{\Delta T} = \frac{T'}{\Delta T'},$$

$$X = \frac{x}{H}, \quad Y = \frac{y}{L}, \quad h* = \frac{H}{L},$$

$$Ra = GrPr, \quad Pr = \frac{v}{\alpha}.$$

Since $h*$ (= $H/L$) $\gg$ 1, Equation 10.145b reduces to

$$U\frac{\partial \phi}{\partial X} = \frac{h*}{Ra}\frac{\partial^2 \phi}{\partial Y^2}. \tag{10.147}$$

The derivative of Equation 10.145a with respect to $Y$, on the other hand, gives

$$U\frac{\partial \phi}{\partial Y} = \frac{d^3U}{dY^2}, \tag{10.148}$$

which implies that $\partial \phi / \partial Y$ is a function of $Y$ only. Therefore, Equation 10.147 yields

$$\frac{\partial \phi}{\partial X} = \text{constant} = \gamma \tag{10.149}$$

or the temperature distribution must then be of the form

$$\phi(X, Y) = \bar{T}(Y) + \gamma X. \tag{10.150}$$

Experimental investigations mentioned earlier indicate, for both high and low Prandtl number fluids, that the temperature gradient $\gamma$ increases with increasing Rayleigh number, from a value of zero in the conduction regime to an asymptotic value of about 0.5 in the boundary-layer regime. Figure 10.17 shows the results obtained by Elder [41] for paraffin.

It is to be noted that, since the walls are isothermal, a solution of the form (10.150) is strictly valid only at $X = 0$. Under these conditions, Equations 10.145a and 10.147 can be written as

$$\frac{d^2U}{dY^2} + \bar{T}(Y) = -\gamma X + \frac{1}{Grh*}\frac{d\bar{p}}{dX}, \tag{10.151a}$$

$$\frac{d^2\bar{T}}{dY^2} - 64\lambda^4 U(Y) = 0, \tag{10.151b}$$

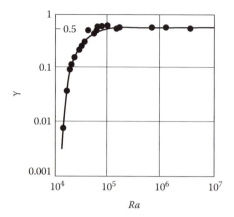

**FIGURE 10.17**
Vertical temperature gradient. (From Elder, J. M., *J. Fluid Mech.*, 23, 77, 1965.)

where

$$\lambda^4 = \frac{\gamma Ra}{64h^*}. \tag{10.151c}$$

The left-hand side of Equation 10.151a is a function of $Y$ only and the right-hand side is a function of $X$ only. Hence, Equation 10.151a can be rewritten in two parts as

$$\frac{d^2U}{dY^2} + \bar{T}(Y) = C, \tag{10.152a}$$

$$\frac{1}{Grh^*}\frac{d\bar{p}}{dX} - \gamma X = C, \tag{10.152b}$$

where $C$ is a constant.

Around $x = 0$, the boundary conditions become

$$U(0) = U(1) = 0, \tag{10.153a,b}$$

$$\bar{T}(0) = \frac{1}{2}, \quad \bar{T}(1) = -\frac{1}{2}. \tag{10.153c,d}$$

The velocity distribution $U(Y)$ also satisfies the following continuity relation:

$$\int_0^1 U(Y)dY = 0. \tag{10.153e}$$

In the following sections, solutions to the conduction and boundary-layer regimes will be given.

### 10.11.2.2 Conduction Regime

In the conduction regime, $\gamma = 0$. Therefore, the earlier formulation reduces to

$$\frac{d^2U}{dY^2} + \bar{T}(Y) = C, \tag{10.154a}$$

$$\frac{d^2\bar{T}}{dY^2} = 0. \tag{10.154b}$$

Under the boundary conditions (10.153), solutions of these equations become

$$\bar{T}(Y) = \frac{1}{2} - Y, \tag{10.154c}$$

$$U(Y) = \frac{Y}{12}(2Y^2 - 3Y + 1). \tag{10.155}$$

Hence, the temperature distribution is linear, whereas the velocity distribution is cubic. These are shown in Figure 10.18.

Heat transfer across the slot can be calculated using Fourier's law, that is,

$$q'' = -k\left(\frac{\partial T}{\partial y}\right)_{y=0} = -k\left(\frac{\partial T}{\partial y}\right)_{y=L}, \tag{10.156}$$

which gives

$$q'' = \frac{k\Delta T}{L}. \tag{10.157}$$

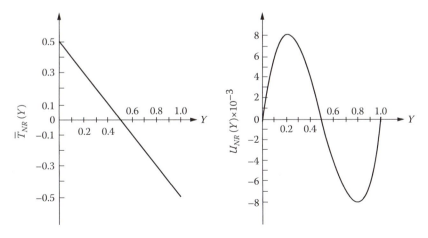

**FIGURE 10.18**
Temperature and velocity distributions in conduction regime.

The Nusselt number defined as $Nu = q''L/k\Delta T$ becomes

$$Nu = 1. \tag{10.158}$$

### 10.11.2.3 Boundary-Layer Regime

Equations 10.151a and b yield

$$\frac{d^4U}{dY^4} + 64\lambda^4 U = 0 \tag{10.159a}$$

and Equation 10.152a gives

$$\bar{T}(Y) = C - \frac{d^2U}{dY^2}. \tag{10.159b}$$

Solutions of these equations, under the boundary conditions (10.153), are

$$\bar{T}(Y) = -\frac{1}{2K}\left[A\sin[\lambda(2Y-1]\cosh[\lambda(2Y-1)] + \cos[\lambda(2Y-1)]\sinh[\lambda(2Y-1)]\right] \tag{10.160a}$$

and

$$U(Y) = -\frac{1}{16\lambda^2 K}\left[A\cos[\lambda(2Y-1)]\sinh[\lambda(2Y-1)], -\sin[\lambda(2Y-1)]\cosh[\lambda(2Y-1)]\right], \tag{10.160b}$$

where

$$A = \frac{\tan(\lambda)}{\tanh(\lambda)}$$

and

$$K = A\sin(\lambda)\cosh(\lambda) + \cos(\lambda)\sinh(\lambda).$$

Equation 10.160 is seen plotted in Figure 10.19 for various values of $\lambda$.

The rate of heat transfer per unit area across the slot, from Equation 10.156, is

$$q'' = \frac{\lambda k \Delta T}{LK} \frac{(A+1)+(A-1)\tan(\lambda)\tanh(\lambda)}{A\tan(\lambda)+\tanh(\lambda)} \tag{10.161}$$

or the Nusselt number becomes

$$Nu = \lambda \frac{(A+1)+(A-1)\tan(\lambda)\tanh(\lambda)}{A\tan(\lambda)+\tanh(\lambda)}. \tag{10.162}$$

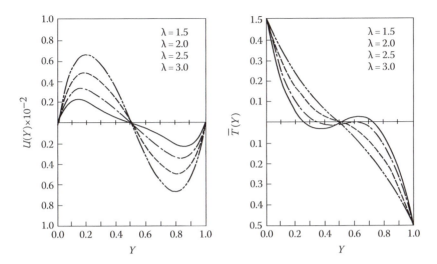

**FIGURE 10.19**
Temperature and velocity distribution in the boundary-layer regime.

Therefore, in the boundary-layer regime, the Nusselt number is a function of

$$\lambda = \frac{\gamma Ra}{64h}.$$

The natural convection in 2D enclosures has been studied by various investigators. For rectangular enclosures with $H/L > 1$ and differentially heated vertical side walls, asymptotic solutions for the boundary-layer regime at high Rayleigh numbers have been given by Gill [42], Bejan [43], and Graebel [44]. Many numerical solutions have been given for vertical enclosures dealing with both insulated and perfectly conducting end walls. The examples are the solutions given by Korpela et al. [45], Schinkel et al. [46], and Bergholz [47]. Heat-transfer characteristics depend on the Rayleigh and Prandtl numbers and the aspect ratio $H/L$:

$$Nu = \phi\left(Ra, Pr\frac{H}{L}\right), \tag{10.163}$$

where $Nu$ is the average Nusselt number.

Several solutions for small aspect ratios ($H/L < 1$) have been given in References 48–50. Natural convection in square enclosure ($H/L = 1$) has also been studied [51–54] with a Boussinesq fluid approximation for $Pr = 0.71$ and Rayleigh numbers of $10^3$–$10^6$. One set of results is shown in Figure 10.20 [53]. The hot wall is on the right. The boundary-layer regimes and the highly stratified core region can be clearly seen. One can also see the penetration of the wall layers near the horizontal wall into the core flow at this Rayleigh number. Because of the geometrical simplicity and well-defined boundary conditions, the square enclosure problem is often used as a standard problem for developing or testing different numerical methods.

Natural convection in rectangular enclosures with tilt angle $\gamma$ (Figure 10.14c) has also been studied extensively [55–57]. Heat-transfer results have been obtained for the various values of Rayleigh number and aspect ratios at different tilt angles. Typical isotherms and streamlines for a tilted enclosure are shown in Figure 10.21 to an air-filled enclosure [57].

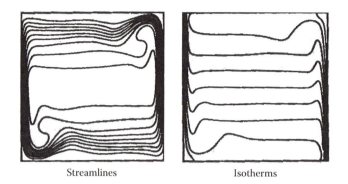

Streamlines                                    Isotherms

**FIGURE 10.20**

Streamlines and isotherms, $Ra = 10^7$, $H/L = 1.0$, $Pr = 0.7$. (From Quon, C., Effects of grid distribution on the computation of high Rayleigh number convection in a differentially heated cavity, in *Numerical Properties and Methodologies in Heat Transfer*, T. M. Shih (Ed.), Hemisphere Publishing Corp., Washington, DC, 1983, pp. 261–281.)

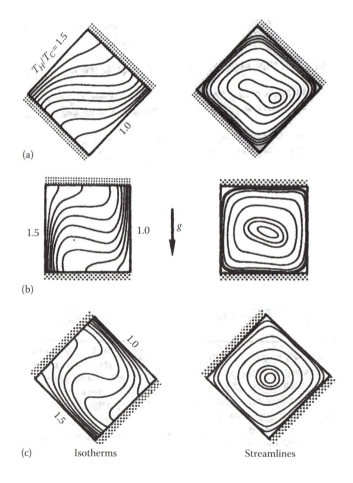

(c)        Isotherms                                    Streamlines

**FIGURE 10.21**

Isotherms and streamlines for tilted square enclosures, $Ra = 10^5$, $H/L = 1.0$. (a) $\gamma = 45°$, (b) $\gamma = 90°$, and (c) $\gamma = 135°$. (From Zhong, Z. Y. et al., Variable-property natural convection in tilted square cavities, in *Numerical Methods in Thermal Problems*, Vol. III, R. W. Lewis, J. A. Johnson, and W. R. Smith (Eds.), Pineridge Press, Swansea, U.K., pp. 968–979, 1983.)

For small values of $\gamma$, heat transfer is by conduction only. It is seen in Figure 10.21a that even at $\gamma = 45°C$, conduction still dominates.

As $\gamma$ increases, convection becomes more important due to increased buoyancy along the isothermal walls and the effect of unstable stratification.

Numerical solutions are given for 2D cylindrical annuli that represent another geometry. For the concentric annulus, streamlines, isotherms, and average heat-transfer data have been obtained by Cho et al. [58] and Farouk and Güçeri [59]. The corresponding solutions for eccentric cylindrical annuli have been obtained by Cho et al. [58] and Prusa and Yoa [60]. These numerical solutions have been verified by experiments [61].

Many experimental techniques are available for the study of natural convection in enclosed spaces [62,63]. A review of these techniques has been given by Hoogendoorn [64]. These experimental studies are important for the validation of the numerical solutions and to provide additional data and information for practically important enclosures.

In practical applications, most of the enclosures are 3D. Fast computing facilities made the numerical solutions possible for such complex enclosures. The use of the laser Doppler velocimeter for 3D flow measurements made an advancement in the development of experimental studies of the 3D enclosures. A review of the numerical and experimental work on 3D enclosures has been given by Yang [35].

## 10.12 Correlations for Natural Convection in Enclosures

Many engineering applications present heat-transfer situations in which natural convection occurs in enclosed spaces. In recent years, natural-convection phenomena in enclosed spaces have stimulated great interest.

Because of the increasing accessibility of large and fast computing facilities and advancement made in experimental techniques, natural convection in various types of enclosed spaces has been investigated. However, in view of the complexity of natural convection in enclosures, the heat-transfer correlations covering all ranges of parameters are still very limited.

Most of the early work on internal natural convection was of experimental or semiempirical nature. In an experimental study of natural-convection heat-transfer coefficients for two parallel, electrically heated vertical plates were measured by Siegel and Norris [65]. The heat input per unit area was substantially uniform and the Grashof number based on plate height was of the order of $10^{10}$.

Bodoia and Osterle [66] investigated the development of free convection between heated vertical plates assuming the fluid to enter the channel with ambient temperature and a flat velocity profile. The basic governing equations were expressed in finite difference forms and solved numerically on a digital computer. Results were presented for the variation of velocity, temperature, and pressure throughout the flow field leading to the establishment of the heat-transfer characteristics and of a "development height" for the channel.

The successful application of natural-convection flows in hollow passages in turbine rotor blades for cooling as demonstrated by Schmidt [67] and the use of this type of heat transfer in some of the many schemes for extracting heat energy from nuclear reactors increased the interest in internal natural convection and pointed out a number of problems for which new answers were needed.

Maslen [68] studied the fully developed laminar flow of a viscous fluid between two vertical parallel planes including the effects of compressibility as well as fluid property variations. In Reference 69, a problem of heating from the following cases where one or both walls are thermally insulated is discussed. Heat generation due to friction and sources is neglected, and stability is investigated.

To understand the effects of confining walls on natural-convection flows, Lighthill [70] analyzed the flow in a closed-end cylindrical tube with the walls at constant temperature and the body force acting toward the closed end. The type of flow in the tube depends primarily on the length–radius ratio for given Prandtl and Rayleigh numbers. For small values of length–radius ratio, the flow is just like free convection about a vertical plate, that is, the effect of the confining walls is negligible if the boundary-layer thickness is much smaller than the tube radius. To study further the stagnation of fluids in closed-end tubes, Ostrach and Thornton [71] considered a configuration identical with Lighthill's model and the temperature was taken to vary along the tube wall.

The study of flow and heat transfer inside a rotating right circular cylindrical of small height that is heated at its lower surface is given in Reference 72.

### 10.12.1 Correlations for Natural Convection between Parallel Walls

An experimental investigation of convective heat transfer in liquids confined by two parallel plates at different temperatures and inclined at various angles with respect to the horizontal has been described in Reference 73. The experiments covered a range of Rayleigh numbers between $5 \times 10^8$ and $7.97 \times 10^8$ and Prandtl numbers between 0.02 and 11.560. The results indicate that the heat-transfer coefficients for all liquids investigated at the various angles, from horizontal to vertical, may be determined from the relationship

$$Nu_L = C(Ra_L)^{1/3}(Pr)^{0.074},$$ (10.164)

where the coefficient $C$ is given as

| $\theta$: | 0° | 30° | 45° | 60° | 90° |
|-----------|-------|-------|-------|-------|-------|
| $C$: | 0.069 | 0.065 | 0.059 | 0.057 | 0.049 |

In all cases, except the vertical case ($\theta = 90°$), the hot plate is below the cold plate. The Nusselt and Rayleigh numbers are defined with the distance between the plates as the characteristic geometric factor.

The Grashof number is defined as

$$Gr_L = \frac{g\beta(T_H - T_c)L^3}{v^2}.$$ (10.165)

For $4 \times 10^4 < Ra < 10^8$, Emery and Chu [74] proposed the following correlation for laminar natural convection confined by two parallel plates:

$$Nu_L = 0.280(Ra_L)^{1/4}\left(\frac{H}{L}\right)^{-1/4}.$$ (10.166)

For turbulent flow, the constant varies considerably, having a value of 0.815 for $Pr = 5$ and a value of 0.355 for $Pr = 1000$. The dependence of $C$ upon the Prandtl number has been found to be weak in laminar flows.

Eckert and Carlson [30] for the case of air ($Pr \cong 0.73$) proposed the following correlation for laminar flow natural convection confined by the two parallel plates:

$$Nu_L = 0.199(Gr_L)^{0.3}\left(\frac{H}{L}\right)^{-1/10}. \tag{10.167}$$

The following correlations for the calculation of natural convection in an enclosed air between vertical walls have been proposed by Jacob [75]:

$$Nu_L = 0.18(Gr_L)^{1/4}\left(\frac{H}{L}\right)^{-1/9} \quad \text{for } 2\times10^3 < Gr_L < 2\times10^4, \tag{10.168}$$

$$Nu_L = 0.065(Gr_L)^{1/3}\left(\frac{H}{L}\right)^{-1/9} \quad \text{for } 2\times10^4 < Gr_L < 11\times10^6. \tag{10.169}$$

The results of numerical computations have been presented for free convection under isothermal wall and constant heat-flux wall boundary conditions for free convection through vertical plane layers in Reference 38. The data for all fluids in the laminar region with an isothermal cold wall and a constant heat-flux hot wall were correlated by

$$Nu_L = 0.42(Ra_L)^{0.25}(Pr)^{0.012}\left(\frac{H}{L}\right)^{-0.30}, \tag{10.170}$$

for $10^4 < Ra_L < 10^7$, $1 < Pr < 20{,}000$, and $10 < H/L < 40$.

For the turbulent regime, the correlation is

$$Nu_L = 0.046(Ra_L)^{1/3}$$

for $10^6 < Ra_L < 10^9$

$1 < Pr < 20$

and

$1 < H/L < 40$. 
$$\tag{10.171}$$

The correlation of Dropkin and Somerscales [73] shows excellent agreement over the turbulent flow regime with the results of this study since neither turbulent correlation is aspect ratio dependent. In Reference 76, 2D laminar natural convection in air contained in a long horizontal rectangular enclosure with isothermal walls at different temperatures has been investigated using numerical techniques. The time-dependent governing differential equations were solved using a method based on that of Crank and Nicholson.

Steady-state solutions were obtained for aspect ratios of 1, 2.5, 10, and 20 and for values of Grashof number, $Gr_L$, covering the range $4 \times 10^3$–$1.4 \times 10^5$. The bounds on the Grashof number for $H/L = 20$ is $8 \times 10^3 < Gr_L < 4 \times 10^4$. The results were correlated with a 3D power law, which yielded

$$Nu = 0.0547(Gr_L)^{0.397}, \quad \text{for } \frac{H}{L} = 1, \tag{10.172}$$

$$Nu_L = 0.155(Gr_L)^{0.315}\left(\frac{H}{L}\right)^{-0.265}, \quad \text{for } 2.5 \le \frac{H}{L} \le 20. \tag{10.173}$$

For the natural convection in enclosed vertical spaces, the heat flux is calculated as

$$q'' = h(T_h - T_c) = Nu_L \frac{k}{L}(T_h - T_c). \tag{10.174}$$

Then the results can be expressed in terms of an effective thermal conductivity defined as

$$q'' = k_e \frac{T_h - T_c}{L}, \tag{10.175}$$

where $k_e = hL = kNu$.

## 10.12.2 Correlations for Spherical and Cylindrical Annuli

An experimental investigation has been described concerning natural convection of air enclosed between two isothermal concentric spheres (at temperatures $T_c$ and $T_h$) of various diameter ratios ranging from 1.19 to 3.14 in Reference 77. Heat-transfer results obtained experimentally were analyzed in terms of the Prandtl number $Pr$, the Grashof number $Gr$, and the gap-radius ratio $L/r_i$ and the following correlation equation was obtained:

$$Nu_L = 0.332(Gr_L)^{0.270}\left(\frac{L}{r_i}\right)^{0.517}, \tag{10.176}$$

where

$$Gr_L = \frac{g\beta(T_h - T_c)L^3}{\nu^2}.$$

To correlate the natural-convection heat transfer, Beckmann [78], Kraussold [79], and Liu et al. [80] defined an *effective thermal conductivity*, which modified the simple conduction solution to account for convection. Applying their concept to the case to isothermal concentric spheres, Bishop et al. [77] defined a different Nusselt number

$$Nu^* = \frac{q_c''L}{4\pi k(\Delta T)r_i r_o} = \frac{k_{eff}}{k} = \frac{Nu}{1 + (L/r_i)}. \tag{10.177}$$

It was observed that $Nu^*$ has only a weak dependence on $L/r_i$. Thus, the form of an overall correlating equation was assumed to be

$$Nu^* = A(Gr_L)^B. \tag{10.178}$$

A least-square technique was used to obtain the values of $A$ and $B$ and the following correlating equation resulted:

$$Nu^* = 0.106(Gr_L)^{0.276}, \tag{10.179}$$

which correlates all of the data with maximum deviations of −13.4% and +15.5%. Equation 10.179 gives better overall correlation for natural convection between isothermal concentric spheres with air in the enclosed space, for

$$0.25 \le \frac{L}{r_i} \le 1.5 \quad \text{and} \quad 2.0 \times 10^4 \le Gr \le 3.6 \times 10^6.$$

### Example 10.3

Air at atmospheric pressure is enclosed between concentric spheres (ID = 3 in., OD = 5 in.). The surface temperatures of the spheres are 120°C and 40°C, respectively. Calculate the heat transfer between spherical surfaces.

### Solution
The average temperature is

$$T_m = \frac{120 + 40}{2} = 80°C.$$

The properties of air at 80°C from Appendix B are

$$k = 0.0299 \text{ W}/(\text{m K}), \quad v = 20.94 \times 10^{-6} \text{ m}^2/\text{s},$$

$$\beta = 2.83 \times 10^{-3} \, 1/\text{K}, \quad Pr = 0.708.$$

The Grashof number is

$$Gr = \frac{\beta g \Delta T L^3}{v^2} = \frac{2.83 \times 10^{-3} \times 9.81 \times (120 - 40)(1 \times 2.54)^3}{(20.94)^2 \times 10^{-12} \times 10^6} = 8.3 \times 10^4.$$

We can use Equation 10.132 to calculate Nusselt number:

$$Nu^* = 0.106(Gr_L)^{0.276} = 0.106(8.3 \times 10^4)^{0.276} = 2.415,$$

which yields

$$k_{eff} = Nu^* k = 2.415 \times 0.0299 = 0.0722 \text{ W}/(\text{m K}).$$

Hence,

$$q'' = \frac{4\pi k_{eff} r_i r_o (T_h - T_c)}{r_o - r_i} = \frac{4\pi \times 0.0722 \times 1.5 \times 2.5 \times 2.54 \times 80}{1 \times 100} = 6.91\,\text{W}.$$

For pure conduction, we have

$$q'' = \frac{4\pi k r_i r_o (T_h - T_c)}{r_o - r_i} = \frac{4\pi \times 0.0299 \times 1.5 \times 2.5 \times 2.54 \times 80}{1 \times 100} = 2.86\,\text{W}.$$

A correlation equation has also been given by Raithby and Hollands [81,82] for differentially heated horizontal annuli formed by concentric and eccentric circular cylinders. They proposed the following correlation for concentric cylinders, which is valid for the range of $10^2 < Ra^*_{cyl} < 10^7$:

$$q'' = \frac{2\pi k_{eff} H}{\ln(r_o/r_i)}(T_H - T_c), \tag{10.180}$$

where

$$\frac{k_{eff}}{k} = 0.386 \left( \frac{Pr}{0.861 + Pr} \right)^{1/4} \left( Ra^*_{cyl} \right)^{1/4}$$

$$\left( Ra^*_{cyl} \right)^{1/4} = \frac{\ln(r_o/r_i)}{(r_o/r_i)^{3/4} \left( D_i^{-3/5} + D_o^{-3/5} \right)^{5/4}} Ra_L^{1/4}$$

$$L = \frac{1}{2}(D_o - D_i), \quad H = \text{length of the cylinder}$$

$$Ra_L = \frac{g\beta(T_H - T_c)L^3}{v^2} Pr$$

A heat-transfer correlation equation for concentric and eccentric spheres has also been given by Raithby and Hollands [81] as follows:

$$q'' = k_{eff} \frac{\pi D_i D_o}{L}(T_H - T_c), \quad \text{for } 10^2 < Ra^*_{sph} < 10^4, \tag{10.181}$$

where

$$k_{eff} = 0.74 \left( \frac{Pr}{0.861 + Pr} \right)^{1/4} \left( Ra^*_{sph} \right)^{1/4}, \tag{10.182a}$$

$$L = \frac{1}{2}(D_o - D_i), \tag{10.182b}$$

and

$$\left( Ra^*_{sph} \right)^{1/4} = \frac{(r_o - r_i)^{1/4} Ra_L^{1/4}}{D_i D_o \left( D_i^{-7/5} + D_0^{-7/5} \right)^{5/4}}. \tag{10.183}$$

All physical properties in the earlier correlations are evaluated at the mean of the hot and cold enclosure-surface temperatures.

Warrington and Powe compiled all experimental data and attempted to provide a generalized correlation for 3D bodies and their enclosures [83]. The experimental data specifically cover concentrically located isothermal spherical, cylindrical, and cubical inner bodies and their isothermal spherical, cylindrical, and cubical enclosures.

Raithby and Holland [82] have compiled heat-transfer data for various enclosures. All these correlations are subject to some uncertainties. In theoretical analysis, certain boundary conditions are difficult. Readers are also referred to Reference 35 for recommended correlations for 2D and 3D enclosures.

## 10.13 Combined Free and Forced Convection

Recently there has been considerable interest in the effect of body forces on forced-convection phenomena. The effect of the gravity field is always present in forced flow heat transfer as a result of the buoyancy forces connected with the temperature differences. Usually they are of a smaller order of magnitude than the external forces and may be neglected. In certain engineering applications, however, this cannot be done. It was, for instance, realized in the early days that the heat exchange in oil coolers is affected markedly by the free-convection currents superimposed on the forced flow.

More recently, applications, in which such large free-convection forces are present that they change the flow patterns even at high velocities, have become important. Forced flow through rotating components is always subject to centrifugal forces. In the presence of temperature differences, these forces create strong free-convection flows.

Cooling of rotating parts such as rotor blades of gas turbines and ramjets attached to the rotor of helicopters are examples in which very large free-convection forces exist. Combined forced and free convection is also commonly encountered in nuclear reactor applications, particularly when dealing with shutdown cooling problems.

It is important to realize that heat transfer in mixed convection can be significantly different from its values in both pure natural and pure forced convection. In a vertical circular duct, the laminar mixed-convection heat-transfer coefficient in buoyancy-assisted flow (usually when the fluid is heated) is higher than its value in pure forced convection. On the other hand, in buoyancy-opposed flow (usually downflow when the fluid is heated), the laminar mixed-convection heat transfer can be lower than that for pure forced flow. In turbulent flow, the heat transfer is often reduced in assisted flow and increased in opposed flow, compared with pure forced convection. An excellent review of mixed convection in internal flow is given by Aung [84], and most of the material of this section is borrowed from his review.

For a horizontal duct, temperature variations in the fluid lead to the possibility of counter-rotating transverse vortices that are superimposed on the streamwise main flow. This so-called secondary flow can also increase the heat transfer significantly. Therefore, buoyancy influences internal forced-convection heat transfer in ways that depend on whether the flow is laminar or turbulent, upflow or downflow, and on duct geometry as well as orientation.

### 10.13.1 Governing Equations for Mixed Convection

The governing equations of convective heat transfer are given in Chapter 2. Let us consider a mixed convection in a vertical circular duct. For a steady flow and heat transfer, in the absence of internal heat generation, with negligible viscous dissipation, no axial diffusion of heat and momentum, and assuming that pressure change is only in the streamwise direction, the equations for continuity, momentum, and energy can be written as

$$\frac{1}{r}\frac{\partial}{\partial r}(\rho r v) + \frac{\partial}{\partial x}(\rho u) = 0, \tag{10.184}$$

$$\rho\left(v\frac{\partial u}{\partial r} + u\frac{\partial u}{\partial x}\right) = -\frac{dp}{dx} \pm g\rho + \frac{1}{r}\frac{\partial}{\partial r}\left(r\mu\frac{\partial u}{\partial r}\right), \tag{10.185}$$

$$v\frac{\partial T}{\partial r} + u\frac{\partial T}{\partial x} = \frac{1}{\rho c_p}\frac{1}{r}\frac{\partial}{\partial r}\left(rk\frac{\partial T}{\partial r}\right). \tag{10.186}$$

The axial coordinate $x$ oriented vertically up, the negative sign in the buoyancy term ($\rho g$) is used when the basic forced convection is directed upward (upflow), while the positive sign is used for downflow.

In the absence of flow and heat transfer, Equation 10.185 can be reduced to

$$-\frac{dp_o}{dx} \pm \rho_i g = 0, \tag{10.187}$$

where $p_o$ is the hydrostatic pressure corresponding to $\rho_i$ and $T_i$, which are inlet density and temperature of the fluid, respectively.

Subtracting Equation 10.187 from Equation 10.185, we get

$$\rho\left(v\frac{\partial u}{\partial r} + u\frac{\partial u}{\partial x}\right) = -\frac{d}{dx}(p - p_o) \pm (\rho - \rho_i)g + \frac{1}{r}\frac{\partial}{\partial r}\left(r\mu\frac{\partial u}{\partial r}\right). \tag{10.188}$$

The Boussinesq approximation of Equation 10.188 by noting that $\rho - \rho_i = \rho_i\,\beta(T - T_i)$ can be expressed as

$$\rho\left(v\frac{\partial u}{\partial r} + u\frac{\partial u}{\partial x}\right) = -\frac{d}{dx}(p - p_o) \pm g\rho_i\beta(T - T_i) + \frac{1}{r}\frac{\partial}{\partial r}\left(r\mu\frac{\partial u}{\partial r}\right). \tag{10.189}$$

If it is assumed that the density variations are negligible in the continuity equation (10.184) and in the momentum equation (10.189) except in the buoyancy term in addition to specific heat, thermal conductivity, and viscosity, the governing equations (10.184), (10.189), and (10.186) in dimensionless form become

$$\frac{\partial V}{\partial R} + \frac{V}{R} + \frac{\partial U}{\partial X} = 0, \tag{10.190}$$

$$\frac{1}{Pr}\left(V\frac{\partial U}{\partial R} + U\frac{\partial U}{\partial X}\right) = -\frac{d\bar{p}}{dx} + \frac{1}{R}\frac{\partial U}{\partial R} + \frac{\partial^2 U}{\partial R^2} \pm \frac{Gr_i}{Re}\theta, \tag{10.191}$$

$$V\frac{\partial \theta}{\partial R} + U\frac{\partial \theta}{\partial X} = \frac{1}{R}\frac{\partial \theta}{\partial R} + \frac{\partial^2 \theta}{\partial R^2}, \tag{10.192}$$

where

$$U = \frac{u}{U_m}, \qquad V = \frac{RePrv}{U_m} \tag{10.193a}$$

$$R = \frac{r}{d/2}, \qquad X = \frac{2x/d}{RePr} \tag{10.193b}$$

$$\bar{P} = \frac{p - p_0}{Pr\rho U_m^2}, \qquad \theta = \frac{T_w - T}{T_w - T_i} \tag{10.193c}$$

The plus and minus signs in Equations 10.189 and 10.191 apply, respectively, to upward and downward flows, when $x$ is oriented vertically up. One can also write the conservation of mass at any cross section of the duct as

$$\int_0^{d/2} 2\pi u \rho r dr = \frac{\pi}{4} d^2 \rho U_m. \tag{10.194}$$

The dependent variables $U$, $V$, and $\theta$ can be obtained as functions of $R$, $X$, $Pr$, and $Gr_i/Re$ and $\bar{P}$ as a function of $X$, $Pr$, and $Gr_i/Re$, where

$$Gr_i = \frac{g\beta(\bar{T}_w - T_i)d^3}{v^2} \tag{10.195}$$

and

$$Re = \frac{\rho U_m d}{\mu}. \tag{10.196}$$

For noncircular ducts, the inside diameter d of the circular duct can be replaced by hydraulic diameter, $D_h$.

From the temperature and velocity distributions, the bulk mean temperature and the local heat-transfer coefficient can be calculated as described in previous chapters (see Chapters 6 and 7). The dimensionless local heat-transfer coefficient, $Nu_x$, will be a function of the distance along the duct, the Prandtl number, and the parameter $Gr_i/Re$. The parameter $Gr_i/Re$ expresses the importance of buoyancy-induced flow relative to forced flow in internal mixed convection.

## 10.13.2 Laminar Mixed Convection in Vertical Ducts

Solutions for combined natural and forced laminar convection in ducts will depend on the duct geometry, the orientation of the duct, and to whether the flow is fully developed or developing. For vertical circular ducts under hydrodynamically and thermally fully developed flow conditions, the effect of buoyancy on the fully developed Nusselt number for both laminar assisted and opposed flows with constant wall heat-flux boundary condition (UHF) is shown in Figure 10.22.

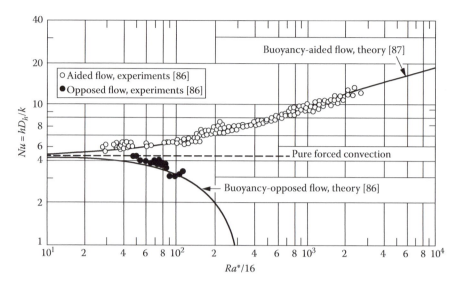

**FIGURE 10.22**
Effect of buoyancy on the fully developed Nusselt number for both laminar assisted and opposed flows in uniformly heated (UHF) vertical circular tubes.

In Figure 10.22, the Rayleigh number is based on the heat flux and is defined as

$$Ra^* = \frac{g\beta(dT_w/dx)d^4}{\alpha v}.$$ (10.197)

As it can be seen in Figure 10.22 as the Rayleigh number increases, the Nusselt number increases above the pure forced-convection value of 4.36 for laminar-assisted flow; but, in buoyancy-opposed flow, the heat transfer decreases with an increase in buoyancy. In Figure 10.22, fluid properties are evaluated at the film temperature.

The dimensionless governing equations, (10.191) through (10.194), can also be solved for thermally developing flow in a duct. In this case, the velocity profile is fully developed upstream of the heated section of the duct. A numerical method can be used with appropriate thermal boundary conditions. At the entrance of the heated section of the duct, the axial velocity distribution is parabolic. The effects of buoyancy cause this profile to change along the duct, but at a large distance from the tube entrance, the profile resumes the fully developed profile for buoyancy-assisted flow [85]. Experiments have indicated that the effect of buoyancy is to cause the thermal entrance length first to decrease with Rayleigh number, then increase at large Rayleigh numbers for buoyancy-assisted flow [85].

Mixed convection for hydrodynamically and thermally developing flow in vertical circular ducts with and without temperature-dependent variables has also been studied [86–88]. For a vertical circular duct of length $L$ with UWT and forced flow having a flat velocity profile at the entrance to the heated section, Jackson et al. [88] recommended the following correlation for the average Nusselt number for the heated section, which is of length $L_x$:

$$Nu = 1.128\left\{Re_b Pr_b \frac{d}{L_x} + \left[3.02\left(Gr_i p_r^2 \frac{d}{L}\right)\right]^{0.4}\right\}^{1/2},$$ (10.198)

which is valid for

$$40.2 \le Gz_b = \frac{Re_b Pr_b d}{L} \le 1710, \tag{10.199}$$

$$1.05 \times 10^5 \le \left( Gr_i Pr \frac{d}{L} \right)_w \le 1.30 \times 10^6. \tag{10.200}$$

The second term in the bracket of Equation 10.198 represents the contribution of natural convection. Equation 10.198 is also valid for combined natural and fully developed laminar forced convection under the same conditions. The experimental results with air are shown in Figure 10.23. On the same figure, also shown is the correlation that applies when a uniform velocity profile exists throughout the test section [88], which is

$$Nu = 1.126(Gr_b)^{1/2} \tag{10.201}$$

with constant properties.

For mixed convection in vertical annuli, the reader may refer to References 84 and 89–103.

Mixed convection in the vertical rectangular duct configuration is also important in practical applications. Early work on mixed convection for hydrodynamically and thermally fully developed flow between vertical parallel plates includes Ostrach [96], Lietzke [97], and Cebeci et al. [98]. Studies have also been conducted by Aung and Worku [99]. For hydrodynamically fully developed flow with constant properties, the governing equations (10.184) through (10.186) for mixed convection in vertical channels reduce to

$$-\frac{d}{dx}(p - p_o) \pm \rho_i g \beta (T - T_i) + \mu \frac{d^2 u}{dy^2} = 0, \tag{10.202}$$

$$\rho c_p u \frac{\partial T}{\partial x} = k \frac{\partial^2 T}{\partial y^2}. \tag{10.203}$$

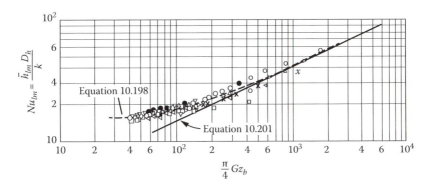

**FIGURE 10.23**
Average heat transfer for hydrodynamically and thermally developing assisted flow in vertical tubes at UWT. (From Jackson, T. W. et al., *Trans. ASME*, 80, 739, 1958.)

The fully developed temperature distribution from Equation 10.203 becomes

$$\frac{\partial^2 T}{\partial y^2} = 0. \tag{10.204}$$

Aung and Worku [99] further assumed that the axial pressure gradient is constant in fully developed flow and solved the momentum equation in the following form:

$$\mu \frac{d^2 u}{dy^2} \pm \rho_i g \beta (T - T_i) + \gamma = 0, \tag{10.205}$$

where $\gamma$ is a constant to be evaluated. The boundary conditions were

$$\text{at } y = 0: \quad u = 0, T = T_c \tag{10.206a,b}$$

$$\text{at } y = b: \quad u = 0, T = T_h \tag{10.206c,d}$$

where $T_c$ and $T_h$ represent temperatures of the channel walls. Aung and Worku [99] obtained the normalized velocity distribution, $u/U_m$, across the duct for various values of dimensionless temperature $(T_c - T_i)/(T_h - T_i)$ with $Gr_2/Re$ as a parameter, where $Gr_2 = g\beta(T_h - T_i)d^3/v^2$.

The information available concerning developing mixed convection between parallel plates is limited. One of the features that distinguish the flow between parallel plates from circular duct flow is the possibility of asymmetric heating on the two walls in the former case. Under certain conditions, this gives rise to reversed flow, even in buoyancy-assisted flows [100]. An analysis of the mixed convection in a channel with symmetric uniform temperature and symmetric uniform-flux heating for hydrodynamically and thermally developing flow has been presented by Yao [101]. Hydrodynamically and thermally developing laminar flows between vertical parallel channels with UWT has also been solved by Aung and Worku [102]. They note that the hydrodynamic development distance is dramatically increased by buoyancy effect.

### 10.13.3 Laminar Mixed Convection in Horizontal Ducts

Under the developing conditions in horizontal ducts, the flow and heat transfer become rotationally asymmetric and are 3D. For steady flow, with constant viscosity and conventional Boussinesq approximation, the governing equations for flow in horizontal tubes can be obtained from Tables 2.1 through 2.3; continuity and energy equations will be the same as in these tables. In the $r$-component of the momentum equation, $f_r = -g\beta(T_w - T)\cos\theta$ and $\theta$-component $f_\theta = g\beta(T_w - T)\sin\theta$ should be added.

Mixed convention in horizontal ducts gives rise to secondary flows, which can cause increase in the friction factor and heat transfer and a decrease in thermal entrance length. A summary of the important correlations is presented in Reference 84.

### 10.13.4 Transition from Laminar to Turbulent Flow

The current understanding of heat transfer in the laminar-to-turbulent flow transition regime is limited. For hydrodynamically fully developed flow in a vertical tube, the transition from laminar to an unstable flow has been studied for water by Scheele et al. [104].

Observations show that for buoyancy-aided flow, the mechanism of transition is similar to that in boundary-layer flow over a flat plate. Here, transition consists of the appearance of regular oscillations that gradually grow in extent and amplitude until the disturbance breaks into fluctuating motion that is characteristic of turbulent flow.

For buoyancy-opposed flow, transition consists of an asymmetric flow that gives rise to reversed flow on one side of the tube. The extent of the reversed flow increases in size as $Gr/Re$ increases (where $Gr = g\beta(T_w - T_c)d^3/v^2$ and $T_c$ is the tube centerline temperature), leading to an eddying flow. Transition to an eddying motion occurs suddenly.

The flow oscillations that accompany the transition process lead to fluctuations in wall temperatures, and these can in turn be used to indicate transition. Using this approach, Hallman [105] found that for buoyancy-assisted, hydrodynamically fully developed flow at the tube entrance with a constant wall heat-flux boundary condition (UHF) in a vertical circular duct, the location of transition depends on the heating rate and the flow rate.

At a constant flow rate, an increase in heating rate causes the point of transition to travel upstream (i.e., down the tube). If the heating rate is held constant, increasing the flow rate moves the transition point downstream. The following correlation may be used to predict the location of transition in a vertical tube [105]:

$$Gr_b Pr = 2664 Gz^{1.83}, \tag{10.207}$$

where

$$Gr_b = \frac{g\beta(\overline{T}_w - T_b)d^3}{v^2}. \tag{10.208}$$

In a horizontal circular duct, the transition to turbulent flow is strongly affected by the presence of secondary flow. With a low initial turbulence level at the inlet to the duct, secondary flow tends to increase the turbulence.

For $ReRa^* < 5 \times 10^5$ with UHF boundary condition, Mori et al. [106] recommend the following correlation when the initial turbulence is low:

$$Re_c = \frac{Re_{co}}{1 + 0.14 \times 10^{-5} ReRa^*}, \tag{10.209}$$

where $Re_{co}$ is the critical Reynolds number without heating. Mori et al. [106] obtained a value of $Re_{co} = 7700$. For high initial turbulence levels, Mori et al. [106] recommended the following correlation for uniformly heated tubes (UHF):

$$Re_c = 128(ReRa^*)^{1/4}. \tag{10.210}$$

## 10.13.5 Turbulent Mixed Convection in Ducts

For vertical upward flow in turbulent mixed convection in heated tubes, fairly well-established criteria for the onset of buoyancy-induced impairment of heat transfer are available. No satisfactory correlation, however, is available yet for the heat transfer. In downflow, a satisfactory correlation now exists. For horizontal circular ducts, experimental evidence indicates that the effect of buoyancy is negligible in turbulent flow.

In vertical ducts, heat transfer is sometimes less than the pure forced-convection value, when natural convection aids forced convection and wherein opposed flow heat transfer is generally larger than the corresponding forced-convection value.

Hall [107] explains the phenomena of heat-transfer impairment in uniformly heated vertical circular ducts by a two-layer model. In this concept, the fluid in the layer close to the heated wall experiences a buoyancy force owing to the reduced density. Acting in the direction of motion, this force tends to decrease the shear stress in the layer away from the wall. Consequently, turbulence production is reduced across the duct, resulting in laminarization. An approximate analysis [107] leads to the following criterion for the onset of buoyancy-induced impairment of heat transfer:

$$\frac{\overline{Gr}}{Re^{2.7}} \leq 10^{-5}, \tag{10.211}$$

where $\overline{Gr} = g(\rho_b - \overline{\rho})d^3/(\overline{\rho}v^2)$. The average density is defined as

$$\overline{\rho} = \frac{1}{T_w - T_b} \int_{T_b}^{T_w} \rho dT. \tag{10.212}$$

In the earlier equation, the over-bar designates the arithmetic mean of the values at the inlet and outlet of the duct.

The thermal impairment in vertical ducts does not persist at higher buoyancy, since the shear stress changes sign and energy inputs to the turbulent motion start to increase, as does the turbulence performance of the tube. Therefore, in buoyancy-aided flow, the heat transfer from the duct is impaired in the low ranges of the Grashof number but recovers and may even be higher than the pure-convection value at high Grashof numbers [108].

For turbulent mixed convection in buoyancy-opposed flow in vertical, uniformly heated tubes (UHF), the heat transfer is generally enhanced over that for pure forced convection. A correlation for buoyancy-opposed flow has been recommended by Jackson and Hall [109]:

$$\frac{Nu}{Nu_F} = \left[1 + 2750\left(\frac{\overline{Gr}}{Re^{2.7}}\right)^{0.91}\right]^{1/3}, \tag{10.213}$$

where

$$\overline{Gr} = \frac{g\beta\rho^2(T_w - T_b)d^3}{2\mu^2}. \tag{10.214}$$

For vertical circular ducts at UWT, in the high Reynolds range, Herbert and Sterns [110] have found that buoyancy effects on the heat transfer are negligible in aided turbulent mixed convection when the Reynolds number exceeds a certain apparent critical value $Re_{ac}$, the value of which can be calculated from the following correlation:

$$Re_{ac} = 3000 + 0.00027 GrPr, \tag{10.215}$$

where the Grashof number is defined by Equation 10.208. Equation 10.215 was based on experiments with water, with $1.8 < Pr < 2.2$ and $2.0 \times 10^7 < Gr < 2.6 \times 10^7$; when $Re$ is greater than $Re_{ac}$, heat transfer is calculated by a correlation for pure forced convection.

When $Re < Re_{ac}$, Herbert and Sterns [110] suggest the following correlation for buoyancy-aided turbulent flow:

$$Nu = 8.5 \times 10^{-2} (GrPr)^{1/3} \tag{10.216}$$

for $4,500 < Re < 15,000$; $3 \times 10^6 < Gr < 30 \times 10^6$ where Grashof number is defined by Equation 10.208.

For buoyancy-opposed turbulent forced convection, the data of Herbert and Sterns [110] indicate that buoyancy effects may be neglected for $Re < 15,000$, and their data are correlated by the following relationship:

$$Nu = 0.56 Re^{0.47} Pr^{0.4}, \tag{10.217}$$

which gives values higher than those for pure forced convection under the constant wall temperature (UWT) boundary condition.

Brown and Gauvin [111] found that for vertical tubes with constant wall temperature boundary condition, in the low Reynolds number range, the turbulent mixed-convection Nusselt number is independent of Reynolds number in both aided and opposed flow, and the heat-transfer coefficient can be found by the following correlation for pure natural convection within ±7%:

$$Nu = 0.13 (GrPr)^{1/3} \quad \text{for } 5 \times 10^6 < Gr < 1 \times 10^7, \tag{10.218}$$

where

$$Gr = \frac{g\beta(T_w - T_{cl})d^3}{\nu^2} \tag{10.219}$$

$T_{cl}$ is the centerline temperature

For opposed flow in the range $378 < Re < 6900$, the Nusselt number in the fully developed turbulent flow regime, which is independent of Reynolds number as noted earlier, is about 45% higher than the value given by Equation 10.218, where the properties are evaluated at the film temperature defined between the wall and centerline temperature.

In horizontal circular ducts, contrary to the case of laminar flow, the influence of buoyancy is completely dominated by turbulent flow. This has been verified by experiments for turbulent flow of air in a horizontal circular duct with constant heat-flux (UHF) boundary condition [106]. Thus, the Nusselt number for turbulent mixed convection in horizontal ducts may be computed by the use of the correlations for pure forced convection in turbulent flow.

## 10.13.6 Flow Regime Maps for Mixed Convection

Metais and Eckert [112] have made a study of the available literature in order to establish the limits of various flow regimes of mixed convection. Figure 10.24 shows the various regimes for combined convection in vertical tubes. The figure contains the results of experiments with UHF as well as UWT. The Grashof number is based on the tube diameter and difference between wall and fluid temperature. The applicable range of Figure 10.24 is $10^{-2} < Pr(d/L) < 1$.

Metais and Eckert have also studied the regimes of combined free and forced convection in horizontal tubes [112]. Figure 10.25 contains the results of the survey for flow through

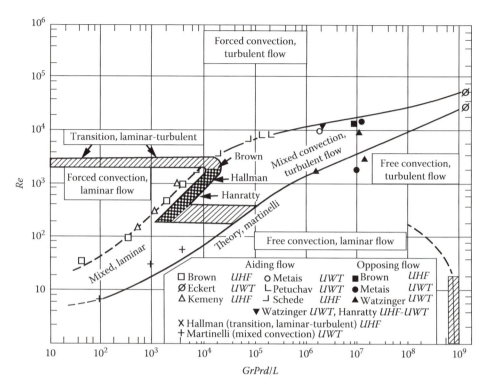

**FIGURE 10.24**
Regimes of free, forced, and mixed-convection flow through vertical circular ducts. (From Metais, B. and Eckert, E. R. G., *J. Heat Transfer*, 86C, 295, 1964.) The results are valid for both upflow and downflow.

a horizontal tube in which the gravitational body forces are, therefore, normal to the pressure gradient. This figure is for UWT. The correlations are found to be equally good regardless of whether the factor $d/L$ is included into the parameter in the abscissa or not. For this reason, both correlations are presented here.

Mixed convection in external flow has also been extensively analyzed for the cases of UWT and uniform surface heat flux (UHF) as boundary conditions. Correlations for mixed convection in laminar-layer flow adjacent to vertical, inclined horizontal flat plates, horizontal cylinders in cross flow, vertical cylinders, spheres, and moving sheets in a quiescent fluid are given [113]. Experimental studies on mixed convection have also been reported for flows along vertical, inclined, and horizontal plates, vertical plates on cross flow, and flows across horizontal cylinders and spheres. In external flow, buoyancy forces can also enhance the surface heat-transfer coefficient when they aid the forced flow and vice versa. The local Nusselt number, $Nu$, for the mixed-convection regime in laminar boundary-layer flow can be correlated by an equation of the following form [114]:

$$Nu^m = Nu_F^m \pm Nu_N^m,$$

(10.220)

where

$Nu_F$ and $Nu_N$ are the Nusselt numbers for pure forced convection and pure free convection
$m$ is a constant

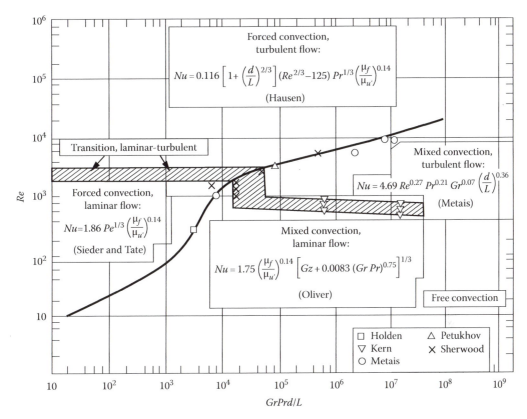

**FIGURE 10.25**
Regimes of free, forced, and mixed-convection flow through horizontal circular ducts with UWT boundary condition. (From Metais, B. and Eckert, E. R. G., *J. Heat Transfer*, 86C, 295, 1964.) Refer to Chapter 9 for recommended correlations for laminar and turbulent forced convection in ducts.

The positive sign is for the buoyancy-assisted case, and the negative is for the buoyancy-opposed case. Equation 10.220 can also be written as

$$\left(\frac{Nu}{Nu_F}\right)^m = 1 \pm \left(\frac{Nu_N}{Nu_F}\right)^m. \tag{10.221}$$

Available mixed-convection results and correlations for various flow configurations for both turbulent and laminar flows under different surface heating conditions (UWT and UHF) are presented in Reference 113.

Some of the results for vertical and inclined plates are presented in Figures 10.26 through 10.28, for a wide range of Prandtl numbers for both the buoyancy-assisting and the buoyancy-opposing flow cases [115–127]. These figures show the region where mixed-convection results deviate from pure free-convection or the pure forced-convection values.

### Example 10.4

Air at atmospheric pressure and 20°C is forced through a horizontal 1 in. diameter tube at an average velocity at 0.3 m/s. Tube wall is maintained at a constant temperature of 140°C. Calculate the heat-transfer coefficient for this situation if tube is 12 in. long.

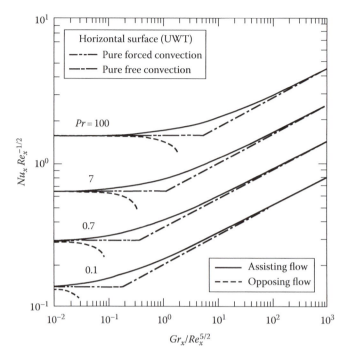

**FIGURE 10.26**
Calculated local Nusselt numbers for flow along a vertical flat plate with UWT. (From Chen, T. S. et al., *ASME J. Heat Transfer*, 108, 835, 1986.)

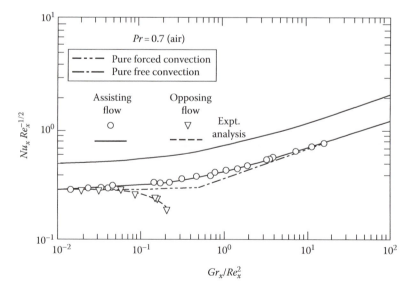

**FIGURE 10.27**
Measured and calculated local Nusselt numbers for air flow along an isothermal vertical flat plate. (From Ramachandran, N. et al., *ASME J. Heat Transfer*, 107, 636, 1985.)

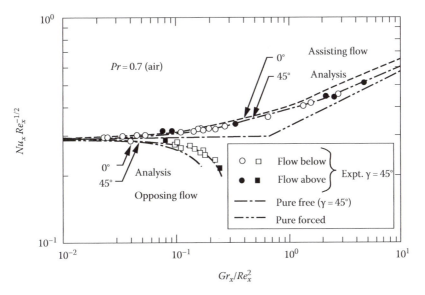

**FIGURE 10.28**
Measured and calculated local Nusselt numbers for air flow along an isothermal inclined flat plate. (From Ramachandran, N. et al., *ASME J. Heat Transfer*, 109, 146, 1987.)

**Solution**
The properties of air should be evaluated at film temperature:

$$T_f = \frac{140 + 20}{2} = 80°C.$$

From Appendix B, we have

$$
\begin{aligned}
&v = 20.94 \times 10^{-6} \text{ m}^2/\text{s} && \beta = 2.83 \times 10^{-3} \text{ 1/K} \\
&k = 0.0299 \text{ W}/(\text{m K}) && \mu = 2.10 \times 10^{-5} \text{ kg}/(\text{m s}) \\
&Pr = 0.708 && \mu_w = 2.35 \times 10^{-5} \text{ kg}/(\text{m s})
\end{aligned}
$$

The Reynolds and Grashof numbers are

$$Re_d = \frac{0.3 \times 0.0254}{20.94 \times 10^{-6}} = 364,$$

$$Gr_d = \frac{g\beta\Delta T d^3}{v^2} = \frac{9.81 \times 2.83 \times 10^{-3} \times 120 \times (0.0254)^3}{(20.94 \times 10^{-6})^2} = 1.24 \times 10^5,$$

and hence

$$Gr_d Pr \frac{d}{L} = 1.24 \times 10^5 \times 0.708 \times \frac{1}{12} = 7.32 \times 10^3,$$

$$Gz = Re_d Pr \frac{d}{L} = 364 \times 0.708 \times \frac{1}{12} = 21.48.$$

From Figure 10.25, we see that we must use mixed-convection laminar flow expression, that is,

$$Nu = 1.75 \left( \frac{\mu_f}{\mu_w} \right)^{0.14} [Gz + 0.0083(GrPr)^{0.75}]^{1/3}$$

$$= 1.75 \left( \frac{2.1 \times 10^{-5}}{2.35 \times 10^{-5}} \right)^{0.14} [21.48 + 0.0083 \times (1.24 \times 10^5 \times 0.708)^{0.75}]^{1/3}$$

$$= 6.88.$$

Hence, the average heat-transfer coefficient is

$$h = \frac{Nuk}{d} = \frac{6.88 \times 0.0299}{0.0254} = 8.1 \text{ W}/(\text{m}^2 \text{ K}).$$

For strictly forced-convection laminar flow, we can use the relation

$$Nu = 1.86(RePr)^{1/3} \left( \frac{\mu_f}{\mu_w} \right)^{0.14} \left( \frac{d}{L} \right)^{1/3}$$

$$= 1.86(364 \times 0.708)^{1/3} \times \left( \frac{2.10 \times 10^{-5}}{2.35 \times 10^{-5}} \right)^{0.14} \times \left( \frac{1}{12} \right)^{1/3}$$

$$= 5.09.$$

Then the average heat-transfer coefficient would be

$$h = \frac{Nuk}{d} = \frac{5.09 \times 0.0299}{0.0254} = 5.99 \text{ W}/(\text{m}^2 \text{ K}).$$

**Example 10.5**

Air at atmospheric pressure and 20°C is forced through a horizontal 3 in. diameter tube at an average velocity of 1.1 m/s. Tube wall is maintained at a constant temperature of 140°C. Calculate the heat-transfer coefficient for this situation if the tube is 12 in. long.

**Solution**

We have the same properties as in Example 10.4. The Reynolds and Grashof numbers are

$$Re = \frac{U_m d}{\nu} = \frac{3 \times 2.54 \times 1.1}{100 \times 20.94 \times 10^{-6}} = 4003,$$

$$Gr = \frac{g\beta\Delta T d^3}{\nu^2} = \frac{9.81 \times 2.83 \times 10^{-3} \times 120(2.54 \times 3)^3}{(20.94)^2 \times 10^{-12} \times 10^6} = 3.36 \times 10^6,$$

and, hence,

$$GrPr\frac{d}{L} = 3.36 \times 10^6 \times 0.708 \times \frac{3}{12} = 5.95 \times 10^5,$$

$$Gz = RePr\frac{d}{L} = 4003 \times 0.708 \times \frac{3}{12} = 708.5.$$

From Figure 10.25, we have mixed-convection turbulent flow. Therefore, we must use

$$Nu = 4.69 Re^{0.27} Pr^{0.21} Gr^{0.07} \left(\frac{d}{L}\right)^{0.36}$$

$$= 4.69(4003)^{0.27}(0.708)^{0.21}(3.36 \times 10^6)^{0.07} \left(\frac{3}{12}\right)^{0.36}$$

$$= 71.09.$$

Hence, the average heat-transfer coefficient is

$$h = \frac{Nuk}{d} = \frac{71.09 \times 0.0299}{3 \times 2.54/100} = 27.89 \text{ W}/(\text{m}^2 \text{ K}).$$

The average heat-transfer coefficient for strictly turbulent flow (Figure 10.25):

$$Nu = 0.116 \left[1 + \left(\frac{d}{L}\right)^{2/3}\right](Re^{2/3} - 125)Pr^{1/3} \left(\frac{\mu_f}{\mu_w}\right)^{0.14}$$

$$= 0.116 \left[1 + \left(\frac{3}{12}\right)^{2/3}\right](4003^{2/3} - 125)0.708^{1/3} \left(\frac{2.10 \times 10^{-5}}{2.35 \times 10^{-5}}\right)^{0.14}$$

$$= 18.038.$$

Hence,

$$h = \frac{Nuk}{d} = \frac{18.038 \times 0.0299}{3 \times 2.54/100} = 7.08 \text{ W}/(\text{m}^2 \text{ K}).$$

## Problems

**10.1** In most of the free-convection correlations, the average Nusselt number *Nu* is expressed as

$$Nu = C(Gr \cdot Pr)^n.$$

Show that, by selecting appropriate average property values in these correlations, the average heat-transfer coefficient *h* can be expressed by

$$h = C' \left(\frac{\Delta T}{L}\right)^n,$$

where *C* and *C'* are constants.

**10.2** A 1 m² vertical plate is maintained at 100°C and exposed to air at atmospheric pressure and 20°C. Compare free-convection heat transfer with that from forcing air over the plate at a velocity equal to the maximum velocity that occurs in the free-convection boundary layer. Comment on the results.

**10.3** Derive an expression for the maximum velocity in the laminar free-convection boundary layer on a vertical plate. At what position in the boundary layer does this velocity occur?

**10.4** A vertical wall at 70°C is exposed to air at 20°C. Calculate the Grashof number 60 cm from the lower edge.

**10.5** A transformer and oil bath are contained in a tank of 1 m inside diameter and 1.5 m height. The electrical losses are dissipated to the surroundings by free convection from the outer surface of the tank. It is to be assumed that the area of the base of the tank does not transmit heat to the surroundings. If the air temperature is 26°C, what is the estimated surface temperature, assuming an electrical loss of 1.8 kW?

**10.6** A vertical wall of an oven is 200 cm long and the surface temperature is 200°C. Air temperature inside the oven is 90°C and pressure is atmospheric. Calculate the following:

　a. Local heat-transfer coefficient at the end of wall

　b. Average heat-transfer coefficient over the wall

**10.7** Air at atmospheric pressure and 20°C is forced through a horizontal 2 in. diameter tube at an average velocity of 0.7 m/s. Tube wall is maintained at a constant temperature of 140°C. Calculate combined free and forced-convection heat-transfer coefficient if the tube is 10 in. long.

**10.8** From Equation 10.33, show that if $Pr$ is much greater than unity, then

$$Nu = C(GrPr)^{1/4},$$

and when $Pr$ is much less than unity,

$$Nu = 0.6(GrPr^2)^{1/4}.$$

**10.9** A swimming pool-type nuclear reactor consists of 30 parallel vertical fuel plates 30 cm wide and 70 cm high, spaced a distance of 5 cm apart. Maximum allowable plate temperature and water coolant average temperature between the plates are 120°C and 40°C, respectively. By making necessary assumptions, calculate the power level of the reactor.

**10.10** Air at atmospheric pressure is enclosed between two vertical plates separated by 2 cm. The temperatures of the plates are 60°C and 100°C, respectively. Calculate the heat flux across the space.

**10.11** Show that for an ideal gas $\beta = 1/T$.

**10.12** Obtain the average Nusselt number based on the temperature difference between the plate and ambient at $L/2$. Compare this result with that for the isothermal plate from the last two columns of Table 10.2.

**10.13** Show that Equations 10.70 and 10.71 can be reduced directly to the well-known equations for the constant-property problem by making usual approximations.

**10.14** Show that for the special case of fluids that obey $p = \rho RT$ (where $R$ is the gas constant) and $\rho \upsilon$ = constant, $\rho k$ = constant, and $c_p$ = constant, the variable-property problem given by Equations 10.70 and 10.71 reduces to the constant-property problem.

**10.15** If a flat plate is not parallel to the body force direction but inclined with an angle $\psi$ from it, show that the Nusselt number for free convection on this inclined plate is

$$Nu = \phi(Pr, Gr\cos\psi).$$

**10.16** Obtain the asymptotic values of $Nu$ from Equation 10.96 for $Pr \to \infty$ and $Pr \to 0$.

---

# References

1. Gebhart, H., Jalvria, Y., Mahajan, R. L., and Sammakia, B., *Buoyancy Induced Flows and Transport*, Hemisphere Publishing Corp., New York, 1987.
2. Pohlhausen, E., Der Warmeaustausch Zwischen Festen Korpen und Flussigkeiten mit kleiner Reiburg und kleiner warme-leitung, *Z. Angew. Math. Mech.*, 1, 115, 1911.
3. Schmidt, E. and Beckmann, W., Das Temperatur-und Geschwindigkeitsfeld Von einer Warme Abgebenden Senkrechten Platte bei naturlicher Konvektion, *Forsch-Ing. Wes.*, 1, 391, 1930.
4. Ostrach, S., Aɴ analysis of laminar free-convection flow and heat transfer about a flat plate parallel to the direction of the generating body force, NACA Report 1111, 1953.
5. McAdams, W. H., *Heat Transmission*, 3rd edn., McGraw-Hill, New York, 1954.
6. Saunders, O. A., Natural convection in liquids, *Proc. R. Soc. London A*, 172, 55–71, 1939.
7. Le Fevre, E. J., Laminar free convection from a vertical plane surface, *Proceedings of the 9th International Congress of Applied Mechanics*, Brussels, Belgium, Vol. 4, p. 168, 1956.
8. Lorenz, L., Veber das Leitungsvermogen der Metalle fur Warme Und Electricitat, *Annele der Physik Und Chemie*, 13, 582, 1881.
9. Sparrow, E. M. and Gregg, J. L., Laminar free convection from a vertical flat plate with uniform surface heat flux, *Trans. ASME*, 78, 435–440, 1956.
10. Rich, B. R., An investigation of heat transfer from an inclined flat plate in free convection, *Trans. ASME*, 75, 489–499, 1953.
11. Vliet, G. C., Natural convection local heat transfer on constant heat flux inclined surfaces, *J. Heat Transfer*, 9, 511–516, 1969.
12. Fujii, T. and Imura, H., Natural convection from a plate with arbitrary inclination, *Int. J. Heat Mass Transfer*, 15, 755–767, 1972.
13. Pern, L. and Gebhart, B., Natural convection boundary layer flow over horizontal and slightly inclined surfaces, *Int. J. Heat Mass Transfer*, 16, 1131–1146, 1972.
14. Sparrow, E. M. and Gregg, J. L., The variable fluid property problem in free convection, *Trans. ASME*, 80, 879–886, 1958.
15. Sparrow, E. M., Laminar free convection on a vertical plate with prescribed nonuniform wall heat flux or prescribed nonuniform wall temperature, NACA TN 3508, 1955.
16. Churchill, S. W. and Chu, H. H. S., Correlating equations for laminar and turbulent free convection from a vertical plate, *Int. J. Heat Mass Transfer*, 18, 1323–1329, 1975.
17. Braun, W. H. and Heighway, J. E., An integral method for natural-convection flows at high and low Prandtl numbers, NASA, Note D.292, 1960.
18. Jaluria, Y., Basics of natural convection, in *Handbook of Single-Phase Convective Heat Transfer*, S. Kakaç, R. K. Shah, and W. Aung (Eds.), John Wiley & Sons, New York, Chapter 12, 1987.
19. Eckert, E. R. G. and Jackson, T. W., Analysis turbulent free convection boundary layer on a flat plate, NACA Report 1015, 1951.
20. Ruckstein, E. and Felski, J. D., Turbulent natural convection at high Prandtl numbers, *Trans. ASME, J. Heat Transfer*, 102, 773–775, 1980.
21. Bayley, F. J., An analysis of turbulent free convection heat transfer, *Inst. Mech. Eng.*, 169, 361–370, 1955.

22. Warner, C. Y. and Arpaci, V. S., An investigation of turbulent natural convection in air at low pressure along a vertical heated plate, *Int. J. Heat Mass Transfer*, 11, 397–406, 1968.

23. Vliet, G. C. and Liu, D. C., An experimental study of turbulent natural convection, *J. Heat Transfer*, 91, 517–531, 1969.

24. Vliet, G. C. and Rose, D. C., Turbulent natural convection on upward and downward facing inclined heat flux surfaces, *J. Heat Transfer*, 97, 549–555, 1975.

25. Yuge, T., Experiments on heat transfer from spheres including combined natural and forced convection, *J. Heat Transfer*, 82C, 214–220, 1960.

26. Özisik, M. N., *Heat Transfer: A Basic Approach*, McGraw-Hill, New York, 1985.

27. Incropera, F. P. and Dewitt, D. P., *Introduction to Heat Transfer*, John Wiley & Sons, New York, 1985.

28. Chavez, A., Holografische Untersuchung an Einspritzstrahlen-Fluiddynamik und Wärmeübergang durch Kondensation, Dissertation, Technische Universität München, Germany, 1991.

29. Eckert, E. R. G. and Soehngen, E., Studies on heat transfer in laminar free convection with the Zehnder–Mach Interferometer, UASF Technical Report 5747, 1948.

30. Eckert, E. R. G. and Carlson, W. O., Natural convection in an air layer enclosed between two vertical plates with different temperatures, *Int. J. Heat Mass Transfer*, 2, 106–120, 1961.

31. Chávez, A. and Mayinger, F., Evaluation of pulsed laser holograms of spray droplets using digital image processing, *Proceedings of the 2nd International Congress on Particle Sizing*, Tempe, AZ, E. Dan Hirleman (Ed.), pp. 462–471, 1990.

32. Holman, J. P., Gartrell, H. E., and Soehngen, E. E., An interferometric method of studying boundary layer oscillations, *J. Heat Transfer, Ser. C*, 80(1), 10, 1960.

33. Hauf, W. and Grigull, U., Optical methods in heat transfer, in *Advances in Heat Transfer*, Vol. 6, Academic Press, New York, pp. 134–360, 1970.

34. Sariarslan, Y., On the effect of time variation of film coefficient upon cooling of a plate by natural convection and radiation, Master of Science thesis, Middle East Technical University, Ankara, Turkey, 1976.

35. Yang, K. T., Natural convection in enclosures, in *Handbook of Single-Phase Convective Heat Transfer*, S. Kakaç, R. K. Shah, and W. Aung (Eds.), John Wiley & Sons, New York, Chapter 13, 1987.

36. Gebhait, B., Jaluria, Y., Mahajan, R. L., and Sammakia, B., *Buoyancy-Induced Flows and Transport*, Hemisphere Publishing Corp., New York, 1988.

37. Chan, Y. L. and Tien, C. L., A numerical study of two-dimensional laminar natural convection in shallow open cavities, *Int. J. Heat Mass Transfer*, 28, 603–612, 1985.

38. MacGregor, R. K. and Emery, A. P., Free convection through vertical plane layers—Moderate and high Prandtl number fluids, *J. Heat Transfer*, 91, 391, 1969.

39. Vest, C. M. and Arpaci, V. S., Stability of natural convection in a vertical slot, *J. Fluid Mech.*, 36, 1–15, 1969.

40. Batchelor, G. K., Heat transfer by free convection across a closed cavity between vertical boundaries at different temperatures, *Q. J. Appl. Math.*, 12, 209–233, 1954.

41. Elder, J. M., Laminar free convection in a vertical slot, *J. Fluid Mech.*, 23, 77–98, 1965.

42. Gill, A. E., The boundary layer regime for convection in a rectangular cavity, *J. Fluid Mech.*, 26, 515–536, 1966.

43. Bejan, A., Note on Gill's solution for free convection in a vertical enclosure, *J. Fluid Mech.*, 90, 561–568, 1979.

44. Graebel, W. P., The influence of Prandtl number on free convection in a rectangular cavity, *Int. J. Heat Mass Transfer*, 24, 125–131, 1981.

45. Korpela, S. A., Lee, Y., and Drummond, J. E., Heat transfer through a double pane window, *J. Heat Transfer*, 104, 539–544, 1982.

46. Schinkel, W. M. M., Linhorst, S. J. M., and Hougendoom, D. J., The stratification in natural convection in vertical enclosures, *J. Heat Transfer*, 105, 267–272, 1983.

47. Bergholz, R. F., Instability of steady natural convection in a vertical fluid layer, *J. Fluid Mech.*, 84, 743, 1978.

48. Cormack, D. E., Leal, L. G., and Seinfield, J. H., Natural convection in a shallow cavity with differentially heated end walls, Pt. 2, numerical solution, *J. Fluid Mech.*, 65, 231–246, 1974.

49. Shiralkar, G. S. and Tien, C. L., A numerical study of laminar natural convection in shallow cavities, *J. Heat Transfer*, 103, 226–231, 1981.

50. Tichy, J. and Gadgil, A., High Rayleigh number laminar convection in low aspect ratio enclosures with adiabatic horizontal walls and differentially heated vertical walls, *J. Heat Transfer*, 104, 103–110, 1982.

51. Jones, I. P. and Thompson, C. P. (Eds.), Numerical solution for a comparison problem on natural convection in an enclosed cavity, AERE-R-9955, HMSO, 1981.

52. de Vahl Davis, G. and Jones, I. P., Natural convection in a square cavity—A comparison exercise, *Int. J. Numer. Methods Fluids*, 3, 227–249, 1983.

53. Quon, C., Effects of grid distribution on the computation of high Rayleigh number convection in a differentially heated cavity, in *Numerical Properties and Methodologies in Heat Transfer*, T. M. Shih (Ed.), Hemisphere Publishing Corp., Washington, DC, pp. 261–281, 1983.

54. Markatos, N. C. and Pericleous, K. A., Laminar and turbulent natural convection in an enclosed cavity, *Int. J. Heat Mass Transfer*, 27, 755–772, 1984.

55. Catton, I., Natural convection in enclosures, *Heat Transfer*, 6, 13–43, 1978.

56. Arnold, J. N., Catton, I., and Edwards, D. K., Experimental investigation of natural convection in inclined rectangular regions of differing aspects ratios, *J. Heat Transfer*, 98, 67–71, 1976.

57. Zhong, Z. Y., Lloyd, J. R., and Yang, K. T., Variable-property natural convection in tilted square cavities, in *Numerical Methods in Thermal Problems*, Vol. III, R. W. Lewis, J. A. Johnson, and W. R. Smith (Eds.), Pineridge Press, Swansea, U.K., pp. 968–979, 1983.

58. Cho, C. H., Chang, K. S., and Park, K. H., Numerical simulation of natural convection in concentric and eccentric horizontal cylindrical annuli, *J. Heat Transfer*, 104, 624–630, 1982.

59. Farouk, B. and Güçeri, S. I., Laminar and turbulent natural convection in the annulus between horizontal concentric cylinders, *J. Heat Transfer*, 104, 631–636, 1982.

60. Prusa, J. and Yao, L. S., Natural convection heat transfer between eccentric horizontal cylinders, *J. Heat Transfer*, 105, 108–116, 1983.

61. Kuehn, T. H. and Goldstein, R. J., An experimental study of natural convection heat transfer in concentric and eccentric horizontal cylindrical annuli, *J. Heat Transfer*, 100, 635–640, 1978.

62. Elder, J. W., Laminar free convection in a vertical slot, *J. Fluid Mech.*, 23, 77–98, 1965.

63. Elder, J. W., Turbulent free convection in a vertical slot, *J. Fluid Mech.*, 23, 99–111, 1965.

64. Hoogendoorn, C. J., Experimental methods in natural convection, in *Natural Convection Fundamentals and Applications*, S. Kakaç, W. Aung, and R. Viskanta (Eds.), Hemisphere Publishing Corp., Washington, DC, pp. 674–696, 1985.

65. Siegel, R. and Norris, R. H., Tests of free convection in a partially enclosed space between two heated vertical plates, *Trans. ASME*, 79, 663–673, 1957.

66. Bodoia, J. R. and Osterle, J. F., The development of free convection between heated vertical plates, *Trans. ASME*, C84, 40–43, 1962.

67. Schmidt, E. H. W., Heat transmission by natural convection at high centrifugal acceleration in water-cooled gas turbine blades, *Proceedings of the General Discussion on Heat Transfer*, London, U.K., 1951.

68. Maslen, S. H., On fully developed channel flows, NACA Technical Note 4319, 1958.

69. Ostrach, S., On the flow, heat transfer, and stability of viscous fluids subject to body forces and heated from below in vertical channels, in *50 Jahre Grenz-Schichtforschung*, H. Gortler (Ed.), Vieweg, Braunschweig, Germany, pp. 226–235, 1955.

70. Lighthill, M. J., Theoretical consideration on free convection in tubes, *Q. J. Mech. Appl. Math.*, 6, 398–439, 1953.

71. Ostrach, S. and Thornton, P. R., On the stagnation of natural-convection flows in closed-end tubes, *Trans. ASME*, 80, 363–367, 1958.

72. Ostrach, S. and Braun, W. H., Natural convection inside a flat rotating container, NACA Technical Note 4223, 1958.

73. Dropkin, D. and Somerscales, E., Heat transfer by natural convection in liquids confined between two parallel plates which are inclined at various angles with respect to horizontal, *J. Heat Transfer*, 87, 77, 1965.

74. Emery, A. and Chu, N. C., Heat transfer across vertical layers, *J. Heat Transfer*, *Trans. ASME*, *Ser. C*, 87, 110–116, 1965.

75. Jacob, M., *Heat Transfer*, John Wiley & Sons, New York, p. 538, 1958.

76. Newell, M. E. and Schmidt, F. W., Heat transfer by laminar natural convection within rectangular enclosures, *J. Heat Transfer*, 92, 159–168, 1970.

77. Bishop, E. N., Mack, R. L., and Scanlon, J. A., Heat transfer by natural convection between concentric spheres, *Int. J. Heat Mass Transfer*, 9, 649, 1966.

78. Beckmann, W., Die Warmeiibertragung in zylindrischen Gasschichten bei naturlicher Konvection, *Forsch. Geb. Ing. Wes.*, 2, 165–178, 213–217, 407, 1931.

79. Kraussold, H., Die Warmeabgabe von zylindrischen Flussinkeitschicten bei naturlichen Konvection, *Forsch. Geb. Ing. Wes.*, 5, 186–191, 1934.

80. Liu, C. Y., Mueller, W. K., and Landis, F., Natural convection heat transfer in long horizontal cylindrical annuli, *Proceedings of the International Developments in Heat Transfer*, Part V, Boulder, CO, ASME, New York, Paper 117, pp. 976–984, 1961.

81. Raithby, G. D. and Hollands, K. G. T., A generalized method of obtaining approximate solutions to laminar and turbulent free convection problems, in *Advances in Heat Transfer*, Vol. 11, T. F. Irvine and J. P. Hartnett (Eds.), Academic Press, New York, pp. 265–315, 1975.

82. Raithby, G. D. and Holland, K. G. T., Natural CONVECTION, in *Handbook of Heat Transfer Fundamentals*, 2nd edn., W. M. Rohsenow, J. P. Hartnett, and E. N. Ganic (Eds.), McGraw-Hill, New York, Chapter 6, pp. 57–62, 1985.

83. Warrington Jr., R. O. and Powe, R. E., The transfer of heat by natural convection between bodies and their enclosures, *Int. J. Heat Mass Transfer*, 28, 319–330, 1985.

84. Aung, W., Mixed convection in internal flow, in *Handbook of Single-Phase Forced Convection*, S. Kakaç, R. K. Shah, and W. Aung (Eds.), John Wiley & Sons, New York, Chapter 15, 1987.

85. Marner, W. J. and McMillan, H. K., Combined free and forced laminar convection in a vertical tube with constant wall temperature, *J. Heat Transfer*, 92, 559–562, 1970.

86. Lawrence, W. T. and Chato, J. C., Heat transfer effects on the developing laminar flow inside vertical tubes, *J. Heat Transfer*, 88, 215–222, 1966.

87. Zeldin, B. and Schmidt, F. W., Developing FLOW with combined forced–free convection in an isothermal vertical tube, *J. Heat Transfer*, 94, 211–223, 1972.

88. Jackson, T. W., Harrison, W. B., and Boteler, W. C., Combined free and forced convection in a constant-temperature vertical tube, *Trans. ASME*, 80, 739–745, 1958.

89. Rokerya, M. S. and Iqbal, M., Effects of viscous dissipation on combined free and forced convection through vertical concentric annuli, *Int. J. Heat Mass Transfer*, 14, 491–494, 1971.

90. Sherwin, K. and Wallis, J. D., A theoretical study of combined natural and forced laminar convection for developing flow down vertical annuli, *Heat Transfer 1970*, Vol. IV, Paper NC 3.9, 1970.

91. Sherwin, K. and Wallis, J. D., A study of laminar convection for flow down vertical annuli, *Proc. Inst. Mech. Eng.* (3H, Paper 34) 182, 330–335, 1967–1968.

92. Sherwin, K. and Wallis, J. D., Combined natural and forced laminar convection for upflow through heated vertical annuli, *Proc. Inst. Mech. Eng.*, *Symposium on Heat and Mass Transfer by Combined Forced and Natural Convection*, Paper CI12/71, Manchester, U.K., 1971.

93. Shumway, R. W. and McEligot, D. M., Heated laminar gas flow in annuli with temperature-dependent transport properties, *Nucl. Sci. Eng.*, 46, 394–407, 1971.

94. Malik, M. R. and Pletcher, R. H., Calculation of variable property heat transfer in ducts of annular cross section, *Numer. Heat Transfer*, 3, 241–257, 1980.

95. El-Shaarawi, M. A. I. and Sarhan, A., Combined forced-free laminar convection in the entry region of a vertical annulus with a rotating inner cylinder, *Int. J. Heat Mass Transfer*, 25(2), 175–186, 1982.

96. Ostrach, S., Combined natural and forced-convection laminar flow heat transfer of fluids with and without heat sources in channels with linearly wall temperatures, NACA Technical Note 3141, 1954.

97. Lietzke, A. F., Theoretical and experimental investigation of heat transfer by laminar natural convection between parallel plates, NACA Report 1223, 1954.

98. Cebeci, T., Khattab, A. A., and LaMont, R., Combined natural and forced convection in vertical ducts, *Heat Transfer*, 2, 419–424, 1982.

99. Aung, W. and Worku, G., Theory of fully developed combined convection including flow reversal, *J. Heat Transfer*, 108, 485–488, 1986.

100. Sparrow, E. M., Chrysler, G. M., and Azevedo, L. F., Observed flow reversals and measured–predicted Nusselt numbers for natural convection in a one-sided heated vertical channel, *J. Heat Transfer*, 106(2), 325–332, 1984.

101. Yao, L. S., Free and forced convection in the entry region of a heated vertical channel, *Int. J. Heat Mass Transfer*, 26(1), 65–72, 1983.

102. Aung, W. and Worku, G., Developing flow and flow reversal in a vertical channel with asymmetric wall temperatures, *J. Heat Transfer*, 108, 299–307, 1986.

103. Mori, Y. and Futagami, K., Forced convective heat transfer in uniformly heated horizontal tubes (second report, theoretical study), *Int. J. Heat Mass Transfer*, 10, 1801–1813, 1967.

104. Scheele, G. F., Rosen, E. M., and Hanratty, T. J., Effect of natural convection on transition to turbulence in vertical pipes, *Can. J. Chem. Eng.*, 38, 67–73, 1960.

105. Hallman, T. M., Experimental study of combined forced and laminar convection in a vertical tube, NASA TN D-l 104, 1961.

106. Mori, Y., Futagami, K., Tokuda, S., and Nakamura, M., Forced convective heat transfer in uniformly heated horizontal tubes, 1st report—Experimental study on the effect of buoyancy, *Int. J. Heat Mass Transfer*, 9, 453–463, 1966.

107. Hall, W. B., Heat transfer near the critical point, *Adv. Heat Transfer*, 7, 1–86, 1971.

108. Fewster, J., Heat transfer to supercritical pressure fluids, Dissertation, University of Manchester, U.K., 1975.

109. Jackson, J. D. and Hall, W. B., Influences of Buoyancy on heat transfer to fluids flowing in vertical tubes under turbulent conditions, in *Turbulent Forced Convection in Channels and Bundles*, S. Kakaç and D. B. Spalding (Eds.), Hemisphere Publishing Corp., pp. 613–673, 1979.

110. Herbert, L. S. and Sterns, U. J., Heat transfer in vertical tubes—Interaction of forced and free convection, *Chem. Eng. J.*, 4, 46–52, 1972.

111. Brown, C. K. and Gauvin, W. H., Combined free-and-forced convection, *Can. J. Chem. Eng.*, 43, 306–318, 1965.

112. Metais, B. and Eckert, E. R. G., Forced mixed and free convection regimes, *J. Heat Transfer*, 86C, 295–296, 1964.

113. Chen, T. S. and Armaly, B. F., Mixed convection in external flow, in *Handbook of Single-Phase Forced Convection*, S. Kakaç, R. K. Shah, and W. Aung (Eds.), John Wiley & Sons, New York, 1987, Chapter 14.

114. Churchill, S. W., A comprehensive correlating equation for laminar, assisting, forced and free convection, *AIChE J.*, 23, 10–16, 1977.

115. Chen, T. S., Armarly, B. F., and Ramachandran, N., Correlations for laminar mixed convection flows on vertical, inclined, and horizontal flat plates, *ASME J. Heat Transfer*, 108, 835–840, 1986.

116. Ramachandran, N., Armaly, B. F., and Chen, T. S., Measurements and predictions of laminar mixed convection flow adjacent to a vertical surface, *ASME J. Heat Transfer*, 107, 636–641, 1985.

117. Ramachandran, N., Armaly, B. F., and Chen, T. S., Measurements of laminar mixed convection from an inclined surface, *ASME J. Heat Transfer*, 109, 146–150, 1987.

118. Hallman, T. M., Combined forced and free-laminar heat transfer in a vertical tubes with uniform internal heat generation, *Trans. ASME*, 78(8), 1831–1841, 1956.

119. Sherwin, K., Laminar convection in uniformly heated vertical concentric annuli, *Br. Chem. Eng.*, 13(11), 1580–1585, 1968.

120. Maitra, D. and Sabba Raju, K., Combined free and forced convection laminar heat transfer in a vertical annulus, *J. Heat Transfer*, 97, 135–137, 1975.
121. Zaki, G. M., El-Genk, M. S., William, T. E., and Philbin, J. S., Experimental heat transfer studies for water in an annulus at low Reynolds number, in *Fundamentals of Forced and Mixed Convection*, Vol. 42, F. A. Kulacki and R. D. Boyd (Eds.), HTD, ASME, New York, pp. 113–120, 1985.
122. Morcos, S. M. and Bergles, A. E., Experimental investigation of combined forced and free laminar convection in horizontal tubes, *J. Heat Transfer*, 97, 212–219, 1975.
123. Depew, C. A. and August, S. E., Heat transfer due to combined free and forced convection in a horizontal and isothermal tube, *J. Heat Transfer*, 93, 380–384, 1971.
124. Jackson, T. W., Spurlock, J. M., and Purdy, K. R., Combined free and force convection in a constant temperature horizontal tube, *AIChE J.*, 7, 38–45, 1961.
125. Yousef, W. W. and Tarasuk, J. D., Free convection effects on laminar forced convective heat transfer in a horizontal isothermal tube, *J. Heat Transfer*, 104, 145–152, 1982.
126. Hattori, N., Combined free and forced-convection heat transfer for fully-developed laminar flow in horizontal concentric annuli (numerical analysis), *JSME Trans.*, 45, 227–239, 1979.
127. Osborne, D. G. and Incropera, F. P., Laminar, mixed convection heat transfer for flow between horizontal parallel plates with asymmetric heating, *Int. J. Heat Mass Transfer*, 28(1), 207–217, 1985.

# 11

## Heat Transfer in High-Speed Flow

## Nomenclature

| | |
|---|---|
| $Cf$ | average skin friction coefficient |
| $C_{f_x}$ | local skin friction coefficient $= 2\tau_w/\rho U_\infty$ |
| $c_p$ | specific heat at constant pressure, J/(kg K) |
| $c_v$ | specific heat at constant volume, J/(kg K) |
| $Ec$ | Eckert number $= U_\infty^2/(T_w - T_\infty)c_p$ |
| $f$ | dimensionless stream function |
| $h$ | heat-transfer coefficient, W/(m² K) |
| $i$ | enthalpy per unit mass, J/kg |
| $k$ | thermal conductivity, W/(m K) |
| $L$ | length, m |
| $M$ | Mach number |
| $Nu$ | average Nusselt number $= hL/k$ |
| $Nu_x$ | local Nusselt number $= h_x/k$ |
| $P$ | pressure, Pa |
| $Pr$ | Prandtl number $= c_p\mu/k$ |
| $q$ | heat flow rate, W |
| $q''$ | heat flux, W/m² |
| $\dot{q}$ | heat generation, W/m³ |
| $R$ | gas constant, J/(kg K) |
| $Re_L$ | Reynolds number based on $L = \rho U_\infty L/\mu$ |
| $Re_x$ | Reynolds number based on $x$, $= \rho U_\infty x/\mu$ |
| $r$ | recovery factor $= (T_{aw} - T_\infty)/(T_s - T_\infty)$ |
| $St$ | Stanton number $= Nu/RePr$ |
| $T$ | temperature, °C, K |
| $T^*$ | reference temperature, °C, K |
| $U_\infty$ | free-stream velocity, m/s |
| $\mathcal{U}$ | internal energy |
| $u$ | velocity component in $x$ direction, m/s |
| $V$ | magnitude of velocity vector, m/s |
| $v$ | velocity component in $y$ direction, m/s |
| $x$ | rectangular coordinate, m |
| $y$ | rectangular coordinate normal to surface, m |

**Greek Symbols**

$\gamma$     ratio of specific heats = $c_p/c_v$
$\eta$     similarity variable
$\Theta$     dimensionless temperature function, Equation 11.39
$\mu$     dynamic viscosity, Pa s
$\nu$     kinematic viscosity, m²/s
$\rho$     density, kg/m³
$\tau$     shear stress, Pa

**Subscripts**

$a$     adiabatic
$aw$     adiabatic wall
$f$     film temperature
$l$     laminar
$s$     stagnation conditions
$t$     turbulent
$w$     wall conditions
$\infty$     free-stream conditions

## 11.1 Introduction

As mentioned before in the previous chapters, the main factors governing the value of heat-transfer coefficients are as follows:

1. Shape of the system
2. Rate of flow of the fluid
3. Physical properties of the fluid

Secondary effects include the following:

1. Entrance effect
2. Distribution of the heat flux over the transfer surface
3. Effect of variations in physical properties of the fluid (the magnitude of the heat flux)
4. Effect of high fluid velocity
5. Effect of the roughness (smoothness) of the heat-transfer surface

We discussed all these effects in the previous chapters except that of the high velocity. Especially with gases, when the velocities are higher than the speed of sound (supersonic) and in some cases a large multiple of the sound velocity (hypersonic), temperature increases due to frictional dissipation effects become important. The blades of gas turbines, the skins of high-speed aircraft, missiles, and satellites are, for example, exposed to gas streams of very high velocities. The study of heat transfer in such practical cases and determination of the high temperatures reached due to frictional effects are of prime importance.

The energy conversion by internal friction is the main object of this chapter. Heating due to the frictional effects may cause large temperature gradients within the boundary layers and therefore surfaces have to be cooled sufficiently so that structures are not destroyed. Properties are also no longer constant since these large temperature differences cause very large property variations near the boundaries.

The temperature variations and heat-transfer processes in high temperature and high-speed flow are very complicated as a consequence of the processes such as ionization and dissociation of gas molecules into atoms.

Only the basic principles involved and some engineering correlations will be discussed in this chapter, and we shall consider the high-speed flows only.

## 11.2 Stagnation Temperature

The state of a fluid at any point in a flow field is generally described by the velocity components and by two state properties; the easiest state properties to measure are the temperature and the pressure.

It is useful to distinguish between the two types of state properties. One type is represented by the pressure and the temperature that are, in principle, measured by instruments moving along with the flowing fluid; properties measured in this way are called *static* properties.

The second characteristic state of a fluid stream is the one that is obtained by isentropically decreasing the velocity to zero. This state is known as *total* properties.

If a thermometer travels with a flowing gas with no relative motion between the two, it will register the average energy of the molecules that collide with it in the course of their normal random motion. If the thermometer is at rest, it will encounter, at least on the side facing upstream, molecules having a velocity due to not only their random motion but also to the velocity of flow and will register a higher temperature. If the ratio of gas velocity to the speed of sound in the gas (i.e., the Mach number) exceeds 0.2, the difference between the two values registered by the thermometer may become important.

Assume that a compressible fluid is brought to rest on the surface of an obstacle as illustrated in Figure 11.1 and the process is adiabatic and reversible, that is, no heat transfer and frictionless flow. From the flow energy equation we have

$$\mathcal{U}_\infty + \frac{p_\infty}{\rho_\infty} + \frac{1}{2}V_\infty^2 = \mathcal{U}_s + \frac{p_s}{\rho_s}, \tag{11.1}$$

where

$\mathcal{U}_\infty$, $p_\infty$, $\rho_\infty$, and $V_\infty$ are the internal energy, pressure, density, and velocity of the fluid in the undisturbed stream, respectively

$\mathcal{U}_s$, $p_s$, and $p_s$ are the values at the stagnation point

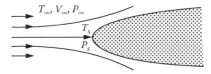

**FIGURE 11.1**
Stagnation temperature and pressure.

Equation 11.1 can also be written in terms of enthalpies as

$$i_\infty + \frac{1}{2}V_\infty^2 = i_s. \tag{11.2}$$

Assuming that the specific heat is constant, from the definition of enthalpy, we have $Ai = cpAT$ (for a perfect gas). Then Equation 11.2 gives

$$T_s - T_\infty = \frac{V_\infty^2}{2c_p}, \tag{11.3}$$

where $T_s$ is called the *stagnation temperature* and it plays an important role in studying heat transfer in high-velocity streams. The temperature rise $T_s - T_\infty$ is due to the work done on the fluid as a result of increase in pressure when the fluid is brought to rest isentropically.

The Mach number $M$, an important parameter, is defined as the ratio of the local velocity to the speed of sound, $a$, defined as

$$M = \frac{V}{a} \tag{11.4}$$

with

$$a = \left\{ \frac{\partial p}{\partial \rho} \right\}_s^{1/2}, \tag{11.5}$$

where the subscript $s$ indicates an isentropic process. For a perfect gas, $pv^\gamma = $ constant, and thus Equation 11.5 gives

$$a = \sqrt{\gamma R T_\infty}, \tag{11.6}$$

where $\gamma = c_p/c_v$ and $R$ is the gas constant. For air, for example, approximate values are

$$R = 287 \, \text{J/(kg K)} \quad \text{and} \quad \gamma = 1.4.$$

Equation 11.3 can be written as

$$\frac{T_s}{T_\infty} = 1 + \frac{V_\infty^2}{2c_p T_\infty}. \tag{11.7}$$

Since for a perfect gas $c_p = \gamma R/(\gamma - 1)$, Equation 11.7 becomes

$$\frac{T_s}{T_\infty} = 1 + \frac{\gamma - 1}{2}M^2. \tag{11.8}$$

The assumption of motionless adiabatic flow is reasonable for the streamline that leads to the stagnation point. It is certainly unreasonable for regions close to the body and downstream of this point. Dissipation of mechanical energy into heat and heat transfer through the gas must be considered in such regions.

**Example 11.1**

A flat plate is placed in a wind tunnel where the free-stream velocity of a gas varies from 30 to 600 m/s. Also, $a = 300$ m/s, and $\gamma = 1.4$. Calculate the difference between stagnation temperature and free-stream temperature for various values of free-stream velocity.

**Solution**

The temperature rise is calculated from Equation 11.7. The results are

$$V_\infty \text{ (m/s)}: 30 \quad 60 \quad 90 \quad 150 \quad 300 \quad 600$$

$$T_s - T_\infty \text{ (K)}: 0.6 \quad 2.4 \quad 5.4 \quad 15 \quad 60 \quad 240$$

## 11.3 Adiabatic Wall Temperature and Recovery Factor

As we have already discussed, the formation of boundary layer over a surface is due to the viscous nature of the fluid flowing over the surface. The boundary-layer energy equation we have obtained for 2D boundary layers includes the term $\mu(\partial u/\partial y)^2$ that accounts for the heat generation due to fluid friction. In fact, the complete energy equation includes a series of such terms. This process of heat generation is irreversible.

In order to explain the effect of viscous dissipation, we shall first take a grossly simplified view that the compressibility effect may be neglected and the properties such as viscosity, thermal conductivity, and specific heat are constant and then consider the flow between two parallel flat plates, one of which is stationary and the other moving in its own plane with a constant velocity $U_\infty$, as illustrated in Figure 11.2 (*Couette flow*).

It is assumed that the dimensions of the plate are very large compared to the distance $d$ between the plates. Then the flow will be fully developed, that is, $u = u(y)$. Further, if $\partial p/\partial x = 0$, then the momentum Equation 2.33a yields

$$\frac{d^2u}{dy^2} = 0. \tag{11.9}$$

Equation 11.9 with the boundary conditions $u(0) = 0$ and $u(d) = U_\infty$ gives

$$u = \frac{U_\infty}{d} y. \tag{11.10}$$

Hence, the velocity profile is linear between the plates.

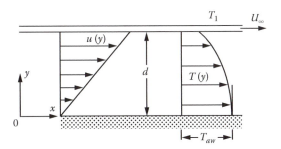

**FIGURE 11.2**
Couette flow.

Let us assume that the lower plate is insulated and the upper plate is kept at constant temperature $T_1$. Since the temperature does not vary with $x$, the energy Equation 2.62 reduces to

$$k\frac{d^2T}{dy^2} = -\mu\left(\frac{du}{dy}\right)^2. \tag{11.11}$$

The boundary conditions are

$$\text{at } y = 0 : \frac{dT}{dy} = 0 \tag{11.12a}$$

$$\text{at } y = d : T = T_1 \tag{11.12b}$$

Hence, there is no convection and the temperature profile is determined by the conduction of heat generated by viscous dissipation from the fluid to the upper boundary. Substituting $\partial u/\partial y$ from Equation 11.10 into Equation 11.11, we get

$$k\frac{d^2T}{dy^2} = -\mu\left(\frac{U_\infty}{d}\right)^2. \tag{11.13}$$

Integrating Equation 11.13 twice and using the boundary conditions (11.12a,b), we obtain

$$T(y) - T_1 = \mu\frac{U_\infty^2}{2kd^2}(d^2 - y^2), \tag{11.14}$$

which is shown plotted on Figure 11.2. Thus, the increase in the temperature of the wall that is perfectly insulated is given by

$$T_{aw} - T_1 = T(0) - T_1 = \frac{\mu U_\infty^2}{2k}. \tag{11.15}$$

The temperature $T_{aw}$ is called the *adiabatic wall temperature*. It should be noted that although the velocity of the lower wall is zero, $T_{aw}$ is not the stagnation temperature $T_s$ that is defined under isentropic conditions. The stagnation temperature $T_s$ corresponding to velocity $U_\infty$ is

$$T_s - T_1 = \frac{U_\infty^2}{2c_p}. \tag{11.16}$$

The ratio of Equation 11.15 to Equation 11.16 is called the *recovery factor r*,

$$r = \frac{T_{aw} - T_1}{U_\infty^2/2c_p} = \frac{T_{aw} - T_1}{T_s - T_1}, \tag{11.17}$$

and for Couette flow

$$\frac{T_{aw} - T_1}{T_s - T_1} = Pr. \tag{11.18}$$

Hence, for the Couette flow, the recovery factor is simply $r = Pr$. Equation 11.15 shows that the kinetic energy of the moving plate $\left(U_\infty^2/2\right)$ is partially converted into thermal energy and causes an adiabatic temperature rise $T_{aw} - T_1$. As seen from Equation 11.17, the recovery factor is the ratio of this adiabatic temperature rise to the fluid kinetic energy at the upper plate that is the stagnation temperature rise.

If the Prandtl number is unity, the excess of adiabatic temperature over the nonstationary plate temperature is the same for this process as for the adiabatic compression processes considered earlier. In practice, the Prandtl number for gases is generally less than unity and we define a recovery factor $r$ for boundary-layer flow as

$$r = \frac{T_{aw} - T_\infty}{T_s - T_\infty}.\qquad(11.19)$$

## 11.4 Governing Equations in High-Velocity Flow

We obtain the following system of boundary-layer equations expressing heat transfer between compressible flow (perfect gas) and a plane boundary:

$$\frac{\partial(\rho u)}{\partial x} + \frac{\partial(\rho v)}{\partial y} = 0 \qquad(11.20)$$

$$\rho\left(u\frac{\partial u}{\partial x} + v\frac{\partial u}{\partial y}\right) = -\frac{\partial p}{\partial x} + \frac{\partial}{\partial y}\left(\mu\frac{\partial u}{\partial y}\right) \qquad(11.21)$$

$$\rho c_p\left(u\frac{\partial T}{\partial x} + v\frac{\partial T}{\partial y}\right) = \frac{\partial}{\partial y}\left(k\frac{\partial T}{\partial y}\right) + \mu\left(\frac{\partial u}{\partial y}\right)^2 + \left(u\frac{\partial p}{\partial x}\right) \qquad(11.22)$$

These equations are valid for variable properties; therefore, the following additional equations should also be introduced:

$$p = \rho RT, \quad \mu = \mu(T), \quad k = k(T) \qquad(11.23a,b,c)$$

Substituting the value of $\partial p/\partial x$ from Equation 11.21 into Equation 11.22, we get

$$\rho c_p\left(u\frac{\partial T}{\partial x} + v\frac{\partial T}{\partial y}\right) + \rho\left(u^2\frac{\partial u}{\partial x} + uv\frac{\partial u}{\partial y}\right) = \frac{\partial}{\partial y}\left(k\frac{\partial T}{\partial y}\right) + \frac{\partial}{\partial y}\left(\mu u\frac{\partial u}{\partial y}\right) \qquad(11.24a)$$

$$\rho c_p\left(u\frac{\partial T}{\partial x} + v\frac{\partial T}{\partial y}\right) + \rho\left(u\frac{\partial(u^2/2)}{\partial x} + v\frac{\partial(u^2/2)}{\partial y}\right) = \frac{\partial}{\partial y}\left(k\frac{\partial T}{\partial y}\right) + \frac{\partial}{\partial y}\left(\mu\frac{\partial(u^2/2)}{\partial y}\right) \qquad(11.24b)$$

Rearrangement of the terms yields

$$\rho c_p \left( u \frac{\partial}{\partial x}\left( T + \frac{u^2}{2c_p} \right) + v \frac{\partial}{\partial y}\left( T + \frac{u^2}{2c_p} \right) \right) = \frac{\partial}{\partial y}\left[ k \frac{\partial}{\partial y}\left( T + \frac{\mu u^2}{2k} \right) \right] \qquad (11.24c)$$

or in terms of the stagnation temperature, we have

$$\rho c_p \left( u \frac{\partial T_s}{\partial x} + v \frac{\partial T_s}{\partial y} \right) = \frac{\partial}{\partial y}\left\{ k \frac{\partial}{\partial y}\left[ T_s + (Pr - 1)\frac{u^2}{2c_p} \right] \right\}. \qquad (11.25)$$

We shall now discuss the constant-property flow. For such a flow, Equation 11.25 becomes

$$\rho c_p \left( u \frac{\partial T_s}{\partial x} + v \frac{\partial T_s}{\partial y} \right) = k \frac{\partial^2 T_s}{\partial y^2} + k(Pr - 1)\frac{\partial^2}{\partial y^2}\left( \frac{u^2}{2c_p} \right). \qquad (11.26)$$

Thus, if the Prandtl number is unity, the solution proceeds in exactly the same manner as for the low-speed flows, except that we replace the temperature $T$ by the total (or stagnation) temperature $(T + u^2/2c_p)$.

The boundary-layer energy equation with constant properties for incompressible flow is

$$\rho c_p \left( u \frac{\partial T}{\partial x} + v \frac{\partial T}{\partial y} \right) = k \frac{\partial^2 T}{\partial y^2} + \mu \left( \frac{\partial u}{\partial y} \right)^2. \qquad (11.27)$$

With the assumption of constant properties, the energy Equation 11.27 is still linear; the dissipation term in Equation 11.27 must be regarded as a distribution of heat source, the magnitude of which is determined by the fluid flow problem. Consequently, the solution to heat transfer at high velocities for Prandtl numbers different from unity can be obtained as the superposition of the following two solutions:

1. We take as a reference solution the one that is for an adiabatic wall. For the boundary-layer flow, this is a solution of the equation

$$\rho c_p \left( u \frac{\partial T}{\partial x} + v \frac{\partial T}{\partial y} \right) = k \frac{\partial^2 T}{\partial y^2} + \dot{q}(x, y), \qquad (11.28)$$

where $\dot{q}(x, y)$ is the source due to the dissipation term. The boundary condition is zero heat flux at the wall, and we denote this solution by $T_1(x, y) - T_\infty = \psi_1(x, y)$.

2. To solution (1), we add a solution of the equation

$$\rho c_p \left( u \frac{\partial T}{\partial x} + v \frac{\partial T}{\partial y} \right) = k \frac{\partial^2 T}{\partial y^2} \qquad (11.29)$$

with the boundary conditions of specified wall temperature or heat flux at the boundary, and we denote this solution by $T_2(x, y) - T_\infty = \psi_2(x, y)$.

By adding these two solutions, we obtain the complete solution of Equation 11.28 under the specified wall temperature or heat flux.

## 11.5 Thermal Boundary Layer over a Flat Plate in High-Speed Flow

In this section, we shall analyze the same heat-transfer problem in incompressible laminar boundary layer over the flat plate we discussed in Chapter 4 by including the viscous dissipation term in the formulation of the problem. Especially for gases, as the velocity increases, the conversion of mechanical energy to thermal energy due to the effects of viscosity becomes increasingly important. Thus, we shall consider an incompressible fluid at temperature $T_\infty$ flowing with a steady and uniform velocity $U_\infty$ parallel to an isothermal flat plate at temperature $T_w$ ($\neq T_\infty$). If we restrict our solution to high subsonic and low supersonic flows ($M < 2$) and modest temperature differences, the effects of the variation in the properties may be neglected. Hence, we may take the properties of the fluid to be constant. Assuming that the flow in the laminar boundary layer over the plate is 2D, the boundary-layer equations including viscous dissipation become

$$\frac{\partial u}{\partial x} + \frac{\partial v}{\partial y} = 0 \tag{11.30}$$

$$u \frac{\partial u}{\partial x} + v \frac{\partial v}{\partial y} = v \frac{\partial^2 u}{\partial y^2} \tag{11.31}$$

$$u \frac{\partial T}{\partial x} + v \frac{\partial T}{\partial y} = \alpha \frac{\partial^2 T}{\partial y^2} + \frac{\mu}{\rho c_p} \left( \frac{\partial u}{\partial y} \right)^2 \tag{11.32}$$

with the following boundary conditions:

$$\text{at } y = 0: u = v = 0, \quad T = T_w, \quad \text{or} \quad \frac{\partial T}{\partial y} = 0 \tag{11.33a,b,c}$$

$$\text{as } y \to \infty: u = U_\infty \quad \text{and} \quad T = T_\infty \tag{11.33d,e}$$

$$\text{at } x = 0: u = U_\infty \quad \text{and} \quad T = T_\infty \tag{11.33f,g}$$

Since the properties were assumed to be constant, the velocity field will be independent of the temperature field. Therefore, the solutions we obtained for the velocity field in Section 4.2 will be applicable, that is,

$$u = U_\infty f'(\eta) \tag{11.34a}$$

$$v = \frac{1}{2} \sqrt{\frac{v U_\infty}{x}} \left[ \eta f'(\eta) - f(\eta) \right], \tag{11.34b}$$

with the similarity variable

$$\eta = y \sqrt{\frac{U_\infty}{vx}}, \tag{11.35}$$

where the function $f(\eta)$ is the solution of the following differential equation:

$$f''' + \frac{1}{2} ff'' = 0 \tag{11.36}$$

with the boundary conditions that $f(0) = f'(0) = 0$ and $f'(\infty) = 1$. The solution of Equation 11.36 was given in Chapter 4.

Since the velocity components are known, the temperature distribution in the thermal boundary layer over the plate can be evaluated by solving Equation 11.32 under the boundary conditions (11.33c, e, g). After substituting the earlier velocity components in the boundary-layer energy Equation 11.32 and using the similarity variable (11.35), it can be shown that the temperature distribution $T(\eta)$ satisfies the following equation:

$$\frac{d^2T}{d\eta^2} + \frac{1}{2} Prf \frac{dT}{d\eta} = -Pr \frac{U_\infty^2}{c_p} (f'')^2 \tag{11.37}$$

with the following boundary conditions of specific wall temperature or adiabatic wall:

$$\text{at } \eta = 0 : T = T_w \quad \text{or} \quad \frac{dT}{d\eta} = 0 \tag{11.38a}$$

$$\text{as } \eta \to \infty : T = T_\infty \tag{11.38b}$$

Before we present the solution for $T(\eta, Pr)$, let us consider the following two auxiliary solutions:

1. *Isothermal flat plate with negligible viscous dissipation.* Consider the same problem but assume that the viscous dissipation is negligible. Let the temperature distribution in the boundary layer in this case be $T_1(\eta)$. Then the formulation of the problem is given by the differential equation

$$\Theta'' + \frac{1}{2} Prf\Theta' = 0 \tag{11.39}$$

with the boundary conditions

$$\text{at } \eta = 0 : \Theta = 0 \tag{11.40a}$$

$$\text{as } \eta \to \infty : \Theta = 1 \tag{11.40b}$$

where we have defined

$$\Theta(\eta, Pr) = \frac{T_1(\eta) - T_w}{T_\infty - T_w}. \tag{11.41}$$

The solution of Equation 11.39 with the boundary conditions (11.40) was discussed in Chapter 4 and it is given by Equation 4.58.

2. *Adiabatic flat plate including viscous dissipation.* Consider now the same problem but assume that the plate is adiabatic, that is, insulated. Let the temperature distribution in the boundary layer in this case be $T_a(\eta)$. The formulation of the problem is then given by the differential equation

$$\Theta_a'' + \frac{1}{2} Prf\,\Theta_a' = -2Pr(f'')^2]$$  (11.42)

with the boundary conditions

$$\text{at } \eta = 0 : \Theta_a' = 0$$  (11.43a)

$$\text{as } \eta \rightarrow \infty : \Theta_a = 1$$  (11.43b)

where we have defined

$$\Theta_a(\eta) = \Theta_a(\eta, Pr) = \frac{T_a(\eta) - T_\infty}{U_\infty^2 / 2c_p}$$  (11.44)

The solution of Equation 11.42 with the boundary conditions (11.43) was obtained by Pohlhausen [1] and it is given by

$$\Theta_a(\eta, Pr) = 2Pr \int_\eta^\infty (f'')^{pr} \left[ \int_0^\eta (f'')^{2-Pr} d\eta \right] d\eta.$$  (11.45)

Again with $f(\eta)$ known (at least in numerical form) from the Blasius solution, Equation 11.45 can now be, in principle, evaluated by numerical methods. But for $Pr = 1$, direct integration is possible and we have

$$\Theta_a(\eta) = 1 - (f')^2 \quad \text{for } Pr = 1.$$  (11.46)

It can be seen from Equation 11.45 that the *adiabatic wall temperature $T_{aw}$*, which is the temperature assumed by the wall owing to frictional heating, will be given by

$$T_{aw} - T_\infty = \frac{U_\infty^2}{2c_p} \Theta_a(0, Pr) = r \frac{U_\infty^2}{2c_p},$$  (11.47)

where $r = \Theta_a(0, Pr)$ is by definition the recovery factor and is a function of the Prandtl number only for the laminar flat-plate flow, and the values of $r$ can be evaluated from Equation 11.45. For the constant-property laminar boundary-layer flow, the variation of the recovery factor with the Prandtl number is shown in Figure 11.3 [2]. The recovery factor can be approximated with sufficient accuracy by the following expressions:

$$r = (Pr)^{1/2}, \quad 0.5 \le Pr \le 47$$  (11.48)

$$r = 1.9(Pr)^{1/3}, \quad Pr > 47$$  (11.49)

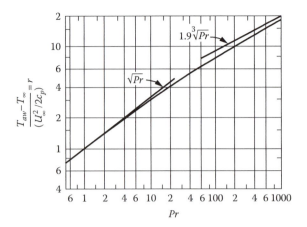

**FIGURE 11.3**
Dependence of adiabatic wall temperature on the Prandtl number for a laminar boundary layer on a flat plate. (Adapted from Schlichting, H., _Boundary-Layer Theory_, translated into English by Kestin, J., 7th edn., McGraw-Hill, New York, 1979. With permission.)

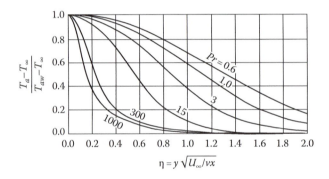

**FIGURE 11.4**
Dimensionless temperature profiles in the laminar boundary layer on an adiabatic flat plate. (Adapted from Schlichting, H., _Boundary-Layer Theory_, translated into English by Kestin, J., 7th edn., McGraw-Hill, New York, 1979. With permission.)

The temperature distribution may also be represented nondimensionally as

$$\frac{T_a(\eta) - T_\infty}{T_{aw} - T_\infty} = \frac{1}{r}\Theta_a(\eta, Pr),$$ (11.50)

which is seen plotted in Figure 11.4 [2].

**_The General Solution._** Since the differential Equation 11.37 and the boundary conditions (11.38) are linear, the solution for $T(\eta)$ of the original problem may be expressed as

$$T(\eta) - T_\infty = C(T_w - T_\infty)\left[1 - \Theta(\eta, Pr)\right] + \frac{U_\infty^2}{2c_p}\Theta_a(\eta, Pr).$$ (11.51)

Equation 11.51 satisfies the differential Equation 11.37 and the boundary condition (11.38b) for any constant C. But for this equation to satisfy the boundary condition (11.38a), the constant should be given by

$$C = 1 - \frac{T_{aw} - T_{\infty}}{T_w - T_{\infty}}. \tag{11.52}$$

Thus, the solution for $T(\eta)$ becomes

$$T(\eta) - T_{\infty} = C(T_w - T_{aw})\left[1 - \Theta(\eta, Pr)\right] + \frac{U_{\infty}^2}{2c_p}\Theta_a(\eta, Pr), \tag{11.53}$$

which satisfies the energy Equation 11.37 and the boundary conditions of (11.38). Equation 11.53 can be expressed in dimensionless form as

$$\frac{T(\eta) - T_{\infty}}{T_w - T_{\infty}} = \left[1 - \frac{1}{2}rEc\right]\left[1 - \Theta(\eta, Pr)\right] + \frac{1}{2}Ec\Theta_a(\eta, Pr), \tag{11.54}$$

where Ec is the Eckert number defined as

$$Ec = \frac{U_{\infty}^2}{c_p(T_w - T_{\infty})}, \tag{11.55}$$

which is proportional to the ratio of the temperature rise of the fluid in an adiabatic compression to the temperature difference between the wall and the fluid at the edge of the boundary. For a perfect gas by combining Equation 11.55 with Equation 11.7, we get

$$Ec = \frac{(\gamma - 1)M_{\infty}^2 T_{\infty}}{T_w - T_{\infty}}. \tag{11.56}$$

Figure 11.5 shows the solution (11.54) for air ($Pr = 0.7$) for various values of the Eckert number. It is interesting to observe that, for $T_w > T_{\infty}$, the plate is heated by the air when $r \cdot Ec > 2$ and cooled when $r \cdot Ec < 2$.

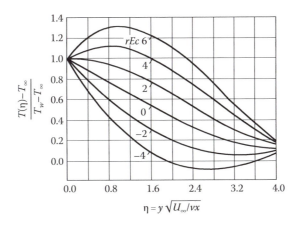

**FIGURE 11.5**
Temperature distribution in a laminar boundary layer over a flat plate for air including frictional heating effects.

The rate of heat transfer per unit area of the plate at any location $x$ can be obtained from Fourier's law of heat conduction, that is,

$$q_w^n = -k\left(\frac{\partial T}{\partial y}\right)_{y=0} = k(T_w - T_{aw})\sqrt{\frac{U_\infty}{vx}}\Theta'(0, Pr), \tag{11.57}$$

because $\Theta'(0, Pr) = 0$. From Equation 4.61, we have $\Theta'(0, Pr) = 0.332Pr^{1/3}$ when $0.6 < Pr < 15$. Hence, Equation 11.57 becomes

$$q_w^n = 0.332kPr^{1/3}\sqrt{\frac{U_\infty}{vx}}(T_w - T_{aw}). \tag{11.58}$$

Heat flux is not proportional to $(T_w - T_\infty)$ and therefore the usual definition of $h$ and $Nu$ based on the temperature difference $(T_w - T_\infty)$ is no longer useful when the viscous dissipation is included in the energy equation. It consequently appears logical to define the heat-transfer coefficient in terms of the difference between the surface temperature and the adiabatic wall temperature, that is,

$$h_x = \frac{q_w^n}{T_w - T_{aw}} = 0.332kPr^{1/3}\sqrt{\frac{U_\infty}{vx}}, \tag{11.59}$$

and the Nusselt number becomes

$$Nu_x = \frac{h_x x}{k} = 0.332kPr^{1/3}Re_x^{1/2}, \tag{11.60}$$

where $Re = U_\infty x/v$. Thus, we see that all low-speed heat-transfer relationships may be used for high-speed flow provided that the temperature difference $(T_w - T_{aw})$ is used in place of usual temperature difference $(T_w - T_\infty)$.

It can also be shown that the average heat-transfer coefficient and the Nusselt number over the entire plate of length $L$ can be obtained as

$$h = 2h_{x=L} \tag{11.61}$$

and

$$Nu = 2Nu_{x=L}. \tag{11.62}$$

Since we have assumed the properties to be constant, the velocity field is independent of the temperature field and therefore the friction coefficient given by Equation 4.37 holds true here, that is,

$$C_{fx} = \frac{0.664}{\sqrt{Re_x}}. \tag{11.63}$$

From Equations 11.60 and 11.63, the following relation (Colburn analogy) between the Stanton number and the friction coefficient is obtained:

$$St_x Pr^{2/3} = \frac{1}{2}C_{fx} \qquad (11.64)$$

The correlations given here assume that the temperature is constant along the surface. Note that Equation 11.64 is valid in general, provided that proper value of the skin friction coefficient is used. Actually, the temperature along any surface subjected to high-velocity flow varies considerably and it is known that such a temperature variation has a marked influence on heat transfer. There are methods that permit a calculation of heat transfer for a variable surface temperature for laminar and turbulent flow conditions, at least in an approximate way.

The relation between skin friction and heat transfer expressed by Equation 11.64 is very useful because it still holds under certain conditions when the property values vary with temperature. The relationships that have been determined for constant-property values can be expected to hold satisfactorily for real fluids or gases as long as the temperature variation throughout the boundary layer is comparatively small. It has been found that this is the case as long as the flow velocities are in the subsonic range and as long as the ratio of wall and free-stream static temperatures is not too large (say, between 0.5 and 1.5).

In the supersonic range and for large temperature differences, the fact that the property values are dependent on temperatures has to be accounted for in a solution of the boundary-layer equations [2].

It was found that the expression (11.59) still holds and in the same way the relationship (11.64) between skin friction and heat transfer is still valid. Only the variation of the friction coefficient for heat transfer has to be taken into account because of the validity of Equation 11.64. Eckert [3] demonstrated that if the specific heat can be assumed constant and all fluid properties are evaluated at the following reference temperature $T^*$, then the low-speed constant-property correlating equations for the $Nu$ number can be used for air for Mach numbers up to 20, the errors being less than a few percent:

$$T^* = T_\infty + 0.50(T_w - T_\infty) + 0.22(T_{aw} - T_\infty) \qquad (11.65)$$

If the specific heat is variable due to very large temperature differences, the energy equation can be solved in terms of total enthalpy and all properties should be evaluated at the temperature corresponding to the reference enthalpy $i^*$ [3]:

$$i^* = 0.50(i_w - i_\infty) + 0.22(i_{aw} - i_e) \qquad (11.66)$$

For variable viscosity and conductivity but constant Prandtl number and specific heat, Rubesin and Johnson [4] have suggested the following expression for the reference temperature:

$$T^* = T_\infty + 0.58(T_w - T_\infty) + 0.19(T_{aw} - T_\infty) \qquad (11.67)$$

On the basis of the foregoing discussion, the following procedure is recommended for the calculation of friction and heat-transfer coefficients in 2D laminar boundary layers along a surface with constant pressure and temperature. Local shearing stress is calculated from the following equation:

$$\tau_w = C_{f_x} \rho \frac{U_\infty^2}{2} \tag{11.68}$$

with the friction factor appearing in this equation given by Equation 11.63. The property values are introduced into both equations at a reference temperature, which may be calculated from Equation 11.65. The heat-transfer coefficient appearing in Equation 11.59 is calculated from the Stanton number, $St_x = h_x/\rho c_p U_\infty$, for essentially constant specific heat.

The density $\rho$ has to be introduced at the adiabatic wall temperature ($T_{aw}$), and the Stanton number is obtainable from the relationship Equation 11.64 into which the Prandtl number has to be introduced again at the reference temperature given by Equation 11.65.

## 11.6 Heat Transfer in 2D Turbulent Boundary Layers

In turbulent boundary layers, the situation is completely different. The present-day understanding of the turbulent exchange mechanism of momentum and energy, especially under conditions of widely varying temperatures, is restricted to such an extent that the exact solutions of the boundary-layer equations cannot be obtained. In any calculation procedure, as indicated in Chapter 7, assumptions have to be made on the exchange mechanisms. In obtaining friction and heat-transfer data, we have to therefore rely heavily on measurements.

The measurements show that the Mach number influences the friction factor. The ratio of the friction factor to the friction factor calculated by assuming constant-property values decreases considerably with increasing Mach number. Reference [5] compared the predictions of different published theories on the change of friction factor with the Mach number. It was found that the results of the various calculations differ to a very high degree.

Faced with this situation, it is appropriate to look for a simple calculation procedure that agrees with the experimental results in the investigated range. Fischer and Norris [6] pointed out that the use of constant-property relations with the reference temperature established for laminar boundary layers gives good agreement for turbulent boundary layers as well. Hence, for a simple calculation procedure, the relationship for friction coefficient in a constant-property fluid must be established first. Several relationships are given in the literature for this purpose for the flat plate. The Blasius equation

$$C_{f_x} = 0.0592 Re_x^{-0.2} \quad \text{for } 5 \times 10^5 < Re_x < 10^7 \tag{11.69}$$

gives good agreement with measured values up to $Re_x = 10^7$.

At higher Reynolds numbers, Schultz-Grunow [7] recommends

$$C_{f_x} = 0.370(\log_{10} Re_x)^{-2.584}, \quad 10^7 < Re_x < 10^9 \tag{11.70}$$

and the average-friction coefficient can be calculated from

$$C_{f_x} = 0.455(\log_{10} Re_L)^{-2.584}, \quad 10^7 < Re_L < 10^9. \tag{11.71}$$

Eckert [3,8] suggested the use of constant-property value relationship together with Equation 11.65 for turbulent as well as for laminar flow because it was found that the normal constant-property relationship agreed with the results of the experiments when the property values were introduced at this reference temperature.

Kuerti [9] suggested the following relationship for the recovery factor in a turbulent boundary layer:

$$r = \sqrt[3]{Pr} \tag{11.72}$$

The simplest way to calculate the heat-transfer coefficient in turbulent flow is again to obtain it from the friction data with the equation

$$St_x Pr^{2/3} = \frac{1}{2} C_{f_x}. \tag{11.73}$$

For small temperature differences and low velocities, Colburn has found that Equation 11.73 is valid for turbulent flow as well as laminar flow. The Stanton number for turbulent flow over an isothermal plate can be calculated by the use of the skin friction coefficient (11.69).

In terms of the local Nusselt number, we obtain

$$Nu_x = 0.0296 Re_x^{4/5} Pr^{1/3} \quad \text{for } 0.6 \le Pr \le 60. \tag{11.74}$$

In practical applications, laminar–turbulent transition may occur on the flat plate and the average heat-transfer coefficient can be defined as

$$h = \frac{1}{L} \int_0^{x_{cr}} h_l dx + \frac{1}{L} \int_{x_{cr}}^{L} h_t dx, \tag{11.75}$$

where $h_l$ and $h_t$ are the heat-transfer coefficients for the laminar and turbulent portions of the plate, respectively, and it is assumed that transition occurs at $x = x_{cr}$. Using Equations 11.60 and 11.74, we obtain

$$Nu = \left[ 0.664 (Re_{cr})^{1/2} + 0.037 (Re_L)^{4/5} - (Re_{cr})^{4/5} \right] Pr^{1/3}. \tag{11.76}$$

If the typical transition Reynolds number of $Re_{cr} = 5 \times 10^5$ is assumed, Equation 11.76 gives

$$Nu = \left[ 0.037 (Re_L)^{4/5} - 871 \right] Pr^{1/3} \tag{11.77}$$

for $0.6 \le Pr \le 60$ and $5 \times 10^5 < Re_L \le 10^8$ (see also Chapter 7).

Therefore, constant-property relationships should be used to calculate the friction factor, the recovery factor, and the heat-transfer coefficient, namely, Equations 11.69 through 11.71 for the friction factor, Equation 11.72 for the recovery factor, and Equations 11.73 through 11.77 and additional correlations given in Chapter 5 under various boundary conditions for the heat-transfer coefficient. Property values should be introduced into these equations at the reference temperature given by Equation 11.65. As stated in Reference [3], it is assumed that $c_p$ is constant.

These recommendations should be checked at large Mach numbers and especially at large temperature differences and high temperatures.

Foregoing results are applicable for smooth surfaces. It is known that roughness increases friction and heat transfer in turbulent flow when the roughness exceeds certain limits.

It should be remembered that the Navier–Stokes equations are based on the assumption that the fluid is a continuum. In flight with increasing altitude, the free-path length of air molecules becomes larger and larger as a consequence of the decreasing air density. This causes deviation from continuum flow, at first, in the immediate vicinity of a solid surface. Therefore, the calculation procedure is to use equations for friction and heat-transfer parameters that have been developed for constant-property fluids and to adapt them to the conditions where properties vary in such a way that those properties are introduced into the equations at a properly determined reference temperature. The effects of temperature-dependent fluid properties on heat transfer have been discussed in Chapter 9.

For very small temperature differences (perhaps <5°C for liquids and <50°C for gases) between the free stream and wall and $M_\infty < 0.3$ [10], all properties can be evaluated at the film temperature defined by Equation 4.70, to obtain reasonably accurate predictions by the use of the equations of this chapter.

Van Driest [11] obtained solutions for skin friction coefficient, Stanton number, velocity, and temperature profiles for compressible laminar boundary layer along a flat plate, for $Pr = 0.75$ as a function of Mach number for various values of $T_w/T_\infty$. For a fixed value of $T_w/T_\infty$, the skin friction coefficient decreases with increasing Mach number. For a given Mach number, lowering the wall temperature increases the skin friction coefficient. Similar results are obtained for the Stanton number [10,11]. The Eckert reference-property method can also be used along with the incompressible correlations for the Nusselt number.

For laminar and turbulent flow over a flat plate, numerical solutions for the arbitrary wall temperature and wall heat flux variation can be obtained. With numerical methods, it is also possible to solve combined conduction and convection problems.

Several correlations have been given for calculating the skin friction and heat-transfer coefficient for compressible turbulent boundary layers on a flat plate. The correlations that were developed by Spalding and Chi [12] and Van Driest [13] have higher accuracy than the rest, according to studies by Hopkins and Keener [14] and Cary and Bertram [15].

### Example 11.2

Air at 20°C and 0.689 × 104 N/m² static values flows over an isothermal flat plate at 600 m/s. The temperature of the plate is 50°C and the length of it is 100 cm. Calculate the following:

  a. The average heat-transfer coefficient for the whole plate
  b. The rate of heat transfer between the plate and air

### Solution

The Reynolds number at the end of the plate can be estimated by evaluating the air properties at the film temperature:

$$T_f = \frac{T_w + T_\infty}{2} = \frac{50 + 20}{2} = 35°\text{C}$$

Since the variation of the absolute viscosity with pressure is negligible, from Appendix B, at 35°C, we have

$$\mu = 1.87 \times 10^{-5} \text{ kg/(m s)}.$$

Density of air at 35°C and $0.689 \times 10^4$ N/m² can be found from the equation of state, that is,

$$\rho = \frac{p}{RT} = \frac{0.689 \times 10^4}{287 \times (273 + 35)} = 0.0779 \text{ kg/m}^3.$$

Hence, the Reynolds number at the end of the plate is

$$Re_L = \frac{\rho U_\infty L}{\mu} = \frac{0.0779 \times 600 \times 1}{1.87 \times 10^5} = 2.5 \times 10^6.$$

Therefore, both laminar and turbulent boundary layers must be considered over the plate. We must now determine the reference temperatures for the two regimes and then evaluate the properties at these temperatures.

The free-stream sound velocity is calculated from Equation 11.5:

$$a = \sqrt{\gamma R T_\infty} = \sqrt{1.4 \times 287 \times (273 + 20)} = 343.11 \text{ m/s}$$

Hence, the free-stream Mach number and stagnation temperature are

$$M_\infty = \frac{U_\infty}{a} = \frac{600}{343.11} = 1.749$$

$$T_s = T_\infty \left[ 1 + \frac{\gamma - 1}{2} M_\infty^2 \right] = 293 \left[ 1 + \frac{1.4 - 1}{2} (1.749)^2 \right] = 472 \text{ K}$$

*Laminar portion.* Let us assume a Prandtl number of 0.7. Hence, we have

$$r = Pr^{1/2} = (0.70)^{1/2} = 0.837$$

and from Equation 11.19, we get

$$T_{aw} = r(T_s - T_\infty) + T_\infty = 0.837(472 - 293) + 293 = 443 \text{ K}.$$

Then the reference temperature for laminar portion is

$$T^* = T_\infty + 0.5(T_w - T_\infty) + 0.22(T_{aw} - T_\infty)$$

$$= 293 + 0.5(323 - 293) + 0.22(443 - 293) = 341 \text{ K} = 68°C$$

From Appendix B, the properties to be used in the laminar portion at 68°C are as follows:

$$\mu^* = 2.04 \times 10^{-5} \text{ kg/(m s)}$$

$$k^* = 0.0291 \text{ W/(m K)}$$

$$Pr^* = 0.7086$$

The Prandtl number is sufficiently close to the assumed value. Density of air at 68°C and $0.689 \times 10^4$ N/m² is

$$\rho^* = \frac{0.689 \times 10^4}{287 \times 341} = 0.070 \text{ kg}/\text{m}^3.$$

We shall assume that the critical Reynolds number is

$$Re_{cr} = \frac{\rho^* U_\infty x_{cr}}{\mu^*} = 5 \times 10^5,$$

which yields

$$x_{cr} = 5 \times 10^5 \frac{2.04 \times 10^{-5}}{0.070 \times 600} = 0.2429 \text{ m}.$$

Hence, the average heat-transfer coefficient over the laminar portion is

$$h = 2h_{x=x_{cr}} = 0.664 \frac{k^*}{x_{cr}} (Pr^*)^{1/3} (Re_{cr})^{1/2}$$

$$= 0.664 \frac{0.0291}{0.2429} \times (0.7086)^{1/3} (5 \times 10^5)^{1/2}$$

$$= 50.15 \text{ W}/(\text{m}^2 \text{ K})$$

*Turbulent portion.* Again let us assume $Pr = 0.7$. Then

$$r = Pr^{1/3} = (0.7)^{1/3} = 0.8879.$$

Therefore,

$$T_{aw} = 0.8879(472 - 293) + 293 = 452 \text{ K}$$

and

$$T^* = 293 + 0.5(323 - 293) + 0.22(452 - 293) = 343 \text{ K} = 70°\text{C}.$$

Hence, the properties to be used in the turbulent portion, from Appendix B, are

$$\mu^* = 2.04 \times 10^{-5} \text{ kg}/(\text{m s})$$

$$k^* = 0.0291 \text{ W}/(\text{m K})$$

$$Pr^* = 0.7085$$

and the density from the equation of state is

$$\rho^* = \frac{0.689 \times 10^4}{287 \times 343} = 0.070 \text{ kg}/\text{m}^3.$$

The local heat-transfer coefficient can be calculated from Equations 11.69 and 11.73, which yields

$$h_x = 0.0296 \rho^* c_p^* U_\infty (Pr^*)^{2/3} \left( \frac{\rho^* U_\infty x}{\mu^*} \right)^{-0.2}.$$

Inserting the numerical values for the properties, we obtain

$$h_x = 86.89(x)^{-0.2} \ W/(m^2 \ K).$$

The average heat-transfer coefficient can be calculated from

$$h = \frac{\int_{x_{cr}}^{L} h_x dx}{\int_{x_{cr}}^{L} dx} = \frac{\int_{0.2429}^{1} 86.89 x^{-0.2} dx}{\int_{0.2429}^{1} dx} = 97.21 \ W/(m^2 \ K).$$

a. Average heat-transfer coefficient for the whole plate is

$$h = \frac{50.15 \times 0.2429 + 97.21 \times 0.7571}{1} = 85.78 \ W/(m^2 \ K).$$

b. The rate of heat transfer between the plate and air per unit depth of the plate, in the laminar portion, is

$$q_1 = hA(T_w - T_{aw})$$
$$= 50.51 \times 0.2429 \times 1 \times (323 - 442)$$
$$= -1449.59 \ W$$

In turbulent portion,

$$q_2 = hA(T_w - T_{aw})$$
$$= 97.21 \times 0.7571 \times 1 \times (323 - 451)$$
$$= -9420.50 \ W$$

Therefore, the total amount of heat transfer is

$$q = q_1 + q_2 = -10871 \ W,$$

so that a total amount of 10 871 W of cooling is required for the laminar and turbulent portions.

## Problems

**11.1** Air at 15°C and atmosphere pressure flows with a velocity of 200 m/s parallel to an insulated flat plate. Determine the adiabatic surface temperature of the plate at 2.5 cm from the leading edge.

**11.2** Air at 15°C and atmospheric pressure flows over an isothermal flat plate at 200 m/s. The temperature of the plate is 64°C, and its length and width are 60 cm and 30 cm, respectively. Determine the magnitude and direction of the heat-transfer rates of the surface.

**11.3** Consider a Couette flow system. The upper plate is maintained at 100°C and moves at a velocity of 250 m/s and the lower plate is insulated. The distance between the plates is 0.025 cm. Assuming no pressure gradient and constant fluid properties, calculate for air and $H_2$:

    a. The recovery factor

    b. The stagnation temperature

**11.4** a. Temperature distribution for laminar flow over an adiabatic flat plate is given by Equation 11.45. Obtain this solution showing every step clearly and then the general solution given by Equation 11.51.

    b. From the *general solution* given by Equation 11.51, show that

$$\theta_a(0) = r.$$

**11.5** Discuss the physical meaning of $Ec > 0$, $Ec < 0$, $Ec = 0$, $rEc = 2$, $rEc > 2$, and $rEc < 2$ in Equation 11.54. In which cases will the wall not be cooled by the stream of air flowing past it?

**11.6** Air at Mach number 3, $2.07 \times 10^4$ N/m² absolute pressure, and –10°C temperature flows past a flat plate. The plate is to be maintained at a constant temperature of 90°C. If the plate is 50 cm long, how much cooling will be required to maintain this temperature?

**11.7** If we define the Nusselt number based on $(T_w - T_\infty)$, show that Equation 11.60 leads to

$$Nu_x = 0.332\sqrt[3]{Pr}\sqrt{Re_x}\left(1 - \frac{1}{2}Ec\phi(Pr)\right),$$

where

$$\phi(Pr) = \sqrt{Pr} \text{ for } 0.5 < Pr < 5.$$

**11.8** Obtain an expression for the local Stanton number in turbulent flow over a flat plate in terms of the Prandtl number and the local Reynolds number for $10^7 < Re_x < 10^9$.

## References

1. Pohlhausen, E., Der Warmeaustausch zwichen festen Koipern und Hussigkeiten mit kleiner Reibung and kleiner Warmeleitung, *Z. Angew, Math. Mech. (ZAMM)*, 1, 115–121, 1921.
2. Schlichting, H., *Boundary-Layer Theory*, translated into English by J. Kestin, 7th edn., McGraw-Hill, New York, 1979.

3. Eckert, E. R. G., Engineering relations for heat transfer and friction in high-velocity laminar and turbulent boundary layer flow over surfaces with constant pressure and temperature, *Trans. ASME*, 78, 1273–1284, 1956.

4. Rubesin, M. W. and Johnson, H. A., Aerodynamic heating and convective heat transfer— Summary of literature survey, *Trans. ASME*, 71, 383–388, 1949.

5. Chapman, D. R. and Kester, R. H., Measurements of turbulent skin friction on cylinders in axial row at subsonic and supersonic velocities, *J. Aeronaut. Sci.*, 20, 44, 1953.

6. Fischer, W. W. and Norris, R., Supersonic convective heat transfer correlation from skin temperature measurement on V-2 rocket in flight, *Trans. ASME*, 71, 457–469, 1949.

7. Schultz-Grunow, F., New frictional resistance law for smooth plates, NACA Technical Memorandum 986, 1941, Transactions of Neus Widerstandsgesetz fur Glatte Platten, *Luf tf ahrf orschung*, 17, 239–246, 1940.

8. Eckert, E. R. G. and Drake, R. M. *Analysis of Heat and Mass Transfer*, McGraw-Hill, New York, 1972.

9. Kuerti, G., *Advances in Applied Mechanics*, Vol. 2, Academic Press, New York, Report No. AL-1866, 1951.

10. Pletcher, R. H., External flow forced convection, in *Handbook of Single-Phase Convection Heat Transfer*, S. Kakaç, R. K. Shah, and W. Aung (Eds.), John Wiley & Sons, New York, Chapter 2, 1987.

11. van Driest, E. R., Investigation of laminar boundary layer in compressible fluid using the crocco method, NACA Technical Note 2587, 1952.

12. Spalding, D. B. and Chi, S. W., The drag of a compressible turbulent boundary layer on a smooth flat plate with and without heat transfer, *Fluid Mech.*, 18, 117–143, 1964.

13. van Driest, E. R., The problem of aerodynamic heating, *Aeronaut. Eng. Rev.*, 15, 26–41, 1956.

14. Hopkins, E. J. and Keener, E. R., Pressure gradients effects on hypersonic skin friction and boundary-layer profiles, *AIAA J.*, 10, 1141, 1972.

15. Cary, A. M. and Bertram, M. H., Engineering prediction of turbulent skin friction and heat transfer in high speed flow, NASA Technical Note D-7507, 1974.

# 12

## Convective Heat Transfer in Microchannels

## Nomenclature

| | |
|---|---|
| $A$ | temperature coefficient, K/m |
| $\bar{A}$ | temperature coefficient $= AL/T_i - T_{s,i}$ |
| $A(\omega)$ | constant defined by Equation 12.79 |
| $a,b$ | constant defined by Equation 12.122 |
| $B(\omega)$ | constant defined by Equation 12.79 |
| $Br$ | Brinkman number $= Ec/Pr = \mu u_m^2/k(T_w - T_i)$ |
| $Br^*$ | Brinkman number for the flow inside microchannel |
| $C$ | Sutherland's coefficient for Equation 12.88, 111 K for air |
| $c$ | specific heat, J/(kg K) |
| $c_{jump}$ | temperature jump coefficient |
| $c_v$ | specific heat at constant volume, J/(kg K) |
| $D_h$ | hydraulic diameter, m |
| $d$ | one-half channel width, m |
| $e$ | average height of the roughness elements, m |
| $F$ | constant defined by Equation 12.116 |
| $F_m$ | tangential momentum accommodation coefficient |
| $F_T$ | thermal accommodation coefficient |
| $f$ | Fanning friction factor defined by Equation 6.33 |
| $f_c$ | constant defined by Equation 12.122 |
| $G_2, G_3$ | constant defined by Equation 12.102 |
| $Gz$ | Graetz number $= 2R(RePr)/L$ |
| $H$ | channel height, m |
| $h$ | heat-transfer coefficient, W/(m² K) |
| $h_x$ | local heat-transfer coefficient, W/(m² K) |
| $h_\varepsilon$ | viscous dissipation heat-transfer coefficient, W/(m² K) |
| $Kn$ | Knudsen number $= \lambda/L$ |
| $k$ | thermal conductivity, W/(m K) |
| $k_r$ | ratio of specific heats of the gas |
| $k_{ref}$ | reference dimensionless thermal conductivity |
| $\bar{k}$ | Boltzmann constant, $1.3806 \times 10^{-23}$ J/K, dimensionless thermal conductivity |
| $\kappa$ | microchannel-dimensionless boundary condition $= (2 - F_T)2\gamma/F_T(\gamma + 1)Pr$ |
| $L$ | characteristic flow dimension, m |
| $L_t$ | thermal and hydrodynamic entry length |
| $Ma$ | Mach number $= U_\infty/U_a$ |
| $N_u$ | average Nusselt number $= hL/k$ |

| $(Nu)_{fd}$ | fully developed Nusselt number |
|---|---|
| $Nu_T^s$ | slip-flow constant wall temperature Nusselt number |
| $Nu_x$ | local Nusselt number $= h_x D_h/k$ |
| $Nu_v$ | viscous dissipation Nusselt number |
| $P$ | pressure, N/m$^2$ |
| $\bar{P}$ | dimensionless pressure |
| $Pe$ | Peclet number $= Re\ Pr$ |
| $Po$ | Poiseuille number |
| $Pr$ | Prandtl number $= c_p\mu/k$ |
| $q$ | heat-transfer rate, W |
| $q''$ | constant wall heat flux, W/m$^2$ |
| $q_w$ | wall heat flux, W/m$^2$ |
| $Q_i$ | incoming stream energy, J |
| $Q_r$ | energy carried by the reflected molecules, J |
| $Q_w$ | energy of the molecules leaving the surface at the wall temperature, J |
| $R$ | channel radius, m |
| $\mathfrak{R}$ | gas constant |
| $Re$ | Reynolds number, $= \rho U_\infty L/\mu$ |
| $Re_x$ | local Reynolds number based on $x$, $= \rho U_\infty x/\mu$ |
| $Re_L$ | Reynolds number based on $L$, $= \rho U_\infty L/\mu$ |
| $r$ | radius coordinate, m |
| $T$ | fluid temperature, K, input temperature, K |
| $T_b$ | bulk temperature, K |
| $T_i$ | inlet temperature, K |
| $T_0$ | reference temperature, at 273 K |
| $T_s$ | surface temperature, K |
| $T_\infty$ | free-stream temperature, K |
| $T_w$ | wall temperature, K |
| $t$ | time, s |
| $U$ | fluid velocity, m/s |
| $U_a$ | speed of sound in the gas, m/s |
| $U_\infty$ | free-stream velocity, m/s |
| $u$ | axial velocity, m/s |
| $u_s$ | slip velocity, m/s |
| $u_m$ | mean velocity, m/s |
| $\bar{u}$ | dimensionless velocity |
| $v$ | velocity component in $y$ direction, m/s |
| $\bar{v}$ | dimensionless velocity |
| $v_r$ | radius velocity, m/s |
| $W$ | microchannel height, m |
| $W_c$ | center-to-center distance of microchannels, m |
| $w$ | velocity component in $z$ direction, m/s |
| $x$ | rectangular coordinate parallel to surface, m |
| $\bar{x}$ | dimensionless rectangular coordinate parallel to surface |
| $x_e^*$ | entrance length, m |
| $x^*$ | developing region of laminar flow in a circular tube |
| $y$ | transverse coordinate, m |
| $\bar{y}$ | transverse coordinate, m |
| $z$ | constant for Equation 12.99 |

## Greek Symbols

| | |
|---|---|
| $\alpha$ | thermal diffusivity $= k/\rho\, c$, m$^2$/s |
| $\alpha^*$ | constant defined by Equation 12.122 |
| $\gamma$ | specific heat ratio, aspect ratio of the short side to the long side of a microchannel |
| $\varepsilon$ | relative roughness defined by Equation 12.81 |
| $\eta$ | dimensionless transverse coordinate |
| $\theta$ | dimensionless temperature profile $= (T - T_\infty/)(T_W - T_\infty)$, side angle |
| $\bar{\theta}$ | constant wall heat-flux dimensionless temperature profile $= T - T_{s,i}/T_i - T_{s,i}$ |
| $\theta^*$ | constant wall heat-flux dimensionless temperature profile $= T - T_i/q''(H/k)$ |
| $\theta_b$ | dimensionless bulk temperature |
| $\theta_m$ | dimensionless mean temperature |
| $\theta_s$ | dimensionless surface temperature |
| $\lambda$ | mean free molecular path, m |
| $\lambda_D$ | Darcy coefficient |
| $\mu$ | dynamic viscosity, Pa s |
| $\mu_f$ | free-stream dynamic viscosity, Pa s |
| $\mu_w$ | dynamic viscosity at wall, Pa s |
| $\bar{\mu}$ | dimensionless dynamic viscosity |
| $\nu$ | kinematic viscosity, m$^2$/s |
| $\xi$ | dimensionless rectangular coordinate |
| $\rho$ | density, kg/m$^3$ |
| $\sigma$ | molecular diameter, m |
| $\tau$ | dimensionless time $= \alpha t/H^2$ |
| $\phi$ | heat dissipation, W/m$^3$ |
| $\omega$ | Verhoff and Fisher coordinate variable, constant given by Equation 12.77 |

## 12.1 Introduction

With the advancement of miniaturized manufacturing, microdevices are successfully employed in a variety of fields such as micro-heat sinks, micro-reactors, micro-motors, micro-biochips, micro-valves, micro-fuel cells, and components in electronic industries. Therefore, a significant number of microdevices have been used widely and continuously. Generally, all devices with dimensions between 1 μm and 1 mm can be called "microdevices," which can be divided into three main categories [1]: MEMS or micro-electromechanical systems (sensors, lab-on-a-chip, etc.), MOEMS or micro-opto-electro-mechanical systems (optical switches, digital, etc.), and MFD or micro-flow devices (micro-heat exchangers, micropumps, etc.). The use of minichannels, microchannels, and microfluidics in general is becoming increasingly important to the biomedical community. However, the transport and manipulation of living cells and biological macromolecules place increasingly critical demands on maintaining system conditions within acceptable ranges [2]. One important application of microchannels is designing and controlling lab-on-a-chip devices, the devices are miniaturized biomedical or chemistry laboratories on a small glass or plastic chip with a network of microchannels, electrodes, sensors, and electrical circuits [3]. The early applications involved micromachined devices such as micropumps, micro-valves, and microsensors [4].

Due to the miniaturization trend especially in electronics, problems associated with overheating of the microdevices have emerged and high-performance heat-transfer devices have developed. Experimental studies in microchannel flow and heat transfer have shown that the macroscopic conventional theories based on the continuum hypothesis cannot be used to explain their phenomena. In other words, there are no generalized solutions for the determination of heat-transfer and flow characteristics in designing microdevices. The convective heat transfer in liquid and gases flows should be considered separately. In microchannel flow, experimental and theoretical results for liquid flows cannot be considered or modified for gaseous flows because the flow regime boundaries are significantly different as well as flow and heat-transfer characteristics. For an example, gas flow in microchannels, the slip flow, and temperature jump have to be considered. Heat-transfer coefficients for liquid flow in microchannels are very high due to their small hydraulic diameters. The high pressure gradients lead to low flow rates. In small cooling devices, the ability of the fluid stream to carry the heat away for a given temperature rise becomes limited, but reducing the flow length of the channels and increasing the liquid flow rate must be matched with their performances [4]. As a result, employing multiple streams with short paths in a microdevice was recommended [4]. One should note that, in the transition and turbulent flow regions, the reduced flow length reduces the pressure drop, multiple inlets enlarging channel area where the heat-transfer rate is higher, and turbulent flow can help to increase heat-transfer coefficients in these regions.

Two basic considerations for the flow in microchannels are as molecular models (a collection of molecules) and continuum models (continuous and indefinitely divisible fluid). The first modeling is deterministic or probabilistic modeling. The second modeling considers fluid properties related with positions and time, so conservation of mass, momentum, and energy are nonlinear partial differential equations, Navier–Stokes equations. When the Navier–Stokes governing equations are applied to use with flow in MEMS, results are often equivocal because of considered parameters such as viscous dissipation, thermal conductivity, mean flow velocity, and temperature distributions.

## 12.2 Definitions in Microchannels

Fluid velocity in microchannels is not very high because of dominant pressure drops occurred in the small channels. According to the low velocity, the Reynolds number is also small. From this matter, the transition velocity range from laminar to turbulent flow occurs earlier, the viscous effects and friction factors are also higher than those obtained from conventional calculations. All these discrepancies affect both the fluid flow and the convective heat transfer in microchannels [5]. There are important distinguished explanations between macroscopic and microscopic points of view presented in this section.

### 12.2.1 Knudsen Number

Rarefied flow is defined as gaseous flow with very low pressure. The interaction between gas molecules and channel walls becomes as frequent as intermolecular collisions, so the boundaries and the molecular structure influence the flow; it is also called rarefaction effects. The average distance travelled by molecules between the collisions is called the mean path. The ratio of the mean free path to the characteristic length is used to represent rarefaction effects and is also called the Knudsen number [6]:

$$Kn = \frac{\lambda}{L}, \tag{12.1}$$

where
  $L$ is a characteristic flow dimension such as channel hydraulic diameter ($D_h$)
  $\lambda$ is the mean free molecular path corresponding to the distance travelled by the molecules between the collision and defined as

$$\lambda = \frac{\bar{k}T}{\sqrt{2}\pi P\sigma^2}. \tag{12.2}$$

The Knudsen number is a well-established criterion to indicate whether fluid flow problem can be solved by the continuum approach. The local Knudsen number is used to determine the degree of rarefaction and the degree of deviation from continuum model. Flow regimes are specified according to Knudsen number as shown in Table 12.1 [7].

In continuum flow regime, no-slip flow where $Kn \leq 0.001$, for gaseous flow in microchannels, continuum assumption that is widely used for macroscopic heat-transfer problems becomes valid. In the slip-flow regime, where $0.001 < Kn \leq 0.1$, continuum model is applicable except in the layer next to the wall, which can be identified as the Knudsen layer. For the Knudsen layer, slip boundary conditions should be considered. If the flow is in the transition regime, where $0.1 < Kn \leq 10$, and the flow continues into free molecular flow regime, molecular approach should be used. In other words, the Boltzmann equation should be considered for atomic-level studies of gaseous flow in transition regime [8]. In the classic continuum flow regime, one may use the compressible Navier–Stokes equations, the ideal gas equation of state, and classic boundary conditions to express the continuity of temperature and velocity between the fluid and the wall [9]. For transition flow and free molecular flow, for the Knudsen numbers higher than unity, a slip-flow model may still be used with the Burnett equations. For higher Knudsen numbers, in the full transition and in the free molecular regimes, the Boltzmann equation must be directly treated by appropriate numerical techniques such as the direct simulation Monte Carlo (DSMC) method or the lattice Boltzmann methods (LBM) [9].

Since change in fluid density inside microdevices can be significant, the compressibility must be taken into account. The Mach number is a dimensionless parameter for

**TABLE 12.1**

Flow Regimes Based on the Knudsen Number

| Regime | Method of Calculation | $Kn$ Range |
|---|---|---|
| Continuum | Navier–Stokes and energy equations with no-slip/no-jump boundary conditions | $Kn \leq 0.001$ |
| Slip flow | Navier–Stokes and energy equations with slip/jump boundary conditions, DSMC[a] | $0.001 < Kn \leq 0.1$ |
| Transition | Boltzmann Transport Equations[a], DSMC[a] | $0.1 < Kn \leq 10$ |
| Free molecule | Boltzmann Transport Equations[a], DSMC[a] | $Kn > 10$ |

*Source:* Gad-El-Hak, M., *J. Fluids Eng.*, 121, 5, 1999.
[a] From Gad-El-Hak [7].

compressible fluid; if $Ma < 0.3$, the flow can be assumed incompressible. According to the kinetic theory, viscosity can be expressed as a function of mean free path $\left(\mu = \rho\lambda\sqrt{2RT/\pi}\right)$. The Knudsen number can also be expressed as [10]

$$Kn = \frac{Ma}{Re}\sqrt{\frac{\pi k_r}{2}},\qquad(12.3)$$

where

the Reynolds number is $Re = \rho U_\infty L/\mu$
the Mach number is $Ma = U_\infty/U_a$.

Note that $U_\infty$ is free-stream velocity and $U_a = \sqrt{k_r \Re T}$ is the speed of sound in the gas. Here, $k_r$ is the ratio of specific heats of the gas and $\Re$ is the gas constant.

### 12.2.2 Velocity Profile

As rarefaction effects, when gas molecules impinge on the wall, gas molecules will be reflected after collision and the molecules leave some of their momentum and create shear stress on the wall. If specular reflection occurs, the tangential momentum remains the same, whereas the normal momentum will be reserved. If diffuse reflection occurs, the tangential momentum will vanish and the tangential momentum balance at the wall yields the slip velocity as [11–13]

$$u_s = 2\frac{2 - F_m}{F_m}\frac{\mu}{\rho u_m}\left(\frac{du}{dy}\right)_w\qquad(12.4)$$

and

$$\mu \cong \frac{1}{2}\rho u_m \lambda.\qquad(12.5)$$

By neglecting thermal creep, the tangential momentum balance at the wall gives the first-order slip velocity in Cartesian coordinate as [10,14,15]

$$u_s = -\frac{2 - F_m}{F_m}\lambda\left(\frac{du}{dy}\right)_w + 3\sqrt{\frac{\Re T}{8\pi}}\frac{\lambda}{T}\left(\frac{\partial T}{\partial x}\right)_w,\qquad(12.6)$$

and in cylindrical coordinate as [15]

$$u_s = -\frac{2 - F_m}{F_m}\lambda\left(\frac{du}{dr}\right)_R + 3\sqrt{\frac{\Re T}{8\pi}}\frac{\lambda}{T}\left(\frac{\partial T}{\partial x}\right)_R,\qquad(12.7)$$

where $F_m$ is the tangential momentum accommodation coefficient. Molecules will be reflected specularly if they collide to an ideal smooth surface, the tangential momentum of the molecules will be conserved and no shear stress will be created in the wall and thus $F_m$

is equal to 0. If diffusive reflection occurs, in other words, the tangential momentum will be lost at the wall, $F_m$ will equal to 1. For real surfaces, molecules reflect from the wall both diffusively and specularly. According to experiments, there is a ratio between diffusive and specular reflections in the range of 0.2–0.8 [13].

For slip flow, the fully developed velocity profile can be obtained from the momentum equations for laminar flow with constant thermophysical properties in Cartesian coordinate as

$$u = \frac{3}{2} u_m \frac{[1-(y/H)^2 + 4Kn]}{1+6Kn}, \tag{12.8}$$

and the fully developed velocity profile for slip flow can be obtained from the momentum equations for laminar flow with constant thermophysical properties in cylindrical coordinate as

$$u = 2u_m \frac{[1-(r/R)^2 + 4Kn]}{1+8Kn}. \tag{12.9}$$

### 12.2.3 Temperature Jump

The difference between the fluid temperature at the wall and the wall temperature, temperature jump, is proposed to be [13]

$$T_s - T_w = c_{jump} \left( \frac{\partial T}{\partial y} \right)_w. \tag{12.10}$$

The thermal accommodation coefficient, $F_T$, is defined as

$$F_T = \frac{Q_i - Q_r}{Q_i - Q_w}, \tag{12.11}$$

where
  $Q_i$ is the incoming stream energy
  $Q_r$ is the energy carried by the reflected molecules
  $Q_w$ is the energy of the molecules leaving the surface at the wall temperature [13].

For a perfect gas, the temperature jump coefficient can be obtained as [13]

$$c_{jump} = \frac{2-F_T}{F_T} \frac{2\gamma}{(\gamma+1)} \frac{\lambda}{Pr}, \tag{12.12}$$

where

$$2\frac{\lambda}{Pr} = \frac{k\sqrt{2\pi\Re T}}{c_v P}. \tag{12.13}$$

Then the temperature jump can also be expressed in Cartesian coordinate as [13]

$$T_s - T_w = -\frac{2 - F_T}{F_T} \frac{2\gamma}{(\gamma + 1)} \frac{\lambda}{Pr} \left( \frac{\partial T}{\partial y} \right)_w, \tag{12.14}$$

and in cylindrical coordinate as

$$T_s - T_w = -\frac{2 - F_T}{F_T} \frac{2\gamma}{(\gamma + 1)} \frac{\lambda}{Pr} \left( \frac{\partial T}{\partial r} \right)_R. \tag{12.15}$$

### 12.2.4 Brinkman Number

The friction between the fluid layers causes viscous heat generation. Although viscous heat generation can be neglected for continuum flow, it is an important factor for fluid flow in microchannels because of ratio between large surface area and volume. The Brinkman number is a dimensionless parameter representing heat that is generated by viscous dissipation. The definition of the Brinkman number varies with the boundary condition at the wall. When constant wall temperature (CWT) boundary condition is considered, the Brinkman number is given by

$$Br = \frac{\mu u_m^2}{k(T_w - T_i)}. \tag{12.16}$$

In addition, if constant wall heat-flux boundary condition is considered, the Brinkman number can be defined as

$$Br = \frac{\mu u_m^2}{q_w d}, \tag{12.17}$$

where $T_w - T_i$ is the wall–fluid temperature difference at a particular axial location. The Brinkman number is used to measure the relative importance at viscous heating to heat conduction in the fluid along the microchannels. The Brinkman number is positive for heating the fluid, and the Brinkman number is negative for cooling the fluid. So, for cooling the gas, the heat-transfer coefficient increases.

The effects of the Brinkman and Knudsen numbers on heat transfer are also illustrated in Table 12.2 [15]. It can be clearly seen that, for both Brinkman numbers, Nusselt numbers decreases with increasing the Knudsen number, which can be explained by the increasing temperature jump.

### 12.2.5 Governing Equations

For steady 2D and incompressible flow with constant thermophysical properties, governing equations can be written in Cartesian coordinates as [16,17]

Continuity equation:

$$\frac{\partial u}{\partial x} + \frac{\partial v}{\partial y} = 0 \tag{12.18}$$

**TABLE 12.2**

Nusselt Numbers for Developed Laminar Flow Where
$q_w$ = Constant and $T_w$ = Constant at $Pr$ = 0.7

| | *T* = Constant | | $q_w$ = Constant | | |
|---|---|---|---|---|---|
| *Kn* | *Br* = 0.00 | *Br* = 0.01 | *Br* = 0.00 | *Br* = 0.01 | *Br* = −0.01 |
| 0.00 | 3.6566 | 9.5985 | 4.3649 | 4.1825 | 4.564 |
| 0.02 | 3.4163 | 7.4270 | 4.1088 | 4.0022 | 4.2212 |
| 0.04 | 3.1706 | 6.0313 | 3.8036 | 3.7398 | 3.8695 |
| 0.06 | 2.9377 | 5.0651 | 3.4992 | 3.4498 | 3.5395 |
| 0.08 | 2.7244 | 4.3594 | 3.2163 | 3.1912 | 3.2419 |
| 0.10 | 2.5323 | 3.8227 | 2.9616 | 2.944 | 2.9784 |

*Source:* Cetin, B. et al., *Int. J. Trans. Phenomena*, 8, 297, 2006.

Momentum equation:

$$u\frac{\partial u}{\partial x}+v\frac{\partial v}{\partial y}=-\frac{1}{\rho}\frac{\partial P}{\partial x}+\upsilon\left(\frac{\partial^2 u}{\partial x^2}+\frac{\partial^2 v}{\partial y^2}\right) \qquad (12.19)$$

$$u\frac{\partial v}{\partial x}+v\frac{\partial v}{\partial y}=-\frac{1}{\rho}\frac{\partial P}{\partial y}+\upsilon\left(\frac{\partial^2 v}{\partial x^2}+\frac{\partial^2 v}{\partial y^2}\right) \qquad (12.20)$$

Energy equation:

$$u\frac{\partial T}{\partial x}+v\frac{\partial T}{\partial y}=\alpha\left(\frac{\partial^2 T}{\partial x^2}+\frac{\partial^2 T}{\partial y^2}\right)+\frac{1}{\rho c}\phi \qquad (12.21)$$

The 2D governing equations can be written in cylindrical coordinates, from Tables 2.1 through 2.3 [15] where $v = v_r$ and $u = w$, as

Continuity equation:

$$\frac{\partial u}{\partial x}+\frac{\partial v}{\partial r}=0 \qquad (12.22)$$

Momentum equation:

$$u\frac{\partial u}{\partial x}+v\frac{\partial u}{\partial r}=-\frac{1}{\rho}\frac{\partial P}{\partial x}+\upsilon\left(\frac{1}{r}\frac{\partial}{\partial r}\left(r\frac{\partial u}{\partial r}\right)+\frac{\partial^2 u}{\partial x^2}\right) \qquad (12.23)$$

$$u\frac{\partial v}{\partial x}+v\frac{\partial v}{\partial r}=-\frac{1}{\rho}\frac{\partial P}{\partial r}+\upsilon\left(\frac{\partial}{\partial r}\left(\frac{1}{r}\frac{\partial}{\partial r}(rv)\right)+\frac{\partial^2 v}{\partial x^2}\right) \qquad (12.24)$$

Energy equation:

$$u\frac{\partial T}{\partial x}+v\frac{\partial T}{\partial r}=\frac{\alpha}{r}\frac{\partial}{\partial r}\left(r\frac{\partial T}{\partial r}\right)+\alpha\frac{\partial^2 T}{\partial x^2}+\phi \tag{12.25}$$

where

$$\phi=\frac{2\mu}{\rho c}\left[\left(\frac{\partial v}{\partial r}\right)^2+\left(\frac{v}{r}\right)^2+\left(\frac{\partial u}{\partial x}\right)^2+\frac{1}{2}\left(\frac{\partial u}{\partial r}+\frac{\partial v}{\partial x}\right)^2\right] \tag{12.26}$$

### 12.2.5.1 Flow through a Parallel-Plate Microchannel

For steady fully developed incompressible laminar flow with constant thermophysical properties, the Cartesian governing equations can be reduced to

$$\frac{\partial u}{\partial x}=\frac{\partial v}{\partial y}=0: \quad u=u(y), \tag{12.27}$$

$$0=-\frac{1}{\rho}\frac{dP}{dx}+\upsilon\left(\frac{d^2 u}{dx^2}\right), \tag{12.28}$$

$$0=-\frac{1}{\rho}\frac{dP}{dy}, \tag{12.29}$$

$$u\frac{dT}{dx}=\alpha\left(\frac{d^2 T}{dy^2}\right)+\frac{\upsilon}{\rho c}\left(\frac{du}{dy}\right)^2. \tag{12.30}$$

Boundary conditions are as follows:

$$\text{at } x=0: \quad T=T_i, \tag{12.31a}$$

$$\text{at } y=H: \quad T=T_s, \tag{12.31b}$$

$$\text{at } y=0: \quad \frac{\partial T}{\partial y}=0, \tag{12.31c}$$

where $T_s$ is the slip temperature of the fluid, which is different from the wall temperature.

From the velocity profile in Equation 12.8, the velocity profile can be rewritten for flow between a parallel-plate microchannel, distance between the plates defined as $d$, as

$$u=\frac{3}{2}u_m\left[1-\left(\frac{y}{d}\right)^2\right]. \tag{12.32}$$

### 12.2.5.2 Flow through a Microtube

For steady fully developed incompressible laminar flow with constant thermophysical properties, the cylindrical governing equations can be reduced to

$$\frac{\partial u}{\partial x} = \frac{\partial v}{\partial r} = 0: \quad u = u(r), \tag{12.33}$$

$$\frac{1}{r}\frac{d}{dr}\left(r\frac{du}{dr}\right) = \frac{1}{\mu}\frac{dP}{dx}, \tag{12.34}$$

$$u\frac{\partial v}{\partial x} + v\frac{\partial v}{\partial r} = -\frac{1}{\rho}\frac{\partial P}{\partial r} + \upsilon\left(\frac{\partial}{\partial r}\left(\frac{1}{r}\frac{\partial}{\partial r}(rv)\right) + \frac{\partial^2 v}{\partial x^2}\right), \tag{12.35}$$

$$u\frac{\partial T}{\partial x} = \frac{\alpha}{r}\frac{\partial}{\partial r}\left(r\frac{\partial T}{\partial r}\right) + \frac{\upsilon}{c}\left(\frac{du}{dx}\right)^2. \tag{12.36}$$

Boundary conditions are as follows:

$$\text{at } x = 0: \quad T = T_i, \tag{12.37a}$$

$$\text{at } r = R: \quad T = T_s \quad \text{and} \quad u = u_s, \tag{12.37b}$$

$$\text{at } r = 0: \quad \frac{\partial T}{\partial r} = 0 \quad \text{and} \quad u = U, \tag{12.37c}$$

where $u_s$ and $U$ are the slip velocity and finite velocity of the fluid.

From the velocity profile in Equation 12.9, the velocity profile can be rewritten for flow through a microtube, a radius of the microtube defined as $R$, as

$$u = 2u_m\left(1 - \left(\frac{r}{R}\right)^2\right). \tag{12.38}$$

## 12.3 Convective Heat Transfer for Gaseous Flow in Microchannels

To understand microchannel flow and heat transfer in microchannels and microtubes, the typical domains, Figures 12.1 and 12.2, are considered for both coordinates, Cartesian and cylindrical coordinates.

There is an unheated section at the inlets to be able to have a fully developed velocity profile. Since they have a high length to diameter ratio ($L/D \sim 100$), the 2D, incompressible, constant thermophysical property, hydrodynamically developed, thermally developing and single-phase laminar flow in microtubes and in microchannels was considered

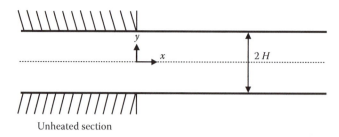

**FIGURE 12.1**
Cartesian coordinate schematic of hydrodynamically fully developed flow.

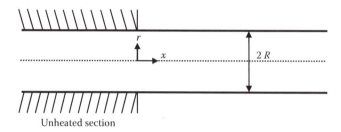

**FIGURE 12.2**
Cylindrical coordinate schematic of hydrodynamically fully developed flow.

with negligible axial conduction for both uniform wall temperature (UWT) and uniform wall heat-flux thermal boundary conditions by imposing the rarefaction effects into the boundary conditions and by including the viscous dissipation term, which is important for microchannels. The fully developed velocity profiles are determined by solving the momentum equations, Equations 12.30 and 12.36, by using the slip velocity profiles for both geometries as in Equations 12.8 and 12.9.

### 12.3.1 Constant Wall Temperature

For CWT, the boundary conditions are (12.31a), (12.31b), and (12.31c) for Cartesian coordinate and (12.37a), (12.37b), and (12.37c) for cylindrical coordinate. The dimensionless energy equation can be expressed for the microchannels as

$$\frac{\bar{u}}{4}\frac{d\theta}{d\xi} = \left(\frac{d^2\theta}{d\eta^2}\right) + \frac{8Br}{(1+8Kn)^2}\eta^2, \qquad (12.39)$$

and for the microtubes as

$$\frac{\bar{u}}{2}\frac{d\theta}{d\xi} = \left(\frac{d^2\theta}{d\eta^2}\right) + \frac{1}{\eta}\frac{\partial\theta}{\partial\eta} + \frac{16Br}{(1+8Kn)^2}\eta^2, \qquad (12.40)$$

where

$$\theta = \frac{T - T_w}{T_i - T_w}, \qquad (12.41a)$$

$$\bar{u} = \frac{u}{u_m}, \tag{12.41b}$$

$$\eta = \begin{cases} \dfrac{r}{R} \\ \dfrac{y}{H} \end{cases}, \tag{12.41c}$$

$$\xi = \begin{cases} \dfrac{x}{HRePr} & \text{for microchannel} \\ \dfrac{x}{RRePr} & \text{for microtube} \end{cases}, \tag{12.41d}$$

$$Br = \frac{\mu u_m^2}{k(T_i - T_w)}. \tag{12.41e}$$

The dimensionless boundary conditions can be expressed for the microchannels as

$$\text{at } \eta = 0: \quad \frac{\partial \theta}{\partial \eta} = 0, \tag{12.42a}$$

$$\text{at } \eta = 1: \quad \theta_s = -2\kappa Kn \left( \frac{\partial \theta}{\partial \eta} \right)_{\eta=1}, \tag{12.42b}$$

$$\text{at } \xi = 0: \theta = 1 \tag{12.42c}$$

where

$$\text{at } \kappa = \frac{2 - F_T}{F_T} \frac{2\gamma}{\gamma + 1} \frac{1}{Pr}. \tag{12.43}$$

The Nusselt number is determined for CWT in the microchannel as

$$Nu_x = \frac{h_x D_h}{k} = -\frac{4 \left( \partial \theta / \partial \eta \right)_{\eta=1}}{\theta_m}, \tag{12.44}$$

and in the microtube as

$$Nu_x = \frac{h_x D_h}{k} = -\frac{2 \left( \partial \theta / \partial \eta \right)_{\eta=1}}{\theta_m}, \tag{12.45}$$

where $\theta_m$ is the dimensionless mean temperature, and, for microchannel and microtube, respectively, they are defined as

$$\theta_m = \int_0^1 \left(\frac{u}{u_m}\right)\theta(\eta,\xi)d\eta, \tag{12.46}$$

$$\theta_m = 2\int_0^1 \left(\frac{u}{u_m}\right)\theta(\eta,\xi)\eta d\eta. \tag{12.47}$$

### 12.3.2 Constant Wall Heat Flux

For constant wall heat flux, the modified definition of dimensionless temperature for flow inside microchannel is

$$\theta^* = \frac{T - T_i}{q''(H/k)}, \tag{12.48}$$

and inside microtube is

$$\theta^* = \frac{T - T_i}{q''(R/k)}. \tag{12.49}$$

The Brinkman number for the flow inside microchannel is

$$Br^* = \frac{\mu u_m^2}{2q''H}, \tag{12.50}$$

and for the flow inside microtube is

$$Br^* = \frac{\mu u_m^2}{2q''R}. \tag{12.51}$$

The dimensionless energy equation can be expressed for the microchannels as

$$\frac{\bar{u}}{4}\frac{d\theta^*}{d\xi} = \left(\frac{d^2\theta^*}{d\eta^2}\right) + \frac{18Br^*}{(1+8Kn)^2}\eta^2, \tag{12.52}$$

and for the microtubes as

$$\frac{\bar{u}}{2}\frac{d\theta^*}{d\xi} = \left(\frac{d^2\theta^*}{d\eta^2}\right) + \frac{1}{\eta}\frac{\partial\theta^*}{\partial\eta} + \frac{32Br^*}{(1+8Kn)^2}\eta^2. \tag{12.53}$$

The dimensionless boundary conditions can be expressed for the microchannels as

$$\text{at } \eta = 0: \quad \frac{\partial \theta^*}{\partial \eta} = 0, \tag{12.54a}$$

$$\text{at } \eta = 1: \quad \frac{\partial \theta^*}{\partial \eta} = 1, \tag{12.54b}$$

$$\text{at } \xi = 0: \quad \theta^* = 0. \tag{12.54c}$$

The Nusselt number is determined for constant wall heat flux in the microchannel as

$$Nu_x = \frac{h_x D_h}{k} = \frac{4}{\theta_m - \theta_s - 2\kappa Kn}, \tag{12.55}$$

and in the microtube as

$$Nu_x = \frac{h_x D}{k} = \frac{2}{\theta_m - \theta_s - 2\kappa Kn}, \tag{12.56}$$

where $\theta_m$ is the dimensionless mean temperature, defined in Equations 12.46 and 12.47, and $\theta_s$ is the dimensionless fluid temperature at the surface. To determine the local Nusselt number, the energy equation, Equations 12.39, 12.40, 12.52, and 12.53, should be solved with indicated boundary conditions to obtain the temperature distribution. However, for the flow inside microchannel, there are additional terms in the fully developed velocity profile, at the wall thermal boundary conditions and viscous heating. These extensions in the problem introduce some difficulties in the analytical solution. The analytical solution was obtained for CWT as in several works [12,18–20], for constant wall heat flux in some literature works [21,22] without viscous heating, and for both thermal boundary conditions with unit thermal accommodation coefficient with viscous heating [23,24] in microchannels. The steady-state problem is transformed into a transient problem; elliptic partial differential equation is transformed into a parabolic differential equation, by defining the dimensionless time as

$$\tau = \begin{cases} \dfrac{\alpha t}{H^2} & \text{for microchannel} \\[2ex] \dfrac{\alpha t}{R^2} & \text{for microtube} \end{cases}, \tag{12.57}$$

where an appropriate time step is

$$\Delta \tau = 0.4 \Delta \eta^2. \tag{12.58}$$

Velocity gradient at the wall decreases with increasing rarefaction, which leads to a reduction in the friction factor for flows in microchannels. Large slip on the wall increases convection along the surface, on the other hand a large temperature jump decrease the heat transfer by reducing the temperature gradient at the wall. Neglecting temperature jump results in the overestimation of the heat-transfer coefficient [15]. From the previous governing equations and boundary conditions, one can solve the governing equations by applying the boundary conditions numerically. Results are shown in Tables 12.3 through 12.6 [15].

For fixed κ parameter, the deviation from the continuum increases with increasing rarefaction for both CWT and constant wall heat-flux cases without viscous heating. For fixed $Kn$ number, the deviation from the continuum increases and $Nu$ decreases gradually with increasing κ parameter without viscous heating. The fully developed $Nu$ number decreases or increases compared to the continuum fully developed $Nu$ number values depending on the $Kn$ number and κ values, without viscous heating. For viscous heating and CWT case, even for small $Br$ numbers, there is a deviation from the $Br = 0$ case. Local $Nu$ number value experiences a jump in magnitude for positive $Br$ number and a singular point for a negative $Br$ number. The magnitude of the $Br$ number and the temperature jump affects the axial location of the jump and the singular point. As $Br$ increases, the points of $Nu$ number get closer to the entrance. With increasing temperature jump, the axial locations of the points move far away from the entrance. The increase in the temperature jump decreases the jump in $Nu$ number. The effect of viscous heating should be considered even for small $Br$ numbers with large length over diameter ratios, $(L/2H)$, which is the case for flows in microchannels. Moreover, the length of the microchannel is a critical choice to avoid the

**TABLE 12.3**

Nusselt Number for Fully Developed Flow in Microchannel with CWT

| | $Br = 0$ without Viscous Dissipation | | | | $Br \neq 0$ with Viscous Dissipation | | | |
|---|---|---|---|---|---|---|---|---|
| $Kn$ | $\kappa = 0$ | $\kappa = 0.5$ | $\kappa = 1.667$ | $\kappa = 10$ | $\kappa = 0$ | $\kappa = 0.5$ | $\kappa = 1.667$ | $\kappa = 10$ |
| 0 | 7.541 | 7.541 | 7.541 | 7.541 | 17.497 | 17.497 | 17.497 | 17.497 |
| 0.02 | 7.739 | 7.478 | 6.925 | 4.476 | 17.735 | 16.290 | 13.688 | 6.394 |
| 0.04 | 7.905 | 7.382 | 6.374 | 3.145 | 17.931 | 15.205 | 11.222 | 3.909 |
| 0.06 | 8.048 | 7.264 | 5.882 | 2.415 | 18.096 | 14.233 | 9.49 | 2.813 |
| 0.08 | 8.172 | 7.131 | 5.445 | 1.957 | 18.236 | 13.363 | 8.229 | 2.189 |
| 0.10 | 8.279 | 6.989 | 5.058 | 1.643 | 18.358 | 12.583 | 7.256 | 1.792 |

*Source:* Cetin, B. et al., *Int. J. Trans. Phenomena*, 8, 297, 2006.

**TABLE 12.4**

Nusselt Number for Fully Developed Flow in Microtube with CWT

| | $Br = 0$ without Viscous Dissipation | | | | $Br \neq 0$ with Viscous Dissipation | | | |
|---|---|---|---|---|---|---|---|---|
| $Kn$ | $\kappa = 0$ | $\kappa = 0.5$ | $\kappa = 1.667$ | $\kappa = 10$ | $\kappa = 0$ | $\kappa = 0.5$ | $\kappa = 1.667$ | $\kappa = 10$ |
| 0 | 3.656 | 3.656 | 3.656 | 3.656 | 9.598 | 9.598 | 9.598 | 9.598 |
| 0.02 | 3.855 | 3.739 | 3.488 | 2.291 | 9.871 | 8.984 | 7.427 | 3.316 |
| 0.04 | 4.020 | 3.778 | 3.292 | 1.624 | 10.088 | 8.394 | 6.031 | 1.995 |
| 0.06 | 4.160 | 3.785 | 3.087 | 1.247 | 10.264 | 7.848 | 5.065 | 1.423 |
| 0.08 | 4.279 | 3.767 | 2.887 | 1.008 | 10.411 | 7.350 | 4.359 | 1.115 |
| 0.10 | 4.382 | 3.732 | 2.697 | 0.844 | 10.535 | 6.900 | 3.822 | 0.912 |

*Source:* Cetin, B. et al., *Int. J. Trans. Phenomena*, 8, 297, 2006.

**TABLE 12.5**

Nusselt Number with Different *Br* for Fully Developed
Flow in Microtube with Constant Wall Heat Flux

| *Br* | *Kn* | $\kappa = 0$ | $\kappa = 0.5$ | $\kappa = 1.667$ | $\kappa = 10$ |
|------|------|------|------|------|------|
| 0 | 0.00 | 4.364 | 4.364 | 4.364 | 4.364 |
| | 0.02 | 4.710 | 4.498 | 4.071 | 2.425 |
| | 0.04 | 4.998 | 4.544 | 3.749 | 1.666 |
| | 0.06 | 5.241 | 4.529 | 3.438 | 1.265 |
| | 0.08 | 5.448 | 4.473 | 3.155 | 1.017 |
| | 0.10 | 5.627 | 4.391 | 2.903 | 0.849 |
| 0.1 | 0.00 | 3.038 | 3.038 | 3.038 | 3.038 |
| | 0.02 | 3.598 | 3.468 | 3.209 | 2.091 |
| | 0.04 | 4.058 | 3.753 | 3.194 | 1.547 |
| | 0.06 | 4.446 | 3.923 | 3.077 | 1.212 |
| | 0.08 | 4.771 | 4.007 | 2.916 | 0.911 |
| | 0.10 | 5.046 | 4.029 | 2.741 | 0.835 |
| −0.1 | 0.00 | 7.743 | 7.743 | 7.743 | 7.743 |
| | 0.02 | 6.835 | 6.397 | 5.566 | 2.888 |
| | 0.04 | 6.505 | 5.756 | 4.537 | 1.806 |
| | 0.06 | 6.382 | 5.356 | 3.895 | 1.322 |
| | 0.08 | 6.348 | 5.063 | 3.438 | 1.044 |
| | 0.10 | 6.359 | 4.825 | 3.087 | 0.864 |

*Source:* Cetin, B. et al., *Int. J. Trans. Phenomena*, 8, 297, 2006.

**TABLE 12.6**

Nusselt Number with Different *Br* for Fully
Developed Flow in Microchannel with CWT

| *Br* | *Kn* | $\kappa = 0$ | $\kappa = 0.5$ | $\kappa = 1.667$ | $\kappa = 10$ |
|------|------|------|------|------|------|
| 0 | 0.00 | 8.235 | 8.235 | 8.235 | 8.235 |
| | 0.02 | 8.555 | 8.204 | 7.487 | 4.611 |
| | 0.04 | 8.825 | 8.110 | 6.819 | 3.192 |
| | 0.06 | 9.057 | 7.973 | 6.233 | 2.437 |
| | 0.08 | 9.257 | 7.811 | 5.724 | 1.968 |
| | 0.10 | 9.432 | 7.632 | 5.281 | 1.650 |
| 0.1 | 0.00 | 6.250 | 6.250 | 6.250 | 6.250 |
| | 0.02 | 6.857 | 6.630 | 6.154 | 4.068 |
| | 0.04 | 7.366 | 6.860 | 5.913 | 2.978 |
| | 0.06 | 7.794 | 6.978 | 5.608 | 2.335 |
| | 0.08 | 8.157 | 7.013 | 5.283 | 1.914 |
| | 0.10 | 8.467 | 6.988 | 4.964 | 1.618 |
| −0.1 | 0.00 | 12.069 | 12.069 | 12.069 | 12.069 |
| | 0.02 | 11.370 | 10.759 | 9.559 | 5.321 |
| | 0.04 | 11.007 | 9.915 | 8.052 | 3.438 |
| | 0.06 | 10.808 | 9.300 | 7.016 | 2.548 |
| | 0.08 | 10.700 | 8.814 | 6.254 | 2.027 |
| | 0.10 | 10.645 | 8.407 | 5.640 | 1.684 |

*Source:* Cetin, B. et al., *Int. J. Trans. Phenomena*, 8, 297, 2006.

misleading designs for negative *Br* numbers. In the viscous dissipation, $Br > 0$, with UWT case, *Nu* number decreases with increase in *Kn* number and *Nu* number has larger values for the case with viscous heating than without viscous heating case [15].

For constant wall heat-flux case with viscous heating, the deviation is proportional to the magnitude of the *Br* number. For small *Br* numbers, the effect can be neglected. $Br < 0$ means that fluid is being cooled, the Nusselt number takes higher values, and lower values for $Br > 0$. For fully developed conditions, at given $Br > 0$, *Nu* number decreases as *Kn* increases. For given *Kn* number with the increase in *Pr* number, increase in κ parameter, *Nu* number decreases. This is due to the fact that the temperature jump reduces heat transfer; as *Kn* increases, the temperature jump also increases.

## 12.4 Effects of Temperature Jump

In Equation 12.7, thermal creep, which accounts for fluid flow induced by the temperature gradient, is neglected; this term is second order in the Knudsen number [25–27]. Therefore, for the moderate temperature gradients, the second term is negligible for low Knudsen numbers, which is the case for slip-flow regime. Neglecting thermal creep, as was done in Equation 12.7, temperature jump can be expressed as in Equation 12.15 With coordinates shown in Figure 12.3 [27], the fully developed velocity profile can be determined by solving the equation of motion for a steady and fully developed laminar flow of a viscous fluid in a tube [13] with the use of the slip velocity, Equation 12.2, at the wall and the symmetry condition at the centerline of the tube as in Equation 12.9.

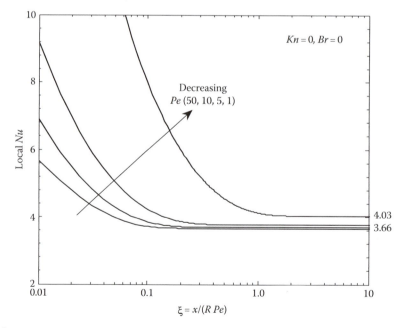

**FIGURE 12.3**

Variation of local *Nu* as a function of dimensionless axial coordinate for different *Pe* numbers ($Kn = 0$). (From Cetin, B. et al., *Int. Commun. Heat Mass Transfer*, 35, 535, 2008.)

### 12.4.1 Constant Wall Temperature

For CWT, the boundary conditions are also (12.37a) through (12.37c) for cylindrical coordinate. The dimensionless energy equation can be expressed for the microchannels as in Equation 12.40 by using the fully developed velocity profile for slip flow from Equation 12.9. Then, we can obtain another dimensionless form as

$$\frac{Gz(1-\eta^2+4Kn)}{2(1+8Kn)}\frac{\partial\theta}{\partial\xi}=\frac{1}{\eta}\frac{\partial}{\partial\eta}\left(\eta\frac{\partial\theta}{\partial\eta}\right)+\frac{16Br}{(1+8Kn)^2}\eta^2, \tag{12.59}$$

where $Gz = (2R/L)RePr$ and the boundary conditions for cylindrical coordinate are as

$$\text{at } \eta=0: \quad \frac{\partial\theta}{\partial\eta}=0, \tag{12.60a}$$

$$\text{at } \eta=1: \quad \theta=0, \tag{12.60b}$$

$$\text{at } \xi=0: \quad \theta=1. \tag{12.60c}$$

Note that Equations 12.60a and 12.60c are the same as Equations 12.42a and 12.42c, but Equation 12.42b is replaced with Equation 12.60b. Heat transfer from the wall to the fluid by convection can be written in terms of bulk temperature as

$$q''=h_x(T_w-T_b), \tag{12.61}$$

where $T_b$ the bulk temperature of the fluid along the tube for an incompressible fluid with constant specific heat is defined as

$$T_b=\frac{\int_A TudA}{\int_A udA}. \tag{12.62}$$

Heat flux at the wall from Equation 12.61 can also be written using Fourier's law of heat conduction as

$$q''=k\left.\frac{\partial T}{\partial r}\right|_{r=R}. \tag{12.63}$$

The Nusselt number can be obtained for the CWT as [8]

$$Nu_x=\frac{h_xD}{k}=-\frac{2(\partial\theta/\partial\eta)|_{\eta=1}}{\theta_m-(4\gamma/\gamma+1)(Kn/Pr)(\partial\theta/\partial\eta)|_{\eta=1}}, \tag{12.64}$$

where $\theta_b$ is the dimensionless bulk temperature, which can also be obtained from Equation 12.47.

## 12.4.2 Constant Wall Heat Flux

For this condition, the energy equation is the same as Equation 12.36, and the boundary conditions become

$$\text{at } x = 0: \quad T = T_i, \tag{12.65a}$$

$$\text{at } r = R: \quad q_w = k\frac{\partial T}{\partial r}, \tag{12.65b}$$

$$\text{at } r = 0: \quad \frac{\partial T}{\partial r} = 0. \tag{12.65c}$$

Also note that Equations 12.65a and 12.65c are the same as Equations 12.37a and 12.37c, but Equation 12.37b is replaced with Equation 12.60b. The energy equation can be made dimensionless by using Equation 12.49 as

$$\frac{Gz(1-\eta^2+4Kn)}{2(1+8Kn)}\frac{\partial\theta^*}{\partial\xi} = \frac{1}{\eta}\frac{\partial}{\partial\eta}\left(\eta\frac{\partial\theta^*}{\partial\eta}\right) + \frac{32Br^*}{(1+8Kn)^2}\eta^2, \tag{12.66}$$

where $Br^*$ is defined as in Equation 12.51. The dimensionless boundary conditions can be expressed as the same as in Equations 12.54a through 12.54c. Similarly, we can derive an expression for the local Nusselt number as in the CWT case [8]:

$$Nu_x = \frac{h_x D}{k} = \frac{2}{\theta_s^* + (4\gamma/\gamma+1)(Kn/Pr) - \theta_b^*}. \tag{12.67}$$

## 12.4.3 Linear Wall Temperature

For this condition, Equation 12.36 can be used as the energy equation, and the boundary conditions become

$$\text{at } x = 0: \quad T = T_i, \tag{12.68a}$$

$$\text{at } r = R: \quad T = T_s(x) = T_{s,i} + Ax, \tag{12.68b}$$

$$\text{at } r = 0: \quad \frac{\partial T}{\partial r} = 0, \tag{12.68c}$$

where $T_{s,i}$ is the fluid temperature at the inlet surface or the inlet wall. Note that Equations 12.68a and 12.68c are the same as Equations 12.37a and 12.37c, but Equation 12.37b is replaced with Equation 12.68b. The temperature jump as in Equation 12.15 can also be applied with the previous boundary condition as

$$T_s(x) - T_w(x) = -\frac{2-F_T}{F_T}\frac{2\gamma}{(\gamma+1)}\frac{\lambda}{Pr}\left(\frac{\partial T}{\partial r}\right)_R. \tag{12.69}$$

For the linear wall temperature, the modified dimensionless temperature for the flow is

$$\bar{\theta} = \frac{T - T_{s,i}}{T_i - T_{s,i}}. \tag{12.70}$$

The dimensionless form of the energy equation can be written as follows:

$$\frac{Gz(1 - \eta^2 + 4Kn)}{2(1 + 8Kn)} \frac{\partial \bar{\theta}}{\partial \xi} = \frac{1}{\eta} \frac{\partial}{\partial \eta} \left( \eta \frac{\partial \bar{\theta}}{\partial \eta} \right) + \frac{16Br}{(1 + 8Kn)^2} \eta^2, \tag{12.71}$$

$$\text{at } \eta = 0: \quad \frac{\partial \bar{\theta}}{\partial \eta} = 0, \tag{12.72a}$$

$$\text{at } \eta = 1: \quad \bar{\theta} = \bar{A}\xi, \tag{12.72b}$$

$$\text{at } \xi = 0: \quad \bar{\theta} = 1, \tag{12.72c}$$

where $Br$ and $\bar{A}$ are defined by

$$Br = \frac{\mu u_m^2}{k(T_i - T_{s,i})}, \tag{12.73}$$

$$\bar{A} = \frac{AL}{T_i - T_{s,i}}. \tag{12.74}$$

Note that Equations 12.72a and 12.72c are similar to Equations 12.60a and 12.60c, but Equation 12.60b is replaced with Equation 12.72b. The local Nusselt number considering the temperature jump can also be obtained as

$$Nu_x = \frac{h_x D}{k} = -\frac{2(\partial\theta/\partial\eta)\big|_{\eta=1}}{\bar{\theta}_b - (4\gamma/(\gamma+1))(Kn/Pr)(\partial\theta/\partial\eta)\big|_{\eta=1} - \bar{A}\xi}. \tag{12.75}$$

In general, for the previous three boundary conditions stated, velocity slip and temperature jump affect the heat transfer in opposite ways: a large slip on the wall will increase the convection along the surface. On the other hand, a large temperature jump will decrease the heat transfer by reducing the temperature gradient at the wall. Therefore, neglecting temperature jump will result in the overestimation of the heat-transfer coefficient [26].

Table 12.7 [26] shows the typical Nusselt number for both CWT, constant wall heat-flux boundary, and linear wall temperature conditions with considering temperature jump. For CWT boundary condition considering temperature jump, the fully developed Nusselt number decreases as $Kn$ increases and is a function of the Prandtl number. Prandtl number, the increase of which increases heat transfer, is important, since it directly influences the magnitude of the temperature jump. On the other hand, without considering temperature jump and for CWT boundary condition, the fully developed Nusselt number increases

**TABLE 12.7**

Effect of Temperature Jump on the Fully Developed
Nusselt Number Values at $Br = 0$

| Knudsen Number | Nusselt Number | | |
|---|---|---|---|
| | $T_w$ = Constant | $q_w$ = Constant | $T_w = T_s(x) =$ $T_{s,i} + Ax$ |
| $Pr = 0.6$ | | | |
| 0.00 | 3.6566 | 4.3649 | 4.3654 |
| 0.02 | 3.3527 | 4.0205 | 4.0215 |
| 0.04 | 3.0627 | 3.6548 | 3.6582 |
| 0.06 | 2.8006 | 3.3126 | 3.3270 |
| 0.08 | 2.5689 | 3.0081 | 3.0229 |
| 0.10 | 2.3659 | 2.7425 | 2.7427 |
| 0.12 | 2.1882 | 2.5125 | 2.5264 |
| $Pr = 0.7$ | | | |
| 0.00 | 3.6566 | 4.3649 | 4.3654 |
| 0.02 | 3.4163 | 4.1088 | 4.1125 |
| 0.04 | 3.1706 | 3.8036 | 3.8043 |
| 0.06 | 2.9377 | 3.4992 | 3.5009 |
| 0.08 | 2.7244 | 3.2163 | 3.2296 |
| 0.10 | 2.5323 | 2.9616 | 2.9691 |
| 0.12 | 2.3604 | 2.7354 | 2.7391 |
| $Pr = 0.8$ | | | |
| 0.00 | 3.6566 | 4.3649 | 4.3654 |
| 0.02 | 3.4657 | 4.1788 | 4.1790 |
| 0.04 | 3.2567 | 3.9243 | 3.9246 |
| 0.06 | 3.0497 | 3.6544 | 3.6616 |
| 0.08 | 2.8540 | 3.3932 | 3.3933 |
| 0.10 | 2.6733 | 3.1510 | 3.1526 |
| 0.12 | 2.5084 | 2.9309 | 2.9411 |

*Source:*   Sun, W. et al., *Int. J. Therm. Sci.*, 46, 1084, 2007.

as *Kn* increases and keeps unchanged as *Pr* increases. Still, *Kn* has more influence with the presence of viscous dissipation. In the thermal entrance region to the fully developed region, the system first reaches the fully developed condition as if there is no viscous heating. Then, at some point, *Nu* makes a jump to its final value. As *Br* increases, the jump occurs at a shorter distance from the entrance. Since the wall temperature is constant, the *Nu* values all converge to the same fully developed value, although the Nusselt number is greater for larger values of the Brinkman number in the thermal entrance region.

For constant heat-flux (CHF) boundary conditions with temperature jump, the fully developed Nusselt number decreases as *Kn* increases and increases as *Pr* increases. Since the definition of the Brinkman number is different for the CHF boundary condition case, a positive *Br* means that the heat is transferred to the fluid from the wall as opposed to the constant temperature case. The behavior of the system under linear wall temperature boundary conditions is similar to the behavior of the system under CHF boundary conditions. For both these cases, the Nusselt number approaches to the value of the CHF under the same conditions. This section can ensure that neglecting the temperature jump consideration can overpredict the Nusselt number.

## 12.5 Effects of Viscous Dissipation

For microchannel gas flow, the classical approach used for thermally developing flow is not consistent with the presence of fluid axial heat conduction [27]. Therefore, mathematically more consistent boundary conditions should be used in the analysis as stated by Shah and London [28], as shown in Figure 12.3. The fully developed velocity profile can be determined by solving momentum equations [17] together with the slip velocity boundary condition at the wall for the hydrodynamically fully developed, constant-property fluid flow in microtubes, as in Equation 12.9. 2D energy equation for hydrodynamically fully developed, thermally developing flow with constant properties can be written as in Equation 12.25 with boundary conditions as shown in Equations 12.37b and 12.37c, as well as

$$\text{at } r = R: \quad T = T_s, \tag{12.76a}$$

$$\text{at } r = 0: \quad \frac{\partial T}{\partial r} = 0, \tag{12.76b}$$

$$\text{at } x \rightarrow \infty: \quad \frac{\partial T}{\partial x} = 0, \tag{12.76c}$$

$$\text{at } x \rightarrow -\infty: \quad T = T_i, \tag{12.76d}$$

where $T_i$ and $T_s$ can represent the temperature of the gas at the inlet and at the surface, respectively. For CWT case, the energy equation can be nondimensionalized by the following dimensionless quantities as in Equations 12.41a through 12.41d. Since the boundary conditions at infinity are difficult to handle in numerical analysis, a coordinate transformation is performed [27]. By defining a new coordinate variable, $\omega$, suggested by Verhoff and Fisher [29], and Yang and Ebadian [30],

$$\xi = \tan\left(\frac{\pi}{2}\omega\right), \tag{12.77}$$

the energy equation and the boundary conditions become

$$\left[\frac{\bar{u}}{2}A(\omega) + \frac{B(\omega)}{Pe^2}\right]\frac{\partial\theta}{\partial\omega} = \frac{\partial^2\theta}{\partial\eta^2} + \frac{1}{\eta}\frac{\partial\theta}{\partial\eta} + \frac{A^2(\omega)}{Pe^2}\frac{\partial^2\theta}{\partial\omega^2} + \frac{16Br}{(1+8Kn)^2}\eta^2, \tag{12.78}$$

$$\text{at } \eta = 1: \quad \theta_s = -2\kappa Kn\left(\frac{\partial\theta}{\partial\eta}\right)_{\eta=1}, \tag{12.79a}$$

$$\text{at } \eta = 0: \quad \frac{\partial\theta}{\partial\eta} = 0, \tag{12.79b}$$

$$\text{at } \omega = 1: \quad \frac{\partial\theta}{\partial\omega} = 0, \tag{12.79c}$$

$$\text{at } \omega = -1: \quad \theta = 1, \tag{12.79d}$$

where

$$A(\omega) = \frac{2}{\pi}\cos^2\left(\frac{\pi}{2}\omega\right),$$                                    (12.79e)

$$B(\omega) = \frac{4}{\pi}\cos^3\left(\frac{\pi}{2}\omega\right)\sin\left(\frac{\pi}{2}\omega\right).$$          (12.79f)

The Nusselt number in Equation 12.45 is determined as

$$Nu_\varepsilon = \frac{h_\varepsilon D}{k} = -\frac{2(\partial\theta/\partial\eta)_{\eta=1}}{\theta_m}.$$       (12.80)

The temperature field is determined by solving the energy equation numerically. One can vary $Kn$ between 0 and 0.1, which are the applicability limits of the slip-flow regime. Parameter $\kappa$ is taken as 0 and 1.667. When $\kappa = 0$ that is a fictitious case, one can see the effect of the slip velocity in the absence of temperature jump. $\kappa = 1.667$ is a typical value for air, which is the working fluid in many engineering applications. To see the effect from higher $\kappa$ value, $\kappa = 10$ can be considered. $Pe$ number can vary between 1 and 50. One can consider $Pe$ as high as 1000 to demonstrate $Pe \to \infty$ case. From numerical simulations, one can obtain results as in Table 12.8 [15,27,28,31] and Figure 12.3. One can find that local $Nu$ tends to increase with decreasing $Pe$ as well as the fully developed $Nu$. This means the thermal entrance length also increases with decreasing $Pe$ due to finite axial conduction. All local $Nu$ tends to reach infinity at the origin, $x = 0$, for all $Pe$ numbers because the temperature gradient at the wall is infinite. The mean $Nu$ value in the entrance region increases with decreasing $Pe$ number. The thermal lengths for different $Nu$ numbers are approximately equal for the same $Pe$ numbers. In other words, the effect of $Pe$ number on the thermal entrance length is independent of the $Kn$ number. $Pe$ number only affects the thermal entrance length. By increasing $Pe$ number, the fully developed $Nu$ number increases, so does the mean $Nu$ number [27].

**TABLE 12.8**

Comparison of Fully Developed Nusselt Number for Flow in Microtube with CWT

| $K$ | $Pe=1$ | | $Pe=2$ | | $Pe=5$ | | $Pe=10$ | | $Pe\to\infty$ | | $Kn$ |
|---|---|---|---|---|---|---|---|---|---|---|---|
| | $Nu_\infty$ | $Nu_\infty^*$ | $Nu_\infty$ | $Nu_\infty^*$ | $Nu_\infty$ | $Nu_\infty^{**}$ | $Nu_\infty$ | $Nu_\infty^*$ | $Nu_\infty$ | $Nu_\infty^{***}$ | |
| 0 | 4.028 | 4.303 | 3.922 | 3.925 | 3.767 | 3.767 | 3.965 | 3.697 | 3.656 | 3.656 | 0.00 |
| 1.667 | 4.028 | 4.303 | 3.922 | 3.925 | 3.767 | 3.767 | 3.965 | 3.697 | 3.656 | 3.656 | 0.00 |
| 10 | 4.028 | 4.303 | 3.922 | 3.925 | 3.767 | 3.767 | 3.965 | 3.697 | 3.656 | 3.656 | 0.00 |
| 0 | 4.358 | — | 4.270 | — | 4.131 | — | 4.061 | — | 4.020 | 4.020 | 0.04 |
| 1.667 | 3.604 | | 3.517 | | 3.387 | | 3.325 | | 3.292 | 3.292 | 0.04 |
| 10 | 1.706 | | 1.678 | | 1.643 | | 1.630 | | 1.624 | 1.624 | 0.04 |
| 0 | 4.585 | — | 4.509 | — | 4.386 | — | 4.319 | — | 4.279 | 4.279 | 0.08 |
| 1.667 | 3.093 | | 3.036 | | 2.949 | | 2.909 | | 2.887 | 2.887 | 0.08 |
| 10 | 1.029 | | 1.021 | | 1.012 | | 1.009 | | 1.008 | 1.008 | 0.08 |

*Note:* $Nu_\infty$ are simulated results [27], $Nu_\infty^*$ are results from Shah and London [28], $Nu_\infty^{**}$ are results from Lahjomri and Qubarra [31], and $Nu_\infty^{***}$ are results from Cetin et al. [15].

## 12.6 Effects of Channel Roughness

To study channel roughness effect, the 2D laminar, viscous, incompressible, single-phase flow at steady state between parallel plates can be used to consider, and CWT condition can be assumed. The roughness may be modeled by adjusting triangular obstructions along the channel walls. The flow can be assumed to have uniform velocity and temperature at the channel inlet and in the slip-flow regime inside the channel. Channel surface roughness should be characterized by the relative roughness, $\varepsilon$, which is defined as

$$\varepsilon = \frac{e}{D_h},\tag{12.81}$$

where $e$ is the average height of the roughness elements. In the typical study, relative roughness values of 1.325% and 2.0% are considered for both continuum where $Kn = 0$ and slip-flow regimes where $0.001 < Kn < 0.1$, and relative roughness value of 2.65% can also be considered only for the continuum regime to study the effect of higher roughness value. Spacing and base angles of the triangular roughness elements can be specified according to degree of roughness, such as small angles and big spacing for low levels of roughness. Some literatures [32,33] also considered triangles as the roughness geometry for incompressible flow with no-slip boundary conditions and compressible flow.

Governing equations are the continuity, momentum, and energy equations, which can be written in vectorial form, when body forces are neglected, as in Equations 2.12, 2.34, and 2.52 [17]. The rarefaction effect is accounted for by imposing the slip velocity and temperature jump boundary conditions at the wall, Equations 12.6 and 12.15. To obtain the nondimensional governing equations, one can define dimensionless parameters similar to in Equation 12.41. Then, the boundary conditions in nondimensional form can be defined as in Equation 12.42.

The local Nusselt number can be defined by the use of nondimensional parameters as

$$Nu_x = \frac{D_h}{\theta_m}\left(\frac{\partial \theta}{\partial \eta}\right)_w.\tag{12.82}$$

The numerical methods can be applied to the previous governing equations and boundary conditions. The typical solutions for 45° triangular roughness with spacing, five times their height, are shown in Table 12.9 [34]; neglecting axial conduction in rough channels should yield underestimated results even at relatively large $Pe$ values with or without viscous dissipation [34]. For the continuum case, $Kn = 0$, when viscous dissipation is neglected and axial conduction is included, the triangular-shaped roughness reduces the local $Nu$ between the consecutive roughness elements compared to smooth channel fully developed values, due to lower gradient regions and increases at the peaks of the roughness elements. However, reduction is dominant at the rough section. For rarefied flow, surface roughness has an increasing effect on $Nu$, which is more obvious at low $Kn$ values and proportional to the relative roughness height. Maximum $Nu$ is observed at $Kn = 0.02$ among the considered $Kn$ values. As the flow becomes more rarefied, magnitude of this increase in $Nu$ becomes smaller [34]. When viscous dissipation is included, an increase in $Nu$ is observed both in smooth and rough channels, compared to cases where it is neglected. As the relative height of the roughness elements is increased, also the effect of viscous dissipation is found to be more significant at low-rarefied flow. Even in low velocity flows, surface roughness increases $Nu$ significantly in the presence of viscous dissipation as expressed in Table 12.3.

**TABLE 12.9**

Channel and Rough Section Averaged Nusselt Number Values for Various Mean
Velocities and Corresponding Brinkman Number Values at $Kn = 0.0665$ and $D_h = 1 \times 10^{-6}$

| $u_m$ (m/s) | $Br \times 10^7$ | Channel Averaged $Nu$ | | | Rough Section Averaged $Nu$ | | |
| --- | --- | --- | --- | --- | --- | --- | --- |
| | | Smooth | $\varepsilon = 1.325\%$ | $\varepsilon = 2.0\%$ | Smooth | $\varepsilon = 1.325\%$ | $\varepsilon = 2.0\%$ |
| 0.05 | 2 | 0.972 | 1.025 | 1.043 | 0.905 | 1.003 | 1.036 |
| 0.10 | 7 | 0.996 | 1.062 | 1.089 | 0.950 | 1.074 | 1.125 |
| 1.00 | 700 | 1.082 | 1.212 | 1.26 | 1.118 | 1.364 | 1.460 |
| 3.00 | 6,400 | 1.124 | 1.276 | 1.331 | 1.176 | 1.463 | 1.565 |
| 5.00 | 17,800 | 1.143 | 1.301 | 1.356 | 1.189 | 1.483 | 1.586 |
| 7.00 | 34,800 | 1.155 | 1.316 | 1.371 | 1.194 | 1.490 | 1.593 |
| 9.00 | 57,500 | 1.165 | 1.326 | 1.382 | 1.196 | 1.493 | 1.596 |
| 10.00 | 71,000 | 1.169 | 1.330 | 1.385 | 1.196 | 1.494 | 1.597 |
| 12.00 | 10,200,000 | 1.176 | 1.338 | 1.393 | 1.197 | 1.496 | 1.599 |
| 15.00 | 16,000,000 | 1.184 | 1.346 | 1.402 | 1.198 | 1.497 | 1.600 |
| 20.00 | 28,400,000 | 1.195 | 1.358 | 1.414 | 1.199 | 1.499 | 1.602 |
| 25.00 | 44,400,000 | 1.204 | 1.367 | 1.423 | 1.199 | 1.500 | 1.603 |
| 30.00 | 63,900,000 | 1.211 | 1.374 | 1.430 | 1.199 | 1.500 | 1.604 |

*Source:* Turgay, M.B. et al., Roughness effect on the heat transfer coefficient for gaseous flow in microchannels, *Proceedings of the International Heat Transfer Conference (IHTC14)*, Washington, DC, IHTC14-23142, August 8–13, 2010.

## 12.7 Effects of Variable Fluid Properties

When fluid properties are varied, thermal conductivity and viscosity, the steady-state, 2D, incompressible, laminar, and single-phase gas flow can be considered. The velocity profile and temperature jump develop as in Equations 12.7 and 12.15. In Equation 12.7, the thermal accommodation coefficient, $F_T$, is defined as in Equation 12.11 and may take a value in the range 0.0–1.0, depending on the gas and solid surface, the gas temperature and pressure, and the temperature difference between the gas and the surface, and is determined experimentally. Using the slip velocity boundary condition, the fully developed velocity profile is Equation 12.9 [26], where the Knudsen number from Equation 12.1 can be taken as $Kn = \lambda/2R$.

### 12.7.1 Constant Wall Temperature

For CWT boundary condition, the nondimensional energy equation and the boundary conditions for the flow inside a microtube, including axial conduction and viscous dissipation, can be rearranged from Equations 12.36 and 12.37 for cylindrical coordinate as

$$\frac{\bar{u}}{2}\frac{\partial \theta}{\partial \xi} = \frac{1}{\eta}\frac{\partial}{\partial \eta}\left(\eta \frac{\partial \theta}{\partial \eta}\right) + \frac{1}{Pe^2}\frac{\partial^2 \theta}{d\xi^2} + Br\left(\frac{\partial \bar{u}}{\partial \eta}\right)^2, \tag{12.83}$$

with all boundary conditions from Equations 12.42 and 12.43.

The energy equation can also be solved numerically [26] and analytically, using general eigenfunction expansion [35] as in the previous section. Using the temperature distribution, the local Nusselt number may be determined as in Equation 12.64 where $\theta_m$ is the nondimensional mean temperature defined by Equation 12.47.

### 12.7.2 Constant Wall Heat Flux

In this case, the nondimensional energy equation and the boundary conditions become [26]

$$\frac{Gz(1-\eta^2+4Kn)}{2(1+8Kn)}\frac{\partial\theta^*}{\partial\xi}=\frac{1}{\eta}\frac{\partial}{\partial\eta}\left(\eta\frac{\partial\theta^*}{\partial\eta}\right)+\frac{1}{Pe^2}\frac{\partial^2\theta^*}{\partial\xi^2}+\frac{32Br^*}{(1+8Kn)^2}\eta^2, \tag{12.84}$$

$$\text{at } \eta=0: \quad \frac{\partial\theta^*}{\partial\eta}=0, \tag{12.85a}$$

$$\text{at } \eta=1: \quad \frac{\partial\theta^*}{\partial\eta}=1, \tag{12.85b}$$

$$\text{at } \xi=0: \quad \theta^*=1. \tag{12.85c}$$

The boundary conditions, Equation 12.85, are almost the same as (12.54), excepting (12.85c).

The nondimensional terms have been used are as in Equations 12.49 and 12.51. The temperature profile is determined numerically, and using the temperature distribution, the local Nusselt number may be determined as [26]

$$Nu_x=\frac{h_xD}{k}=\frac{2}{\theta_s^*+(4\gamma/(\gamma+1))(Kn/Pr)-\theta_m^*}. \tag{12.86}$$

The variation of thermal conductivity, $k$, can be expressed by using a sixth-degree polynomial. The dimensional thermal conductivity is then nondimensionalized with a reference value $k_{ref}$. The equation for the polynomial is given in the following:

$$k(T)=(1.035\times10^{-18}T^6)-(3.447\times10^{-15}T^5)+(3.627\times10^{-12}T^4)$$

$$-(1.071\times10^{-10}T^3)-(2.985\times10^{-6}T^2)+(4.178\times10^{-3}T)-2.212\times10^{-3}. \tag{12.87}$$

Variation of viscosity is modeled using Sutherland's formula in this study. Sutherland's formula is based on the kinetic theory. The formula is given in the following:

$$\mu(T)=\mu_0\frac{T_0+C}{T+C}\left(\frac{T}{T_0}\right)^{3/2}, \tag{12.88}$$

where
$T_0$ is the reference temperature, at 273 K
$C$ is Sutherland's coefficient and equal to 111 K for air
$T$ is the input temperature

The nondimensional $x$- and $y$-momentum and energy equations for unsteady flow are given in the following:

Momentum equation:

$$\frac{\partial \bar{u}}{\partial t} + \frac{\partial (\bar{u}^2)}{\partial x} + \frac{\partial (\bar{u}\bar{v})}{\partial y} = -\frac{\partial \bar{P}}{\partial x} + \frac{1}{Re}\left( \frac{\partial^2 \bar{u}}{\partial x^2} + \frac{\partial^2 \bar{u}}{\partial y^2} + 2\frac{\partial \bar{\mu}}{\partial x}\frac{\partial \bar{u}}{\partial x} + \frac{\partial \bar{\mu}}{\partial y}\frac{\partial \bar{v}}{\partial x} + \frac{\partial \bar{\mu}}{\partial y}\frac{\partial \bar{u}}{\partial y} \right) \tag{12.89}$$

$$\frac{\partial \bar{v}}{\partial t} + \frac{\partial (\bar{v}^2)}{\partial y} + \frac{\partial (\bar{u}\bar{v})}{\partial x} = -\frac{\partial \bar{P}}{\partial y} + \frac{1}{Re}\left( \frac{\partial^2 \bar{v}}{\partial x^2} + \frac{\partial^2 \bar{v}}{\partial y^2} + 2\frac{\partial \bar{\mu}}{\partial y}\frac{\partial \bar{v}}{\partial y} + \frac{\partial \bar{\mu}}{\partial x}\frac{\partial \bar{v}}{\partial x} + \frac{\partial \bar{\mu}}{\partial x}\frac{\partial \bar{u}}{\partial y} \right) \tag{12.90}$$

Energy equation:

$$\frac{\partial \theta}{\partial t} + \bar{u}\frac{\partial \theta}{\partial x} + \bar{v}\frac{\partial \theta}{\partial y} = \frac{1}{RePr}\left( \frac{\partial \bar{k}}{\partial x}\frac{\partial \theta}{\partial x} + \bar{k}\frac{\partial^2 \theta}{\partial x^2} + \frac{\partial \bar{k}}{\partial y}\frac{\partial \theta}{\partial y} + \bar{k}\frac{\partial^2 \theta}{\partial y^2} \right)$$

$$+ \frac{Br}{Re\,Pr}\left\{ 2\bar{\mu}\left[ \left(\frac{\partial \bar{u}}{\partial x}\right)^2 + \left(\frac{\partial \bar{v}}{\partial y}\right)^2 + \left(\frac{\partial \bar{u}}{\partial y} + \frac{\partial \bar{v}}{\partial x}\right) \right] \right\} \tag{12.91}$$

where

$$\bar{k} = \frac{k(T)}{k_{ref}} \tag{12.92}$$

$$\bar{\mu} = \frac{\mu(T)}{\mu_0}. \tag{12.93}$$

Thus, the variation of thermal conductivity and viscosity with temperature, as shown in Equations 12.87 and 12.88, respectively, is accounted for. On the other hand, in the definitions of the Peclet and Brinkman numbers, the viscosity and thermal conductivity at the reference temperature are utilized; thus, $Pe$ and $Br$ are not temperature-variable. Energy and momentum equations are solved in a coupled manner to account for the viscosity variation. Coupled solutions are made for pressure and velocity for investigating simultaneously developing flow. Variation of specific heat, $c$, and density, $\rho$, with temperature is not included, since these properties vary in negligible amounts within the studied temperature range [36].

Results should be compared with constant-property results for microchannels, with the effects of rarefaction and viscous dissipation, and be obtained numerically as in Table 12.10 [36,37] and Figure 12.4 [36]. For high values of rarefaction, high $Kn$ and temperature jump, high $\kappa$, the effect of axial conduction is negligible. However, for lower rarefaction and temperature jump values, as $Pe$ decreases, axial conduction effect increases, the fully developed $Nu$ increases more significantly. It may be concluded from these observations that the effect of axial conduction should not be neglected for low-rarefied flows and with low values of temperature jump [36]. Regardless of the effect of axial conduction, for a given $Kn$ and $\kappa$ value, the flow reaches the same fully developed $Nu$ value for all values of $Br$.

**TABLE 12.10**

Laminar Flow Fully Developed $Nu$ Values for Both CWT ($Nu_T$) and Constant Wall Heat-Flux ($Nu_q$) Cases (and $Nu_T^*$ and $Nu_q^*$ Are Results from Tunc and Bayazitoglu [37] at $Pr = 0.6$ [36])

| $Kn$ | $Nu_T^*$ | $Nu_T$ | $Nu_q^*$ | $Nu_q$ |
|------|----------|--------|----------|--------|
| 0.00 | 3.6751 | 3.6566 | 4.3627 | 4.3649 |
| 0.02 | 3.3675 | 3.3527 | 3.9801 | 4.0205 |
| 0.04 | 3.0745 | 3.0627 | 3.5984 | 3.6548 |
| 0.06 | 2.8101 | 2.8006 | 3.2519 | 3.3126 |
| 0.08 | 2.5767 | 2.5689 | 2.9487 | 3.0081 |
| 0.10 | 2.3723 | 2.3659 | 2.6868 | 2.7425 |

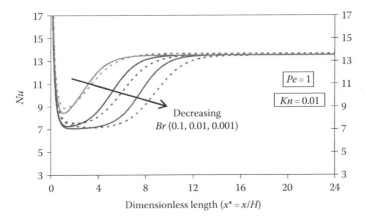

**FIGURE 12.4**
Variation of the local Nusselt numbers with axial position for constant (solid lines) and variable (dashed lines) property solutions for different Brinkman numbers. (From Kakaç, S. et al., *Heat Mass Transfer*, 47, 879, 2011.)

When the fluid is cooled, $Br > 0$ for CWT and $Br < 0$ for constant wall heat flux, $Nu$ takes higher values. The increase in fully developed $Nu$ value with the added effect of viscous dissipation suggests that this effect should not be neglected for long channels. Since microconduits have high length-to-diameter ratios, even for low values of $Br$, viscous dissipation effect must be considered [36].

In general, for CWT and constant wall heat-flux conditions, velocity slip and temperature jump affect the heat transfer in opposite ways: a large slip on the wall will increase the convection along the surface. On the other hand, a large temperature jump will decrease the heat transfer by reducing the temperature gradient at the wall. Therefore, neglecting temperature jump will result in the overestimation of the heat-transfer coefficient. When viscous dissipation is neglected, the effect of axial conduction should be included for $Pe < 100$. When viscous dissipation is included in the analysis, axial conduction is significant for $Pe < 100$ for short channels. The variation of thermophysical properties affects the temperature profile. For both fluid heating and cooling cases, the variation in local $Nu$ due to temperature-variable properties is significant in the developing region [36]. The fully developed $Nu$ is almost invariant for constant and variable properties cases due to reduced temperature gradients in this region. In addition, the amounts of axial conduction, viscous dissipation, and property variation do not significantly affect the fully developed Nusselt number values. Property variation effects should be included in numerical solutions of airflow in microchannels when the entry length is comparable with the channel length. However, constant-property assumption will yield satisfactory results for long channels, in which entry length is negligible compared to the overall channel length.

## 12.8 Empirical Correlations for Gaseous Forced Convection in Microchannels

### 12.8.1 Nusselt Number Correlations

#### 12.8.1.1 Laminar and Turbulent Regimes

For the engineering applications, the prediction of the heat-transfer coefficient and the friction factor are important. From the early 1980s, several researches have been conducted [38]. Wu and Little [39] tested the heat-transfer characteristics of gases flowing in the trapezoidal silicon/glass microchannels of widths 130–300 μm and depths of 30–60 μm, involving both laminar and turbulent flow regimes. The experimental studies found that average Nusselt numbers are higher than those predicted by the conventional correlations for fully developed laminar flows and for fully developed turbulent flows. They also indicated that transition from laminar to turbulent flow occurs at low Reynolds numbers of 400–900 depending on the various test configurations. Therefore, it is reported that the heat transfer is improved because of early transition in flow regime. As a result of their research, a Nusselt number correlation was proposed for trapezoidal microchannels as

$$Nu = 0.00222 Pr^{0.4} Re^{1.08} \quad \text{for } Re > 3000. \tag{12.94}$$

Choi et al. [40] conducted experimental data on the Nusselt numbers for both turbulent and laminar regimes of nitrogen gas, for Reynolds numbers between the range of 50 and 20,000. In this study, microtubes with diameters ranging from 3 to 81 μm and all flows were fully developed for both hydraulically and thermally. They found that their experimental results differed from the conventional theory, the conventional correlations. Although the heat-transfer coefficients in the conventional correlations were independent from the Reynolds number for fully established laminar flow, it was reported by Choi et al. [40,74] that heat-transfer coefficients depended on the Reynolds number in laminar flow. Moreover, for turbulent flow in microtubes, heat-transfer coefficients, measured from the experiments, were larger than those predicted by conventional correlations. Two new correlations for the average Nusselt number in laminar and turbulent regime were proposed:

$$Nu = 0.000972 Re^{1.17} Pr^{1/3} \quad \text{for } Re < 2000, \tag{12.95}$$

$$Nu = 3.82 \times 10^{-6} Re^{1.96} Pr^{1/3} \quad \text{for } 2,500 < Re < 20,000. \tag{12.96}$$

In this study, it was also reported that the ratio of micro- to macroturbulent Nusselt numbers was obtained as a function of the Reynolds number; $Nu_{micro}/Nu_{macro} = 0.000166 Re^{1.16}$. However, those correlations proposed by Choi et al. [40] are not in agreement with the correlation proposed by Wu and Little [39].

Yu et al. [41] investigated the heat transfer in microtubes with diameters 19, 52, and 102 μm. They used dry nitrogen gas and water as their working fluids in the experiments. They studied gaseous flow in microtubes for Reynolds numbers between the range of 6,000 and 20,000 and for Prandtl numbers between the range of 0.7 and 5. It was reported that the heat-transfer coefficient was enhanced in turbulent flow regime. As a result, they proposed a correlation for the Nusselt number in microtubes as in the following form:

$$Nu = 0.007 Re^{1.2} Pr^{0.2} \quad \text{for } 6000 < Re < 20000. \tag{12.97}$$

Peng and Peterson [18] studied the effect of dimensions of the microchannel geometry on heat transfer for laminar and turbulent regimes and reported that Nusselt number depends on Reynolds number, Prandtl number, and the microchannel aspect ratio. Peng and Peterson determined that geometric configuration of the microchannel was a critical parameter for heat-transfer phenomena. They also proposed two correlations for laminar and turbulent regime:

$$Nu = 0.1165(D_h/W_c)^{0.81}(H/W)^{0.79} Re^{0.62} Pr^{1/3} \quad \text{for } Re < 2200, \tag{12.98}$$

$$Nu = 0.072(D_h/W_c)^{1.15}(1 - 2.421(z - 0.5)^2) Re^{0.8} Pr^{1/3} \quad \text{for } Re > 2200, \tag{12.99}$$

where
   $H$ and $W$ are the height and width of the microchannels, respectively
   $W_c$ is the center-to-center distance of microchannels

$z = \min(H,W)/\max(H,W)$, for $z = 0.5$ value, turbulent convective heat transfer is maximum regardless of $H/W$ or $W/H$.

Ameel et al. [42] studied hydrodynamically fully developed laminar gaseous flow in microtubes analytically. They solved the problem under CHF boundary condition. They used two approximations in order to simplify the problem. First, temperature jump boundary condition was not taken into account directly. Secondly, the effect of momentum and thermal accommodation coefficients were assumed to be united. The 2D and steady-state flow was considered to be incompressible with constant physical properties. In addition, viscous effects were neglected. They developed a relation for the Nusselt number by considering the previous work of Baron et al. [20]:

$$Nu = \frac{48(2\beta - 1)^2}{(24\beta^2 - 16\beta + 3)\left[1 + \dfrac{24\gamma(\beta - 1)(2\beta - 1)^2}{(24\beta^2 - 16\beta + 3)(\gamma + 1)Pr}\right]}. \tag{12.100}$$

In Equation 12.100, $\beta$ is defined as $\beta = 1 + 4Kn$. They determined that the Nusselt number decreases with increasing Knudsen number. It can be said that rarefaction effects decrease the Nusselt number. Moreover, they developed a relation between Knudsen number and the entrance length as

$$x_e^* = 0.0828 + 0.14 Kn^{0.69}. \tag{12.101}$$

It was also reported that with the increasing rarefaction effects thermal length increases too. For $Kn = 0$, Equation 12.100 reduces to $Nu = 4.364$, which is the Nusselt number under fully developed conditions.

**Example 12.1**

Using the *Nu* number expression given by Equation 12.100, calculate the fully developed heat-transfer coefficient of air for CWT and constant wall heat-flux boundary conditions for air in engineering applications with *Pr* equals to 0.7, the ratio between $c_p$ and $c_v$ equals to 1.4, and CHF boundary conditions when *Kn* equals to 0.01 and 0.1.

**Solution**
Properties of air, from Appendix B (Table B.1), are

$$k = 0.0271 \text{ W/m K.}$$

The Nusselt correlation for CHF condition is

$$Nu = \frac{48(2\beta - 1)^2}{(24\beta^2 - 16\beta + 3)\left[1 + \dfrac{24\gamma(\beta - 1)(2\beta - 1)^2}{(24\beta^2 - 16\beta + 3)(\gamma + 1)Pr}\right]}.$$

For *Kn* = 0.01,

$$\beta = 1 + (4 \times Kn) = 1 + (4 \times 0.01) = 1.04,$$

$$Nu = \frac{48 \times (2 \times 1.04 - 1)^2}{(24 \times 1.04^2 - 16 \times 1.04 + 3) \times \left[1 + \dfrac{24 \times 1.4 \times (1.04 - 1) \times (2 \times 1.04 - 1)^2}{(24 \times 1.04^2 - 16 \times 1.04 + 3) \times (1.4 + 1) \times 0.7}\right]}.$$

$$= 4.2250$$

So the heat-transfer coefficient for the CHF condition is

$$h = \frac{Nu \times k}{d} = \frac{4.2250 \times 0.0271}{0.24 \times 10^{-3}} = 447.07 \text{ W/m}^2 \text{ K.}$$

For *Kn* = 0.1,

$$\beta = 1 + (4 \times Kn) = 1 + (4 \times 0.1) = 1.4,$$

$$Nu = \frac{48 \times (2 \times 1.4 - 1)^2}{(24 \times 1.4^2 - 16 \times 1.4 + 3) \times \left[1 + \dfrac{24 \times 1.4 \times (1.4 - 1) \times (2 \times 1.4 - 1)^2}{(24 \times 1.4^2 - 16 \times 1.4 + 3) \times (1.4 + 1) \times 0.7}\right]}.$$

$$= 2.9037$$

So the heat-transfer coefficient for the CHF condition is

$$h = \frac{Nu \times k}{d} = \frac{2.9037 \times 0.0271}{0.24 \times 10^{-3}} = 327.88 \text{ W/m}^2 \text{ K.}$$

Yu and Ameel [43] used the same integral transform technique with Bayazitoglu et al. [6] obtaining the Nusselt number for rectangular microchannels. In their solutions, they used uniform temperature boundary condition. They also included slip-flow effects, but viscous effects were neglected. It was reported that the Knudsen number, Prandtl number, aspect ratio, and temperature jump may cause the Nusselt number to deviate from the conventional value. By integrating the studies of Bayazitoglu et al. [6] and Yu and Ameel [43], the results for the fully developed Nusselt number under uniform heat flux (UHF) or uniform temperature boundary conditions were given for different geometries in Table 5.3, in the calculations, viscous effects were neglected ($Br = 0$).

Renksizbulut et al. [44] studied rarefied gas flow and heat transfer in the entrance region of rectangular microchannels. They made investigations in the slip-flow regime numerically. Both velocity slip and temperature jump boundary conditions were considered in the study. It was reported that the Nusselt number decreases in the entrance region due to rarefaction effects. For different channel, aspect ratios and $Kn \leq 0.1$ with the range of $0.1 \leq Re \leq 1000$ and $Pr = 1$. Under these conditions, they also modified a correlation of earlier study of Renksizbulut et al. [45] for trapezoidal and rectangular microchannels under fully developed conditions.

$$(Nu)_{fd} = \left[ 2.87\left(\frac{90°}{\theta}\right)^{-0.26} + 4.8\exp\left(-3.9\gamma\left(\frac{90°}{\theta}\right)^{0.21}\right)\right] G_2 G_3, \qquad (12.102a)$$

$$G_2 = 1 + 0.075(1+\gamma)\exp(-0.45Re), \qquad (12.102b)$$

$$G_3 = 1 - 1.75Kn^{0.64}(1 - 0.72\tanh(2\gamma)), \qquad (12.102c)$$

where
  $\theta$ is the side angle
  $\gamma$ is defined as the aspect ratio of the short side to the long side of the channel

For fully developed flows, an expression for the friction coefficient was also obtained as

$$f\,Re = 13.9\left(\frac{90°}{\theta}\right) + 10.4\exp\left[-3.25\gamma\left(\frac{90°}{\theta}\right)^{0.23}\right], \qquad (12.103)$$

where $f$ is the Fanning friction factor defined as in Equation 6.33.

Table 12.11 [8,38] shows the fully developed Nusselt number values for rectangular and circular channels for various values of the Knudson number under two boundary conditions.

### 12.8.1.2 Laminar to Turbulent Transition Regime

As mentioned in the earlier sections in this chapter, the molecular approach is not enough to express the transition flow regime, and the effect of the Knudsen number becomes more important. The Boltzmann equation should be used in order to investigate the atomic-level flows in transition regime. Boltzmann equation can be solved by several methods such as molecular dynamics (MD), DSMC. In addition, the simplified Boltzmann equation can be solved by using lattice Boltzmann method (LBM). However, the current computational methods (MD and DSMC) cannot provide an effective solution for transition flow regime [10]. Thermal creep is also an important phenomenon for transition flow regime [14].

Hadjiconstantinou et al. [46] investigated the convective heat-transfer characteristics of a gaseous flow by using a 2D model in microchannels. They focused on monoatomic gases and used the simplest monoatomic gas model, the hard sphere gas. The simulations were performed by using DSMC for $0.02 < Kn < 2$. They found a weak dependence

**TABLE 12.11**

Nusselt Number for Different Geometries Subject to Slip Flow

| Br = 0.0 | | $Kn = 0.00$ | | $Kn = 0.04$ | | $Kn = 0.08$ | | $Kn = 0.12$ | |
|---|---|---|---|---|---|---|---|---|---|
| | | $Nu_T$ | $Nu_q$ | $Nu_T$ | $Nu_q$ | $Nu_T$ | $Nu_q$ | $Nu_T$ | $Nu_q$ |
| Cylindrical | | 3.67 | 4.36 | 3.18 | 3.75 | 2.73 | 3.16 | 2.37 | 2.68 |
| | $\gamma = 1$ | 2.98 | 3.10 | 2.71 | 2.85 | 2.44 | 2.53 | 2.17 | 2.24 |
| | $\gamma = 0.84$ | 3.00 | 3.09 | 2.73 | 2.82 | 2.46 | 2.48 | 2.19 | 2.17 |
| Rectangular aspect ratio $\gamma = a/b$ | $\gamma = 0.75$ | 3.05 | 3.08 | 2.77 | 2.81 | 2.49 | 2.44 | 2.22 | 2.12 |
| | $\gamma = 0.5$ | 3.39 | 3.03 | 2.92 | 2.71 | 2.55 | 2.26 | 2.24 | 2.18 |
| | $\gamma = 0.25$ | 4.44 | 2.93 | 3.55 | 2.42 | 2.89 | 1.81 | 2.44 | 1.68 |
| | $\gamma = 0.125$ | 5.59 | 2.85 | 4.30 | 1.92 | 3.47 | 1.25 | 2.80 | 1.12 |
| Two parallel plates | | 7.54 | 8.23 | 6.26 | 6.82 | 5.29 | 5.72 | 4.56 | 4.89 |

*Source:* Bayazitoglu, Y. and Kakaç, S., Flow regimes in microchannel single-phase gaseous fluid flow, in *Microscale Heat Transfer-Fundamentals and Applications*, S. Kakaç et al., (Eds.), Kluwer Academic Publishers, Dordrecht, the Netherlands, pp. 75–92, 2005; Kakaç, S. et al., *Heat Exchangers: Selection, Rating, and Thermal Design*, 3rd edn., CRC Press, Boca Raton, FL, 2012.

of the Nusselt number on the Peclet number and introduced a dependence of the Nusselt number on the Peclet number as

$$Nu_T^s(Pe, Kn = 0) = 8.11742(1 - 0.0154295Pe + 0.0017359Pe^2), \qquad (12.104)$$

where $Nu_T^s$ represents the slip-flow CWT Nusselt number.

Several experimental researches have also been conducted in order to understand the behaviors in the transition flow regime. Wu and Little [47] tested the gaseous transition flow regime, occurring at the Reynolds number ranging from 1000 to 3000 in silicon and glass microchannels. They considered silicon microchannels being smooth and glass microchannels having high value of relative roughness. Therefore, the roughness of the microchannel had impacts on the transition. Acosta et al. [48] carried on experiments of isothermal gas flows in rectangular microchannels with very small aspect ratios, $0.019 < \gamma < 0.05$, in the transition regime at the Reynolds number about 2770. The results indicated very similar behaviors with respect to conventional transition. Some literature surveys, experimental and numerical results, for the transition gaseous flow in microchannels were summarized in Table 12.12 [38].

Microscale heat transfer has been an important issue for researchers since there are more miniature equipments. The conclusion drawn for microscale gaseous flow can be summarized as follows [38]:

1. The behavior of heat transfer in microchannels is very different when compared with that of conventional-sized situation.

2. Convective heat transfer in microchannels is highly dependent on the Knudsen number, Prandtl number, Brinkman number, size and geometry of microchannels.

3. Slip velocity and temperature jump affects heat transfer in microchannels while velocity slip increases the heat transfer, temperature jump reduces the heat transfer by reducing the temperature gradient at the wall. The magnitude of temperature jump is directly dependent on the Prandtl number.

**TABLE 12.12**

Experimental Data on the Laminar to Turbulent Flow Transition
in Microchannels

| Literatures | Cross Section | $\varepsilon/D_h$ | $\gamma$ | Critical Reynolds |
|---|---|---|---|---|
| Wu and Little [47] | Trap. | — | 0.444 | 1000–3000 |
| Acosta et al. [48] | Rect. | — | 0.019, 0.033, 0.05 | 2770 |
| Choi et al. [40] | Circ. | — | — | 500–2000 |
| Li et al. [49] | Circ. | — | — | 2300 |
| Li et al. [50] | Circ. | 0.04 | — | 1700–2000 |
| Yang et al. [51] | Circ. | — | — | 1200–3800 |

*Source:* Kakaç, S. et al., *Heat Exchangers: Selection, Rating, and Thermal Design*, 3rd edn.,
CRC Press, Boca Raton, FL, 2012.

4. Rarefaction effects play an important role especially in gaseous flows in microchannels, and the degree of rarefaction is defined by the Knudsen number. The
Knudsen number also determines the degree of deviation from conventional
model. As the Knudsen number increases, a reduction in the Nusselt number and
heat transfer occurs.

5. Viscous heat dissipation is a significant factor for microchannel heat transfer. The
viscous heat dissipation effects are defined by the Brinkman number.

6. In microchannels, the Nusselt number can be correlated as a function of the
Brinkman, Reynolds, Prandtl, and geometric parameters of microchannels.

7. According to experimental review for correlations of the Nusselt number, in the
laminar regime, the Nusselt number increases with the Reynolds number [39,40].
However, in different study, it is stated that the Nusselt number decreases with
increasing the Reynolds number in laminar regime [18].

8. In microchannels, the friction factor can be correlated as a function of Reynolds
number and geometric parameters of microchannels.

9. Experimental review for friction number can be summarized that in several studies, the friction factor for fully developed flow is found to be lower than conventional value [41,52–56]. However, in other studies the friction factor is found to be
lower than conventional value [39,49].

10. For gaseous fully developed laminar flow, friction factor decreases with the
increasing Knudsen number [56,57].

11. In several experimental works, it was concluded that the friction factor depends
on the material and relative roughness of microchannel [39,41,52–54].

12. The critical Reynolds number for transition regime is affected by the fluid temperature, velocity, and geometric parameters of microchannel.

13. Axial conduction, as thermal creep, should be considered for transition flow
regime.

14. In transition flow regime, the critical Reynolds numbers depend on wall roughness in microchannels [39].

15. The critical Reynolds numbers decrease with the microchannel hydraulic diameter [40,58].

### 12.8.2 Friction Factor Correlations

The product of the Reynolds number and the friction factor should be constant for laminar flow and represented as the Poiseuille number:

$$Po = 4f \, Re, \tag{12.105a}$$

and for laminar flow in circular tubes,

$$fRe = 16. \tag{12.105b}$$

The Poiseuille number can be alternately defined with the Darcy coefficient

$$Po = \lambda_D \, Re. \tag{12.106}$$

The frictional pressure drop can be obtained by using the friction factors, and the Nusselt numbers for fully developed laminar flow for different duct geometries are given in Tables 6.1 and 12.13 [8,38]. The frictional pressure drop can be calculated from Equation 6.33.

Various experimental investigations, in microchannels, reported the variation of $fRe$ for different Reynolds numbers, exhibit increasing and decreasing trends, both in laminar and turbulent flow regimes. From Wu and Little [39], as mentioned earlier, nitrogen, hydrogen, and argon gases were used as working fluids. They found the friction factor to be 10%–30% larger than those predicted by the Moody chart. A result of great relative roughness and asymmetric distribution of the surface roughness on the microchannel walls causes the

**TABLE 12.13**

Nusselt Number for Ellipse Subject to Slip Flow

| Duct Shape | Ellipse Major Axis/ Minor Axis ($a/b$) | $Nu_H$ | $aNu_T$ | $Po$ |
|---|---|---|---|---|
| | 1 | 4.36 | 3.66 | 64.00 |
| | 2 | 4.56 | 3.74 | 67.28 |
| | 4 | 4.88 | 3.79 | 72.96 |
| | 8 | 5.09 | 3.72 | 76.60 |
| | 16 | 5.18 | 3.65 | 78.16 |

*Source:* Bayazitoglu, Y. and Kakaç, S., Flow regimes in microchannel single-phase gaseous fluid flow, in *Microscale Heat Transfer-Fundamentals and Applications*, S. Kakaç et al., (Eds.), Kluwer Academic Publishers, Dordrecht, the Netherlands, pp. 75–92, 2005; Kakaç, S. et al., *Heat Exchangers: Selection, Rating, and Thermal Design*, 3rd edn., CRC Press, Boca Raton, FL, 2012.

*Note:* $Nu = hD_h/k$; $Re = \rho u_m D_h/\mu$; $Nu_H$ is the Nusselt number under a CHF boundary condition, constant axial heat flux, and circumferential temperature; $Nu_T$ is Nusselt number under a CWT boundary condition.

different predictions. They proposed correlations for various Reynolds numbers and flow regimes in order to calculate friction factors:

$$f = (110 \pm 8)Re^{-1} \quad \text{for } Re \le 900, \tag{12.107a}$$

$$f = 0.165(3.48 - \log Re)^{2.4} + (0.081 \pm 0.007) \quad \text{for } 900 < Re < 3000, \tag{12.107b}$$

$$f = (0.195 \pm 0.017)Re^{-0.11} \quad \text{for } 3,000 < Re < 15,000. \tag{12.107c}$$

The pressure drop can also be calculated from Equation 6.33. Acosta et al. [48] studied friction factors for rectangular microchannels with a hydraulic diameter, ranging between 368.9 and 990.4 µm. The rectangular channels used in the investigations had small aspect ratios ($0.019 < \gamma < 0.05$). All channels with different aspect ratios showed similar friction factor values. Therefore, as conclusion, this behavior was consistent with the conventional asymptotic behavior of the Poiseuille number (*fRe*) for channels with small aspect ratios.

Pfhaler et al. [52–54] investigated the friction factor for liquid and gas flows in microchannels, 1.6–3.4 µm hydraulic diameters (the rectangular channels 100 µm wide and 0.5–50 µm deep) with Reynolds numbers varying between 50 and 300. They found that their friction factors were lower than those from the conventional theory. Their friction factor results were opposite to the results of Wu and Little [39]. In this study, it was also reported that the viscosity becomes size dependent at microscales. This work concluded that small friction value for fluids was due to the reduction of viscosity with decreasing size and rarefaction effects.

Choi et al. [40] investigated the friction factor in silica microtubes with diameters of 3, 7, 10, 53, and 81 µm. Nitrogen gas was used as the operating fluid, which was conducted for the Reynolds numbers between the range of 30 and 20,000. It was reported that no variation in friction factor with the wall roughness and the Poiseuille number was lower than the conventional value. They proposed the following correlations for different Reynolds numbers:

$$f = \frac{64}{Re} \left[ 1 + 30 \left( \frac{\nu}{dC_a} \right) \right]^{-1} \quad \text{for } Re < 2300, C_a = 30 \pm 7, \tag{12.108a}$$

$$f = 0.14Re^{-0.192} \quad \text{for } 4,000 < Re < 18,000. \tag{12.108b}$$

Arklic et al. [57] investigated the flow in rectangular silicon microchannel with a hydraulic diameter of 2.6 µm. Helium and nitrogen gases were used as working fluids. This work reported that lower friction factors were obtained in microchannels compared to conventional results for macrochannels. Their models represented the relationship between mass flow rate and pressure by including a slip-flow boundary condition at the wall.

Pong et al. [58] measured the pressure along the 4.5 mm length of rectangular microchannels 5–40 µm wide and 1.2 µm deep for helium and nitrogen flows. They found that pressure distribution in the channel was nonlinear, this fact was caused by a result of rarefaction and compressibility effects as their conclusion. Liu et al. [59] also studied pressure drop along the rectangular microchannels, and helium was also used. The procedure of the Pong et al. [58] was recalled. But this work found that the pressure drop was smaller than the conventional values. These findings were completed by Shih et al. work [60] that used nitrogen and helium as the working fluids. It was concluded that when inlet pressure is lower than 0.25 MPa, the pressure drop can be predicted by slip-flow model.

Yu et al. [41] experimentally studied on silica microtubes with diameters of 19.52 and 102 µm to investigate both gas (nitrogen) and liquid flows in the range of Reynolds numbers 250 and 20,000. They concluded that friction factors were lower than conventional macroscale predictions. They also investigated how the tube roughness affects the flow. They proposed the following correlations in order to calculate friction factor defined as Equation 6.33 for laminar and transition flow regimes.

$$f = 50.13/Re \quad \text{for } Re < 2000, \tag{12.109a}$$

$$f = 0.302/Re^{0.25} \quad \text{for } 2000 < Re < 6000. \tag{12.109b}$$

Harley et al. [55] investigated the flow experimentally in trapezoidal silicon microchannels with 100 µm in width and 0.5–20 µm in depth. Nitrogen, helium, and argon gases were used as working fluids. The hydraulic diameters of the microchannels were in the range between 1.01 and 35.91 µm and the aspect ratios between the range of 0.0053 and 0.161. Compressibility and rarefaction effects on the Poiseuille number were investigated and concluded that friction factors were lower than predicted according to conventional theory. They also explained all these observations by using the first-order slip model for isothermal fully developed flow.

Araki et al. [56] investigated the flow characteristics of nitrogen and helium along three different trapezoidal microchannel geometries with hydraulic diameter differs from 3 and 10 µm and reported that the investigated friction factors were also lower than that predicted by the conventional theory caused by a result of rarefaction effects. Li et al. [49] examined friction factor for the nitrogen gas flow in five different microtubes with diameters from 80 to 166.6 µm. They found a nonlinear distribution of the pressure along microtubes when the Mach number exceeded 0.3. However, the friction factors in this work were higher than those predicted by the conventional theory.

Yang et al. [51] investigated the airflow in circular microtubes with diameters ranging between 173 and 4010 µm and lengths ranging from 15 to 1 mm. From their results, the laminar regime friction factors were in a good agreement with conventional theory, but the turbulent friction factors were lower than those predicted by conventional correlation for macrotubes.

Lalonde et al. [61] experimentally studied the friction factor for airflow in microtube with a diameter of 52.8 µm. The friction factors obtained from the experiments indicated a good agreement with the predictions of conventional theory. Turner et al. [62] experimentally investigated compressible gas flow, nitrogen, helium, and air, in rough and smooth rectangular microchannels with hydraulic diameter ranging between 4 and 100 µm. They only investigated the laminar flow regime for the Reynolds numbers between the ranges of 0.02 and 1000. The friction factors obtained in these experiments were in agreement with the conventional theory and Lalonde et al. [61].

Hsieh et al. [63] examined the characteristics of nitrogen flow in a rectangular microchannel with the length of 24 mm, width of 200 µm, and depth of 50 µm, between the range of the Reynolds number 2.6 and 89.4 and the range of the Knudsen number from 0.001 to 0.02. Their friction factors were lower than that was predicted by the conventional theory due to compressibility and rarefaction effects. Moreover, they proposed the following correlation for friction factor defined as in the form of Equation 6.33:

$$f = 48.44/Re^{1.02} \quad \text{for } 2.6 < Re < 89.4. \tag{12.110}$$

Tang and He [64] also investigated on friction factors in silica microtubes and square microchannels with hydraulic diameters ranging from 50 to 210 µm. The measured friction factors were in a good agreement with conventional predictions, but it was noted that, in smaller microtubes, compressibility, roughness, and rarefaction effects cannot be neglected. Celata et al. [65] investigated compressibility effects of helium flow, along the microtubes with diameters between 30 and 254 µm and high $L/D$ ratios, on the friction factor. They found that the quantitative behavior was predicted by the incompressible theory, and the frictional losses at the inlet and outlet did not affect to the total pressure drop significantly.

Some outstanding observed literatures concerning the friction factors were summarized in Table 12.14 [1,15,38]. It can be noted that there is no general solution for the friction factor behavior in microchannels. It can also be concluded that, at the lower values of hydraulic

**TABLE 12.14**

Experimental Results on Friction Factor for Gas Flows in Microchannels

| Literatures | *fRe* | $D_h$ (µm) | Cross Section | Test Fluids |
|---|---|---|---|---|
| Wu and Little [39] | ↑↑ | 55.8–72.4 | Trap. | $N_2$, $H_2$, Ar |
| Acosta et al. [48] | ≈ | 368.9–990.4 | Rect. | He |
| Phfaler et al. [52–54] | ↓↓ | 1.6–65 | Rect.-Trap | $N_2$, He |
| Choi et al. [40] | ↓↓ | 3–81 | Circ. | $N_2$ |
| Arkilic et al. [57] | ↓↓ | 2.6 | Rect. | He |
| Pong et al. [58] | ↓↓ | 1.94–2.33 | Rect. | $N_2$, He |
| Liu et al. [59] | ↓↓ | 2.33 | Rect. | $N_2$, He |
| Shih et al. [60] | ↓↓ | 2.33 | Rect. | $N_2$, He |
| Yu et al. [43] | ↓↓ | 19–102 | Circ. | $N_2$ |
| Harley et al. [55] | ↓↓ | 1.01–35.91 | Rect.-Trap. | $N_2$, He, Ar |
| Araki et al. [56] | ↓↓ | 3–10 | Trap. | $N_2$, He |
| Li et al. [49] | ↑↑ | 128.8–179.8 | Circ. | $N_2$ |
| Yang et al. [51] | ↓↓ | 173–4010 | Circ. | Air |
| Lalonde et al. [61] | ≈ | 58.2 | Circ. | Air |
| Turner et al. [62] | ≈ | 4–100 | Rect. | Air, $N_2$, He |
| Hsieh et al. [63] | ↓↓ | 80 | Rect. | $N_2$ |
| Tang et al. [64] | ≈ | 50–201 | Rect.-Circ. | $N_2$ |
| Celata et al. [65] | ≈ | 30–254 | Circ. | He |
| Kohl et al. [66] | ≈ | 24.9–99.8 | Rect. | Air |

*Source:* Morini, G.L., *Int. J. Therm. Sci.*, 43, 631, 2004; Cetin, B. et al., *Int. J. Trans. Phenomena*, 8, 297, 2006; Kakaç, S. et al., *Heat Exchangers: Selection, Rating, and Thermal Design*, 3rd edn., CRC Press, Boca Raton, FL, 2012.

*Note:* ↑↑ represents higher *fRe* than the conventional theory; ↓↓ represents lower *fRe* than the conventional theory; ≈ represents agreed *fRe* with the conventional theory.

diameters, there is no agreement with the conventional theory due to the rarefaction effects. For transition flow regime, at the Reynolds number equal to 2300, Li et al. [49] conducted experiments with gases in order to calculate the friction factors for circular microtubes. Their results were in a good agreement with the conventional theory. In their continued work [50], they concluded that the transition flow regime occurred at Reynolds numbers between 1700 and 2000.

## 12.9 Empirical Correlations for Liquid Forced Convection in Microchannels

There are more applications of the liquid flow in microchannels for cooling of electronics and microdevices. Both theoretical and experimental investigations have been conducted. Convective heat transfer in liquid flowing through microchannels has been extensively experimented to obtain the characteristics in the laminar, transitional, and turbulent regimes. The geometric parameters of individual rectangular microchannels, namely, the hydraulic diameter and the aspect ratio, and the geometry of the microchannel plate have significant influence on the single-phase convective heat-transfer characteristics. In microchannels, the flow transition point and range are functions of the heating rate or the wall temperature conditions. The transitions are also a direct result of the large liquid temperature rise in the microchannels, which causes significant liquid thermophysical property variations and significant increases in the relevant flow parameters such as the Reynolds number. Hence, the transition point and range are affected by the liquid temperature, velocity, and geometric parameters of the microchannel.

Some of the correlations aimed to predict the friction factor and Nusselt number for liquid flow in microchannels. The mechanism by which heat is transferred in an energy conversion system is complex. However, three basic and distinct modes of heat transfer have been classified: conduction, convection, and radiation [17]. Convection is the heat-transfer mechanism that occurs in a fluid by the mixing of one portion of the fluid with another portion due to gross movements of the mass of fluid. Although the actual process of energy transfer from one fluid particle or molecule to another is still heat conduction, the energy may be transported from one point in space to another by the displacement of the fluid itself. Hence, forced convection heat transfer with or without phase change [67] can be considered. For the simple analysis, the flow without phase change, or the single-phase liquid flow, should be investigated. The thermal and hydrodynamic entry lengths for the flow in microducts can be expressed as

$$\frac{L_t}{D_h} = 0.056 RePr, \tag{12.111}$$

$$\frac{L_h}{D_h} = 0.056 Re. \tag{12.112}$$

The local heat transfer in the developing region of a laminar flow in a circular tube is given by the following equations [43]:

$$Nu_x = 4.363 + 8.68(10^3 x^*)^{-0.506} e^{-41x^*}, \tag{12.113}$$

$$x^* = \frac{x/D_h}{RePr}. \tag{12.114}$$

Most of the studies reported in the open literature indicate that the improvement in the experimental techniques for the majority of the macrochannel correlations is applicable to the studies in microchannels. Gnielinski [68] developed the Nusselt number and friction factor correlation for the transition regime from laminar to turbulent:

$$Nu = \left. \frac{(f/8)(Re - 1000)Pr}{1 + 12.7(f/8)^{1/2}(Pr^{2/3} - 1)} \middle| \begin{array}{l} 3000 < Re < 5 \times 10^6 \\ 0.5 < Pr < 2000 \end{array} \right. \tag{12.115}$$
$$f = [1.82 \log(Re) - 1.64]^{-2}$$

Adams et al. [17,69] modified the Gnielinski correlation, which was a generalized correlation for the Nusselt number for turbulent flow in small-diameter tubes (0.76–1.09 mm):

$$Nu_{Adams} = Nu_{Gnielinski}(1 + F) \left\{ \begin{array}{l} F = 7.6 \times 10^{-6} Re\left(1 - \left(\dfrac{d}{D_0}\right)^2\right) \\ 2,600 < Re < 23,000 \\ 1.53 < Pr < 6.43 \\ D_0 = 1.164 \text{ mm is the reference diameter} \end{array} \right. \tag{12.116}$$

Tsuzuki et al. [70] found the Nusselt number correlation for water in their experimental study:

$$Nu = 0.253 Re^{0.597} Pr^{0.349} \left\{ \begin{array}{l} 100 < Re < 1500 \\ 2 < Pr < 11 \end{array} \right., \tag{12.117}$$

$$Nu = 0.207 Re^{0.627} Pr^{0.34} \left\{ \begin{array}{l} 1,500 < Re < 15,000 \\ 1 < Pr < 3 \end{array} \right., \tag{12.118}$$

and friction factor correlation was given by White [71] as

$$f = 0.0791 Re^{-1/4} \left\{ 10^4 < Re < 10^5 \right.. \tag{12.119}$$

Jiang et al. [72] found correlations for friction factor for large relative roughness with respect to their experiments:

$$f = \frac{1639}{Re^{1.48}} \{ Re < 600, \tag{12.120}$$

$$f = \frac{5.45}{Re^{0.55}} \{ 600 < Re < 2800. \tag{12.121}$$

The following correlations are also given for turbulent flow regime [73]:

$$f = (1.0875 - 0.1125\alpha^*) f_c \left\{ \begin{array}{l} 4 \times 10^3 < Re < 10 \\ f_c = 0.00128 + 0.1143 Re^{-0.311} \\ \alpha^* = \dfrac{a}{b} \end{array} \right. . \tag{12.122}$$

### Example 12.2

Water flows at a temperature of 30°C through a microchannel with a hydraulic diameter of 0.03 mm, and the length of the channel is 10 mm. The Reynolds number for flow is 400. By the use of correlation in Equation 6.33 (a), calculate the friction coefficient, the Poiseuille number, and the Darcy coefficient, and (b) by the use of Equation 6.33, calculate the frictional pressure drop in bar. Kinematic viscosity of water at 30°C can be taken as $v$ and equals to $0.8315 \times 10^{-6}$ m$^2$/s.

**Solution**
From properties of water at $T = 30$°C,

$$\rho = 995.65 \text{ kg/m}^3,$$

$$\mu = 0.0007973 \text{ Pa s},$$

$$k = 0.6155 \text{ W/ m K},.$$

$$Pr = 5.43,$$

$$v = 0.8315 \text{ m}^2/\text{s}.$$

The water velocity can be found from the given Reynolds number:

$$Re = 400 = \frac{D_h \times u_m}{v},$$

$$500 = \frac{0.03 \times 10^{-3} \times u_m}{0.8315 \times 10^{-6}},$$

$$u_m = 11.1 \text{ m/s}.$$

Since $Re = 400 < 600$, the Jiang et al. [47] correlations, Equation 12.120, can be used to calculate the friction factor:

$$f = \frac{1639}{Re^{1.48}} \quad \text{for } Re < 600$$

$$= \frac{1639}{400^{1.48}} = 0.231.$$

$$\lambda_{Dr} = 4f = 0.9238$$

For laminar flow, the Poiseuille number is introduced, Equation 12.105a, as

$$Po = fRe$$

$$= 0.231 \times 400$$

$$= 92.38.$$

The Poiseuille number can be defined with the Darcy coefficient, Equation 12.106, as

$$Po = \lambda_D Re,$$

$$92.38 = \lambda_D \times 400,$$

$$\lambda_D = 92.38/400 = 0.231.$$

The frictional pressure drop, Equation 6.33, can be obtained as

$$\Delta p = \frac{\lambda_{Dr} L \rho u_m^2}{2 D_h}$$

$$= \frac{0.9238 \times 0.01 \times 995.65 \times 11.1^2}{2 \times 0.03 \times 10^{-3}}.$$

$$= 18842.893 \text{ kPa}$$

$$= 188.43 \text{ bar}$$

Choi et al. [74] found the Nusselt number correlations Equation 12.108 for laminar and turbulent flow in microchannels. Yu et al. [41] also found a Nusselt number correlation as in Equation 12.110 for turbulent flow in microchannels:

$$Nu = 0.000972 Re^{1.17} Pr^{1/3} \{ Re < 2000, \tag{12.123}$$

$$Nu = 3.82 \times 10^{-6} Re^{1.96} Pr^{1/3} \{ 2,500 < Re < 20,000, \tag{12.124}$$

$$Nu = 0.007 Re^{1.2} Pr^{0.2} \{ 6,000 < Re < 20,000. \tag{12.125}$$

Bejan [75] recommended the equation given by Gnielinski, Equation 12.110 over the traditional Dittus–Boelter equation with limited errors about ±10%:

$$Nu = 0.0.12(Re^{0.87} - 280)Pr^{0.4} \begin{cases} 3 \times 10^3 \le Re \le 10^6 \\ 1.5 \le Pr < 500 \end{cases}.$$  (12.126)

Park and Punch [76] proposed a new correlation for the Nusselt number with respect to their experimental data:

$$Nu = 0.015Br^{-0.22}Re^{0.62}Pr^{0.33} \begin{cases} 69 < Re < 800 \\ 4.44 < Pr < 5.69 \\ 106\,\mu m < D_h < 307\,\mu m \\ Br < 0.0006 \end{cases}.$$  (12.127)

Lee et al. [77] examined heat transfer in rectangular microchannels in their study and found a new interval for the Sieder–Tate correlation given by [73]

$$Nu = 1.86\left(\frac{RePrd}{L}\right)^{1/3}\left(\frac{\mu_f}{\mu_w}\right)^{0.14} \{Re < 2200.$$  (12.128)

Wu and Little [39] presented a Nusselt number correlation for turbulent liquid nitrogen flow in microchannels:

$$Nu = 0.0022Re^{1.09}Pr^{0.4}\{Re > 3000.$$  (12.129)

Wang and Peng [78] reported a Nusselt number correlation with respect to their experimental study:

$$Nu = 0.00805Re^{0.8}Pr^{1/3}\{Re > 1500.$$  (12.130)

Additional correlations were given in two literature [79,80] as

$$Nu = 0.00805Re^{4/5}Pr^{1/3},$$  (12.131)

for fully developed turbulent flow $1000 < Re < 1500$,

$$Nu = 0.038Re^{0.62}Pr^{1/3},$$  (12.132)

for laminar flow $Re < 700$, and for turbulent $Re > 700$,

$$Nu = 0.072Re^{0.8}Pr^{1/3}.$$  (12.133)

**Example 12.3**

Water flows at a temperature of 30°C through a rectangular microchannel, 0.03 mm × 320 μm, used to cool a chip with 10 mm length of the channel. The water flow rate is 21.6 × $10^{-6}$ kg/s. Determine (1) the heat-transfer coefficient for the CHF and CWT boundary conditions and (2) the friction factor and the pressure drop in the channel.

**Solution**

From properties of water at $T = 30°C$,

$$\rho = 995.65 \text{ kg/m}^3,$$

$$\mu = 0.0007973 \text{ Pa} \cdot \text{s},$$

$$c_p = 4.1798 \text{ kJ/kg K},$$

$$k = 0.6155 \text{ W/m K},$$

$$Pr = 5.43.$$

The hydraulic diameter can be calculated as

$$D_h = \frac{2 \times a \times b}{a + b} = \frac{2 \times 40 \times 320}{40 + 320} = 54.86 \text{ μm},$$

and the Reynolds number can be obtained as

$$Re = \frac{4 \times \dot{m}}{\pi \times \mu \times D_h} = \frac{4 \times 21.6 \times 10^{-6}}{\pi \times 0.0007973 \times 54.86 \times 10^{-6}} = 628.54.$$

To check the hydrodynamically fully developed flow condition, $L_h$ has to be found:

$$L_h = 0.056 \times Re \times D_h = 0.056 \times 628.54 \times 71.11 \times 10^{-6}$$

$$= 1.93 \times 10^{-3} \text{ m}$$

$$= 1.93 \text{ mm} < 10 \text{ mm},$$

so the flow is hydrodynamic fully developed.

To check the thermal fully developed flow condition, $L_t$ has to be found:

$$L_t = 0.05 \times Re \times Pr \times D_h = 0.05 \times 628.54 \times 5.43 \times 71.11 \times 10^{-6}$$

$$= 9.36 \times 10^{-3} \text{ m}$$

$$= 9.36 \text{ mm} < 10 \text{ mm},$$

so the flow is thermally fully developed.

Choi et al. [41] correlation, Equation 12.95, can be used to calculate the average Nusselt number:

$$Nu = 0.000972 Re^{1.17} Pr^{1/3} \quad \text{for } Re < 2000$$

$$= 0.0384 \times 628.54^{1.17} \times 5.43^{1/3}$$

$$= 3.21,$$

$$h = \frac{Nu_\infty \times k}{d} = \frac{3.21 \times 0.6155}{54.86 \times 10^{-6}} = 36028.43 \text{ W/m}^2 \text{ K}.$$

Jiang et al. [72] correlations, Equation 12.121, can be used to calculate the friction factor for rough surfaces:

$$f = \frac{5.45}{Re^{0.55}} \quad \text{for } 600 < Re < 2800$$

$$= \frac{5.45}{628.54^{0.55}} = 0.158.$$

The pressure drop can be calculated from Equation 6.33:

$$\Delta p = 4f \times \frac{L}{D_h} \times \frac{\rho \times u_m^2}{2}$$

$$= 4 \times 0.158 \times \frac{0.010}{54.86 \times 10^{-6}} \times \frac{995.65 \times 9.17^2}{2}.$$

$$= 4.83 \text{ MPa}$$

$$= 48.3 \text{ bar}$$

For microchannels, the relative roughness values are expected to be higher than the limit of 0.05 used in the Moody diagram. Celeta et al. [81] measured the friction factor for refrigerant, R-114, flowing in a channel 130 μm in diameter with the Reynolds number varied from 100 to 8000 and the relative channel surface roughness of 2.65%. Their experimental results showed that, for laminar flow where $Re < 583$, the friction factor was in good agreement with the Poiseuille theory, but for higher values of the Reynolds number where the transition occurred in the $Re$ range of 1881–2479, the experimental data were higher than the Poiseuille theory. Satish et al. [2,4] considered the effect of cross-sectional area reduction due to protrusive roughness elements and proposed the use of the constricted flow area in calculating the friction factor, using a constricted diameter, $D_{cf} = D - 2\varepsilon$, in a modified Moody diagram [64]. In the turbulent region, a constant value of a friction factor was above $\varepsilon/D_{cf} > 0.03$. In the turbulent fully rough region, $0.03 \leq \varepsilon/D_{cf} \leq 0.05$, the friction factor can be calculated as

$$f_{Darcy,cf} = \lambda_D = 0.042. \tag{12.134}$$

Fanning friction factor can be calculated from the Darcy coefficient as follows:

$$\lambda_D = 4f = 0.042, \tag{12.135a}$$

$$f = \frac{0.042}{4}, \tag{12.135b}$$

$$f = 0.0105. \tag{12.135c}$$

For relative roughness, values that were higher than 0.05 were not compatible with Equation 12.118 because of the reason that experimental data were not available beyond $\varepsilon/D_{cf} > 0.03$. Constricted friction factors can be seen in Table 12.15 [38] for different cases [38].

**TABLE 12.15**

Constricted Friction Factors

| Flow Region | $f_{cf}$ | Interval |
|---|---|---|
| Laminar region | $f_{cf} = \dfrac{Po}{Re_{cf}}$ | $0 \leq \varepsilon/D_{h,cf} \leq 0.15$ |
| Fully developed turbulent region | $f_{cf} = \dfrac{0.25\left[\log_{10}\left(\left((\varepsilon/D_{h,cf})/3.7\right)+\left(5.74/Re_{cf}^{0.9}\right)\right)\right]^{-2}}{4}$ | $0 \leq \varepsilon/D_{h,cf} \leq 0.03$ |

*Source:* Kakaç, S. et al., *Heat Exchangers: Selection, Rating, and Thermal Design*, 3rd edn., CRC Press, Boca Raton, FL, 2012.

## Problems

**12.1** Repeat Problem 6.5; consider both hydrodynamically and thermally fully developed steady laminar flow of a constant-property fluid through a circular microchannel of CWT. Obtain an expression for the Nusselt number by solving Equation 12.36 iteratively, first starting with the temperature distribution for the constant wall heat-flux case and then repeating the procedure until the Nusselt number reaches a limit.

**12.2** Consider both hydrodynamically and thermally fully developed steady laminar flow of a constant-property fluid through a circular microchannel with a CHF condition maintained at the microchannel wall. Neglecting axial conduction of heat and assuming that the velocity profile can be approximated to be uniform across the entire flow area of the microchannel or slug flow, obtain an expression for the Nusselt number. Compare the result with Equation 6.68 and explain the reason for the difference on a physical basis.

**12.3** Consider a fully developed steady laminar flow of a constant-property fluid through a circular microchannel with a CHF condition maintained at the microchannel wall. Neglecting axial conduction of heat, obtain an expression for the Nusselt number with the effect of viscous dissipation included in the analysis. How does this frictional heating affect the Nusselt number in the microchannel?

**12.4** Consider the fully developed flow of a very viscous fluid in a circular microchannel of radius $r_0$. Obtain an expression for the Nusselt number if the boundary condition is given as

$$at = r = r_0 : T = T_w < T_m,$$

where $T_m$ is the microchannel mean fluid temperature.

**12.5** Evaluate the values of the local Nusselt number from Section 12.4 and compare the values with values from Equation 6.113 at $(x/d)/(RePr) = 0.0025, 0.05, 0.2,$ and $0.5$ along the pipe.

**12.6** Consider the flow of a constant-property fluid at a mass flow rate m in an electrically heated microtube of diameter $d$ and length $L$. The heat flux to the fluid along the length of the microtube is given as

$$q_w'' = q_0'' \sin \frac{\pi x}{L},$$

where $q_w''$ is a given constant. Determine the variation of the tube surface temperature along the length of the microtube. Assume that the heat-transfer coefficient is constant and known.

**12.7** Calculate the *Nu* number expression given by Equation 12.97 of water with temperature equals to 80°C and when *Re* equals to 6,000 and 20,000.

**12.8** Using the *Nu* number expression given by Equation 12.100, calculate the fully developed heat-transfer coefficient of dry air at atmospheric pressure for CWT and constant wall heat-flux boundary conditions for air in engineering applications with temperature equals to 20°C and CHF boundary conditions when *Kn* equals to 0.01 and 0.1.

**12.9** Air flows at a temperature of 30°C through a microchannel with a hydraulic diameter of 0.03 mm, and the length of the channel is 10 mm. The Reynolds number for flow is 400. By the use of correlation in Equation 6.33, (a) calculate the friction coefficient, the Poiseuille number, and the Darcy coefficient, and (b) by the use of Equation 6.33, calculate the frictional pressure drop in bar.

---

# References

1. Morini, G. L., Single-phase convective heat transfer in microchannels: A review of experimental results, *Int. J. Therm. Sci.*, 43, 631–651, 2004.
2. Kandlikar, S. G. and King, M. R., Introduction, in *Heat Transfer and Fluid Flow in Minichannels and Microchannels*, S. G. Kandlikar, S. Garimella, D. Li, S. Colin, and M. R. King (Eds.), Elsevier Science, London, U.K., pp. 1–7, 2006.
3. Li, D., Single phase electrokinetic flow in microchannels, in *Heat Transfer and Fluid Flow in Minichannels and Microchannels*, S. G. Kandlikar, S. Garimella, D. Li, S. Colin, and M. R. King (Eds.), Elsevier Science, London, U.K., pp. 137–174, 2006.
4. Kandlikar, S. G., Single-phase liquid flow in minichannels and microchannels, in *Heat Transfer and Fluid Flow in Minichannels and Microchannels*, S. G. Kandlikar, S. Garimella, D. Li, S. Colin, and M. R. King (Eds.), Elsevier Science, London, U.K., pp. 87–136, 2006.
5. Zhang, M. Z., *Nano/Microscale Heat Transfer*, McGraw-Hill, New York, 2007.
6. Bayazitoglu, Y., Tunc, G., Wilson, K., and Tjahjono, I., Convective heat transfer for single-phase gases in microchannel slip flow: Analytical solutions, in *Microscale Heat Transfer- Fundamentals and Applications*, S. Kakaç, L. Vasiliev, Y. Bayazitoglu, and Y. Yener (Eds.), Kluwer Academic Publishers, Dordrecht, the Netherlands, pp. 125–148, 2005.
7. Gad-El-Hak, M., The fluid mechanics of microdevices-the Freshman scholar lecture, *J. Fluids Eng.*, 121, 5–33, 1999.
8. Bayazitoglu, Y. and Kakaç, S., Flow regimes in microchannel single-phase gaseous fluid flow, in *Microscale Heat Transfer-Fundamentals and Applications*, S. Kakaç, L. Vasiliev, Y. Bayazitoglu, and Y. Yener (Eds.), Kluwer Academic Publishers, Dordrecht, the Netherlands, pp. 75–92, 2005.
9. Colin, S., Single-phase gas flow in microchannels, in *microchannel single-phase gaseous fluid flow*, in *Microscale Heat Transfer-Fundamentals and Applications*, S. Kakaç, L. Vasiliev, Y. Bayazitoglu, and Y. Yener (Eds.), Kluwer Academic Publishers, Dordrecht, the Netherlands, pp. 9–86, 2005.
10. Gad-El-Hak, M., Momentum and heat transfer in MEMS, *Congrs Francais de Thermique*, SFT, Grenoble, France, 3–6 June, Elsevier, Paris, 2003.
11. Kennard, E. H., *Kinetic Theory of Gases*, McGraw-Hill Book Company, Inc., New York, 1938.
12. Tuckerman, D. B. and Pease, R. F. W., High performance heat sinking for VLSI, *IEEE Electron Dev. Lett.*, 2(5), 126–129, 1981.

13. Yener, Y., Kakaç, S., Avelino, M., and Okutucu, T., Single-phase gas flow in microchannels: A state-of-the-art review, in *Microscale Heat Transfer-Fundamentals and Applications*, S. Kakaç, L. Vasiliev, Y. Bayazitoglu, and Y. Yener (Eds.), Kluwer Academic Publishers, Dordrecht, the Netherlands, pp. 1–24, 2005.
14. Beskok, A., Trimmer, W., and Karniadakis, G., Rarefaction compressibility and thermal creep effects in gas microflows, *Proc. ASME DSC*, 57(2), 877–892, 1995.
15. Cetin, B., Yuncu, H., and Kakaç, S., Gaseous flow in microconduits with viscous dissipation, *Int. J. Trans. Phenomena*, 8, 297–315, 2006.
16. Kakaç, S. and Yener, Y., *Heat Conduction*, 4th edn., Taylor & Francis Group, 2008.
17. Kakaç, S. and Yener, Y., *Convective Heat Transfer*, 2nd edn., CRC Press, Boca Raton, FL, 1995.
18. Peng, X. F. and Peterson, G. P., Convective heat transfer and flow friction for water flow in microchannels, *Int. J. Heat Mass Transfer*, 39(12), 2599–2608, 1996.
19. Chen, Y. T., Kang, S. W., Tuh, W. C., and Hsiao, T. H., Experimental investigation of fluid flow and heat transfer in microchannels, *Tamkang J. Sci. Eng.*, 7(1), 11–16, 2004.
20. Barron, R. F., Wang, X. M., Warrington, R. O., and Ameel, T. A., The Graetz problem extended to slip flow, *Int. J. Heat Mass Transfer*, 40(8), 1817–1823, 1997.
21. Peng, X. F. and Peterson, G. P., The effect of thermofluid and geometrical parameters on convection of liquids through rectangular microchannels, *Int. J. Heat Mass Transfer*, 38(4), 755–758, 1995.
22. Wang, B. X. and Peng, X. F., Experimental investigation of heat transfer in flat plates with rectangular microchannels, *Int. J. Heat Mass Transfer*, 38(1), 127–137, 1995.
23. Harms, T. M., Kazmierczak, M. J., and Gerner, F. M., Developing convective heat transfer in deep rectangular microchannels, *Int. J. Heat Fluid Flow*, 20, 149–157, 1999.
24. Wu, H. Y. and Cheng, P., An experimental study of convective heat transfer in silicon microchannels with different surface conditions, *Int. J. Heat Mass Transfer*, 46, 2547–2556, 2003.
25. Beskok, A., Karniadakis, G. M., and Trimmer, W., Rarefaction and compressibility effects in gas microflows, *J. Fluid Engrg.*, 118, 448–456, 1996.
26. Sun, W., Kakaç, S., and Yazicioglu, A. G., A numerical study of single-phase convective heat transfer in microtubes for slip flow, *Int. J. Therm. Sci.*, 46, 1084–1094, 2007.
27. Cetin, B., Yazicioglu, A. G., and Kakaç, S., Fluid flow in microtubes with axial conduction including rarefaction and viscous dissipation, *Int. Commun. Heat Mass Transfer*, 35, 535–544, 2008.
28. Shah, R. K. and London, A. L., Laminar flow forced convection in ducts, in *Laminar Flow Forced Convection in Ducts*, *Advances in Heat Transfer*, T. F. Irvine Jr. and J. P. Hartnett (Eds.), Academic Press, New York, pp. 78–152, 1978.
29. Verhoff, F. H. and Fisher, D. P., A numerical solution of the Graetz problem with axial conduction, *J. Heat Transfer*, 95, 12–134, 1973.
30. Yang, G. and Ebadian, M., Thermal radiation and laminar forced convection in the entrance region of a pipe with axial conduction and radiation, *Int. J. Heat Fluid Flow*, 12(3), 202–209, 1991.
31. Lahjomri, J. and Qubarra, A., Analytical solution of the Graetz problem with axial conduction, *J. Heat Transfer*, 121, 1078–1083, 1999.
32. Croce, G. and D'Agaro, P., Numerical analysis of roughness effect on microtube heat transfer, *Superlattices Microstruct.*, 35, 601–616, 2004.
33. Croce, G., D'Agaro, P., and Filippo, A., Compressibility and rarefaction effects on pressure drop in rough microchannels, *Heat Transfer Eng.*, 28(8–9), 688–695, 2007.
34. Turgay, M. B., Yazicioglu, A. G., and Kakaç, S., Roughness effect on the heat transfer coefficient for gaseous flow in microchannels, *Proceedings of the International Heat Transfer Conference (IHTC14)*, Washington, DC, IHTC14-23142, August 8–13, 2010.
35. Barisik, M., Analytical solution for single phase microtube heat transfer including axial conduction and viscous dissipation, Thesis, Master of Science in Mechanical Engineering, Middle East Technical University, Ankara, Turkey, 2008.
36. Kakaç, S., Yazicioglu, A. G., and Gozukara, A. C., Effect of variable thermal conductivity and viscosity on single phase convective heat transfer in slip flow, *Heat Mass Transfer*, 47, 879–891, 2011.

37. Tunc, G. and Bayazitoglu, Y., Heat transfer in microtubes with viscous dissipation, *Int. J. Heat Mass Transfer*, 44, 2395–2403, 2001.

38. Kakaç, S., Liu, H., and Pramuanjaroenkij, A., *Heat Exchangers: Selection, Rating, and Thermal Design*, 3rd edn., CRC Press, Boca Raton, FL, 2012.

39. Wu, P. Y. and Little, W. A., Measurement of heat transfer characteristics of gas flow in fine channels heat exchangers used for microminiature refrigerators, *Cryogenics*, 24(5), 415–420, 1985.

40. Choi, S. B., Barron, R. F., and Warrington, R. O., Fluid flow and heat transfer in microtubes, in *Micromechanical Sensors, Actuators and Systems*, ASME, Atlanta, GA, DSC-32, 123–134, 1991.

41. Yu, D., Warrington, R. O., Barron, R., and Ameel, T., An experimental and theoretical investigation of fluid flow and heat transfer in microtubes, *Proceedings of ASME/JSME Thermal Engineering Joint Conference*, Honolulu, HI, pp. 523–530, 1995.

42. Ameel, T. A., Wang, X., Barron, R. F., and Warrington, R. O., Laminar forced convection in a circular tube with constant heat flux and slip flow, *Microscale Thermophys. Eng.*, 1(4), 303–320, 1997.

43. Yu, S. and Ameel, T., Slip-flow heat transfer in rectangular microchannels, *Int. J. Heat Mass Transfer*, 44, 4225–4234, 2001.

44. Renksizbulut, M., Niazmand, H., and Tercan, G., Slip-flow and heat transfer in rectangular microchannels with constant wall temperature, *Int. J. Therm. Sci.*, 44, 870–881, 2006.

45. Renksizbulut, M. and Niazmand, H., Laminar flow and heat transfer in the entrance region of trapezoidal channels with constant wall temperature, *J. Heat Transfer*, 128, 63–74, 2006.

46. Hadjiconstantinou, N. G. and Simsek, O., Constant-wall-temperature Nusselt number in micro and nano-channels, *Trans. ASME*, 57(2), 877–892, 2002.

47. Wu, P. and Little, W. A., Measurement of friction factors of the flow of gases in very fine channels used for microminiature Joule-Thompson refrigerators, *Cryogenics*, 23, 3–17, 2003.

48. Acosta, R. E., Muller, R. H., and Tobias, W. C., Transport process in narrow (capillary) channels, *AICHE J.*, 31, 473–482, 1985.

49. Li, Z. X., Du, D. X., and Guo, Z. Y., Characteristics of frictional resistance for gas flows in microtubes, *Proceedings of Symposium on Energy Engineering in the 21st Century*, Hong Kong, 2, 658–664, 2000.

50. Li, Z. X., Du, D. X., and Guo, Z. Y., Experimental study on flow characteristics of liquid in circular microtubes, *Microscale Thermophys. Eng.*, 7, 253–265, 2003.

51. Yang, C. Y., Chien, H. T., Lu, S. R., and Shyu, R. J., Friction characteristics of water, R-134a and air in small tubes, *Proceedings of the International Conference on Heat Transfer and Transport Phenomena in Microscale*, Alberta, Canada, pp. 168–174, 2000.

52. Pfalher, J., Harley, J., Bau, H. H., and Zemel, N., Liquid transport in micron and submicron channels, *Sens. Actuators A*, 21–23, 431–434, 1990.

53. Pfalher, J., Harley, J., Bau, H. H., and Zemel, N., *Liquid and Gas Transport in Small Channels*, ASME, Atlanta, GA, DSC-31, 149–157, 1990.

54. Pfalher, J., Harley, J., Bau, H. H., and Zemel, N., Gas and liquid flow in small channels, in *Micromechanical Sensors, Actuators and Systems*, ASME, Atlanta, GA, DSC-32, 49–60, 1991.

55. Harley, J., Huang, Y., Bau, H. H., and Zemel, J. N., Gas flow in microchannels, *J. Fluid Mech.*, 284, 257–274, 1995.

56. Araki, T., Soo, K. M., Hiroshi, I., and Kenjiro, S., An experimental investigation of gaseous flow characteristics in microchannels, *Proceedings of the International Conference on Heat Transfer and Transport Phenomena in Microscale*, Alberta, Canada, pp. 155–161, 2000.

57. Arkilic, E. B., Breuer, K. S., and Schmidt, M. A., Gaseous flow in microchannels, in *Application of Microfabrication to Fluid Mechanics*, ASME, New York, FED-197, 57–66, 1994.

58. Pong, K., Ho, C., Liu, J., and Tai, Y., Non-linear pressure distribution in uniform microchannels, in *Application of Microfabrication to Fluid Mechanics*, ASME, New York, FED-197, 51–56, 1994.

59. Liu, J., Tai, Y. C., and Ho, C. M., MEMS for pressure distribution studies of gaseous flows through uniform microchannels, *Proceedings of Eighth Annual International Workshop MEMS*, IEEE, pp. 209–215, 1995.

60. Shih, J. C., Ho, C. M., Liu, J., and Tai, Y. C., *Monatomic and Polyatomic Gas Flow Through Uniform Microchannels*, ASME, Atlanta, GA, DSC-59, 197–203, 1996.

61. Lalonde, P., Colin, S., and Caen, R., Mesure de Debit de Gaz dans les Microsystems, *Mec. Ind.*, 2, 355–362, 2001.
62. Turner, S. E., Sun, H., Faghri, M., and Gregory, O. J., Compressible gas flow through smooth and rough microchannels, *Proc. IMECE ASME 2001*, HTD-24144, 2001.
63. Hsieh, S. S., Tsai, H. H., Lin, C. Y., Huang, C. F., and Chien, C. M., Gas flow in long microchannel, *Int. J. Heat Mass Transfer*, 47, 3877–3887, 2004.
64. Tang, G. H. and He, Y. L., An experimental investigation of gaseous flow characteristics in microchannels, *Proceedings of Second International Conference on Microchannels and Minichannels*, Rochester, NY, pp. 359–366, 2004.
65. Celata, G. P., Cumo, M., McPhail, S. J., Tesfagabir, L., and Zummo, G., Experimental study on compressibility effects in microtubes, *Proceedings of the XXIII UIT Italian National Conference*, Parma, Italy, pp. 53–60, 2005.
66. Morini, G. L., Lorenzini, M., and Salvigni, S., Friction characteristics of compressible gas flows in microtubes, *Exp. Therm. Fluid Sci.*, 30, 773–744, 2006.
67. Stephan, K., *Heat Transfer in Condensation and Boiling*, Springer-Verlag, Berlin, Germany, 1992.
68. Gnielinski, V., New equations for heat transfer in turbulent pipe and channel flow, *Int. Chem. Eng.*, 16, 359–368, 1976.
69. Adams, T. M., Abdel-Khalik, S. I., Jeter, S. M., and Qureshi, Z. H., An experimental investigation of single-phase forced convection in microchannels, *Int. J. Heat Mass Transfer*, 41(6–7), 851–857, 1998.
70. Tsuzuki, N., Utamura, M., and Ngo, T. L., Nusselt number correlations for a microchannel heat exchanger hot water supplier with S-shaped fins, *Appl. Therm. Eng.*, 29, 3299–3308, 2009.
71. White, F. M., *Viscous Fluid Flow*, McGraw Hill, New York, 1991.
72. Jiang, P. X., Fan, M. H., Si, G. S., and Ren, Z. P., Thermal-hydraulic performance of small scale microchannel and porous-media heat exchangers, *Int. J. Heat Mass Transfer*, 44, 1039–1051, 2001.
73. Kakaç, S., Shah, R. K., and Aung, W., *Handbook of Single Phase Convective Heat Transfer*, John Wiley & Sons, New York, 1987.
74. Choi, S. B., Barron, R. F., and Warrington, R. O., Fluid flow and heat transfer, in *Micromechanical Sensors, Actuator and Systems*, ASME, Atlanta, GA, DSC-32, 123–134, 1991.
75. Bejan, A., *Heat Transfer*, John Wiley & Sons, Hoboken, NJ, 1993.
76. Park, H. S. and Punch, J., Friction factor and heat transfer in multiple microchannels with uniform flow distribution, *Int. J. Heat Mass Transfer*, 51, 4435–4443, 2008.
77. Lee, P. S., Garimella, S. V., and Liu, D., Investigation of heat transfer in rectangular microchannels, *Int. J. Heat Mass Transfer*, 48, 1688–1704, 2005.
78. Wang, B. X. and Peng, X. F., Experimental investigation of liquid forced convection heat transfer through microchannels, *Int. J. Heat Mass Transfer*, 37(1), 73–82, 1994.
79. Peng, X. F., Peterson, G. P., and Wang, B. X., Heat transfer characteristics of water flowing through microchannels, *Exp. Heat Transfer*, 7, 265–283, 1994.
80. Sobhan, C. B. and Peterson, G. P., *Microscale and Nanoscale Heat Transfer*, CRC Press, Boca Raton, FL, 2008.
81. Celata, G. P., Cumo, M., Gulielmi, M., and Zummo, G., Experimental investigation of hydraulic and single phase heat transfer in 0.130 mm capillary tube, *Nanoscale Microscale Thermophys. Eng.*, 6(2), 85–97, 2002.

# 13

## Enhancement of Convective Heat Transfer with Nanofluids

### Nomenclature

| | |
|---|---|
| $a$ | coefficient in Equation 5.40 |
| $Br$ | Brinkman number $= \mu u^2 / k(T_w - T_b)$ |
| $b$ | coefficient in Equation 5.40 |
| CWT | constant wall temperature condition |
| CHF | constant heat flux |
| $c_p$ | specific heat at constant pressure, J/kg·K |
| $D_x$ | thermal dispersion coefficient in the axial direction |
| $d$ | tube diameter or distance, m |
| $d_p$ | nanoparticle diameter, m |
| $f$ | Fanning friction factor defined by Equation 6.33 |
| $h$ | heat-transfer coefficient, W/m²·K |
| $k$ | thermal conductivity, W/m·K |
| $L$ | characteristic length, m |
| $M$ | molecular weight, kg/mol |
| $N_A$ | Avogadro constant, $6.023 \times 10^{23}$/mol |
| $Nu$ | Nusselt number $= hd/k$ |
| $P_{as}$ | aspect ratio of nanotube |
| $Pe$ | Peclet number $= RePr = ud/\alpha$ |
| $Pr$ | Prandtl number $= \mu c_p / k$ |
| $p$ | pitch ratio |
| $q_w$ | heat flux, W/m² |
| $Re$ | Reynolds number $= \rho u_m d / \mu$ |
| $r$ | radius, m |
| $T$ | temperature, K |
| $t$ | nanolayer thickness, m |
| $u, U$ | velocity, m/s |
| $u_m$ | mean flow velocity, m/s |
| $x$ | $x$-direction distance, m |
| $y$ | $y$-direction distance, m |

### Greek Symbols

| | |
|---|---|
| $\alpha$ | thermal diffusivity, m²/s |
| $\beta$ | constant defined by Equation 13.20 |

| | |
|---|---|
| $\gamma$ | constant defined by Equation 13.19 |
| $\delta_v^+$ | dimensionless thickness of the laminar sublayer |
| $\eta$ | dimensionless radius |
| $\Theta$ | constant defined by Equation 13.22 |
| $\theta$ | dimensionless temperature |
| $\kappa_B$ | Boltzmann constant, $1.3807 \times 10^{-23}$ J/K |
| $\lambda$ | mean free molecule path, m |
| $\mu$ | dynamics viscosity, Pa s |
| $v$ | kinematic viscosity, m$^2$/s |
| $\xi$ | dimensionless distance |
| $\rho$ | density, kg/m$^3$ |
| $\phi$ | particle volume fraction |
| $\psi$ | sphericity |

## Subscripts

| | |
|---|---|
| $b$ | bulk mean |
| $cf$ | constricted flow |
| $cl$ | cluster |
| $d$ | dispersed |
| $eff$ | effective |
| $f$ | base fluid |
| $fd$ | fully developed |
| $i$ | inlet |
| $k$ | nanoparticles with a slip velocity relative |
| $m$ | mixture |
| $nf$ | nanofluid |
| $o$ | outlet or tube radius in m |
| $p$ | nanoparticle |
| $r$ | $r$-direction |
| $t$ | thermal |
| $w$ | wall |
| $x$ | local |

## 13.1 Introduction

Nanofluid is envisioned to describe a fluid in which nanometer-sized particles are suspended in conventional heat-transfer basic fluids. Conventional heat-transfer fluids, including oil, water, and ethylene glycol mixture, are poor heat-transfer fluids, since the thermal conductivity of these fluids plays an important role on the heat-transfer coefficient between the heat-transfer medium and the heat-transfer surface. Therefore, numerous methods have been taken to improve the thermal conductivity of these fluids by suspending nano-/micro- or larger-sized particle materials in liquids. Nanofluids are defined as mixture between solid nanoparticles and base fluid. The nanoparticles with typical length scales of 1–100 nm with high thermal conductivity are suspended in the fluid; generally, low thermal conductivity have been shown to enhance effective thermal conductivity and the convective heat-transfer coefficient of the base fluid. The thermal conductivity of the particle materials, metallic or nonmetallic such as $Al_2O_3$, CuO, Cu, SiO, TiO, are typically

**TABLE 13.1**

Thermal Conductivities of Various Solids and Liquids

| Solids/Liquids | Materials | Thermal Conductivity (W/m K) |
|---|---|---|
| Metallic solids | Silver (Ag) | 429 |
| | Copper (Cu) | 401 |
| | Aluminum (Al) | 237 |
| | Copper oxide (CuO) | 20 |
| Nonmetallic solids | Diamond | 3300 |
| | Carbon nanotubes (CNT) | 3000 |
| | Silicon (Si) | 148 |
| | Alumina (Al$_2$O$_3$) | 40 |
| Metallic liquids | Sodium (Na) @ 644 K | 72.3 |
| Nonmetallic liquids | Water | 0.613 |
| | Ethylene glycol (EG) | 0.253 |
| | Engine oil (EO) | 0.145 |

*Source:* Eastman, J. et al., Enhanced thermal conductivity through the development of nanofluids, *Proceedings of the Symposium on Nanophase and Nanocomposite Materials II*, Boston, MA, Vol. 447, pp. 3–11, 1997.

significantly higher than the base fluids even at low concentrations, resulting in significant increases in the heat-transfer coefficient (Table 13.1).

Therefore, the effective thermal conductivity and heat-transfer coefficient of nanofluids are expected to enhance the heat transfer compared with conventional heat-transfer liquids. Heat-transfer coefficient is the determining factor in forced convection cooling–heating applications of heat-exchange equipments including engines and engine systems. Such enhancement mainly depends upon factors such as particle volume concentration, particle material, particle size, particle shape, base fluid material temperature, and additives. Nanoparticles used in nanofluids have been made out of many materials by physical and chemical synthesis processes. Typical physical methods include the mechanical grinding method and the inert-gas-condensation technique [2]. Current processes specifically for making metal nanoparticles include mechanical milling, inert-gas-condensation technique, chemical precipitation, chemical vapor deposition, microemulsions, spray pyrolysis, and thermal spraying. Nanoparticles in most materials discussed are most commonly produced in the form of powders [3].

By increasing the thermal conductivity of the working fluid, significant improvement in thermal performance can be obtained. In the first period of nanofluid researches, nanofluids were investigated by Masuda et al. [4] and Choi [5]. Experimental research has shown that the thermal conductivity enhancements obtained with nanofluids exceed the enhancements obtained by using conventional suspensions [1,6–11]. In addition, nanoparticles fluidize easily in the flow due to their small sizes, and this eliminates the problems of clogging of channels, erosion on the channel walls, and sedimentation. In this period, the studies on nanofluids were mainly focused on the measurements of thermal conductivity. More experimental and numerical studies regarding the convective heat transfer of nanofluids have been developed continuously. Additionally, from experimental studies in the forced convection of nanofluids in tubes, most studies indicate that the heat-transfer coefficient enhancements observed with nanofluids exceed the associated enhancements obtained in thermal conductivity [12–18]. So accurate prediction of the heat-transfer enhancement with nanofluids is necessary for the utilization and application of nanofluids in theoretical and practical applications. One can start with the simplest method to analyze nanofluid flow and heat transfer by applying single-phase approaches. In the simple approach, the nanofluid can be treated as a single-phase fluid, and the effects of nanoparticles are taken into account only through the usage of the thermophysical properties of the nanofluid in the associated calculations [19]. A numerical study following this approach can be performed with the laminar and turbulent flow of nanofluids such as $Al_2O_3$/water and $Al_2O_3$/ethylene glycol nanofluids because there are available literatures concerning these nanofluids to validate the numerical solutions. Maïga et al. [19] introduced their numerical results of the nanofluids in the straight circular tube under constant wall heat-flux boundary and showed that ethylene glycol-based nanofluids provide better heat-transfer enhancement when compared to the water-based nanofluids. The single-phase approach can be modified by utilizing a thermal dispersion model as proposed by Xuan and Roetzel [20] and Heris et al. [21], the model takes the effect of improved thermal transport due to the random motion of nanoparticles into account. The researchers concluded that the heat-transfer enhancement obtained with nanofluids increases with decreasing particle size and increasing particle volume fraction.

Further analysis showed that nanofluid flow can also be examined by utilizing a two-phase approach [22]; in this analysis, the force interactions between nanoparticles and the fluid matrix can be taken into account. One can use the same flow configuration as in the single-phase approach, laminar flow of $Al_2O_3$/water nanofluid inside a straight circular tube, to consider under constant wall heat-flux boundary condition. In one of these analyses, Bianco et al. [22] concluded that the single-phase analysis provides close results to those of the two-phase approach as long as the variation of nanofluid thermophysical properties with temperature is taken into account. To provide a clear point of view, nanofluid thermophysical properties should be discussed as they play emphasis roles in the analysis of nanofluid heat-transfer enhancement. Each nanofluid property may not vary with the same character, properties, and relationships. This following section will provide details of properties that are varied with other properties.

### 13.1.1 Density

The following expression [14] can be used to determine nanofluids density:

$$\rho_{nf} = \phi\rho_p + (1-\phi)\rho_f. \tag{13.1}$$

where
  $\phi$ is particle volume fraction
  subscripts *nf*, *p*, and *f* correspond to nanofluid, particle, and base fluid, respectively

### 13.1.2 Specific Heat

Two expressions can be used to determine the specific heat of nanofluids [14,20]:

$$c_{p,nf} = \phi c_{p,p} + (1-\phi)c_{p,f} \tag{13.2}$$

and

$$(\rho c_p)_{nf} = \phi(\rho c_p)_p + (1-\phi)(\rho c_p)_f. \tag{13.3}$$

Equation 13.3 is theoretically more consistent since specific heat is a mass-specific quantity whose effect depends on the density of the components of a mixture.

### 13.1.3 Viscosity

Nanofluid viscosity is an important parameter for practical applications since it directly affects the pressure drop in forced convection. Generally, the increase in the viscosity by the addition of nanoparticles to the base fluid is significant, depending on parameters such as particle volume fraction, particle size, temperature, and extent of clustering; increasing particle volume fraction increases viscosity. An expression for determining the dynamic viscosity of dilute suspensions that contain spherical particles can be considered as [23] with neglecting particle interactions, the well-known Einstein's formula [24],

$$\mu_{nf} = (1 + 2.5\phi)\mu_f, \tag{13.4}$$

and Einstein's equation was extended by Brinkman [25] at the maximum particle volume fraction [25]:

$$\mu_{nf} = \frac{1}{(1-\phi)^{2.5}} \mu_f. \tag{13.5}$$

For the viscosity of the nanofluids, an empirical correlation was proposed by Nguyen et al. [26] as

$$\mu_{nf} = (1 + 2.5\phi + 150\phi^2)\mu_f, \tag{13.6}$$

which is based on the room temperature experimental data of $Al_2O_3$/water nanofluids with a particle size of 36 nm.

The viscosity of the nanofluid, water as a base fluid and 1 vol.% Cu, for turbulent forced convection in a tube under the constant wall heat-flux boundary condition was proposed by Behzadmehr et al. [27] as a mixture viscosity:

$$\mu_m = \sum_{k=1}^{n} \phi_k \mu_k, \tag{13.7}$$

where $k$ represents the nanoparticles with a defined condition as the particles flow with a slip velocity relative to the velocity of the base fluid phase. As well as, for other nanofluid applications, the viscosity equations can be found in Table 13.2.

### 13.1.4 Thermal Conductivity

For the determination of nanofluid thermal conductivity, there are many theoretical models in literatures. In addition, experimental data available in the literatures for different types of nanofluids can also be used. The detailed literature review about the thermal conductivity of nanofluids are presented by Kakaç and Pramuanjaroenkij [28] and

**TABLE 13.2**

Viscosity Models for Nanofluids

| Nanofluid Applications | Viscosity Equations |
| --- | --- |
| TiO$_2$/water nanofluids | $\mu_{nf} = 13.47e^{35.98\phi}\mu_f$ |
| Al$_2$O$_3$/water nanofluids | $\mu_{nf} = (1 + 7.3\phi + 123\phi^2)\mu_f$ |
| Al$_2$O$_3$/ethylene glycol nanofluids | $\mu_{nf} = (1 - 0.19\phi + 306\phi^2)\mu_f$ |
| CuO-based nanofluids | $\mu_{eff} = \mu_f + 5\times10^4 \beta\rho_f\phi\sqrt{\dfrac{\kappa_B T}{2\rho_p r_p}}[(-134.63 + 1722.3\phi) + (0.4705 - 6.04\phi)T]$ |
| | where |
| | $\beta = \begin{cases} 0.0137(100\phi)^{-0.8229} \text{ for } \phi < 0.01 \\ 0.0011(100\phi)^{-0.7272} \text{ for } \phi > 0.01 \end{cases}$ |
| | $\kappa_B$ is the Boltzmann constant, $T$ is the temperature in K, $r_p$ is the nanoparticle radius in m |
| CuO/water nanofluids | $\mu_{nf} = \exp[-(2.8751 + 53.548\phi - 107.12\phi^2) + (1078.3 + 15857\phi + 20587\phi^2)(1/T)]$ |

*Source:* Yu, W. et al., *Review and Assessment of Nanofluid Technology for Transportation and Other Applications*, ANL/ESD/07-9, Argonne National Laboratory, Argonne, IL, 2007.

Özerinç et al. [29], which summarized many theoretical models, and the latter work compared their predictions with experimental data available in their literatures:

a. *Maxwell model.* The Maxwell equation expresses an effective thermal conductivity equation for nanofluids consisting of spherical particles [30]:

$$k_{nf} = \frac{k_p + 2k_f + 2(k_p - k_f)\phi}{k_p + 2k_f - (k_p - k_f)\phi}k_f, \tag{13.8}$$

where
   *nf*, *p*, and *f* represent nanofluid, nanoparticles, and base fluid, respectively
   $\phi$ is the volume fraction of particles in the mixture

Note that the effect of the size and shape of the particles and particle interactions were not included in the equation.

b. *Hamilton and Crosser model.* The following equation covers nonspherical particles and introduces the shape factor (*n*) as experimental parameters for different types of particle materials:

$$k_{nf} = \frac{k_p + (n-1)k_f - (n-1)\phi(k_f - k_p)}{k_p + (n-1)k_f + \phi(k_f - k_p)}k_f, \tag{13.9}$$

where *n* is the empirical shape factor and it is defined as

$$n = \frac{3}{\psi}, \tag{13.10}$$

where $\psi$ is the sphericity. Sphericity is the ratio of the surface area of a sphere with a volume equal to that of the particle to the surface area of the particle; $n = 3$ for a sphere [31]. Since the empirical shape factor, *n*, is defined as in Equation 13.10, $n = 3$ for a sphere, and in that case Equation 13.9 becomes identical to Equation 13.8.

c. *Brownian motion model.* The Brownian motion is the random motion of particles suspended in a fluid. Brownian dynamics simulation is used to determine the effective thermal conductivity of nanofluids [32] as

$$k_{nf} = \phi k_p + (1-\phi)k_f, \tag{13.11}$$

where $k_p$ is not simply the bulk thermal conductivity of the nanoparticles but it also includes the effect of the Brownian motion of the nanoparticles on the thermal conductivity.

The thermal conductivity of nanofluids can be assumed as the combination of two parts [33]:

$$k_{nf} = k_{static} + k_{Brownian}, \tag{13.12}$$

where

$k_{static}$ represents the thermal conductivity enhancement due to the higher thermal conductivity of the nanoparticles such as the first two models

$k_{Brownian}$ takes the effect of the Brownian motion into account, the effect of fluid particles moving with nanoparticles around them:

$$k_{Brownian} = 5 \times 10^4 \beta \phi \rho_f c_{p,f} \sqrt{\frac{\kappa_B T}{\rho_p d_p}} f, \tag{13.13}$$

where

$\rho_p$ and $\rho_f$ are the density of nanoparticles and base fluid, respectively
$T$ is the temperature in K

As well as, $c_{p,f}$ is specific heat capacity of base fluid and $\beta$ is a constant obtained from Table 13.3 [33]. Note that the interactions between nanoparticles and fluid volumes moving around them were not considered. Another parameter, $f$, was introduced to the model in order to increase the temperature dependency of the model. Both $f$ and $\beta$ were determined by utilizing available experimental data:

$$f = (-134.63 + 1722.3\phi) + (0.4705 - 6.04\phi)T, \tag{13.14}$$

which is obtained by using the results of the study of Das et al. [8] for CuO nanofluids. For other nanofluids, $f$ can be taken as 1 [34].

**TABLE 13.3**

β Values for Different Nanoparticles to Be Used in Equation 13.13

| Type of Particles | β | Particle Volume Fraction |
|---|---|---|
| Au-citrate, Ag-citrate, and CuO | $0.0137(100\phi)^{-0.8229}$ | $\phi < 1\%$ |
| CuO | $0.0011(100\phi)^{-0.7272}$ | $\phi > 1\%$ |
| Al$_2$O$_3$ | $0.0017(100\phi)^{-0.0841}$ | $\phi > 1\%$ |

*Source:* Koo, J. and Kleinstreuer, C., *J. Nanopart. Res.*, 6(6), 577, 2004.

d. *Clustering model.* Clustering can result in fast transport of heat along relatively large distances since heat can be conducted much faster by solid particles when compared to liquid matrix [35–39]. The resulting thermal conductivity ratio expression is

$$\frac{k_{nf}}{k_f} = \frac{(k_{cl} + 2k_f) + 2\phi_{cl}(k_{cl} - k_f)}{(k_{cl} + 2k_f) - \phi_{cl}(k_{cl} - k_f)} \tag{13.15}$$

where
$k_{cl}$ is the thermal conductivity of the clusters
$\phi_{cl}$ is the particle volume fraction of the clusters

It was shown that the effective thermal conductivity increased with increasing cluster size. Fiber-shaped nanoparticles are more effective in thermal conductivity enhancement when compared to spherical particles. It should be noted that excessive clustering of nanoparticles may result in sedimentation, which adversely affects the thermal conductivity [36]. Therefore, there should be an optimum level of clustering for maximum thermal conductivity enhancement [29].

The thermal conductivity study of nanofluids by considering the Brownian motion and clustering of nanoparticles proposes an equation to predict the thermal conductivity of nanofluids [40]:

$$\frac{k_{nf}}{k_f} = \frac{k_p + 2k_f - 2\phi(k_f - k_p)}{k_p + 2k_f + \phi(k_f - k_p)} + \frac{\rho_p \phi c_{p,p}}{2k_f} \sqrt{\frac{\kappa_B T}{3\pi r_{cl}\mu_f}} \tag{13.16}$$

where
$r_{cl}$ is the apparent radius of the nanoparticle clusters, which should be determined by experiment
$T$ is temperature in K
$\mu_f$ is the dynamic viscosity of the base fluid

The second term on the right-hand side of Equation 13.16 adds the effect of the random motion of the nanoparticles into account.

e. *Liquid layering model.* Liquid molecules formed as layered structures around solid surfaces were expected to enhance the effective thermal conductivity than the normal liquid matrix [41,42]. The modified Maxwell model considering the liquid layer was obtained as

$$k_{nf} = \frac{k_{pe} + 2k_f + 2(k_{pe} - k_f)(1+\beta)^3 \phi}{k_{pe} + 2k_f - (k_{pe} - k_f)(1+\beta)^3 \phi} k_f, \tag{13.17}$$

where $k_{pe}$ is the thermal conductivity of the equivalent nanoparticle:

$$k_{pe} = \frac{[2(1-\gamma) + (1+\beta)^3(1+2\gamma)]\gamma}{-(1-\gamma) + (1+\beta)^3(1+2\gamma)} k_p, \tag{13.18}$$

where

$$\gamma = \frac{k_l}{k_p}, \tag{13.19}$$

and $k_l$ is thermal conductivity of the nanolayer. $\beta$ is defined as

$$\beta = \frac{t}{r_p}, \tag{13.20}$$

where
  $t$ is nanolayer thickness
  $r_p$ is the nanoparticle radius

Equation 13.18 is also known as the Yu and Choi model [41,42].

A nanolayer was modeled as a spherical shell with thickness $t$ around the nanoparticle, by assuming that the thermal conductivity changes linearly across the radial direction:

$$\frac{k_{nf} - k_f}{k_f} = 3\Theta\phi_T + \frac{3\Theta^2\phi_T^2}{1 - \Theta\phi_T}, \tag{13.21}$$

where

$$\Theta = \frac{\beta_{lf}[(1+\gamma)^3 - (\beta_{pl}/\beta_{fl})]}{(1+\gamma)^3 + 2\beta_{lf}\beta_{pl}}, \tag{13.22}$$

and

$$\beta_{lf} = \frac{k_l - k_f}{k_l + 2k_f}, \tag{13.23a}$$

$$\beta_{pl} = \frac{k_p - k_l}{k_p + 2k_l}, \tag{13.23b}$$

$$\beta_{fl} = \frac{k_f - k_l}{k_f + 2k_l}, \tag{13.23c}$$

where
  $\phi_T$ is the total volume fraction of nanoparticles and nanolayers
  $k_l$ is the thermal conductivity of the nanolayer
  $\phi_T$ can be determined using

$$\phi_T = \phi(1+\gamma)^3, \tag{13.24}$$

where

$$\gamma = \frac{t}{r_p}.$$                                                                                     (13.25)

$k_l$ was defined as

$$k_l = \frac{k_f M^2}{(M-\gamma)\ln(1+M)+\gamma M},$$                                           (13.26)

where

$$M = \varepsilon_p(1+\gamma)-1,$$                                                                   (13.27)

$$\varepsilon_p = \frac{k_p}{k_f}.$$                                                                       (13.28)

By using Bruggeman's effective media theory, the effective thermal conductivity of the nanofluid was alternatively determined by Bruggeman and Xue et al. as [43,44]

$$\left(1-\frac{\phi}{\alpha}\right)\frac{k_{nf}-k_f}{2k_{nf}+k_f}+\frac{\phi}{\alpha}\frac{(k_{nf}-k_l)(2k_l+k_p)-\alpha(k_p-k_l)(2k_l+k_{nf})}{(2k_{nf}+k_l)(2k_l+k_p)+2\alpha(k_p-k_l)(k_l-k_{nf})}=0,$$     (13.29)

where subscript $l$ refers to nanolayer. $\alpha$ is defined as

$$\alpha = \left(\frac{r_p}{r_p+t}\right)^3,$$                                                             (13.30)

where $t$ is the thickness of the nanolayer.

### Example 13.1

Compare three important models: Maxwell, Hamilton and Crosser, and Yu and Choi, for the $Al_2O_3$/water nanofluid at 50°C by assuming the nanoparticles radius of 2 nm with 5 W/m K for thermal conductivity of the nanolayer with a thickness of 2 nm. The comparison is based on varied volume fractions from 0 to 0.1 vol.%.

### Solution

Properties of water, from Appendix B (Table B.2), are

$$\rho_f = 988 \text{ kg/m}^3$$

$$\mu_f = 54.8\times10^{-5} \text{ kg/m s}$$

$$c_{p,f} = 4.1806 \text{ kJ/kg K}$$

$$k_f = 0.641 \text{ W/m K}$$

$$Pr_f = 3.57$$

From the Table 13.1 and given parameters, one can obtain

$$k_p = 40 \text{ W/m K}$$

$$r_p = t = 2 \text{ nm}$$

So the effective thermal conductivity ratio can be compared as in the following figure.

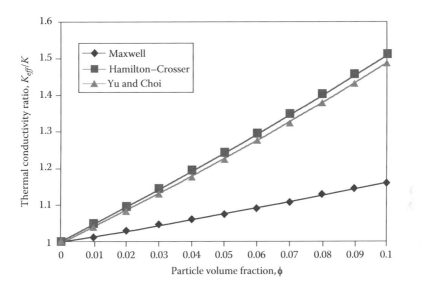

## Example 13.2

Calculate the thermal conductivity of nanofluid, $Al_2O_3$ nanoparticles with a particle radius of 100 nm dispersed in water at 273.15 K and 2% particle volume fraction, by considering a nanolayer as a spherical shell from Equation 5.68, $t$ from Equation 5.67 and letting $k_l = k_p$. Properties of water can be obtained from the water properties at pressure equals to 1 bar.

### Solution
Properties of alumina ($Al_2O_3$), from Table 13.1, are

$$k_p = 40 \text{ W/m K}$$

$$\phi = 0.02$$

$$r_p = 100 \times 10^{-9} \text{ m}$$

$$N_A = 6.023 \times 10^{23} \text{ mol}^{-1}$$

From properties of water at $P = 1$ bar (100 kPa) and $T = 273.15$ K (0°C),

$$k_f = 0.552 \text{ W/m K}$$

$$\rho_f = 999.8 \text{ kg/m}^3$$

$$M_f = 18 \times 10^{-3} \text{ kg/mol}$$

A nanolayer was modeled as a spherical shell by considering these following parameters and letting $k_l = k_p$. From

$$\beta_{lf} = \frac{k_l - k_f}{k_l + 2k_f} = \frac{40 - 0.552}{40 + (2)(0.552)} = 0.96$$

$$\beta_{pl} = \frac{k_p - k_l}{k_p + 2k_l} = \frac{40 - 40}{40 + (2)(40)} = 0$$

$$\beta_{fl} = \frac{k_f - k_l}{k_f + 2k_l} = \frac{0.552 - 40}{0.552 + (2)(40)} = -0.49$$

$$t = \frac{1}{\sqrt{3}} \left( \frac{4M_f}{\rho_f N_A} \right)^{1/3} = \frac{1}{\sqrt{3}} \left( \frac{(4)(18 \times 10^{-3})}{(999.8)(6.023 \times 10^{23})} \right)^{1/3}$$

$$= 2.8443 \times 10^{-10} \text{ m}$$

$$\beta_L = \frac{t}{r_p} = \frac{(2.8443 \times 10^{-10})}{(100 \times 10^{-9})} = 0.0028443$$

$$\phi_T = \phi(1 + \beta_L)^3 = (0.02)(1 + 0.0028443)^3 = 0.02017$$

$$\Theta = \frac{\beta_{lf}[(1 + \beta_L)^3 - (\beta_{pl}/\beta_{fl})]}{(1 + \beta_L)^3 + 2\beta_{lf}\beta_{pl}} = \frac{(0.96)[(1 + 0.0028443)^3 - ((0)/(-0.49))]}{(1 + 0.0028443)^3 + (2)(0.96)(0)}$$

$$= 0.96$$

$$\frac{k_{nf} - k_f}{k_f} = 3\Theta\phi_T + \frac{3\Theta^2\phi_T^2}{1 - \Theta\phi_T}$$

$$\frac{k_{nf} - 0.552}{0.552} = (3)(0.96)(0.02017) + \frac{(3)(0.96)^2(0.02017)^2}{1 - (0.96)(0.02017)}$$

$$k_{nf} = (0.552)\left[ (3)(0.96)(0.02017) + \frac{(3)(0.96)^2(0.02017)^2}{1 - (0.96)(0.02017)} \right] + (0.552)$$

$$k_{nf} = 0.585 \text{ W/m K}$$

The nanofluid can increase the thermal conductivity up to 6%.

f. *Yamada and Ota model for carbon nanotube.* The normalized effective thermal conductivity for carbon nanotube nanofluid was given as

$$k_{nf} = \frac{[k_p/k_f + n - n\phi(1 - k_p/k_f)]}{[k_p/k_f + n + \phi(1 - k_p/k_f)]} k_f, \qquad (13.31)$$

where $n$ is the empirical shape factor defined as

$$n = 2\phi^{0.2}\left(\frac{L_p}{d_p}\right),$$ (13.32)

For carbon nanotube, $L_p$ and $d_p$ are length and diameter of the cylindrical particles. The use of this model, carbon nanotube nanofluid, gives higher enhancement for the convective heat transfer.

## 13.2 Nanofluid Convective Heat-Transfer Modeling

The enhancement of the heat-transfer coefficient is a better indicator than the thermal conductivity enhancement for nanofluids used in the heat-transfer processes. Heris et al. [16] set up experiments with nanofluids, $Al_2O_3$/water and CuO/water under laminar flow up to turbulent flow. This literature found more heat-transfer enhancement as high as 40% with $Al_2O_3$ particles while the thermal conductivity enhancement was less than 15% [3]. Heat-transfer experimental results are available from Pak and Cho [14], Xuan and Li [45], Yang et al. [46], Heris et al. [16], Ding et al. [47], Ma et al. [48], Chen et al. [49], and Kulkarni et al. [50]. Pak and Cho [14] performed experiments on turbulent heat-transfer performance and turbulent frictions of two nanofluids, $\gamma$-$Al_2O_3$/water and $TiO_2$/water.

Some computer simulations dealt with the effective thermal conductivity of nanofluids [32,51] or effective viscosity [52] and most of them focus on the heat transfer of nanofluids [19,53–77]. A complete understanding about the heat-transfer enhancement in forced convection in laminar and turbulent flows with nanofluids is necessary for their practical applications to enhance heat-transfer rate. Nanofluids, in nature, are multicomponent fluids, so it is also treated as either a two-phase homogeneous flow with no slip between nanoparticles and the fluid, which are also in thermal equilibrium; or it is treated with a slip between the particles and the base fluid with thermal equilibrium. Most forced convection flows depend on both the Reynolds and Prandtl numbers, but for the case of nanofluids, additional parameters are included to take the thermal properties of all the constituents into account.

### 13.2.1 Numerical Analysis

From nanofluid thermophysical properties, it is expected that the heat-transfer coefficient of the nanofluid depends on the thermal conductivity and the heat capacity of the base fluid and nanoparticles, flow pattern, the Reynolds and Prandtl numbers, temperature, the volume fraction of the suspended particles, the dimensions, and shape of the particles. The size of the dispersed particles presents some difficulty in analyzing the interaction between the fluid and the solid particles during energy transfer. Many researchers have suggested that in fact the Brownian motion is one of the factors in the enhancement of heat transfer. This random motion of ultra-fine particles would create a slip velocity between the solid particles and the fluid medium. Xuan and Roetzel [6] also suggest including small perturbations in the temperature and velocity formulation to account for the Brownian motion [28]. Maïga et al. [19] investigated laminar and turbulent nanofluid flow inside circular tubes. $Al_2O_3$/water and $Al_2O_3$/ethylene glycol nanofluids were considered under the constant wall heat-flux boundary condition. This work assumed nanofluids as single-phase fluids, and the effect of nanoparticles was taken into account only through the substitution of the nanofluid thermophysical properties into the governing equations; as results, it was concluded that $Al_2O_3$/ethylene glycol nanofluid provides higher enhancement when compared to $Al_2O_3$/water nanofluid. Another single-phase analysis of nanofluid heat transfer was made by Heris et al. [21]. This group performed a numerical analysis that simulates their experimental study [13] by utilizing the thermal dispersion model proposed by Xuan and Roetzel [20]. In the analysis, they did not take the variation of thermal conductivity with temperature into account. Additionally, they assumed uniform thermal dispersion throughout the domain. This analysis considered the effect of particle volume fraction and particle size on heat transfer and concluded that increasing particle volume fraction and decreasing particle size increase the heat-transfer enhancement.

There are also numerical studies that consider two-phase approach in the literatures [28,34]. One of these studies, Behzadmehr et al. [27], investigated the flow of a nanofluid in a circular tube at turbulent regime, a numerical solution was made for the constant wall heat-flux boundary condition, and the difference between the velocities of the nanoparticles and fluid molecules was taken into account. The numerical results were compared with a previous experimental study of Cu/water nanofluids [12], and good agreement was observed. The researchers also compared the results of single-phase assumption with experimental data, and it was seen that single-phase approach, in some cases, underpredicts the associated Nusselt number values. Bianco et al. [22] considered the laminar flow of $Al_2O_3$/water nanofluid under constant wall heat-flux boundary condition, the problem was analyzed by using both single-phase and two-phase approaches, and results indicated that taking the variation of thermophysical properties with temperature into account results in higher enhancement values. In addition, they noted that the difference between the results of single-phase and two-phase approaches is small, especially when temperature dependence of thermophysical properties is taken into account. This is an important result that can be considered as an indication of the fact that the single-phase assumption provides acceptable results [34]. There are many other numerical studies about nanofluids in the literatures [28,29,34,78] providing a summary of those numerical studies and also review some theoretical studies regarding the convective heat transfer of nanofluids.

The governing equations under the specified boundary conditions can be solved. In this case, the equation of conservation (mass, momentum, and energy) that are well known for single-phase flow can be extended for nanofluids. If the microconvective and microdiffusion of the suspended particles (hydrodynamic dispersion) are neglected, these two approaches will result in less heat-transfer coefficients than the experimental findings. Solutions to governing equations can be given assuming the nanofluid is compressible with no slip between the particles and the fluid, but they are in thermal equilibrium. Under such conditions, the general conservation equations in the vectorial form can be written as [79]

*Conservation of Mass*

$$div(\rho\bar{v}) = 0 \qquad\qquad (2.12)$$

*Conservation of Momentum*

$$div(\rho\bar{v}\ \bar{v}) = -gradP + \mu\nabla^2\bar{v} \qquad\qquad (2.34)$$

*Conservation of Energy*

$$div(\rho\bar{v}\ c_pT) = div(K \cdot grad\,T) \qquad\qquad (2.52)$$

These equations can be simplified depending on the required solution. Because of the several effects, the slip velocity between the ultra-fine nanoparticles and the fluid may not be zero.

For 2D fully developed nanofluid flow in heated tube, the energy equation in the presence of heat dissipation can be written as

$$u\frac{\partial T}{\partial x} + v\frac{\partial T}{\partial r} = \frac{\alpha}{r}\frac{\partial}{\partial r}\left(r\frac{\partial T}{\partial r}\right) + \alpha\frac{\partial^2 T}{\partial x^2} + \phi, \tag{13.33}$$

where

$$\phi = \frac{2\mu}{\rho c}\left[\left(\frac{\partial v}{\partial r}\right)^2 + \left(\frac{v}{r}\right)^2 + \left(\frac{\partial u}{\partial x}\right)^2 + \frac{1}{2}\left(\frac{\partial u}{\partial r} + \frac{\partial v}{\partial x}\right)^2\right]. \tag{13.34}$$

In the simplest case, there is only flow in an axial direction or $u = U$ and $v = v_r = 0$, so

$$u\frac{\partial T}{\partial x} = \frac{\alpha}{r}\frac{\partial}{\partial r}\left(r\frac{\partial T}{\partial r}\right) + \alpha\frac{\partial^2 T}{\partial x^2} + \frac{2\mu}{\rho c}\left[\frac{1}{2}\left(\frac{\partial u}{\partial r}\right)^2\right]. \tag{13.35}$$

For nanofluid flow, the thermal dispersion coefficient in the axial direction, $D_x$, takes into account for the contribution of the hydrodynamic dispersion and the irregular movement of the nanoparticles. Since the thermophysical properties of nanofluids including $D_x$ are considered, the classical thermal diffusivity is replaced with the effective apparent thermal diffusivity of nanofluids, $\alpha_{nf}^*$, in Equation 13.35 as

$$u\frac{\partial T}{\partial x} = \frac{\alpha_{nf}^*}{r}\frac{\partial}{\partial r}\left(r\frac{\partial T}{\partial r}\right) + \alpha_{nf}^*\frac{\partial^2 T}{\partial x^2} + \frac{2\mu_{nf}}{(\rho c)_{nf}}\left[\frac{1}{2}\left(\frac{\partial u}{\partial r}\right)^2\right], \tag{13.36}$$

where

$$\alpha_{nf}^* = \alpha_{nf} + \frac{D_x}{(\rho c_p)_{nf}} \tag{13.37}$$

or

$$u\frac{\partial T}{\partial x} = \left(\alpha_{nf} + \frac{D_x}{(\rho c_p)_{nf}}\right)\frac{\partial^2 T}{\partial x^2} + \frac{1}{r}\frac{\partial}{\partial r}\left[\left(\alpha_{nf} + \frac{D_x}{(\rho c_p)_{nf}}\right)r\frac{\partial T}{\partial r}\right] + \left(\frac{\mu}{\rho c_p}\right)_{nf}\left(\frac{du}{dr}\right)^2. \tag{13.38}$$

The apparent thermal conductivity can also be defined as

$$\bar{k} = k_{nf} + D. \tag{13.39}$$

For fully developed, compressible laminar flow, the velocity profile is parabolic as

$$\frac{u}{u_m} = 2\left(1 - \frac{r^2}{r_0^2}\right), \tag{13.40}$$

where ū is the average velocity in the axial direction. If the heat conduction in the axial direction is neglected, then Equation 13.29 can be written as

$$\frac{u}{\alpha}\frac{\partial T}{\partial x} = \frac{1}{r}\frac{\partial}{\partial r}\left[r\frac{\partial T}{\partial r}\right], \tag{6.47}$$

or the nondimensional energy equation can be written as

$$\frac{\bar{u}}{\alpha}\frac{\partial \theta}{\partial \xi} = \frac{1}{\eta}\frac{\partial}{\partial \eta}\left[\eta\frac{\partial \theta}{\partial \eta}\right], \tag{13.41}$$

where dimensionless variables are defined as Equation 6.92. The energy equation in any forms must be solved for nanofluid-forced convection with the given boundary conditions of constant wall temperature (CWT) or constant heat flux (CHF) with a constant inlet temperature. By substituting the velocity profile, Equation 13.40 into Equation 13.41,

$$(1-\eta^2)\frac{\partial \theta}{\partial \xi} = \frac{1}{\eta}\frac{\partial}{\partial \eta}\left[\eta\frac{\partial \theta}{\partial \eta}\right]. \tag{6.94}$$

The classical Graetz solution can also be used with Equation 6.94 for a plug flow, $u = $ constant, for the CWT boundary conditions as also given in [79,80]:

$$Nu_x = \frac{h_x d}{k} = \frac{\sum_{m=1}^{\infty} e^{-4\lambda_m^2(x/d)/Pe}}{\sum_{m=1}^{\infty}(1/\lambda_m^2)e^{-4\lambda_m^2(x/d)/Pe}}; \quad Pe = \frac{u_m d}{\alpha_{nf}^*}. \tag{6.91}$$

The classical Graetz problem can be extended for CHF, CWT, and the linear wall temperature boundary conditions with parabolic velocity profile [79]. Using the parabolic velocity profile, the local Nusselt numbers can be obtained from

$$Nu_x = \frac{h_x d}{k_{nf}} = \frac{\sum_{n=1}^{\infty} A_n e^{-\lambda_n^2 \xi}}{\sum_{n=1}^{\infty}(A_n/\lambda_n^2)e^{-\lambda_n^2 \xi}}\text{(constant wall temperature)} \tag{6.113}$$

$$Nu_x = \frac{h_x d}{k_{nf}} = \left[\frac{11}{48} - \frac{1}{2}\sum_{n=1}^{\infty}\frac{e^{-\beta_n^2 \xi}}{A_n \beta_n^4}\right]^{-1}\text{(constant wall heat flux)} \tag{6.125}$$

$$Nu_x = \frac{h_x d}{k_{nf}} = \frac{\frac{1}{2} + 4\sum_{n=1}^{\infty}\frac{C_n}{2}\frac{R_n'(1)}{\lambda_n^4}e^{-\lambda_n^2 \xi^{-1}}}{\frac{88}{768} + 8\sum_{n=1}^{\infty}\frac{C_n}{2}\frac{R_n'(1)}{\lambda_n^4}e^{-\lambda_n^2 \xi}}\text{(linear wall temperature)}, \tag{6.127}$$

where $\xi = (x/r_0)/Pe$, $Pe = (u_m d)/\alpha_{nf}$, and eigenvalues and coefficients can be found in Section 13.6.9. From Equations 6.91, 6.113, 6.125, and 6.127, asymptotic values of the Nusselt numbers are obtained. The convective heat-transfer coefficient and the effective thermal conductivity for fully developed laminar-flow conditions and different boundary conditions

**TABLE 13.4**

Convective Heat Transfer and Associated Enhancement Ratios of the Fully Developed Conditions of the Base Fluid and Nanofluid

| Conditions | Details | Convective Heat-Transfer Coefficient, $h$ (W/m² K) (Percent Increase in Convective Heat-Transfer Coefficient with Respect to Pure Water) | | |
|---|---|---|---|---|
| | | 0.6% Al₂O₃/Water | 1.8% Al₂O₃/Water | Pure Water |
| 1 | Slug flow with CWT boundary condition | 177.8 (2.80%) | 187.6 (8.51%) | 172.9 |
| 2 | Parabolic velocity profile with CWT boundary condition | 112.4 (2.80%) | 118.6 (8.51%) | 109.3 |
| 3 | Parabolic velocity profile with constant wall heat-flux boundary condition | 134.1 (2.80%) | 141.6 (8.51%) | 130.5 |
| 4 | Parabolic velocity profile with linear wall temperature boundary condition | 134.1 (2.80%) | 141.6 (8.51%) | 130.5 |

*Source:* Kakaç, S. and Pramuanjaroenkij, A., *Int. J. Heat Mass Transfer*, 52, 3187, 2009.

**TABLE 13.5**

Thermal Conductivity and Associated Enhancement Ratios of the Base Fluid and Nanofluid

| Details | 0.6% Al₂O₃/Water | 1.8% Al₂O₃/Water | Pure Water |
|---|---|---|---|
| Thermal conductivity (W/m K) | 0.615 | 0.649 | 0.598 |
| Percent increase in thermal conductivity with respect to pure water | 2.80% | 8.51% | — |

*Source:* Kakaç, S. and Pramuanjaroenkij, A., *Int. J. Heat Mass Transfer*, 52, 3187, 2009.

are given in Tables 13.4 and 13.5 [28], respectively. Noted that the thermal dispersion coefficient, $D_r$, is neglected and the Nusselt number is calculated from $Nu_x = h_x d/k_{nf}$, which underestimates the heat-transfer enhancement since other effects on the enhancement with nanofluids that will be clarified in the following sections.

## 13.2.2 Conventional Correlations with Nanofluid Properties

There are many correlations available in the literature for laminar and turbulent flow in channels under various boundary conditions. One of the most commonly used empirical correlations for the determination of convective heat transfer in laminar-flow regime inside circular tubes with constant heat flow boundary condition is the Shah correlation [81]; the local Nusselt number enhancement ratio is defined as

$$\frac{Nu_{x,nf}}{Nu_{x,f}} = \frac{1.302\xi_{nf}^{-1/3}-1}{1.302\xi_{f}^{-1/3}-1} \quad \text{for } \xi \le 5\times10^{-5}$$

$$\frac{1.302\xi_{nf}^{-1/3}-0.5}{1.302\xi_{f}^{-1/3}-0.5} \quad \text{for } 5\times10^{-5} < \xi \le 1.5\times10^{-3}$$

$$\frac{4.364+0.263\xi_{nf}^{-0.506}e^{-41\xi_{nf}}}{4.364+0.263\xi_{f}^{-0.506}e^{-41\xi_{f}}} \quad \text{for } \xi > 1.5\times10^{-3}, \tag{13.42}$$

where $\xi$ is also defined as in Equation 6.92. In Equation 13.6, as $\xi_f$ and $\xi_{nf}$ increase, $Nu_{x,f}$ and $Nu_{x,nf}$ decrease. Since the Peclet number is inversely proportional to $\xi$, the enhancement of the nanofluid thermal conductivity decreases the Nusselt number enhancement ratio, whereas the enhancement of the nanofluid volumetric heat capacity increases the Nusselt number enhancement ratio.

To obtain the average heat-transfer coefficient enhancement ratio, one should consider the definition of the Nusselt number for the nanofluid and the base fluid:

$$Nu_{nf} = \frac{h_{nf}d}{k_{nf}}, \tag{13.43}$$

$$Nu_f = \frac{h_f d}{k_f}. \tag{13.44}$$

By combining Equations 13.43 and 13.44, the average heat-transfer coefficient enhancement ratio can be obtained as

$$\frac{h_{nf}}{h_f} = \frac{k_{nf}}{k_f} \frac{Nu_{nf}}{Nu_f}. \tag{13.45}$$

The average Nusselt number enhancement ratio can be obtained from Equation 13.45 by integrating the numerator and denominator of Equation 13.42 along the tube length. The heat-transfer performance of the nanofluid flow inside a circular tube can be obtained by using Equations 13.42 and 13.45. Figure 13.1 [12,81,82] presents comparison of the

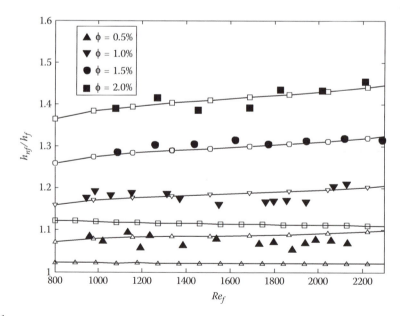

**FIGURE 13.1**
Comparison of the predictions of the correlations with experimental data. Solid lines and dashed lines indicate the predictions of the Li and Xuan correlation [12] and the Shah correlation [81], respectively. Markers indicate experimental data of Li and Xuan [12]. The Shah correlation is plotted for only two particle volume fractions for clarity. $Re_f$: base fluid Reynolds number [82].

predictions of the correlations with experimental data. From Figure 13.1, the Shah correlation significantly underpredicts the experimental data. According to the experimental data, the average heat-transfer coefficient enhancement ratio slightly increases with the Reynolds number, and the Shah correlation cannot predict this increase because of specifying constant enhancement for nearly all the Reynolds numbers in consideration. When it comes to the Li and Xuan correlation, it is seen that the experimental data are correctly predicted by this correlation. Therefore, for the constant heat flow boundary condition, the application of the classical Shah correlation for the nanofluid heat-transfer analysis can be considered as invalid [82].

One of the empirical correlations for the determination of convective heat transfer with nanofluid in laminar-flow regime [12] inside circular tubes, for $800 < Re_{nf} < 2300$ and $0.3\% < \phi < 2\%$, is

$$Nu_{nf} = 0.4328 \left( 1 + 11.285 \phi^{0.754} Pe_d^{0.218} \right) Re_{nf}^{0.333} Pr_{nf}^{0.4}. \tag{13.46}$$

An average Nusselt number of forced convection fluid flow inside a straight circular tube can be determined by using the Sieder–Tate correlation [83] under CWT boundary conditions:

$$Nu = 1.86 \left( Pe \frac{d}{L} \right)^{1/3} \left( \frac{\mu_b}{\mu_w} \right)^{0.14}, \tag{13.47}$$

where $L$ is tube length. Subscripts $b$ and $w$ indicate that the viscosity should be calculated at the bulk mean temperature and wall temperature, respectively. Bulk mean temperature is defined as

$$T_b = \frac{T_i + T_o}{2} \tag{13.48}$$

where
   $T_i$ is inlet temperature
   $T_o$ is outlet temperature

The average Nusselt enhancement ratio by applying Equation 13.47 is

$$\frac{Nu_{nf}}{Nu_f} = \left( \frac{Pe_{nf}}{Pe_f} \right)^{1/3} = \left( \frac{k_f}{k_{nf}} \frac{\rho_{nf}}{\rho_f} \frac{c_{p,nf}}{c_{p,f}} \right)^{1/3} \tag{13.49}$$

and

$$\frac{h_{nf}}{h_f} = \left( \frac{\rho_{nf}}{\rho_f} \frac{c_{p,nf}}{c_{p,f}} \right)^{1/3} \left( \frac{k_{nf}}{k_f} \right)^{2/3}. \tag{13.50}$$

Equation 13.49 shows that the enhancement in the volumetric heat capacity of the nanofluid increases the average Nusselt number enhancement ratio, whereas the enhancement in the thermal conductivity of the nanofluid decreases it. From Equation 13.46, those constants

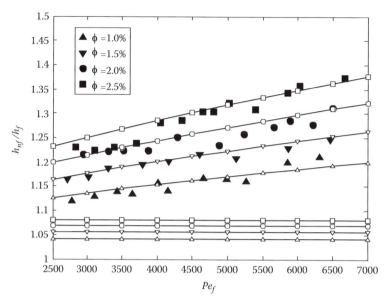

**FIGURE 13.2**
Comparison of the predictions of the correlations with experimental data. Solid lines and dashed lines indicate the predictions of the Li and Xuan correlation [12] and the Sieder–Tate correlation [83], respectively. Markers indicate experimental data of Heris et al. [13]. $Pe_f$: base fluid Peclet number [82].

are determined by fitting the experimental data of Heris et al. [13], valid in the range of the experimental data of $2500 < Pe_{nf} < 3500$ and $0.2\% < \phi < 2.5\%$, the correlation becomes

$$Nu_{nf} = 0.37\left(1 + 58\phi^{0.75}Pe_d^{0.72}\right)Re_{nf}^{0.333}Pr_{nf}^{0.4}. \tag{13.51}$$

From Figure 13.2 [12,13,82,83], the Sieder–Tate correlation and the Li and Xuan correlation are compared with the experimental data of Heris et al. [13], in order to examine the accuracy of the approaches. Predictions of the Sieder–Tate and the Li and Xuan correlations are determined by using Equations 13.50 and 13.51, respectively. Density and specific heat of the nanofluids are calculated according to Equations 13.1 and 13.3 [82]. It can be seen that the application of the classical correlations with nanofluid properties underestimates heat-transfer enhancement.

### 13.2.3 Analysis of Convective Heat Transfer of Nanofluids in Laminar Flow

In the present analysis, the nanofluid flow is considered as a single-phase incompressible flow. The simplest approach for the single-phase assumption is the direct usage of the governing equations of pure fluid flow with the thermophysical properties of the nanofluid in consideration. Thermal dispersion occurs due to the random motion of the nanoparticles in the flow, and it improves heat transport. In order to take the effect of thermal dispersion into account, the energy equation for the forced convection heat transfer of pure fluids is modified by replacing the thermal conductivity terms with an effective thermal conductivity, which is defined as follows [20]:

$$k_{eff} = k_{nf} + k_d, \tag{13.52}$$

where $k_d$ is called the dispersed thermal conductivity calculated by using the following expression [20]:

$$k_d = C(\rho c_p)_{nf} u_x \phi d_p r_0. \tag{13.53a}$$

where

C is an empirical constant determined by matching experimental data
$u_x$ is the local flow velocity
$d_p$ is the particle diameter
$r_0$ is the tube diameter

It should be noted that the dispersed thermal conductivity varies in radial direction due to the local axial velocity term in Equation 13.53, which is taken into account in the present numerical analysis. One can calculate density and specific heat of the nanofluids by using Equations 13.1 and 13.3, respectively. Viscosity of the nanofluids is determined according to the correlation in Equation 13.6 [26]. This correlation is based on the room temperature experimental data of $Al_2O_3$/water nanofluids with a particle size of 36 nm.

Özerinç et al. [29] provided a detailed literature review of nanofluid thermal conductivity and compared the predictions of thermal conductivity models with experimental data available in the literature. To determine the thermal conductivity, a temperature-dependent empirical correlation proposed by Chon et al. [9],

$$k_d = C(\rho c_p)_{nf} u_x \phi d_p r_0, \tag{13.53b}$$

$$\frac{k_{nf}}{k_f} = 1 + 64.7 \phi^{0.7460} \left( \frac{d_f}{d_p} \right)^{0.3690} \left( \frac{k_p}{k_f} \right)^{0.7476} Pr^{0.9955} Re^{1.2321}, \tag{13.54}$$

where

$$Pr = \frac{\mu_f}{\rho_f \alpha_f}, \tag{13.55}$$

$$Re = \frac{\rho_f V_{Br} d_p}{\mu_f}, \tag{13.56}$$

$$V_{Br} = \frac{\kappa_B T}{3\pi \mu_f d_p \lambda_f}, \tag{13.57}$$

noted that $V_{Br}$ is the Brownian velocity of nanoparticles. Assuming constant thermal conductivity in the numerical analysis throughout the flow domain may not be an accurate approach for the case of nanofluids, the variation of nanofluid thermal conductivity with temperature is taken into account in the present analysis:

a. *Laminar flow with nanofluids.* The flow is hydrodynamically fully developed as shown in Figure 13.3 [84]. Then the velocity distribution can be expressed as

$$u_x = 2u_m \left( 1 - \frac{r^2}{r_0^2} \right), \tag{13.58}$$

where
$u_m$ is mean flow velocity
$r$ is radial position

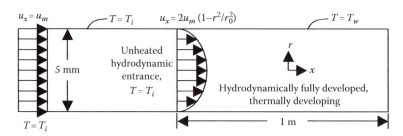

**FIGURE 13.3**
Schematic of hydrodynamically fully developed flow [84].

Since the flow is axisymmetric, all derivatives in angular direction are zero. The energy equation for nanofluids becomes

$$(\rho c)_{nf}\left(\frac{\partial T}{\partial t} + u_x \frac{\partial T}{\partial x}\right) = \frac{1}{r}\frac{\partial}{\partial r}\left(k_{eff} r \frac{\partial T}{\partial r}\right) + \frac{\partial}{\partial x}\left(k_{eff}\frac{\partial T}{\partial x}\right) + \mu_{nf}\left(\frac{\partial u_x}{\partial r}\right)^2, \tag{13.59}$$

Then, the nondimensional energy equation can be written as

$$\frac{\partial \theta}{\partial t^*} + Pe_{nf}(1 - r^{*2})\frac{\partial \theta}{\partial x^*} = \frac{1}{r^*}\frac{\partial}{\partial r^*}\left(k^* r^* \frac{\partial \theta}{\partial r^*}\right) + \frac{\partial}{\partial x^*}\left(k^* \frac{\partial \theta}{\partial x^*}\right) + 16 r^{*2} Br_{nf} \tag{13.60}$$

where

$$\theta = \frac{T - T_w}{T_i - T_w}, \tag{13.61a}$$

$$x^* = \frac{x}{r_0}, \tag{13.61b}$$

$$r^* = \frac{r}{r_0}, \tag{13.61c}$$

$$t^* = \frac{\alpha_{nf,b} t}{r_0^2}, \tag{13.61d}$$

$$k^* = \frac{k_{eff,T}}{k_{eff,b}}, \tag{13.61e}$$

$$Pe_{nf} = \frac{u_m d}{\alpha_{nf,b}}, \tag{13.61f}$$

$$Br_{nf} = \frac{\mu_{nf,b} u_m^2}{k_{nf,b}(T_i - T_w)}, \tag{13.61g}$$

Since $Br_{nf}$ is the nanofluid Brinkman number, on the order of $10^{-7}$, therefore viscous dissipation is negligible. Equation 13.60 becomes

$$\frac{\partial \theta}{\partial t^*} + Pe_{nf}(1-r^{*2})\frac{\partial \theta}{\partial x^*} = \frac{1}{r^*}\frac{\partial}{\partial r^*}\left(k^* r^* \frac{\partial \theta}{\partial r^*}\right) + \frac{\partial}{\partial x^*}\left(k^* \frac{\partial \theta}{\partial x^*}\right). \qquad (13.62)$$

*CWT boundary condition.* For CWT boundary condition, associated boundary conditions are as follows:

$$\text{at } r^* = 0 : \frac{\partial \theta}{\partial r^*} = 0 \qquad (13.63a)$$

$$\text{at } r^* = 1 : \theta = 0 \qquad (13.63b)$$

$$x^* = 0 : \theta = 1 \qquad (13.63c)$$

*Constant wall heat-flux boundary condition.* For constant wall heat-flux boundary condition, nondimensional parameters are the same as the expressions provided for CWT boundary condition as shown in Equation 13.61; the only difference is in the definition of the nondimensional temperature, $\theta$:

$$\theta = \frac{k_{nf}(T - T_i)}{q''_w r_0}, \qquad (13.64)$$

where $q''_w$ is the wall heat flux. Resulting nondimensional energy equation is the same as the expression provided in Equation 13.42. The associated boundary conditions are

$$\text{at } r^* = 0 : \frac{\partial \theta}{\partial r^*} = 0 \qquad (13.65a)$$

$$\text{at } r^* = 1 : \frac{\partial \theta}{\partial r^*} = \frac{1}{k^*} \qquad (13.65b)$$

$$\text{at } x^* = 0 : \theta = 0 \qquad (13.65c)$$

*Determination of heat-transfer coefficient and the Nusselt number.* The local heat-transfer coefficient can be determined by using the following expression as

$$h_{nf,x} = \frac{k_{eff,w,x}(\partial T/\partial r)_{x,r_0}}{T_{w,x} - T_{m,x}}, \qquad (13.66)$$

where $T_{m,x}$ and $T_{w,x}$ are mean temperature and wall temperature at axial location $x$, respectively. The former can be determined by calculating the following expression numerically as

$$T_{m,x} = \frac{\int_0^0 2r u_x(x,r)T(x,r)dr}{u_m r_0^2}. \qquad (13.67)$$

$k_{eff,w,x}$ is the effective thermal conductivity at the wall, and $(\partial T/\partial r)$ can be calculated by using a finite difference formulation. The Nusselt number is determined as follows:

$$Nu_x = \frac{h_x d}{k_{eff,w,x}}. \tag{13.68}$$

Average values of the heat-transfer coefficient and the Nusselt number are determined by integrating the local values along the tube and dividing the result by the tube length.

One can obtain numerical solutions by using numerical methods such as finite different method and finite volume method. In Figures 13.4 [13,82] and 13.5 [21,82], numerical results obtained from the finite difference method and utilized by using C programming language and the experimental data of Heris et al. [13] are compared in terms of the variation of average heat-transfer coefficient enhancement ratio with respect to the Peclet number for different particle volume fractions. When the figures were examined, it is seen that there is good agreement between the experimental data and the numerical results with thermal dispersion. On the other hand, the analysis without thermal dispersion underpredicts the experimental data. The small discrepancies between experimental data and the solution with thermal dispersion might be explained by the fact that the particle volume fraction of a nanofluid may unexpectedly affect the thermal conductivity due to the complicated variation of clustering characteristics with particle volume fraction.

Variation of nanofluid thermal conductivity with temperature and variation of dispersed thermal conductivity with local axial velocity are also taken into account. Results as shown in Figure 13.5 were considered with the same problem as in Figure 13.4, but the numerical solutions obtained from the finite different

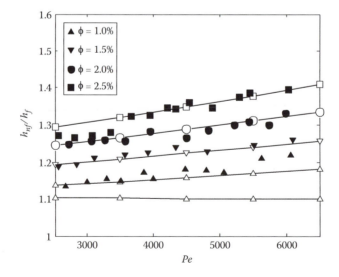

**FIGURE 13.4**
Comparison of the numerical results with experimental data of Heris et al. [13]. Boundary condition is CWT. Lines and markers indicate numerical results and experimental data, respectively. Dashed line is the numerical solution neglecting the thermal dispersion (only 1 vol.% case is shown for clarity) [82].

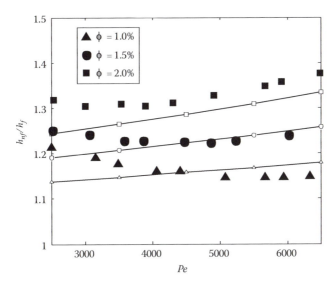

**FIGURE 13.5**
Comparison of the numerical results with numerical data of Heris et al. [21]. Lines and markers indicate the numerical results and study of Heris et al. [21], respectively. $Pe$ is $Pe_{nf}$ for nanofluid and $Pe_f$ for base fluid [82].

method were compared with numerical results obtained from Heris et al. [21], especially for particle volume fractions of 1.0% and 1.5%, which is not in agreement with the experimental data presented in Figure 13.4. For the 2.0 vol.% nanofluid, there is significant difference between the numerical results, and Figure 13.4 shows that the numerical solutions obtained from the finite different method is in very good agreement with the experimental data for 2.0 vol.% nanofluid. Therefore, taking variable thermal conductivity and variable thermal dispersion into account in nanofluid analysis significantly improves the accuracy.

The same flow configuration analyzed numerically in the previous sections is investigated in terms of the axial variation of the local Nusselt number. Figure 13.6 [82,85] shows the associated results for the flow of pure water and $Al_2O_3$/water nanofluid at a Peclet number of 6500, the local Nusselt number is larger for nanofluids throughout the tube due to the thermal dispersion in the flow. Associated values for different particle volume fractions of the $Al_2O_3$/water nanofluid are presented in Tables 13.6 [82] and 13.7 [22,82]. It should be noted that heat-transfer coefficient enhancement ratios are larger than the Nusselt number enhancement ratios since the former shows the combined effect of the Nusselt number enhancement and thermal conductivity enhancement with nanofluids. Figure 13.7 [82,85] shows the associated results for the flow of pure water and $Al_2O_3$/water nanofluid at different Reynolds numbers, the effects from fluid velocity can be observed. The heat-transfer coefficients of the base fluid are lower than these of the nanofluids, and the numerical simulations provide results similar to the experimental results.

For both constant wall heat-flux and CWT boundary conditions, the classical correlations applied to the case of nanofluids by the use of nanofluid properties underpredict the heat-transfer enhancement when compared with the experimental data available in the literatures [13,21,85]; the single-phase approach can provide an accurate prediction of heat-transfer enhancement of nanofluids. Comparison of the

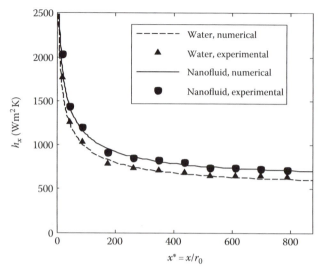

**FIGURE 13.6**
Comparison of the numerical results with experimental data of Kim et al. [85]. Boundary condition is constant wall heat flux. Lines and markers indicate numerical results and experimental data, respectively. $Re_{nf} = Re_f = 1460$ [82].

**TABLE 13.6**

Nusselt Number and Heat-Transfer Coefficient Obtained from the Numerical Solution for Pure Water and $Al_2O_3$/Water Nanofluid with Different Particle Volume Fractions

| Fluid | $Nu_{fd}$ | *Nu* Enhancement Ratio ($Nu_{fd,nf}/Nu_{fd,f}$) | $h_{fd}$ | *h* Enhancement Ratio ($h_{fd,nf}/h_{fd,f}$) |
|---|---|---|---|---|
| $Pe_f = Pe_{nf} = 6500$ | | | | |
| Pure water | 3.66 | — | 480 | — |
| Nanofluid | | | | |
| 1.0 vol.% | 3.77 | 1.030 | 562 | 1.172 |
| 1.5 vol.% | 3.82 | 1.044 | 594 | 1.238 |
| 2.0 vol.% | 3.86 | 1.057 | 624 | 1.300 |
| 2.5 vol.% | 3.91 | 1.069 | 653 | 1.361 |
| $Pe_f = Pe_{nf} = 12,000$ | | | | |
| Pure water | 4.36 | — | 588 | — |
| Nanofluid | | | | |
| 1.0 vol.% | 4.41 | 1.011 | 655 | 1.115 |
| 1.5 vol.% | 4.45 | 1.019 | 697 | 1.187 |
| 2.0 vol.% | 4.48 | 1.027 | 737 | 1.255 |
| 2.5 vol.% | 4.51 | 1.035 | 776 | 1.320 |

*Source:* Özerinç, S. et al., *Int. J. Therm. Sci.*, 62, 138, 2012.
$Pe_f = Pe_{nf} = 6,500$ and $Pe_f = Pe_{nf} = 12,000$.

numerical results with other numerical data available in the literature clearly shows that taking the variation of thermal conductivity with temperature and thermal dispersion in the flow domain into account significantly improves the accuracy of the heat-transfer enhancement predictions. The analysis regarding the fully developed Nusselt number of nanofluid heat transfer showed that the flattening in the radial

**TABLE 13.7**

Average Heat-Transfer Coefficient of Al$_2$O$_3$/Water Nanofluid according to the Numerical Study and the Numerical Study of Bianco et al. [22], $Re_{nf} = 250$

| Al$_2$O$_3$/Water Nanofluid (vol.%) | Numerical Study [82] | Study of Bianco et al. [22] | | | |
|---|---|---|---|---|---|
| | | Single-Phase, Constant k | Single-Phase, Variable k | Two-Phase, Constant k | Two-Phase, Variable k |
| 1.0 | 385 | 364 | 398 | 396 | 421 |
| 4.0 | 450 | 414 | 444 | 422 | 446 |

*Source:* Özerinç, S. et al., *Int. J. Therm. Sci.*, 62, 138, 2012.

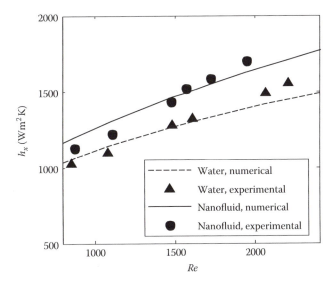

**FIGURE 13.7**
Comparison of the numerical results with experimental data of Kim et al. [85]. Boundary condition is constant wall heat flux. Lines and markers indicate numerical results and experimental data, respectively. $x^* = x/r_0 = 44$. $Re$ is $Re_{nf}$ for nanofluid and $Re_f$ for water [82].

temperature profile improves the Nusselt number. The associated enhancement, which is the result of the thermal dispersion, constitutes a relatively smaller portion of the heat-transfer coefficient enhancement obtained with nanofluids, the larger portion being the result of the thermal conductivity enhancement [82].

b. *Effects of the particle sizes.* For the simplest of the problem in laminar flow, one can neglect the effects of axial conduction in the energy equation, so the energy equation will be reduced to the following form:

$$\frac{Pe_{nf}}{2} u^* \frac{\partial \theta}{\partial x^*} = \frac{1}{r^*} \frac{\partial}{\partial r^*} \left[ r^* k^* \left( 1 + \frac{\varepsilon_h}{\alpha_{nf}} \right) \frac{\partial \theta}{\partial r^*} \right], \tag{13.69}$$

using the thermophysical properties, Equations 13.1, 13.3, and 13.4, with the classical thermal conductivity model from Equation 13.8 and extending equation, Equation 13.9.

The Brownian motion is described as the random movement of nanoparticles in the continuous medium, and it affects the thermal conductivity of the nanofluid since this movement makes a contribution to the heat transport in the nanofluid as in Equation 13.12. At elevated temperatures, the Brownian motion dominates, and the effect of the Brownian motion is introduced as an additional term in thermal conductivity of nanofluids expression by Koo and Kleinstreuer [33] as in Equation 13.13 [33,86].

For $Al_2O_3$ nanoparticles, $f$ is assumed to be 1 as in the literature [34], and from Table 13.3, $\beta$ is also a parameter that differs according to the type of the nanoparticles, which is defined in the following equation for $Al_2O_3$ nanoparticles [33]:

$$\beta = 0.0017(100\phi)^{-0.0841}, \quad \phi > 1\%. \tag{13.70}$$

From the geometry depicted in Figure 13.3, the nondimensional energy equation as in Equation 13.62 can be used to obtain the numerical solutions. The Nusselt number can be obtained according to the dimensionless parameters in Equation 13.62 as

$$Nu = \frac{2}{\theta_w - \theta_b}\left(\frac{\partial\theta}{\partial\eta}\right)_{\eta=1}. \tag{13.71}$$

Dimensionless bulk temperature, $\theta_b$, is determined by calculating the following expression as

$$\theta_b = \frac{\int_0^1 \theta(\xi,\eta)\bar{u}\eta d\eta}{\int_0^1 \bar{u}\eta d\eta}. \tag{13.72}$$

$\theta_b$ can be converted to $T_b$, which can be used to evaluate heat-transfer coefficient. If the length of the tube is long enough to enable nanofluid be thermally fully developed, the Nusselt numbers and heat-transfer coefficients at the tube exit will be similar for different cases. Therefore, the average Nusselt number ($\overline{Nu}$) and average heat-transfer coefficient ($\overline{h}$) given in the following are used for comparison purposes in both laminar and turbulent flow analyses [86]:

$$\overline{h} = \frac{(\rho c_p)_{nf} u_m A(T_{b,in} - T_{b,out})}{\pi DL(T_w - T_b)_{LM}} \tag{13.73}$$

$$\overline{Nu} = \frac{\overline{h}d}{k_{nf}}, \tag{13.74}$$

where $(T_w - T_b)_{LM}$ shows logarithmic mean temperature. Thermophysical properties of the nanofluid are evaluated at the bulk mean temperature, which is the arithmetic mean of inlet and outlet bulk temperatures. $A$ denotes the cross-sectional area of the tube.

Nanoparticles can be assumed to be spherical and to have uniform sizes. Thermophysical properties of alumina can be taken as $k_p$ = 46 W/m K, $\rho_p$ = 3700 kg/m³, and $(c_p)_p$ = 880 J/kg K. Using numerical methods, the thermal conductivity enhancement with the nanofluid with different particle sizes, particle diameters, can be obtained as shown in Figure 13.7. This figure shows that particle diameter increment decreases heat-transfer enhancement, an increase in the Reynolds number has a slight negative effect in heat-transfer enhancement with nanofluids. When volume fraction and particle size can be investigated at once, the heat-transfer enhancement is inversely proportional to the increase in particle size, and smaller particle sizes provide larger enhancement due to effect of the Brownian motion as in Koo and Kleinstreuer model [33]. As a result of the simulations, a considerable increase in enhancement is observed with increasing particle volume fractions for laminar flow, since it increases thermal conductivity of the nanofluid. Lower heat-transfer enhancements are obtained with increasing particle sizes, because of the attenuation in the Brownian motion of the nanoparticles. Particle diameter has a significant effect on enhancement especially in small particle diameter ranges. Figure 13.8 [87] also shows that the heat-transfer enhancement is inversely proportional to the nanoparticle diameter; the decrease in particle diameter of nanofluid increases enhancement.

c. *Heat-transfer analysis with Brownian thermal conductivity.* When heat generation, axial conduction, and viscous dissipation are neglected, 2D energy equation can be written in cylindrical coordinates as

$$\rho c\left(\frac{\partial T}{\partial t} + u_x \frac{\partial T}{\partial x}\right) = \frac{1}{r}\frac{\partial}{\partial r}\left(kr\frac{\partial T}{\partial r}\right). \tag{13.75}$$

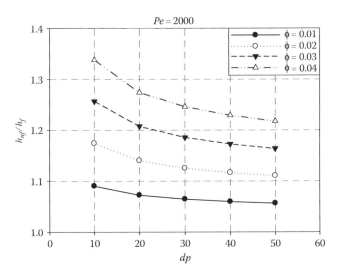

**FIGURE 13.8**
The effect of nanoparticle size with different volume fractions in steady-state region for CWT boundary condition [87].

Because of hydrodynamically fully developed flow assumption that yields radial velocity and axial velocity change to be equal to zero, the velocity profile is determined by solving momentum equation as a parabolic velocity profile:

$$u_x = 2u_m \left(1 - \frac{r^2}{r_0^2}\right), \tag{13.76}$$

where
$u_m$ is mean velocity
$r_0$ is radius of the cylinder

For step change in wall temperature and wall heat-flux conditions, nondimensional temperatures are defined as

$$\theta = \frac{T - T_w}{T_i - T_w} \tag{13.77}$$

and

$$\theta = \frac{k_{nf,b}(T - T_i)}{q''_w r_0}. \tag{13.78}$$

Then, nondimensional energy equation is obtained as

$$\frac{\partial \theta}{\partial t^*} + Pe_{nf}(1 - r^{*2})\frac{\partial \theta}{\partial x^*} = \frac{1}{r^*}\frac{\partial}{\partial r^*}\left(k^* r^* \frac{\partial \theta}{\partial r^*}\right). \tag{13.79}$$

When CWT boundary condition is applied, nondimensional boundary conditions become

$$\text{at } r^* = 0: \frac{\partial \theta}{\partial r^*} = 0 \tag{13.80a}$$

$$\text{at } r^* = 1: \theta = 0 \tag{13.80b}$$

$$\text{at } x^* = 0: \theta = 1 \tag{13.80c}$$

If constant wall heat-flux boundary condition is applied, due to the change in the definition of nondimensional temperature,

$$\text{at } r^* = 0: \frac{\partial \theta}{\partial r^*} = 0 \tag{13.81a}$$

$$\text{at } r^* = 1: \frac{\partial \theta}{\partial r^*} = 1 \tag{13.81b}$$

$$\text{at } x^* = 0: \theta = 0 \tag{13.81c}$$

Nondimensional temperatures calculated by Equation 13.79, then used to evaluate the local Nusselt number ($Nu_x$) and heat-transfer coefficient ($h_{nf,x}$), are defined by

$$Nu_x = \frac{h_{nf,x}d}{k_{nf}}. \tag{13.82}$$

For CWT and CHF boundary conditions, the local Nusselt numbers are calculated as

$$Nu_x = \frac{2}{\theta_b}\frac{\partial\theta}{\partial r^*}\bigg|_{r=r_0} \tag{13.83}$$

and

$$Nu_x = \frac{2}{(\theta_w - \theta_b)}, \tag{13.84}$$

where $\theta_w$ and $\theta_b$ are nondimensional wall and bulk temperatures, respectively. $\theta_b$ can be calculated by

$$\theta_b = \frac{\displaystyle\int_0^1 \theta(x^*,r^*)u^*r^*dr^*}{\displaystyle\int_0^1 u^*r^*dr^*}. \tag{13.85}$$

By using Equations 13.82 through 13.84, local heat-transfer coefficient ($h_{nf,x}$), average heat-transfer coefficient ($h_{ave.}$), and average Nusselt number ($Nu_{ave.}$) can be computed as

$$h_{ave.} = \frac{1}{L}\int_0^L h_{nf,x}\,dx \tag{13.86}$$

$$Nu_{ave.} = \frac{h_{ave.}d}{k_{nf}}. \tag{13.87}$$

When nondimensional energy equation is solved by using finite difference method, the energy equation may be discretized in space with employing second-order backward and central finite difference method for $x$- and $r$-direction respectively, and forward first-order finite difference method may be utilized for time dependency. The dimensionless energy equation can be solved explicitly; thus, stability analysis can be made for maintaining convergence of the code. Dimensionless energy equation is discretized by conservative explicit method as follows:

$$\frac{\theta_{i,j}^{m+1} - \theta_{i,j}^m}{\Delta t^*} + Pe(1 - r^{*2})\frac{3\theta_{i,j}^m - 4\theta_{i-1,j}^m + \theta_{i-2,j}^m}{2\Delta x^*}$$

$$= \frac{1}{r^*(\Delta r^*)^2}\left(k_{j+1/2}^{*m}r_{j+1/2}^{*m}(\theta_{i,j+1}^m - \theta_{i,j}^m) - k_{j-1/2}^{*m}r_{j-1/2}^{*m}(\theta_{i,j}^m - \theta_{i,j-1}^m)\right), \tag{13.88}$$

where $m$, $i$, and $j$ are subscripts that represent dimensionless time, axial, and radial direction nodes, respectively. Time is discretized by forward difference method; axial differentials and radial differentials are discretized by backward and central differencing schemes, respectively.

In transient numerical analyses for step change in CWT, when the inlet Peclet number and particle diameter are taken as 2000 and 20 nm, respectively, thermal conductivity and heat-transfer coefficient of nanofluid can be investigated at various particle volume fractions under transient condition in a circular duct with a length of 1.5 m and a diameter 0.01 m. The flow may be assumed to go into channel with 25°C and step change in wall temperature is 50°C. The higher heat-transfer coefficient values are obtained with the increment of the volume fraction of nanoparticles as it was expected. The step change in wall heat-flux condition can also be investigated numerically. For the hydrodynamically fully developed laminar flow in a circular duct, transient solutions can be carried out.

In the analysis for the same geometry Peclet number is taken as 2000, 20 nm as particle diameter, 25°C as inlet flow temperature, and 3000 W/m² as the wall flux. Table 13.8 [87] shows enhancement of heat transfer at different volume fractions, given data are computed for $Pe = 2000$ (Peclet number), $d_p = 20$ nm (particle diameter) in steady-state region with 25°C inlet flow temperature when wall temperature is taken at 50°C for CWT boundary condition, and for CHF boundary condition, wall heat flux is taken as 3000 W/m². As volume fraction increases, the higher heat-transfer coefficient and enhancement values are obtained. As it has been expected for fully developed laminar flow, the Nusselt number variations with volume fraction are very low. Average heat-transfer coefficients values are higher for the constant wall heat-flux boundary condition due to the larger Nusselt numbers.

Various Peclet numbers and particle volume fractions can be investigated under steady-state condition for both boundary conditions. The enhancement with the Peclet number and nanoparticle diameter variations for CWT and CHF boundary conditions earlier are almost indistinguishable as it can be seen from Tables 13.9 [87] and 13.10 [87]. The effect of the Peclet number on the enhancement of the heat-transfer coefficient is shown in Table 13.9 for both boundary conditions. The enhancement is increasing very low with

**TABLE 13.8**

Effect of Nanoparticle Volume Fraction on Enhancement of the Heat-Transfer Coefficient

| Nanoparticle Volume Fraction | Average Heat-Transfer Coefficient $h_{ave}$ (W/m²·K) | | Nusselt Number ($x^* = 300$) $Nu_x$ | | Enhancement ($h_{nf}/h_f$) of Heat-Transfer Coefficient (%) | |
|---|---|---|---|---|---|---|
| $\Phi$ | CWT | CHF | CWT | CHF | CWT | CHF |
| 0.00 | 264.24 | 325.16 | 3.6261 | 4.3604 | 0.00 | 0.00 |
| 0.01 | 283.37 | 348.59 | 3.6263 | 4.3638 | 7.24 | 7.21 |
| 0.02 | 301.37 | 370.64 | 3.6271 | 4.3669 | 14.05 | 13.99 |
| 0.03 | 319.08 | 392.33 | 3.6275 | 4.3684 | 21.75 | 20.66 |
| 0.04 | 336.67 | 413.89 | 3.6282 | 4.3688 | 27.41 | 27.28 |

*Source:* Tongkratoke, A. et al., Numerical study of nanofluid heat transfer enhancement with mixing thermal conductivity models, *Proceedings of ICHMT International Symposium on Advances in Computational Heat Transfer*, Bath, England, 2012.

CWT is constant wall temperature condition and CHF is constant heat flux.

**TABLE 13.9**

Effect of the Peclet Number on Enhancement of the Heat-Transfer Coefficient

| Nanoparticle Volume Fraction $\phi$ | Enhancement (%) $(h_{nf}/h_f)$ | | Enhancement (%) $(h_{nf}/h_f)$ | | Enhancement (%) $(h_{nf}/h_f)$ | | Enhancement (%) $(h_{nf}/h_f)$ | |
|---|---|---|---|---|---|---|---|---|
| | $Pe = 2000$ | | $Pe = 4000$ | | $Pe = 6000$ | | $Pe = 8000$ | |
| | CWT | CHF | CWT | CHF | CWT | CHF | CWT | CHF |
| 0.01 | 07.24 | 07.21 | 07.28 | 07.28 | 07.29 | 07.30 | 07.29 | 07.30 |
| 0.02 | 14.05 | 13.99 | 14.11 | 14.12 | 14.14 | 14.16 | 14.15 | 14.17 |
| 0.03 | 20.75 | 20.66 | 20.84 | 20.86 | 20.88 | 20.91 | 20.90 | 20.93 |
| 0.04 | 27.41 | 27.29 | 27.53 | 7.55 | 27.58 | 27.62 | 27.60 | 27.64 |

*Source:* Tongkratoke, A. et al., Numerical study of nanofluid heat transfer enhancement with mixing thermal conductivity models, *Proceedings of ICHMT International Symposium on Advances in Computational Heat Transfer*, Bath, England, 2012.

*Note:* CWT is constant wall temperature condition and CHF is constant heat flux.

**TABLE 13.10**

Effect of Particle Diameter on Enhancement of the Heat-Transfer Coefficient

| Nanoparticle Volume Fraction $\phi$ | Enhancement (%) $(h_{nf}/h_f)$ | | Enhancement (%) $(h_{nf}/h_f)$ | | Enhancement (%) $(h_{nf}/h_f)$ | | Enhancement (%) $(h_{nf}/h_f)$ | | Enhancement (%) $(h_{nf}/h_f)$ | |
|---|---|---|---|---|---|---|---|---|---|---|
| | $d_p = 10$ nm | | $d_p = 20$ nm | | $d_p = 30$ nm | | $d_p = 40$ nm | | $d_p = 50$ nm | |
| | CWT | CHF | CWT | CHF | CWT | CHF | CWT | CHF | CWT | CHF |
| 0.01 | 09.04 | 08.99 | 07.24 | 07.21 | 06.45 | 06.42 | 05.97 | 05.94 | 05.65 | 05.62 |
| 0.02 | 17.43 | 17.37 | 14.05 | 13.99 | 12.55 | 12.49 | 11.66 | 11.60 | 11.05 | 10.99 |
| 0.03 | 25.66 | 25.55 | 20.75 | 20.66 | 18.58 | 18.49 | 17.28 | 17.20 | 16.40 | 16.32 |
| 0.04 | 33.80 | 33.66 | 27.41 | 27.29 | 24.58 | 24.46 | 22.90 | 22.79 | 21.75 | 21.65 |

*Source:* Tongkratoke, A. et al., Numerical study of nanofluid heat transfer enhancement with mixing thermal conductivity models, *Proceedings of ICHMT International Symposium on Advances in Computational Heat Transfer*, Bath, England, 2012.

*Note:* CWT is constant wall temperature condition and CHF is constant heat flux.

the higher Peclet numbers on both cases. The constant wall heat-flux boundary condition values are slightly higher than CWT condition for increasing the Peclet number; this can be explained with the behavior at the entrance region of the channel where CHF condition heat-transfer ratios are higher than those of CWT condition (Table 13.9). It also shows that the effect of the Peclet number is very insignificant for hydrodynamically fully developed laminar flow. In Table 13.10, the particle diameter effect on heat-transfer enhancement is shown for different volume fractions. In Table 13.10, the enhancement due to nanoparticle volume fraction increment and particle size decrement is shown, and they are almost same for both boundary conditions. Nanofluids have shown its possibility in enhancing heat-transfer performance above its base fluids. This work presents a numerical study to analyze the nanofluid heat-transfer enhancement using different theoretical models: the effective viscosity and the effective thermal conductivity models. Nanofluid simulation can be performed to help one to understand thermal-fluid properties of nanofluids. The Maxwell and the Brownian motion models can be combined as thermal conductivity models and be applied in the simulation domain alternately, the mixing models. Figure 13.9 [88] shows results from the simulations of $Al_2O_3$/water nanofluid flow with different mixing models under laminar fully developed flow condition through a rectangular pipe.

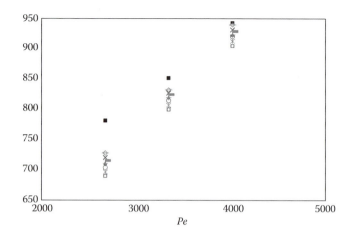

**FIGURE 13.9**
Comparison between the effective thermal conductivity models and experimental data in the prediction of the heat-transfer coefficient at the nanofluid volume fraction at 0.02 [88].

The results showed that different effective thermal conductivity models play an important role, especially at wall surfaces where the convective heat transfer is enhanced effectively. The effective thermal conductivity at the wall region can be increased by increasing nanoparticle amount.

## 13.3 Empirical Correlation for Single-Phase Forced Convection with Nanofluids

a. *Correlation of Yang et al.* This correlation gives the average Nusselt number for the laminar flow inside a straight circular tube under CWT boundary condition [46]:

$$Nu = aRe_{nf}^{b}Pr_{nf}^{1/3}\left(\frac{d}{L}\right)^{1/3}\left(\frac{\mu_b}{\mu_w}\right)^{0.14},$$ 
(13.89)

where
$a = 1.86$
$b = 1/3$

Table 13.11 [46] lists the associated values: the modified values of the constants $a$ and $b$ for the nanofluids.

b. *Correlations of Maïga et al.* The nanofluids considered as single-phase fluids, validity ranges of the correlations are $Re \leq 1000$, $6 \leq Pr \leq 753$, and $\phi \leq 10\%$, the Nusselt numbers are proposed depending on the boundary condition as [67]

$$Nu_{nf} = 0.086Re_{nf}^{0.55}Pr_{nf}^{0.5},\text{ for constant wall heat flux,}$$ 
(13.90)

**TABLE 13.11**

$a$ and $b$ Values to Be Used in Equation 13.18 for Different Nanofluids

| Base Fluid | Nanoparticles | Heated Fluid Temperatures (°C) | A | b |
|---|---|---|---|---|
| A commercial automatic transmission fluid | — | 50 | 1.90 ± 0.13 | 0.33 ± 0.02 |
| A commercial automatic transmission fluid | Graphite 2 wt% | 50 | 1.64 ± 0.08 | 0.33 ± 0.01 |
| A commercial automatic transmission fluid | Graphite 2.5 wt% | 50 | 1.66 ± 0.10 | 0.34 ± 0.02 |
| A commercial automatic transmission fluid | — | 70 | 2.28 ± 0.19 | 0.28 ± 0.02 |
| A commercial automatic transmission fluid | Graphite 2 wt% | 70 | 2.00 ± 0.19 | 0.28 ± 0.02 |
| A commercial automatic transmission fluid | Graphite 2.5 wt% | 70 | 1.88 ± 0.13 | 0.30 ± 0.03 |
| A mixture of two synthetic base oils with some additive packages | — | 50 | 2.09 ± 0.14 | 0.31 ± 0.02 |
| A mixture of two synthetic base oils with some additive packages | Graphite[a] 2 wt% | 50 | 2.17 ± 0.12 | 0.29 ± 0.01 |
| A mixture of two synthetic base oils with some additive packages | Graphite 2 wt% | 50 | 1.81 ± 0.13 | 0.33 ± 0.02 |
| A mixture of two synthetic base oils with some additive packages | — | 70 | 3.11 ± 0.30 | 0.22 ± 0.02 |
| A mixture of two synthetic base oils with some additive packages | Graphite[a] 2 wt% | 70 | 2.92 ± 0.30 | 0.23 ± 0.03 |
| A mixture of two synthetic base oils with some additive packages | Graphite 2 wt% | 70 | 2.66 ± 0.24 | 0.24 ± 0.02 |

*Source:* Yang, Y. et al., *Int. J. Heat Mass Transfer*, 48, 1107, 2005.
[a] From a different source than that used in the other experimental fluids.

$$Nu_{nf} = 0.28Re_{nf}^{0.35}Pr_{nf}^{0.36}, \text{ for CWT.} \qquad (13.91)$$

The fully developed Nusselt number was also proposed for the turbulent flow regime, validity ranges are $10,000 < Re < 500,000$, $6.6 < Pr < 13.9$, and $\phi < 10\%$ [89], as

$$Nu_{fd,nf} = 0.085Re_{nf}^{0.71}Pr_{nf}^{0.35}. \qquad (13.92)$$

c. *Correlation of Buongiorno.* The Nusselt number correlation for the turbulent flow of nanofluids inside straight circular tubes is obtained as [90]

$$Nu_{nf} = \frac{f/8(Re_{nf} - 1000)Pr_{nf}}{1 + \delta_{\upsilon}^{+}\sqrt{f/8}\left(Pr_{\upsilon}^{2/3} - 1\right)}. \qquad (13.93)$$

where $f$ is the friction factor calculated by using a classical correlation for the turbulent flow. $\delta_{\upsilon}^{+}$ is the dimensionless thickness of the laminar sublayer, an empirical constant. $Re_{nf}$ and $Pr_{nf}$ should be calculated at the bulk mean temperature and bulk mean particle volume fraction. $Pr_{\upsilon}$ should be calculated corresponding to the laminar sublayer of the turbulent flow.

d. *Correlations of Velagapudi et al.* A correlation for the turbulent forced convection of nanofluids inside circular tubes was proposed as [91]

$$Nu_{nf} = cRe_{nf}^{m}Pr_{nf}^{n}. \qquad (13.94)$$

The empirical constants in Equation 13.94, namely, $c$, $m$, and $n$, were determined by applying a regression analysis to the experimental data [14,45,92], so

$$Nu_{nf} = 0.0256 Re_{nf}^{0.8} Pr_{nf}^{0.4}, \text{ for Al}_2O_3/\text{water nanofluids,}$$ (13.95)

$$Nu_{nf} = 0.027 Re_{nf}^{0.8} Pr_{nf}^{0.4}, \text{ for Cu / water nanofluids.}$$ (13.96)

A correlation for the nanofluids laminar flow inside circular tubes was proposed as

$$Nu_{nf} = b\left( Re_{nf} Pr_{nf} \frac{d}{L} \right)^{1/3}.$$ (13.97)

The empirical constant $b$ was found to be equal to 1.98 [14,15].

### Example 13.3

Using the experimental correlation with nanofluids given in Equations 13.17, 13.21, and 13.24 for turbulent flow conditions, calculate the Nusselt number under the same nanofluid conditions as the following condition: a hot nanofluid of $Al_2O_3/$water flows through the tube with the Reynolds and Prandtl numbers equal to 10,000 and 7, respectively, and 1% volume fraction.

### Solution

$$Re_{nf} = 10{,}000$$

$$Pr_{nf} = 7$$

$$\phi = 1\%$$

From Equation 13.17: the Pak and Cho correlation

$$Nu_{nf} = 0.021 Re_{nf}^{0.8} Pr_{nf}^{0.5}$$

$$= (0.021)(10{,}000)^{0.8}(7)^{0.5}$$

$$= 88.058$$

From Equation 13.21: the Maïga et al. correlation

$$Nu_{fd,nf} = 0.085 Re_{nf}^{0.71} Pr_{nf}^{0.35}$$

$$= (0.085)(10{,}000)^{0.71}(7)^{0.35}$$

$$= 116.199$$

From Equation 13.24: the Velagapudi et al. correlation

$$Nu_{nf} = 0.0256 Re_{nf}^{0.8} Pr_{nf}^{0.4}$$

$$= (0.0256)(10{,}000)^{0.8}(7)^{0.4}$$

$$= 88.36$$

As one can notice, the Pak and Cho correlation performs the result close to the Velagapudi et al. correlation, then the heat-transfer coefficient can be defined by $h_{nf} = Nu_{nf} k_{nf}/d$.

e. *Correlation of Chun et al.* The effective heat-transfer coefficient correlation for the laminar flow of nanofluids which was $Al_2O_3$/transformer oil inside a double-pipe heat exchanger for the applicability ranges of $100 < Re < 450$ and $0.25\% < \phi < 0.5\%$, was as a following expression [93]:

$$h_{nf} = 1.7 \frac{k_{nf}}{d} Re_{nf}^{0.4}. \tag{13.98}$$

The researchers [93] indicated that the Prandtl number is not an effective parameter for the proposed correlation.

f. *Correlation of Sharma et al.* A correlation for the determination of the Nusselt number was provided, for the transition regime flow of $Al_2O_3(47\ nm)$/water nanofluid inside a tube with twisted tape inserts under constant wall heat-flux boundary condition, as [94]

$$Nu_{nf} = 3.138 \times 10^{-3} Re_{nf} Pr_{nf}^{0.6} \left(1.0 + \frac{H}{d}\right)^{0.03} (1+\phi)^{1.22}. \tag{13.99}$$

The applicability range of the correlation is $0 < H/d < 15$, $3500 < Re < 8500$, $4.5 < Pr < 5.5$, and $35 < T_b < 40$. $H/d$ is called the twist ratio, which is the pitch for 180° rotation divided by the tube inner diameter.

For the same flow configuration, a friction factor correlation was provided as

$$f_{nf} = 172 Re_{nf}^{-0.96} (1.0+\phi)^{2.15} \left(1.0 + \frac{H}{d}\right)^{2.15}. \tag{13.100}$$

Validity range in terms of particle volume fraction is $\phi < 1\%$ and $f$ is the Fanning friction factor defined as in Equation 6.33.

g. *Correlation of Anoop et al.* The convective heat-transfer performance of has been investigated. A local Nusselt number correlation for laminar flow of $Al_2O_3$/water nanofluids inside a circular tube under constant wall heat-flux boundary condition was proposed based on the experimental data as [18]

$$Nu_{x,nf} = 4.36 +$$

$$\left[6.219 \cdot 10^{-3} x_+^{-1.1522} \left(1+\phi^{0.1533}\right) \exp\left(-2.5228 x_+\right)\right] \left[1 + 0.57825 \left(\frac{d_p}{100\ nm}\right)^{-0.2183}\right]. \tag{13.101}$$

$d_p$ is particle size in nm. $x_+$ is defined as follows:

$$x_+ = \frac{x}{d Re_{nf} Pr_{nf}}, \tag{13.102}$$

where $x$ is axial position. The validity range of the correlation is $50 < x/d < 200$, $500 < Re < 2000$.

h. *Correlation of Asirvatham et al.* The local Nusselt number correlation for the convective heat transfer of laminar CuO(40 nm)/water nanofluid, $Re < 2300$ and $\phi = 0.003\%$, inside a circular tube under constant wall heat-flux boundary condition was introduced as [95]

$$Nu_{x,nf} = 0.155 Re_{nf}^{0.59} Pr_{nf}^{0.35} \left(\frac{d}{x}\right)^{0.38}.$$

(13.103)

**Example 13.4**

Nanofluid, CuO/water, flows in a tube. CuO nanoparticles with a particle radius of 100 nm dispersed in water at 273.15 K and 0.003% particle volume fraction by using the Yu and Choi model. CuO density is considered to be 6.31 g/cm³. The specific heat of CuO can be calculated from $c_{p,CuO} = 5.154 \times 10^{-3} T^3$. Properties of water can be obtained from the water properties at pressure equals to 1 bar. Determine the effective thermal conductivity with nanolayer consideration, the density, the specific heat values, and the heat-transfer coefficient enhancement ratio of nanofluid. The enhancement ratio can be calculated by using Equations 13.49 and 13.103. Nanofluid flows with a mean velocity of 0.04 m/s.

**Solution**
Where
$\phi = 0.003\% = 0.003/100 = 0.00003$
$k_p = 20$ W/m K (from Table 13.1; copper oxide (CuO))
$r_p = 100$ nm $= 100 \times 10^{-9}$ m
$N_A = 6.023 \times 10^{23}$/mol

From properties of water at $P = 1$ bar (100 kPa) and $T = 273.15$ K (0°C),

$k_f = 0.552$ W/m K
$\mu_f = 0.0005468$ Pa s
$\rho_f = 999.8$ kg/m³
$M_f = 18$ g/mol $= 18 \times 10^{-3}$ kg/mol
$Pr_f = 13.67$.
$\qquad \varepsilon_p = k_p/k_f$

From    $= 20/0.552$,

$\qquad \varepsilon_p = 36.232$

$$t = \frac{1}{\sqrt{3}} \left(\frac{4M_f}{\rho_f N_A}\right)^{1/3}$$

$$= \frac{1}{\sqrt{3}} \left(\frac{(4)(18 \times 10^{-3})}{(999.8)(6.023 \times 10^{23})}\right)^{1/3}$$

$$t = 2.8443 \times 10^{-10} \text{ m}$$

$$\beta_L = \frac{t}{r_p} = \frac{(2.8443 \times 10^{-10})}{(100 \times 10^{-9})} = 0.0028443.$$

From $Z = \varepsilon_p(1 + \beta_L) - 1$,
$$= (36.232)(1 + 0.0028443) - 1$$
$$Z = 35.335$$

$$k_l = \frac{k_f Z^2}{(Z - \beta_L)\ln(1 + Z) + \beta_L Z}$$

$$= \frac{(0.552)(35.335)^2}{(35.335 - 0.0028443)\ln(1 + 35.335) + (0.0028443)(35.335)}$$

$$k_l = 5.425 \ \text{W/m K}$$

$$\gamma_k = \frac{k_l}{k_p} = \frac{5.425}{20} = 0.271$$

$$k_{pe} = \frac{[2(1 - \gamma_k) + (1 + \beta_L)^3(1 + 2\gamma_k)]\gamma_k}{-(1 - \gamma_k) + (1 + \beta_L)^3(1 + 2\gamma_k)} k_p$$

$$= \frac{[(2)(1 - 0.271) + (1 + 0.0028443)^3(1 + (2)(0.271))](0.271)}{-(1 - 0.271) + (1 + 0.0028443)^3(1 + (2)(0.271))} \quad (20)$$

$$k_{pe} = 19.767 \ \text{W/m K}$$

$$k_{nf} = \frac{k_{pe} + 2k_f + 2(k_{pe} - k_f)(1 + \beta_L)^3 \phi}{k_{pe} + 2k_f - (k_{pe} - k_f)(1 + \beta_L)^3 \phi} k_f$$

$$= \frac{(19.767) + (2)(0.552) + (2)(19.767 - 0.552)(1 + 0.0028443)^3(0.00003)}{(19.767) + (2)(0.552) - (19.767 - 0.552)(1 + 0.0028443)^3(0.00003)} (0.552)$$

$$k_{nf} = 0.552 \ \text{W/m K},$$

where $\rho_p = 6.31 \ \text{g/cm}^3 = 6310 \ \text{kg/m}^3$.
From properties of water at $P = 1$ bar (100 kPa) and $T = 273.15$ K (0°C),

$$\rho_f = 999.8 \ \text{kg/m}^3$$

$$\rho_{nf} = \phi \rho_p + (1 - \phi)\rho_f$$

$$= (0.00003)(6310) + (1 - 0.00003)(999.8)$$

$$\rho_{nf} = 999.9593 \ \text{kg/m}^3$$

$$\rho_p = 6.31 \ \text{g/cm}^3 = 6310 \ \text{kg/m}^3$$

$$c_{p,CuO} = 5.154 \times 10^{-3} T^3$$

$$= (5.154 \times 10^{-3})(273.15)^3$$

$$c_{p,CuO} = 105,038.3833 \text{ J/kg K}$$

$$c_{p,f} = 4.2179 \text{ kJ/kg K} = 4217.9 \text{ J/kg K}$$

$$\rho_f = 999.8 \text{ kg/m}^3$$

$$c_{p,nf} = \phi c_{p,p} + (1-\phi)c_{p,f}$$
$$= (0.02)(105,038.3833) + (1-0.02)(4217.9)$$

$$c_{p,nf} = 4220.925 \text{ J/kg K} = 4.220925 \text{ kJ/kg K}.$$

The specific heat can also be calculated by

$$(\rho c_p)_{nf} = \phi(\rho c_p)_p + (1-\phi)(\rho c_p)_f$$

$$((999.9593)c_p)_{nf} = (0.00003)(6310)(105,038.3833) + (1-0.00003)(999.8)(4217.9)$$

$$((999.9593)c_p)_{nf} = 4,236,814$$

$$c_{p,nf} = \frac{4236814}{999.9593} = 4236.986 \text{ J/kg K} = 4.237 \text{ kJ/kg K},$$

where $k_{nf} = 0.552$ W/m K

$$\rho_{nf} = 999.9593 \text{ kg/m}^3$$

$$c_{p,nf} = 4236.986 \text{ J/kg K}$$

$$\mu_{nf} = (1 + 7.3\phi + 123\phi^2)\mu_{f,b} = 5.47 \times 10^{-4} \text{ Pa s}$$

$$Re_f = \frac{\rho_f u_m d_i}{\mu_f} = \frac{999.8 \times 0.04 \times 0.0254}{5.47 \times 10^{-4}} = 1857.6$$

$$Re_{nf} = \frac{\rho_{nf} u_m d_i}{\mu_{nf}} = \frac{1106.004 \times 0.04 \times 0.0254}{5.47 \times 10^{-4}} = 1857.6$$

$$Pr_{nf} = \frac{\mu_{nf} c_{p,nf}}{k_{nf}} = \frac{5.47 \times 10^{-4} \times 4236.986}{0.552} = 4.198$$

From Equation 13.49,

$$\frac{h_{nf}}{h_f} = \left(\frac{\rho_{nf}}{\rho_f} \frac{c_{p,nf}}{c_{p,f}}\right)^{1/3} \left(\frac{k_{nf}}{k_f}\right)^{2/3}$$

$$= \left(\frac{(999.9593)}{(999.8)} \frac{(4236.986)}{(4217.9)}\right)^{1/3} \left(\frac{0.552}{0.552}\right)^{2/3}$$

$$\frac{h_{nf}}{h_f} = 1.002.$$

From

$$\frac{Nu_{x,nf}}{Nu_{x,f}} = \frac{0.155Re_{nf}^{0.59}Pr_{nf}^{0.35}(d/x)^{0.38}}{0.155Re_{f}^{0.59}Pr_{f}^{0.35}(d/x)^{0.38}} = \frac{Re_{nf}^{0.59}Pr_{nf}^{0.35}}{Re_{f}^{0.59}Pr_{f}^{0.35}} = 0.665,$$

$$\frac{h_{nf}}{h_f} = \frac{Nu_{x,nf}}{Nu_{x,f}} \frac{k_{nf}}{k_f} = 0.665\frac{0.552}{0.552} = 0.665.$$

So the conventional correlation overestimates the Asirvatham et al. correlation [95], which is obtained from the experimental study.

i. *Correlation of Jung et al.* This following correlation [96] is validated for $5 < Re < 300$ and $0.6\% < \phi < 1.8\%$; the $Al_2O_3$(170 nm)/water nanofluid experimentally investigated inside a rectangular microchannel under constant wall heat-flux boundary condition:

$$Nu_{nf} = 0.014\phi^{0.095}Re_{nf}^{0.4}Pr_{nf}^{0.6} \qquad (13.104)$$

and the friction factor can be determined as

$$f_{nf} = \frac{56.9}{Re_{nf}}, \qquad (13.105)$$

where $f$ is the Fanning friction factor defined as in Equation 6.33.

j. *Correlation of Chandrasekar et al.* The determined Nusselt number and friction factor laminar flow of $Al_2O_3$(43 nm)/water nanofluid inside a straight circular tube with wire coil inserts under constant wall heat-flux boundary condition were investigated as [97]

$$Nu_{nf} = 0.279(RePr)_{nf}^{0.558}\left(\frac{p}{d}\right)^{-0.477}(1+\phi)^{134.65} \qquad (13.106)$$

$$f_{nf} = 530.8Re_{nf}^{-0.909}\left(\frac{p}{d}\right)^{-1.388}(1+\phi)^{-512.26}. \qquad (13.107)$$

$p$ is pitch ratio; validity ranges of the correlations are $Re < 2300$, $\phi = 0.1\%$, $2 \leq p/d \leq 3$; and $f$ is the Fanning friction factor defined as in Equation 6.33.

k. *Correlation of Duangthongsuk and Wongwises.* A Nusselt number correlation for the turbulent flow of $TiO_2$(21 nm)/water nanofluid inside a horizontal double-tube counterflow heat exchanger, $\phi \leq 1\%$, and 3000 < Re < 18,000, was obtained as [98]

$$Nu_{nf} = 0.074 Re_{nf}^{0.707} Pr_{nf}^{0.385} \phi^{0.074}.$$  (13.108)

The friction factor for particle volume fractions up to 2% was also proposed as

$$f_{nf} = 0.961 \phi^{0.052} Re_{nf}^{-0.375},$$  (13.109)

where *f* is the Fanning friction factor defined as in Equation 6.33.

l. *Correlation of Wu et al.* A Nusselt number correlation for $Al_2O_3$(56 nm)/water nanofluid flowing inside trapezoidal microchannels, laminar flow with 200 < Re < 1300, $4.6 \leq Pr \leq 5.8$, and $0 \leq \phi \leq 0.26\%$, under constant wall heat-flux boundary condition was provided as [99]

$$Nu_{nf} = 0.566(1 + 100\phi)^{0.57} Re_{nf}^{0.20} Pr_{nf}^{0.18}.$$  (13.110)

A correlation for friction factor, 190 < Re < 1020 and $0 \leq \phi \leq 0.26\%$, was also provided as

$$f_{nf} = 5.43(1 + 100\phi)^{0.18} Re_{nf}^{-0.81},$$  (13.111)

where *f* is the Fanning friction factor defined as in Equation 6.33.

**Example 13.5**

Using the experimental correlation with nanofluids given in Equations 13.42, the Wu et al. correlation, for turbulent flow conditions, calculates the Nusselt number under the same nanofluid conditions as Example 13.2, but a hot nanofluid of $Al_2O_3$/water flows with the Reynolds and Prandtl numbers equals to 10,000 and 7, respectively, and 0.25% volume fraction.

**Solution**

$$Re_{nf} = 10,000$$

$$Pr_{nf} = 7$$

$$\phi = 0.25\%$$

From Example 13.2, the Pak and Cho, Maïga et al., and Velagapudi et al. correlations do not take volume fraction into accounts, so

the Pak and Cho correlation provides $Nu_{nf} = 88.058$

the Maïga et al. correlation provides $Nu_{nf} = 116.199$

the Velagapudi et al. correlation provides $Nu_{nf} = 88.36$

From Equation 13.42: the Wu et al. correlation

$$Nu_{nf} = 0.566(1 + (100\phi))^{0.57} Re_{nf}^{0.2} Pr_{nf}^{0.18}$$

$$= 0.566(1 + (100 \times 0.25))^{0.57} 10,000_{nf}^{0.2} 7_{nf}^{0.18}$$

$$Nu_{nf} = 32.47$$

As one can notice, the Wu et al. correlation underestimated the results from other correlations because the former is used for $Al_2O_3$/water nanofluid flowing inside trapezoidal microchannels, not circular tube. As well as, this correlation is used for laminar flow, so this example can show that using a wrong correlation leads to under-/overestimated results [100].

---

## Problems

**13.1** Determine the thermal conductivities of nanofluid, alumina ($Al_2O_3$)/water, by using two classical models, the Maxwell and the Hamilton and Crosser models, and compare these results with the use of the base fluid:

1. At 5% of particle volume fraction and sphericity value equals to 1
2. At 5% of particle volume fraction and sphericity value equals to 3

**13.2** By considering the combination of static and the Brownian motion parts, determine the thermal conductivity of nanofluid, alumina ($Al_2O_3$) nanoparticles with a mean diameter of 36 nm dispersed in water at 273.15 K and 2% particle volume fraction by using the Maxwell model to calculate the static part. Density of alumina is considered to be 4 $g/cm^3$. Properties of water can be obtained from the water properties at pressure equals to 1 bar.

**13.3** Determine the thermal conductivity of nanofluid in Problem 13.2 when considering the apparent radius of the nanoparticle clusters equal to $r_{cl1} = 10$ nm and $r_{cl2} = 100$ nm, respectively. Compare results from both cluster radius values.

**13.4** Determine the effective thermal conductivity of nanofluid, CuO nanoparticles with a particle radius of 100 nm dispersed in water at 273.15 K and 2% particle volume fraction by using the Yu and Choi model. Properties of water can be obtained from the water properties at pressure equals to 1 bar.

**13.5** Calculate the thermal conductivity of nanofluid, CuO nanoparticles with a particle radius of 100 nm dispersed in water at 273.15 K and 2% particle volume fraction, by considering a nanolayer as a spherical shell from Equation 13.21, $t$ from Equation 13.22 and letting $k_l = k_p$. Properties of water can be obtained from the water properties at pressure equals to 1 bar.

**13.6** Calculate the thermal conductivity of nanofluid in Problem 13.5 by considering the combination of the nanoparticle and nanolayer.

**13.7** Determine the density of nanofluid, CuO/water at 2% and 3% particle volume fractions. Properties of water can be obtained from the water properties at pressure equals to 1 bar. Density of CuO is considered to be 6.31 $g/cm^3$.

**13.8** Compare specific heat values of nanofluid, CuO/water, which are calculated by using (13.2) at 273.15 K and 2% particle volume fractions. Properties of water can be obtained from the water properties at pressure equals to 1 bar. The specific heat of CuO can be calculated from $c_{p,CuO} = 5.154 \times 10^{-3}\, T^3$.

**13.9** Compare viscosity values of nanofluid, CuO/water, which are calculated by using Equations 13.5 and 13.6 at 273.15 K and 2% particle volume fractions. Properties of water can be obtained from the water properties at pressure equals to 1 bar.

**13.10** Using information from Problem 13.9 with following data to calculate the viscosity of the nanofluid at 323.15 K. Density of CuO is considered to be 6.31 g/cm$^3$ and CuO nanoparticles have particle radius of 100 nm.

# References

1. Eastman, J., Choi, S. U. S., Li, S., Thompson, L., and Lee, S., Enhanced thermal conductivity through the development of nanofluids, *Proceedings of the Symposium on Nanophase and Nanocomposite Materials II*, Boston, MA, Vol. 447, pp. 3–11, 1997.
2. Granquist, C. G. and Buhrman, R. A., Ultrafine metal particles, *J. Appl. Phys.*, 47, 2200–2219, 1976.
3. Yu, W., France, D. M., Choi, S. U. S., and Routbort, J. L., *Review and Assessment of Nanofluid Technology for Transportation and Other Applications*, ANL/ESD/07-9, Argonne National Laboratory, Argonne, IL, 2007.
4. Masuda, H., Ebata, A., Teramae, K., and Hishinuma, N., Alteration of thermal conductivity and viscosity of liquid by dispersing ultra-fine particles (dispersion of g-Al$_2$O$_3$, SiO$_2$, and TiO$_2$ ultra-fine particles), *Netsu Bussei*, 4, 227–233, 1993.
5. Choi, S. U. S., Enhancing thermal conductivity of fluids with nanoparticles, in *Developments and Applications of Non-Newtonian Flows*, D. A. Siginer and H. P. Wang (Eds.), ASME, New York, FED-Vol.231/MD, Vol. 66, pp. 99–105, 1995.
6. Eastman, J. A., Choi, S. U. S., Li, S., Yu, W., and Thompson, L. J., anomalously increased effective thermal conductivities of ethylene glycol-based nanofluids containing copper nanoparticles, *Appl. Phys. Lett.*, 78(6), 718–720, 2001.
7. Xie, H., Wang, J., Xi, T., Liu, Y., Ai, F., and Wu, Q., Thermal conductivity enhancement of suspensions containing nanosized alumina particles, *J. Appl. Phys.*, 91(7), 4568–4572, 2002.
8. Das, S. K., Putra, N., Thiesen, P., and Roetzel, W., Temperature dependence of thermal conductivity enhancement for nanofluids, *J. Heat Transfer*, 125(4), 567–574, 2003.
9. Chon, C. H., Kihm, K. D., Lee, S. P., and Choi, S. U. S., Empirical correlation finding the role of temperature and particle size for nanofluid (Al$_2$O$_3$) thermal conductivity enhancement, *Appl. Phys. Lett.*, 87(15), 153107, 2005.
10. Li, C. H. and Peterson, G. P., Experimental investigation of temperature and volume fraction variations on the effective thermal conductivity of nanoparticle suspensions (nanofluids), *J. Appl. Phys.*, 99(8), 084314, 2006.
11. Oh, D., Jain, A., Eaton, J. K., Goodson, K. E., and Lee, J. S., Thermal conductivity measurement and sedimentation detection of aluminum oxide nanofluids by using the 3ω method, *Int. J. Heat Fluid Flow*, 29(5), 1456–1461, 2008.
12. Li, Q. and Xuan, Y., Convective heat transfer and flow characteristics of cu–water nanofluid, *Sci. China Ser. E: Technol. Sci.*, 45(4), 408–416, 2002.
13. Heris, S. Z., Esfahany, M. N., and Etemad, S., Experimental investigation of convective heat transfer of Al$_2$O$_3$/water nanofluid in circular tube, *Int. J. Heat Fluid Flow*, 28(2), 203–210, 2007.
14. Pak, B. C. and Cho, Y. I., Hydrodynamic and heat transfer study of dispersed fluids with submicron metallic oxide particles, *Exp. Heat Transfer*, 11(2), 151–170, 1998.

15. Wen, D. and Ding, Y., Experimental investigation into convective heat transfer of nanofluids at the entrance region under laminar flow conditions, *Int. J. Heat Mass Transfer*, 47(24), 5181–5188, 2004.
16. Heris, S. Z., Etemad, S., and Esfahany, M. N., Experimental investigation of oxide nanofluids laminar flow convective heat transfer, *Int. Commun. Heat Mass Transfer*, 33(4), 529–535, 2006.
17. Hwang, K. S., Jang, S. P., and Choi, S. U. S., Flow and convective heat transfer characteristics of water-based $Al_2O_3$ nanofluids in fully developed laminar flow regime, *Int. J. Heat Mass Transfer*, 52(1–2), 193–199, 2009.
18. Anoop, K., Sundararajan, T., and Das, S. K., Effect of particle size on the convective heat transfer in nanofluid in the developing region, *Int. J. Heat Mass Transfer*, 52(9–10), 2189–2195, 2009.
19. Maïga, S. E. B., Nguyen, C. T., Galanis, N., and Roy, G., Heat transfer behaviours of nanofluids in a uniformly heated tube, *Superlattices Microstruct.*, 35(3–6), 543–557, 2004.
20. Xuan, Y. and Roetzel, W., Conceptions for heat transfer correlation of nanofluids, *Int. J. Heat Mass Transfer*, 43(19), 3701–3707, 2000.
21. Heris, S. Z., Esfahany, M. N., and Etemad, G., Numerical investigation of nanofluid laminar convective heat transfer through a circular tube, *Numer. Heat Transfer A: Appl. Int. J. Comput. Methodol.*, 52(11), 1043–1058, 2007.
22. Bianco, V., Chiacchio, F., Manca, O., and Nardini, S., Numerical investigation of nanofluids forced convection in circular tubes, *Appl. Therm. Eng.*, 29(17–18), 3632–3642, 2009.
23. Einstein, A., A new determination of the molecular dimensions, *Ann. Phys.*, 324(2), 289–306, 1906.
24. Drew, D. A. and Passman, S. L., *Theory of Multicomponent Fluids*, Springer, Berlin, Germany, 1999.
25. Brinkman, H. C. The viscosity of concentrated suspensions and solutions, *J. Chem. Phys.*, 20, 571–581, 1952.
26. Nguyen, C., Desgranges, F., Roy, G., Galanis, N., Maré, T., Boucher, S., and Mintsa, H. A., Temperature and particle-size dependent viscosity data for water-based nanofluids—Hysteresis phenomenon, *Int. J. Heat Fluid Flow*, 28(6), 1492–1506, 2007.
27. Behzadmehr, A., Saffar-Avval, M., and Galanis, N., Prediction of turbulent forced convection of a nanofluid in a tube with uniform heat flux using a two phase approach, *Int. J. Heat Fluid Flow*, 28, 211–219, 2007.
28. Kakaç, S. and Pramuanjaroenkij, A., Review of convective heat transfer enhancement with nanofluids, *Int. J. Heat Mass Transfer*, 52, 3187–3196, 2009.
29. Özerinç, S., Kakaç, S., and Yazıcıoğlu, A. G., Enhanced thermal conductivity of nanofluids: A state-of-the-art review, *Microfluid. Nanofluid.*, 8(2), 145–170, 2010.
30. Maxwell, J. C., *A Treatise on Electricity and Magnetism*, Clarendon Press, Oxford, U.K., 1873.
31. Hamilton, R. L. and Crosser, O. K., Thermal conductivity of heterogeneous two component systems, *Ind. Eng. Chem. Fundam.*, 1, 182–191, 1962.
32. Bhattacharya, P., Saha, S. K., Yadav, A., Phelan, P. E., and Prasher, R. S., Brownian dynamics simulation to determine the effective thermal conductivity of nanofluids, *J. Appl. Phys.*, 95(11), 6492–6494, 2004.
33. Koo, J. and Kleinstreuer, C., A new thermal conductivity model for nanofluids, *J. Nanopart. Res.*, 6(6), 577–588, 2004.
34. Özerinç, S., Heat transfer enhancement with nanofluids, Thesis, Master of Science in Mechanical Engineering, Middle East Technical University, Turkey, May 2010.
35. Wang, B., Zhou, L., and Peng, X., A fractal model for predicting the effective thermal conductivity of liquid with suspension of nanoparticles, *Int. J. Heat Mass Transfer*, 46(14), 2665–2672, 2003.
36. Prasher, R., Phelan, P. E., and Bhattacharya, P., Effect of aggregation kinetics on the thermal conductivity of nanoscale colloidal solutions (nanofluid), *Nano Lett.*, 6(7), 1529–1534, 2006.
37. He, Y., Jin, Y., Chen, H., Ding, Y., Cang, D., and Lu, H., Heat transfer and flow behaviour of aqueous suspensions of $TiO_2$ nanoparticles (nanofluids) flowing upward through a vertical pipe, *Int. J. Heat Mass Transfer*, 50(11–12), 2272–2281, 2007.
38. Evans, W., Prasher, R., Fish, J., Meakin, P., Phelan, P., and Keblinski, P., Effect of aggregation and interfacial thermal resistance on thermal conductivity of nanocomposites and colloidal nanofluids, *Int. J. Heat Mass Transfer*, 51(5–6), 1431–1438, 2008.

39. Nan, C., Birringer, R., Clarke, D. R., and Gleiter, H., Effective thermal conductivity of particulate composites with interfacial thermal resistance, *J. Appl. Phys.*, 81(10), 6692–6699, 1997.

40. Xuan, Y., Li, Q., and Hu, W., Aggregation structure and thermal conductivity of nanofluids, *AIChE J.*, 49(4), 1038–1043, 2003.

41. Yu, C., Richter, A. G., Datta, A., Durbin, M. K., and Dutta, P., Observation of molecular layering in thin liquid films using x-ray reflectivity, *Phys. Rev. Lett.*, 82(11), 2326–2329, 1999.

42. Yu, W. and Choi, S. U. S., The role of interfacial layers in the enhanced thermal conductivity of nanofluids: A renovated maxwell model, *J. Nanopart. Res.*, 5(1), 167–171, 2003.

43. Bruggeman, D. A. G., The calculation of various physical constants of heterogeneous substances. I. The dielectric constants and conductivities of mixtures composed of isotropic substances, *Ann. Phys.*, 416(7), 636–664, 1935.

44. Xue, Q. and Xu, W., A model of thermal conductivity of nanofluids with interfacial shells, *Mater. Chem. Phys.*, 90(2–3), 298–301, 2005.

45. Xuan, Y. and Li, Q., Investigation convective heat transfer and flow features of nanofluids, *J. Heat Transfer*, 125, 151–155, 2002.

46. Yang, Y., Zhang, Z. G., Grulke, E. A., Anderson, W. B., and Wu, G., Heat transfer properties of nanoparticle-in-fluid dispersions (nanofluids) in laminar flow, *Int. J. Heat Mass Transfer*, 48, 1107–1116, 2005.

47. Ding, Y., Alias, H., Wen, D., and Williams, R. A., Heat transfer of aqueous suspensions of carbon nanotubes (CNT nanofluids), *Int. J. Heat Mass Transfer*, 49, 240–250, 2006.

48. Ma, H. B., Wilson, C., Borgmeyer, B., Park, K., and Yu, Q., Effect of nanofluid on the heat transport capability in an oscillatory heat pipe, *Appl. Phys. Lett.*, 88, 143116, 2006.

49. Chen, H., Yang, W., He, Y., Ding, Y., Zhang, L., Tan, C., Lapkin, A. A., and Bavykin, D. V., Heat Transfer behaviour of aqueous suspensions of titanate nanofluids, *Powder Technol.*, 183, 63–72, 2008.

50. Kulkarni, D. P., Namburu, P. K., Bargar, H. E., and Das, D. K., Convective heat transfer and fluid dynamic characteristics of SiO$_2$ ethylene glycol/water nanofluid, *Heat Transfer Eng.*, 29(12), 1027–1035, 2008.

51. Xue, L., Keblinski, P., Phillpot, S. R., Choi, S. U. S., and Eastman, J. A., Effect of liquid layering at the liquid–solid interface on thermal transport, *Int. J. Heat Mass Transfer*, 47, 4277–4284, 2004.

52. Pozhar, L. A., Structure and dynamics of nanofluids: Theory and simulations to calculate viscosity, *Phys. Rev. E*, 61, 1432–1446, 2000.

53. Gupte, S. K., Advani, S. G., and Huq, P., Role of micro-convection due to non-affine motion of particles in a mono-disperse suspension, *Int. J. Heat Mass Transfer*, 38, 2945–2958, 1995.

54. Sato, Y., Deutsch, E., and Simonin, O., Direct numerical simulations of heat transfer by solid particles suspended in homogeneous isotropic turbulence, *Int. J. Heat Fluid Flow*, 19, 187–192, 1998.

55. Ali, A., Vafai, K., and Khaled, A. R. A., Comparative study between parallel and counter flow configurations between air and falling film desiccant in the presence of nanoparticle suspensions, *Int. J. Energy Res.*, 27, 725–745, 2003.

56. Khanafer, K., Vafai, K., and Lightstone, M., Buoyancy-driven heat transfer enhancement in a two-dimensional enclosure utilizing nanofluids, *Int. J. Heat Mass Transfer*, 46, 3639–3653, 2003.

57. Ali, A., Vafai, K., and Khaled, A. R. A., Analysis of heat and mass transfer between air and falling film in a cross flow configuration, *Int. J. Heat Mass Transfer*, 47, 743–755, 2004.

58. Gosselin, L. and da Silva, A. K., Combined heat transfer and power dissipation optimization of nanofluid flows, *Appl. Phys. Lett.*, 85, 4160–4162, 2004.

59. Kim, J., Kang, Y. T., and Choi, C. K., Analysis of convective instability and heat transfer characteristics of nanofluids, *Phys. Fluids*, 16, 2395–2401, 2004.

60. Nguyen, C. T., Roy, G., Gauthier, C., and Galanis, N., Heat transfer enhancement using Al$_2$O$_3$–water nanofluid for an electronic liquid cooling system, *Appl. Therm. Eng.*, 27(8–9), 1501–1506, 2007.

61. Roy, G., Nguyen, C. T., and Lajoie, P. R., Numerical investigation of laminar flow and heat transfer in a radial flow cooling system with the use of nanofluids, *Superlattices Microstruct.*, 35, 497–511, 2004.

62. Shenogin, S., Xue, L., Ozisik, R., Keblinski, P., and Cahill, D. G., Role of thermal boundary resistance on the heat flow in carbon-nanotube composited, *J. Appl. Phys.*, 95, 8136–8144, 2004.
63. Ding, Y. and Wen, D., Particle migration in a flow of nanoparticle suspensions, *Powder Technol.*, 149, 84–92, 2005.
64. Khaled, A. R. A. and Vafai, K., Heat transfer enhancement through control of thermal dispersion effects, *Int. J. Heat Mass Transfer*, 48, 2172–2185, 2005.
65. Koo, J. and Kleinstreuer, C., Laminar nanofluid flow in microheat-sinks, *Int. J. Heat Mass Transfer*, 48, 2652–2661, 2005.
66. Kumar, S. and Murthy, J. Y., A numerical technique for computing effective thermal conductivity of fluid–particle mixtures (Part B), *Num. Heat Transfer*, 47, 555–572, 2005.
67. Maïga, S. E. B., Palm, S. J., Nguyen, C. T., Roy, G., and Galanis, N., Heat transfer enhancement by using nanofluids in forced convection flows, *Int. J. Heat Fluid Flow*, 6, 530–546, 2005.
68. Wen, D. and Ding, Y., Effect of particle migration on heat transfer in suspensions of nanoparticles flowing through minichannels, *Microfluid Nanofluid*, 1, 183–189, 2005.
69. Wen, D. and Ding, Y., Experimental investigation into pool boiling heat transfer of aqueous based C-alumina nanofluids, *J. Nanopart. Res.*, 7, 265–274, 2005.
70. Wen, D. and Ding, Y., Formulation of nanofluids for natural convective heat transfer applications, *Int. J. Heat Fluid Flow*, 26, 855–864, 2005.
71. Xuan, Y. and Yao, Z., Lattice Boltzmann model for nanofluids, *Heat Mass Transfer*, 41, 199–205, 2005.
72. Evans, W., Fish, J., and Keblinski, P., Role of brownian motion hydrodynamics on nanofluid thermal conductivity, *Appl. Phys. Lett.*, 88, 093116, 2006.
73. Jou, R. Y. and Tzeng, S. C., Numerical research of nature convective heat transfer enhancement filled with nanofluids in rectangular enclosures, *Int. Commun. Heat Mass Transfer*, 3, 727–736, 2006.
74. Keblinski, P. and Thomin, J., Hydrodynamic field around a brownian particle, *Phys. Rev. E*, 73, 010502, 2006.
75. Kim, J., Choi, C. K., Kang, Y. T., and Kim, M. G., Effect of thermodiffusion nanoparticles on convective instability in binary nanofluids, *Nanoscale Microscale Thermophys. Eng.*, 10, 29–39, 2006.
76. Mansour, R. B., Galanis, N., and Nguyen, C. T., Effect of uncertainties in physical properties on forced convection heat transfer with nanofluids, *Appl. Therm. Eng.*, 27, 240–249, 2006.
77. Prasher, R., Evans, W., Meakin, P., Fish, J., Phelan, P., and Keblinski, P., Effect of aggregation on thermal conduction in colloidal nanofluids, *Appl. Phys. Lett.*, 89, 143119, 2006.
78. Wang, X. and Mujumdar, A., A review on nanofluids—Part I: Theoretical and numerical investigations, *Braz. J. Chem. Eng.*, 25(4), 613–630, 2008.
79. Kakaç, S. and Yener, Y., *Convective Heat Transfer*, 2nd edn., CRC Press, Boca Raton, FL, 1995.
80. Xuan, Y. and Li, Q., Heat transfer enhancement of nanofluids, *Int. J. Heat Fluid Flow*, 21, 58–64, 2000.
81. Shah, R. K. and London, A. L., *Laminar Flow Forced Convection in Ducts*, Supplement 1 to Advances in Heat Transfer, Academic Press, New York, 1978.
82. Özerinç, S., Yazıcıoğlu, A. G., and Kakaç, S., Numerical analysis of laminar forced convection with temperature-dependent thermal conductivity of nanofluids and thermal dispersion, *Int. J. Therm. Sci.*, 62, 138–148, 2012.
83. Sieder, E. N. and Tate, G. E., Heat transfer and pressure drop of liquids in tubes, *Ind. Eng. Chem.*, 28(12), 1429–1435, 1936.
84. Özerinç, S., Kakaç, S., and Yazıcıoğlu, A. G., Convective heat transfer enhancement with nanofluids: The effect of temperature-variable thermal conductivity, *Proceedings of the ASME 2010 10th Biennial Conference on Engineering Systems Design and Analysis*, Istanbul, ESDA2010-25235, 2010.
85. Kim, D., Kwon, Y., Cho, Y., Li, C., Cheong, S., Hwang, Y., Lee, J., Hong, D., and Moon, S., Convective heat transfer characteristics of nanofluids under laminar and turbulent flow conditions, *Curr. Appl. Phys.*, 9(2), 119–123, 2009.
86. Apaçoğlu, B., Kirez, O., Kakaç, S., and Yazıcıoğlu, A. G., Enhancement of convective heat transfer in laminar and turbulent flows with nanofluids, *Proceedings of VIII Minsk International Seminar "Heat Pipes, Heat Pumps, Refrigerators, Power Sources"*, Minsk, Belarus, pp. 5–17, 2011.

87. Tongkratoke, A., Pramuanjaroenkij, A., Chaengbamrung, A., and Kakaç, S., Numerical study of nanofluid heat transfer enhancement with mixing thermal conductivity models, *Proceedings of ICHMT International Symposium on Advances in Computational Heat Transfer*, Bath, England, 2012.
88. Sert, İ. O., Silindirik Kanallarda Nanoakışkanlarla Laminer Zorlanmış Taşınımla ısı Transferinin Sayısal Analizi, Thesis, Master of Science in Mechanical Engineering, TOBB University of Economics and Technology, Ankara, Turkey, December 2010.
89. Maïga, S. E. B., Nguyen, C. T., Galanis, N., Roy, G., Mare, T., and Coqueux, M., Heat transfer enhancement in turbulent tube flow using $Al_2O_3$ nanoparticle suspension, *Int. J. Numer. Method H*, 16(3), 275–292, 2006.
90. Buongiorno, J., Convective transport in nanofluids, *J. Heat Transfer*, 128(3), 240–250, 2006.
91. Vasu, V., Rama Krishna, K., and Kumar, A., Analytical prediction of forced convective heat transfer of fluids embedded with nanostructured materials (nanofluids), *Pramana*, 69(3), 411–421, 2007.
92. Putra, N., Roetzel, W., and Das, S., Natural convection of nano-fluids, *Heat Mass Transfer*, 39(8), 775–784, 2003.
93. Chun, B., Kang, H., and Kim, S., Effect of alumina nanoparticles in the fluid on heat transfer in double-pipe heat exchanger system, *Korean J. Chem. Eng.*, 25(5), 966–971, 2008.
94. Sharma, K., Sundar, L. S., and Sarma, P., Estimation of heat transfer coefficient and friction factor in the transition flow with low volume concentration of $Al_2O_3$ nanofluid flowing in a circular tube and with twisted tape insert, *Int. Commun. Heat Mass Transfer*, 36(5), 503–507, 2009.
95. Asirvatham, L. G., Vishal, N., Gangatharan, S. K., and Lal, D. M., Experimental study on forced convective heat transfer with low volume fraction of CuO/water nanofluid, *Energies*, 2(1), 97–110, 2009.
96. Jung, J., Oh, H., and Kwak, H., Forced convective heat transfer of nanofluids in microchannels, *Int. J. Heat Mass Transfer*, 52(1–2), 466–472, 2009.
97. Chandrasekar, M., Suresh, S., and Chandra Bose, A., Experimental studies on heat transfer and friction factor characteristics of $Al_2O_3$/water nanofluid in a circular pipe under laminar flow with wire coil inserts, *Exp. Therm. Fluid Sci.*, 34(2), 122–130, 2010.
98. Duangthongsuk, W. and Wongwises, S., An experimental study on the heat transfer performance and pressure drop of $TiO_2$–water nanofluids flowing under a turbulent flow regime, *Int. J. Heat Mass Transfer*, 53(1–3), 334–344, 2010.
99. Wu, X., Wu, H., and Cheng, P., Pressure drop and heat transfer of $Al_2O_3$–$H_2O$ nanofluids through silicon microchannels, *J. Micromech. Microeng.*, 19(10), 105020, 2009.
100. Kakaç, S., Liu, H., and Pramuanjaroenkij, A., *Heat Exchangers: Selection, Rating, and Thermal Design*, CRC Press, Boca Raton, FL, 2012.

# Appendix A: Physical Properties of Metals and Nonmetals

## Nomenclature

| | |
|---|---|
| $c$ | Specific heat, kJ/(kg K) |
| $c_p$ | Specific heat at constant pressure, kJ/(kg K) |
| $c_{pf}$ | Specific heat at constant pressure of saturated liquid, U/(kg K) |
| $c_{pg}$ | Specific heat at constant pressure of saturated vapor, kJ/(kg K) |
| $c_v$ | Specific heat at constant volume, kJ/(kg K) |
| $h$ | Specific enthalpy, kJ/(kg K) |
| $k$ | Thermal conductivity, W(m K) |
| $Pr$ | Prandtl Number, $c_p \mu / k$ |
| $p$ | Pressure, bar |
| $s$ | Specific entropy, kJ/(kg K) |
| $T$ | Temperature, °C, K |
| $\upsilon$ | Specific volume |
| $v_s$ | Velocity of sound, m/s |
| $Z$ | Compressibility factor |

## Greek Symbols

| | |
|---|---|
| $\alpha$ | Thermal diffusivity, m$^2$/s |
| $\beta$ | Coefficient of thermal expansion, 1/K |
| $\gamma$ | Ratio of principal specific heat, $c_p / c_v$ |
| $\varepsilon$ | Emissivity |
| $\mu$ | Viscosity, Pa s |
| $\nu$ | Kinematic viscosity, m$^2$/s |
| $\rho$ | Density, kg/m$^3$ |
| $\sigma$ | Surface tension, N/m |

## Subscripts

| | |
|---|---|
| $f$ | Saturated liquid |
| $g$ | Saturated vapor or gas |

**TABLE A.1**

Thermophysical Properties of Metals

| Metal | Temperature Range $T$, °C | Density $\rho$, g/cm³ | Specific Heat $c$, kJ/(kg K) | Thermal Conductivity $k$, W/(m K) | Emissivity $\varepsilon$ |
|---|---|---|---|---|---|
| Aluminum | 0–400 | 2.72 | 0.895 | 202–250 | 0.04–0.06 (polished) |
| | | | | | 0.07–0.09 (commercial) |
| | | | | | 0.2–0.3 (oxidized) |
| Brass (70% Cu, 30% Zn) | 100–300 | 8.52 | 0.38 | 104–147 | 0.03–0.07 (polished) |
| | | | | | 0.2–0.25 (commercial) |
| | | | | | 0.45–0.55 (oxidized) |
| Bronze (75% Cu, 25% Sn) | 0–100 | 8.67 | 0.34 | 26 | 0.03–0.07 (polished) |
| | | | | | 0.4–0.5 (oxidized) |
| Constantan (60% Cu, 40% Ni) | 0–100 | 8.92 | 0.42 | 22–26 | 0.03–0.06 (polished) |
| | | | | | 0.2–0.4 (oxidized) |
| Copper | 0–600 | 8.95 | 0.38 | 385–350 | 0.02–0.04 (polished) |
| | | | | | 0.1–0.2 (commercial) |
| Iron (C = 4%, cast) | 0–1000 | 7.26 | 0.42 | 52–35 | 0.2–0.25 (polished) |
| | | | | | 0.55–0.65 (oxidized) |
| | | | | | 0.6–0.8 (rusted) |
| Iron (C ≃ 0.5%, wrought) | 0–1000 | 7.85 | 0.46 | 59–35 | 0.3–0.35 (polished) |
| | | | | | 0.9–0.95 (oxidized) |
| Lead | 0–300 | 11.37 | 0.13 | 35–30 | 0.05–0.08 (polished) |
| | | | | | 0.3–0.6 (oxidized) |
| Magnesium | 0–300 | 1.75 | 1.01 | 171–157 | 0.07–0.13 (polished) |
| Mercury | 0–300 | 13.4 | 0.125 | 8–10 | 0.1–0.12 |
| Molybdenum | 0–1000 | 10.22 | 0.251 | 125–99 | 0.06–0.10 (polished) |
| Nickel | 0–400 | 8.9 | 0.45 | 93–59 | 0.05–0.07 (polished) |
| | | | | | 0.35–0.49 (oxidized) |
| Platinum | 0–1000 | 21.4 | 0.24 | 70–75 | 0.05–0.03 (polished) |
| | | | | | 0.07–0.11 (oxidized) |
| Silver | 0–400 | 10.52 | 0.23 | 410–360 | 0.01–0.03 (polished) |
| | | | | | 0.02–0.04 (oxidized) |
| Steel (C ≃ 1%) | 0–1000 | 7.80 | 0.47 | 43–28 | 0.07–0.17 (polished) |
| Steel (Cr ≃ 1%) | 0–1000 | 7.86 | 0.46 | 62–33 | 0.07–0.17 (polished) |
| Steel (Cr 18%, Ni 8%) | 0–1000 | 7.81 | 0.46 | 16–26 | 0.07–0.17 (polished) |
| Tin | 0–200 | 7.3 | 0.23 | 65–57 | 0.04–0.06 (polished) |
| Tungsten | 0–1000 | 19.35 | 0.13 | 166–76 | 0.04–0.08 (polished) |
| | | | | | 0.1–0.2 (filament) |
| Zinc | 0–400 | 7.14 | 0.38 | 112–93 | 0.02–0.03 (polished) |
| | | | | | 0.10–0.11 (oxidized) |
| | | | | | 0.2–0.3 (galvanized) |

*Source:* Kakaç, S. and Yener, Y., *Heat Conduction*, 3rd edn., Taylor & Francis Group, Washington, DC, pp. 343–345, 1993. With permission.

**TABLE A.2**

Thermophysical Properties of Nonmetals

| Metal | Temperature Range $T$, °C | Density $\rho$, g/cm³ | Specific Heat $c$, kJ/(kg K) | Thermal Conductivity $k$, W/(m K) | Emissivity $\varepsilon$ |
|---|---|---|---|---|---|
| Asbestos | 100–1000 | 0.47–0.57 | 0.816 | 0.15–0.22 | 0.93–0.97 |
| Brick, rough red | 100–1000 | 1.76 | 0.84 | 0.38–0.43 | 0.90–0.95 |
| Clay | 0–200 | 1.46 | 0.88 | 1.3 | 0.91 |
| Concrete | 0–200 | 2.1 | 0.88 | 0.81–1.4 | 0.94 |
| Glass, window | 0–600 | 2.2 | 0.84 | 0.78 | 0.94–0.66 |
| Glass wool | 23 | 0.024 | 0.7 | 0.038 | |
| Ice | 0 | 0.91 | 1.9 | 2.2 | 0.97–0.99 |
| Limestone | 100–400 | 2.5 | 0.92 | 1.3 | 0.95–0.80 |
| Marble | 0–100 | 2.60 | 0.79 | 2.07–2.94 | 0.93–0.95 |
| Plasterboard | 0–100 | 1.25 | 0.84 | 0.43 | 0.92 |
| Rubber (hard) | 0–100 | 1.2 | 1.42 | 0.15 | 0.94 |
| Sandstone | 0–300 | 2.24 | 0.71 | 1.83 | 0.83–0.9 |
| Wood (oak) | 0–100 | 0.6–0.8 | 2.4 | 0.17–0.21 | 0.90 |

# Reference

1. Kakaç, S. and Yener, Y., *Heat Conduction*, 3rd edn., Taylor & Francis Group, Washington, DC, 1993.

# Appendix B: Physical Properties of Air, Water, Liquid Metals, and Refrigerants

**TABLE B.1**

Properties of Dry Air at Atmospheric Pressure

| Temperature °C | $\rho$ kg/m³ | $c_p$ kJ/(kg K) | $k$ W/(m K) | $\beta$ × 10³ 1/K | $\mu$ × 10⁵ kg/(m s) | $\nu$ × 10⁶ m²/s | $\alpha$ × 10⁶ m²/s | $Pr$ |
|---|---|---|---|---|---|---|---|---|
| −150 | 2.793 | 1.026 | 0.0120 | 8.21 | 0.870 | 3.11 | 4.19 | 0.74 |
| −100 | 1.980 | 1.009 | 0.0165 | 5.82 | 1.18 | 5.96 | 8.28 | 0.72 |
| −50 | 1.534 | 1.005 | 0.0206 | 4.51 | 1.47 | 9.55 | 13.4 | 0.715 |
| 0 | 1.2930 | 1.005 | 0.0242 | 3.67 | 1.72 | 13.30 | 18.7 | 0.711 |
| 20 | 1.2045 | 1.005 | 0.0257 | 3.43 | 1.82 | 15.11 | 21.4 | 0.713 |
| 40 | 1.1267 | 1.009 | 0.0271 | 3.20 | 1.91 | 16.97 | 23.9 | 0.711 |
| 60 | 1.0595 | 1.009 | 0.0285 | 3.00 | 2.00 | 18.90 | 26.7 | 0.709 |
| 80 | 0.9908 | 1.009 | 0.0299 | 2.83 | 2.10 | 20.94 | 29.6 | 0.708 |
| 100 | 0.9458 | 1.013 | 0.0314 | 2.68 | 2.18 | 23.06 | 32.8 | 0.704 |
| 120 | 0.8980 | 1.013 | 0.0328 | 2.55 | 2.27 | 25.23 | 36.1 | 0.70 |
| 140 | 0.8535 | 1.013 | 0.0343 | 2.43 | 2.35 | 27.55 | 39.7 | 0.694 |
| 160 | 0.8150 | 1.017 | 0.0358 | 2.32 | 2.43 | 29.85 | 43.0 | 0.693 |
| 180 | 0.7785 | 1.022 | 0.0372 | 2.21 | 2.51 | 32.29 | 46.7 | 0.69 |
| 200 | 0.7475 | 1.026 | 0.0386 | 2.11 | 2.58 | 34.63 | 50.5 | 0.685 |
| 250 | 0.6745 | 1.034 | 0.0421 | 1.91 | 2.78 | 41.17 | 60.3 | 0.68 |
| 300 | 0.6157 | 1.047 | 0.0390 | 1.75 | 2.95 | 47.85 | 70.3 | 0.68 |
| 350 | 0.5662 | 1.055 | 0.0485 | 1.61 | 3.12 | 55.05 | 81.1 | 0.68 |
| 400 | 0.5242 | 1.068 | 0.0516 | 1.49 | 3.28 | 62.53 | 91.9 | 0.68 |
| 450 | 0.4875 | 1.080 | 0.0543 | – | 3.44 | 70.54 | 103.1 | 0.685 |
| 500 | 0.4564 | 1.092 | 0.0570 | – | 3.86 | 70.48 | 114.2 | 0.69 |
| 600 | 0.4041 | 1.114 | 0.0621 | – | 3.58 | 95.57 | 138.2 | 0.69 |
| 700 | 0.3625 | 1.135 | 0.0667 | – | 4.12 | 113.7 | 162.2 | 0.70 |
| 800 | 0.3287 | 0.156 | 0.0706 | – | 4.37 | 132.8 | 185.8 | 0.715 |
| 900 | 0.321 | 1.172 | 0.0741 | – | 4.59 | 152.5 | 210 | 0.725 |
| 1000 | 0.277 | 1.185 | 0.0770 | – | 4.80 | 175 | 235 | 0.735 |

**TABLE B.2**

Properties of Water

| Temperature °C | Pressure kgf/cm² | $\rho$ kg/m³ | $c_p$ kJ/(kg K) | $h_{fg}$ kcal/kg | $k$ W/(m k) | $\beta$ ×10³ 1/K | $\mu$ ×10³ kg/(m s) | $\nu$ ×10⁶ m²/s | $\alpha$ ×10⁶ m²/s | $Pr$ |
|---|---|---|---|---|---|---|---|---|---|---|
| 0 | 1 | 999.8 | 4.2179 | 597.3 | 0.552 | −0.07 | 1.792 | 1.795 | 0.131 | 13.67 |
| 10 | 1 | 999.7 | 4.1994 | 591.7 | 0.587 | +0.088 | 1.307 | 1.307 | 0.138 | 9.47 |
| 20 | 1 | 998.2 | 4.1819 | 586.0 | 0.598 | 0.206 | 1.002 | 1.004 | 0.143 | 7.01 |
| 30 | 1 | 995.7 | 4.1785 | 580.4 | 0.614 | 0.303 | 0.797 | 0.801 | 0.148 | 5.43 |
| 40 | 1 | 992.2 | 4.1785 | 574.7 | 0.628 | 0.385 | 0.653 | 0.658 | 0.151 | 4.35 |
| 50 | 1 | 988.0 | 4.1806 | 569.0 | 0.641 | 0.457 | 0.548 | 0.554 | 0.155 | 3.57 |
| 60 | 1 | 983.2 | 4.1844 | 563.2 | 0.652 | 0.523 | 0.467 | 0.475 | 0.158 | 3.00 |
| 70 | 1 | 977.8 | 4.1898 | 557.3 | 0.661 | 0.585 | 0.404 | 0.413 | 0.161 | 2.56 |
| 80 | 1 | 971.8 | 4.1965 | 551.3 | 0.669 | 0.643 | 0.355 | 0.365 | 0.164 | 2.23 |
| 90 | 1 | 965.3 | 4.2053 | 545.2 | 0.676 | 0.698 | 0.315 | 0.326 | 0.166 | 1.96 |
| 100 | 1.0332 | 958.4 | 4.2162 | 539.0 | 0.682 | 0.752 | 0.282 | 0.295 | 0.169 | 1.75 |
| 120 | 2.0245 | 943.1 | 4.245 | 526.1 | 0.686 | 0.860 | 0.235 | 0.2485 | 0.171 | 1.45 |
| 140 | 3.6848 | 926.1 | 4.287 | 512.3 | 0.684 | 0.975 | 0.199 | 0.215 | 0.172 | 1.25 |
| 160 | 6.3023 | 907.4 | 4.341 | 497.4 | 0.682 | 1.098 | 0.172 | 0.1890 | 0.173 | 1.09 |
| 180 | 10.225 | 886.9 | 4.409 | 481.3 | 0.676 | 1.233 | 0.151 | 0.1697 | 0.172 | 0.98 |
| 200 | 15.857 | 864.7 | 4.497 | 463.5 | 0.666 | 1.392 | 0.136 | 0.1579 | 0.171 | 0.92 |
| 220 | 23.659 | 840.3 | 4.610 | 443.7 | 0.653 | 1.597 | 0.125 | 0.1488 | 0.168 | 0.88 |
| 240 | 34.140 | 813.6 | 4.761 | 421.7 | 0.636 | 1.862 | 0.116 | 0.1420 | 0.164 | 0.87 |
| 260 | 47.866 | 784.0 | 4.978 | 396.8 | 0.612 | 2.21 | 0.107 | 0.1365 | 0.157 | 0.87 |
| 280 | 65.457 | 750.7 | 5.309 | 368.5 | 0.581 | 2.70 | 0.0994 | 0.1325 | 0.145 | 0.91 |
| 300 | 87.611 | 712.5 | 5.86 | 335.4 | 0.541 | 3.46 | 0.0935 | 0.1298 | 0.129 | 1.00 |
| 320 | 115.12 | 667.0 | 6.62 | 295.6 | 0.491 | 4.60 | 0.0856 | 0.1282 | 0.111 | 1.15 |
| 340 | 148.96 | 609.5 | 8.37 | 245.3 | 0.430 | 8.25 | 0.0775 | 0.1272 | 0.0844 | 1.5 |
| 360 | 190.42 | 524.5 | 13.4 | 171.9 | 0.349 | – | 0.0683 | 0.1306 | 0.0500 | 2.6 |

**TABLE B.3**

Thermophysical Properties of Steam at 1 Bar Pressure

| $T$ K | $v$ m³/kg | $h$ kJ/kg | $s$ kJ/(kg K) | $c_p$ kJ/(kg K) | $c_v$ kJ/(kg K) | $\gamma$ | $Z$ | $\bar{v}_s$ m/s | $\mu$ $10^{-5}$ Pa s | $k$ W/(m K) | $Pr$ |
|---|---|---|---|---|---|---|---|---|---|---|---|
| 373.15 | 1.679 | 2676.2 | 7.356 | 2.029 | 1.510 | 1.344 | 0.9750 | 472.8 | 1.20 | 0.0248 | 0.982 |
| 400 | 1.827 | 2730.2 | 7.502 | 1.996 | 1.496 | 1.344 | 0.9897 | 490.4 | 1.32 | 0.0268 | 0.980 |
| 450 | 2.063 | 2829.7 | 7.741 | 1.981 | 1.498 | 1.322 | 0.9934 | 520.6 | 1.52 | 0.0311 | 0.968 |
| 500 | 2.298 | 2928.7 | 7.944 | 1.983 | 1.510 | 1.313 | 0.9959 | 540.3 | 1.73 | 0.0358 | 0.958 |
| 550 | 2.531 | 3028 | 8.134 | 2.000 | 1.531 | 1.306 | 0.9971 | 574.2 | 1.94 | 0.0410 | 0.946 |
| 600 | 2.763 | 3129 | 8.309 | 2.024 | 1.557 | 1.300 | 0.9978 | 598.6 | 2.15 | 0.0464 | 0.938 |
| 650 | 2.995 | 3231 | 8.472 | 2.054 | 1.589 | 1.293 | 0.9988 | 621.8 | 2.36 | 0.0521 | 0.930 |
| 700 | 3.227 | 3334 | 8.625 | 2.085 | 1.620 | 1.287 | 0.9989 | 643.9 | 2.57 | 0.0581 | 0.922 |
| 750 | 3.459 | 3439 | 8.770 | 2.118 | 1.653 | 1.281 | 0.9992 | 665.1 | 2.77 | 0.0646 | 0.913 |
| 800 | 3.690 | 3546 | 8.908 | 2.151 | 1.687 | 1.275 | 0.9995 | 685.4 | 2.98 | 0.0710 | 0.903 |
| 850 | 3.921 | 3654 | 9.039 | 2.185 | 1.722 | 1.269 | 0.9996 | 705.1 | 3.18 | 0.0776 | 0.897 |
| 900 | 4.152 | 3764 | 9.165 | 2.219 | 1.756 | 1.264 | 0.9996 | 723.9 | 3.39 | 0.0843 | 0.892 |
| 950 | 4.383 | 3876 | 9.286 | 2.253 | 1.791 | 1.258 | 0.9997 | 742.2 | 3.59 | 0.0912 | 0.886 |
| 1000 | 4.614 | 3990 | 9.402 | 2.286 | 1.823 | 1.254 | 0.9998 | 760.1 | 3.78 | 0.0981 | 0.881 |
| 1100 | 5.076 | 4223 | 9.625 | 2.36 | | | 0.9999 | 794.3 | 4.13 | 0.113 | 0.858 |
| 1200 | 5.538 | 4463 | 9.384 | 2.43 | | | 1.0000 | 826.8 | 4.48 | 0.130 | 0.837 |
| 1300 | 5.999 | 4711 | 10.032 | 2.51 | | | 1.0000 | 857.9 | 4.77 | 0.144 | 0.826 |
| 1400 | 6.461 | 4965 | 10.221 | 2.58 | | | 1.0000 | 887.9 | 5.06 | 0.160 | 0.816 |
| 1500 | 6.924 | 5227 | 10.402 | 2.65 | | | 1.0002 | 916.9 | 5.35 | 0.18 | 0.788 |
| 1600 | 7.386 | 5497 | 10.576 | 2.73 | | | 1.0004 | 945.0 | 5.65 | 0.21 | 0.735 |
| 1800 | 8.316 | 6068 | 10.912 | 3.02 | | | 1.0011 | 999.4 | 6.19 | 0.33 | 0.567 |
| 2000 | 9.263 | 6706 | 11.248 | 3.79 | | | 1.0036 | 1051.0 | 6.70 | 0.57 | 0.445 |

*Source:* Kakaç, S., Ed., *Boilers, Evaporators and Condensers*, Wiley, New York, p. 822, 1991. With permission.

**TABLE B.4**

Thermophysical Properties of Water Steam at High Pressures

| T K | v m³/kg | h kJ/kg | s kJ/(kg K) | $c_p$ kJ/(kg K) | $c_v$ W/(kg K) | γ | Z | $\bar{v}_s$ m/s | μ Pa s | k W/(m K) | Pr |
|---|---|---|---|---|---|---|---|---|---|---|---|
| | | | | *P* = 10 bar | | | | | | | |
| 300 | 1.003. − 3ᵃ | 113.4 | 0.392 | 4.18 | 4.13 | 1.01 | 0.0072 | 1500 | 8.57. − 4 | 0.615 | 5.82 |
| 350 | 1.027. − 3 | 322.5 | 1.037 | 4.19 | 3.89 | 1.08 | 0.0064 | 1552 | 3.70. − 4 | 0.668 | 2.32 |
| 400 | 1.067. − 3 | 533.4 | 1.600 | 4.25 | 3.65 | 1.17 | 0.0058 | 1509 | 2.17. − 4 | 0.689 | 1.34 |
| 450 | 1.123. − 3 | 749.0 | 2.109 | 4.39 | 3.44 | 1.28 | 0.0054 | 1399 | 1.51. − 4 | 0.677 | 0.981 |
| 500 | 0.221 | 2891 | 6.823 | 2.29 | 1.68 | 1.36 | 0.957 | 535.7 | 1.71. − 5 | 0.038 | 1.028 |
| 600 | 0.271 | 3109 | 7.223 | 2.13 | 1.61 | 1.32 | 0.987 | 592.5 | 2.15. − 5 | 0.047 | 0.963 |
| 800 | 0.367 | 3537 | 7.837 | 2.18 | 1.70 | 1.28 | 0.994 | 686.2 | 2.99. − 5 | 0.072 | 0.908 |
| 1000 | 0.460 | 3984 | 8.336 | 2.30 | 1.83 | 1.26 | 0.997 | 759.4 | 3.78. − 5 | 0.099 | 0.881 |
| 1500 | 0.692 | 5224 | 9.337 | 2.66 | | | 1.000 | 917.2 | 5.35. − 5 | 0.18 | 0.80 |
| 2000 | 0.925 | 6649 | 10.154 | 3.29 | | | 1.002 | 1050 | 6.70. − 5 | 0.39 | 0.57 |
| | | | | *P* = 50 bar | | | | | | | |
| 300 | 1.001. − 3 | 117.1 | 0.391 | 4.16 | 4.11 | 1.01 | 0.0362 | 1508 | 8.55. − 4 | 0.618 | 5.76 |
| 350 | 1.025. − 3 | 325.6 | 1.034 | 4.18 | 3.88 | 1.08 | 0.0317 | 1561 | 3.71. − 4 | 0.671 | 2.31 |
| 400 | 1.064. − 3 | 536.0 | 1.596 | 4.24 | 3.64 | 1.16 | 0.0288 | 1519 | 2.18. − 4 | 0.691 | 1.34 |
| 450 | 1.120. − 3 | 751.4 | 2.103 | 4.37 | 3.43 | 1.27 | 0.0270 | 1437 | 1.52. − 4 | 0.681 | 0.975 |
| 500 | 1.200. − 3 | 976.1 | 2.575 | 4.64 | 3.25 | 1.43 | 0.0260 | 1246 | 1.19. − 4 | 0.645 | 0.856 |
| 600 | 0.0490 | 3013 | 6.350 | 2.85 | 1.94 | 1.47 | 0.885 | 560.5 | 2.14. − 5 | 0.054 | 1.129 |
| 800 | 0.0713 | 3496 | 7.049 | 2.31 | 1.74 | 1.32 | 0.966 | 674.5 | 3.03. − 5 | 0.075 | 0.929 |
| 1000 | 0.0911 | 3961 | 7.575 | 2.35 | 1.85 | 1.27 | 0.987 | 756.5 | 3.81. − 5 | 0.102 | 0.880 |
| 1500 | 0.1384 | 5214 | 8.589 | 2.66 | | | 1.000 | 918.8 | 5.37. − 5 | 0.18 | 0.81 |
| 2000 | 0.1850 | 6626 | 9.398 | 3.12 | | | 1.002 | 1053 | 6.70. − 5 | 0.33 | 0.64 |
| | | | | *P* = 100 bar | | | | | | | |
| 300 | 9.99. − 4 | 121.8 | 0.390 | 4.15 | 4.09 | 1.01 | 0.0722 | 1516 | 8.52. − 4 | 0.622 | 5.69 |
| 350 | 1.022. − 3 | 329.6 | 1.031 | 4.17 | 3.87 | 1.08 | 0.0633 | 1571 | 3.73. − 4 | 0.675 | 2.31 |
| 400 | 1.061. − 3 | 539.6 | 1.590 | 4.23 | 3.64 | 1.16 | 0.0575 | 1532 | 2.20. − 4 | 0.694 | 1.34 |

| | | | | | | | | | | | |
|---|---|---|---|---|---|---|---|---|---|---|---|
| 450 | 1.116. − 3 | 754.1 | 2.097 | 4.35 | 3.43 | 1.27 | 0.0537 | 1452 | 1.53. − 4 | 0.685 | 0.975 |
| 500 | 1.193. − 3 | 977.3 | 2.567 | 4.60 | 3.24 | 1.42 | 0.0517 | 1269 | 1.21. − 4 | 0.651 | 0.853 |
| 600 | 0.0201 | 2820 | 5.775 | 5.22 | 2.64 | 1.97 | 0.726 | 502.3 | 2.14. − 5 | 0.073 | 1.74 |
| 800 | 0.0343 | 3442 | 6.685 | 2.52 | 1.82 | 1.38 | 0.929 | 662.4 | 3.08. − 5 | 0.081 | 0.960 |
| 1000 | 0.0449 | 3935 | 7.233 | 2.44 | 1.88 | 1.30 | 0.973 | 753.3 | 3.85. − 5 | 0.107 | 0.876 |
| 1500 | 0.0692 | 5203 | 8.262 | 2.68 | | | 1.000 | 921.1 | 5.37. − 5 | 0.18 | 0.82 |
| 2000 | 0.0926 | 6616 | 9.073 | 3.08 | | | 1.003 | 1057 | 6.70. − 5 | 0.31 | 0.67 |
| **P = 250 bar** | | | | | | | | | | | |
| 300 | 9.93 − 3 | 135.3 | 0.385 | 4.12 | 4.06 | 1.02 | 0.1792 | 1542 | 8.48. − 4 | 0.634 | 5.50 |
| 350 | 1.016. − 3 | 341.7 | 1.022 | 4.14 | 3.84 | 1.08 | 0.1572 | 1599 | 3.78. − 4 | 0.686 | 2.28 |
| 400 | 1.053. − 3 | 550.1 | 1.578 | 4.20 | 3.62 | 1.16 | 0.1426 | 1568 | 2.24. − 4 | 0.704 | 1.33 |
| 450 | 1.105. − 3 | 762.4 | 2.078 | 4.30 | 3.41 | 1.26 | 0.1330 | 1496 | 1.57. − 4 | 0.696 | 0.969 |
| 500 | 1.175. − 3 | 981.9 | 2.541 | 4.50 | | | 0.1273 | 1331 | 1.24. − 4 | 0.666 | 0.838 |
| 600 | 1.454. − 3 | 1479 | 3.443 | 5.88 | 4.22 | 1.40 | 0.1313 | 896.9 | 8.63. − 5 | 0.532 | 0.952 |
| 800 | 0.0120 | 3261 | 6.086 | 3.41 | | | 0.813 | 627.3 | 3.29. − 5 | 0.109 | 1.03 |
| 1000 | 0.0173 | 3845 | 6.741 | 2.69 | 1.97 | 1.36 | 0.935 | 745.9 | 3.98. − 5 | 0.125 | 0.856 |
| 1500 | 0.0277 | 5186 | 7.827 | 2.73 | | | 1.000 | 929.1 | 5.40. − 5 | 0.18 | 0.819 |
| 2000 | 0.0372 | 6608 | 8.642 | 3.04 | | | 1.008 | 1068 | | | |
| **P = 500 bar** | | | | | | | | | | | |
| 300 | 9.83. − 4 | 157.7 | 0.378 | 4.06 | 3.98 | 1.02 | 0.3549 | 1583 | 8.45. − 4 | 0.650 | 5.28 |
| 350 | 1.005. − 3 | 361.8 | 1.007 | 4.10 | 3.81 | 1.08 | 0.3112 | 1644 | 3.87. − 4 | 0.700 | 2.27 |
| 400 | 1.041. − 3 | 567.8 | 1.557 | 4.14 | 3.59 | 1.15 | 0.2820 | 1623 | 2.31. − 4 | 0.719 | 1.33 |
| 450 | 1.088. − 3 | 776.9 | 2.050 | 4.23 | 3.39 | 1.25 | 0.2618 | 1561 | 1.62. − 4 | 0.714 | 0.960 |
| 500 | 1.151. − 3 | 991.5 | 2.502 | 4.37 | | | 0.2493 | 1418 | 1.29. − 4 | 0.689 | 0.822 |
| 600 | 1.362. − 3 | 1456 | 3.346 | 5.08 | 3.72 | 1.37 | 0.2459 | 1080 | 9.34. − 5 | 0.588 | 0.808 |
| 800 | 4.576. − 3 | 2895 | 5.937 | 5.84 | 2.79 | 2.10 | 0.620 | 597.8 | 4.04. − 5 | 0.178 | 1.33 |
| 1000 | 8.102. − 3 | 3697 | 6.302 | 3.17 | 1.81 | 1.76 | 0.878 | 742.1 | 4.28. − 5 | 0.150 | 0.905 |
| 1500 | 0.0139 | 5157 | 7.484 | 2.82 | | | 1.004 | 943.6 | | | |
| 2000 | 0.0188 | 6595 | 8.310 | 3.04 | | | 1.018 | 1086 | | | |

*Source:* Kakaç, S., Ed., *Boilers, Evaporators and Condensers*, Wiley, New York, pp. 823–824, 1991. With permission.
[a] The notation 1.003. − 3 signifies $1.003 \times 10^{-3}$.

**TABLE B.5**

Properties of Liquid Metals

| Liquid Metal and Melting Point | Temperature °C | $k$ W/(m K) | $\rho$ kg/m³ | $c$ kJ/(kg K) | $\mu$ × 10⁴ kg/(m s) |
|---|---|---|---|---|---|
| Bismuth (288°C) | 315 | 16.40 | 10,000 | 0.1444 | 16.22 |
| | 538 | 15.70 | 9,730 | 0.1545 | 10.97 |
| | 760 | 15.70 | 9,450 | 0.1645 | 7.89 |
| Lead (327°C) | 371 | 18.26 | 10,500 | 0.159 | 24.0 |
| | 482 | 19.77 | 10,400 | 0.155 | 19.25 |
| | 704 | – | 10,130 | – | 13.69 |
| Mercury (−39°C) | 10 | 8.14 | 13,550 | 0.138 | 15.92 |
| | 149 | 11.63 | 13,200 | 0.138 | 10.97 |
| | 315 | 14.07 | 12,800 | 0.134 | 8.64 |
| Potassium (64°C) | 315 | 45.0 | 804 | 0.80 | 3.72 |
| | 427 | 39.5 | 740 | 0.75 | 1.78 |
| | 704 | 33.1 | 674 | 0.75 | 1.28 |
| Sodium (98°C) | 93 | 86.0 | 930 | 1.38 | 7.0 |
| | 371 | 72.5 | 860 | 1.29 | 2.81 |
| | 704 | 59.8 | 776 | 1.26 | 1.78 |
| 56% Na, 44% K (19°C) | 93 | 25.6 | 885 | 1.13 | 5.78 |
| | 371 | 27.6 | 820 | 1.06 | 2.36 |
| | 704 | 28.8 | 723 | 1.04 | 5.94 |
| 22% Na, 78% K (−11°C) | 93 | 24.4 | 850 | 0.95 | 4.94 |
| | 400 | 26.7 | 775 | 0.88 | 2.07 |
| | 760 | – | 690 | 0.88 | 1.46 |
| 44.5% Pb, 55.5% Bi (125°C) | 315 | 9.0 | 10,500 | 0.147 | – |
| | 371 | 11.9 | 10,220 | 0.47 | 15.36 |
| | 650 | – | 9,820 | – | 11.47 |

**TABLE B.6**

Thermophysical Properties of Saturated Refrigerant 12

| $P$ bar | $T$ K | $v_f$ $10^{-4}$ m³/kg | $v_g$ m³/kg | $h_f$ kJ/kg | $h_g$ kJ/kg | $s_f$ kJ/(kg K) | $s_g$ kJ/(kg K) | | |
|---|---|---|---|---|---|---|---|---|---|
| 0.10 | 200.1 | 6.217 | 1.365 | 334.8 | 518.1 | 3.724 | 4.640 | | |
| 0.15 | 206.3 | 6.282 | 0.936 | 340.1 | 521.0 | 3.750 | 4.627 | | |
| 0.20 | 211.1 | 6.332 | 0.716 | 344.1 | 523.2 | 3.769 | 4.618 | | |
| 0.25 | 214.9 | 6.374 | 0.582 | 347.4 | 525.0 | 3.785 | 4.611 | | |
| 0.30 | 218.2 | 6.411 | 0.491 | 350.2 | 526.5 | 3.798 | 4.606 | | |
| 0.4 | 223.5 | 6.437 | 0.376 | 354.9 | 529.1 | 3.819 | 4.598 | | |
| 0.5 | 227.9 | 6.525 | 0.306 | 358.8 | 531.2 | 3.836 | 4.592 | | |
| 0.6 | 231.7 | 6.570 | 0.254 | 362.1 | 532.9 | 3.850 | 4.588 | | |
| 0.8 | 237.9 | 6.648 | 0.198 | 367.6 | 535.8 | 3.874 | 4.581 | | |
| 1.0 | 243.0 | 6.719 | 0.160 | 372.1 | 538.2 | 3.893 | 4.576 | | |
| 1.5 | 253.0 | 6.859 | 0.110 | 381.2 | 542.9 | 3.929 | 4.568 | | |
| 2.0 | 260.6 | 6.970 | 0.0840 | 388.2 | 546.4 | 3.956 | 4.563 | | |
| 2.5 | 266.9 | 7.067 | 0.0681 | 394.0 | 549.2 | 3.978 | 4.560 | | |
| 3.0 | 272.3 | 7.183 | 0.0573 | 399.1 | 551.6 | 3.997 | 4.557 | | |
| 4.0 | 281.3 | 7.307 | 0.0435 | 407.6 | 555.6 | 4.027 | 4.553 | | |
| 5.0 | 288.8 | 7.444 | 0.0351 | 414.8 | 558.8 | 4.052 | 4.551 | | |
| 6.0 | 295.2 | 7.571 | 0.0294 | 421.1 | 561.5 | 4.073 | 4.549 | | |
| 8.0 | 306.0 | 7.804 | 0.0221 | 431.8 | 565.7 | 4.108 | 4.546 | | |
| 10 | 314.9 | 8.022 | 0.0176 | 440.8 | 569.0 | 4.137 | 4.544 | | |
| 15 | 332.6 | 8.548 | 0.0114 | 459.3 | 574.5 | 4.193 | 4.539 | | |
| 20 | 346.3 | 9.96 | 0.0082 | 474.8 | 577.5 | 4.237 | 4.534 | | |
| 25 | 357.5 | 9.715 | 0.0062 | 488.7 | 578.5 | 4.275 | 4.527 | | |
| 30 | 367.2 | 10.47 | 0.0048 | 502.0 | 577.6 | 4.311 | 4.517 | | |
| 35 | 375.7 | 11.49 | 0.0036 | 515.9 | 574.1 | 4.347 | 4.502 | | |
| 40 | 383.3 | 13.45 | 0.0025 | 532.7 | 564.1 | 4.389 | 4.471 | | |
| 41.2[a] | 385.0 | 17.92 | 0.0018 | 548.3 | 548.3 | 4.429 | 4.429 | | |

| $P$ bar | $c_{pf}$ kJ/(kg K) | $c_{pg}$ kJ/(kg K) | $u_f$ $10^{-4}$ Pa s | $\mu_g$ $10^{-5}$ Pa s | $k_f$ W/(m K) | $k_g$ W/(m K) | $Pr_f$ | $Pr_g$ | $\sigma$ N/m |
|---|---|---|---|---|---|---|---|---|---|
| 0.10 | 0.855 | | 6.16 | | 0.105 | 0.0050 | 5.01 | | |
| 0.15 | 0.861 | | 5.61 | | 0.103 | 0.0053 | 4.69 | | |
| 0.20 | 0.865 | | 5.28 | | 0.101 | 0.0055 | 4.52 | | |
| 0.25 | 0.868 | | 4.99 | | 0.099 | 0.0056 | 4.38 | | |
| 0.30 | 0.872 | | 4.79 | | 0.098 | 0.0057 | 4.26 | | |
| 0.4 | 0.876 | | 4.48 | | 0.097 | 0.0060 | 4.05 | | 0.0189 |
| 0.5 | 0.880 | 0.545 | 4.25 | 1.00 | 0.095 | 0.0062 | 3.94 | 0.89 | 0.0182 |
| 0.6 | 0.884 | 0.552 | 4.08 | 1.02 | 0.094 | 0.0063 | 3.84 | 0.88 | 0.0176 |
| 0.8 | 0.889 | 0.564 | 3.81 | 1.04 | 0.091 | 0.0066 | 3.72 | 0.88 | 0.0167 |
| 1.0 | 0.894 | 0.574 | 3.59 | 1.06 | 0.089 | 0.0069 | 3.61 | 0.88 | 0.0159 |
| 1.5 | 0.905 | 0.600 | 3.23 | 1.10 | 0.086 | 0.0074 | 3.40 | 0.89 | 0.0145 |
| 2.0 | 0.914 | 0.613 | 2.95 | 1.13 | 0.083 | 0.0077 | 3.25 | 0.90 | 0.0134 |
| 2.5 | 0.922 | 0.626 | 2.78 | 1.15 | 0.081 | 0.0081 | 3.16 | 0.91 | 0.0125 |
| 3.0 | 0.930 | 0.640 | 2.62 | 1.18 | 0.079 | 0.0083 | 3.08 | 0.91 | 0.0118 |
| 4.0 | 0.944 | 0.663 | 2.40 | 1.22 | 0.075 | 0.0088 | 3.02 | 0.92 | 0.0106 |

*(continued)*

**TABLE B.6 (continued)**

Thermophysical Properties of Saturated Refrigerant 12

| $P$ bar | $c_{pf}$ kJ/(kg K) | $c_{pg}$ kJ/(kg K) | $u_f$ $10^{-4}$ Pa s | $\mu_g$ $10^{-5}$ Pa s | $k_f$ W/(m K) | $k_g$ W/(m K) | $Pr_f$ | $Pr_g$ | $\sigma$ N/m |
|---|---|---|---|---|---|---|---|---|---|
| 5.0 | 0.957 | 0.683 | 2.24 | 1.25 | 0.073 | 0.0092 | 2.94 | 0.93 | 0.0096 |
| 6.0 | 0.969 | 0.702 | 2.13 | 1.28 | 0.070 | 0.0095 | 2.95 | 0.95 | 0.0087 |
| 8.0 | 0.995 | 0.737 | 1.96 | 1.33 | 0.066 | 0.0101 | 2.95 | 0.97 | 0.0074 |
| 10 | 1.023 | 0.769 | 1.88 | 1.38 | 0.063 | 0.0107 | 3.05 | 1.01 | 0.0063 |
| 15 | 1.102 | 0.865 | 1.67 | 1.50 | 0.057 | 0.0117 | 3.23 | 1.11 | 0.0042 |
| 20 | 1.234 | 0.969 | 1.49 | 1.69 | 0.053 | 0.0126 | 3.47 | 1.30 | 0.0029 |
| 25 | 1.36 | 1.19 | 1.33 | | 0.047 | 0.0134 | 3.84 | | 0.0019 |
| 30 | 1.52 | 1.60 | 1.16 | | 0.042 | 0.014 | 4.2 | | 0.0009 |
| 35 | 1.73 | 2.5 | | | 0.037 | 0.016 | | | 0.0005 |
| 40 | | | | | | | | | 0.0001 |
| 41.2 | | | | | | | | | 0.0000 |

*Source:* Kakaç, S., Ed., *Boilers, Evaporators and Condensers*, Wiley, New York, pp. 809–810, 1991. With permission.

[a] Critical point.

**TABLE B.7**

Thermophysical Properties of Refrigerant 12 at 1 Bar Pressure

| $T$ K | $\upsilon$ m³/kg | $h$ kJ/kg | $s$ kJ/(kg K) | $\mu$ $10^{-5}$ Pa s | $c_p$ kJ/(kg K) | $k$ W/(m K) | $Pr$ |
|---|---|---|---|---|---|---|---|
| 300 | 0.2024 | 572.1 | 4.701 | 1.26 | 0.614 | 0.0097 | 0.798 |
| 320 | 0.2167 | 584.5 | 4.741 | 1.34 | 0.631 | 0.0107 | 0.788 |
| 340 | 0.2309 | 597.3 | 4.780 | 1.42 | 0.647 | 0.0118 | 0.775 |
| 360 | 0.2450 | 610.3 | 4.817 | 1.49 | 0.661 | 0.0129 | 0.760 |
| 380 | 0.2590 | 623.7 | 4.853 | 1.56 | 0.674 | 0.0140 | 0.745 |
| 400 | 0.2730 | 637.3 | 4.890 | 1.62 | 0.684 | 0.0151 | 0.730 |
| 420 | 0.2870 | 651.2 | 4.924 | 1.67 | 0.694 | 0.0162 | 0.715 |
| 440 | 0.3009 | 665.3 | 4.956 | 1.72 | 0.705 | 0.0173 | 0.703 |
| 460 | 0.3148 | 697.7 | 4.987 | 1.78 | 0.716 | 0.0184 | 0.693 |
| 480 | 0.3288 | 694.3 | 5.018 | 1.84 | 0.727 | 0.0196 | 0.683 |
| 500 | 0.3427 | 709.0 | 5.048 | 1.90 | 0.739 | 0.0208 | 0.674 |

*Source:* Kakaç, S., Ed., *Boilers, Evaporators and Condensers*, Wiley, New York, p. 811, 1991. With permission.

**TABLE B.8**

Thermophysical Properties of Saturated Refrigerant 22

| T K | P bar | $v_f$ m³/kg | $v_g$ m³/kg | $h_f$ kJ/kg | $h_g$ kJ/kg | $s_f$ kJ/(kg K) | $s_g$ kJ/(kg K) | $c_{pf}$ kJ/(kg K) | $c_{pg}$ kJ/(kg K) |
|---|---|---|---|---|---|---|---|---|---|
| 150 | 0.0017 | 6.209. − 4[a] | 83.40 | 268.2 | 547.3 | 3.355 | 5.215 | 1.059 | |
| 160 | 0.0054 | 6.293. − 4 | 28.20 | 278.2 | 552.1 | 3.430 | 5.141 | 1.058 | |
| 170 | 0.0150 | 6.381. − 4 | 10.85 | 288.3 | 557.0 | 3.494 | 5.075 | 1.057 | |
| 180 | 0.0369 | 6.474. − 4 | 4.673 | 298.7 | 561.9 | 3.551 | 5.013 | 1.058 | |
| 190 | 0.0821 | 6.573. − 4 | 2.225 | 308.6 | 566.8 | 3.605 | 4.963 | 1.060 | |
| 200 | 0.1662 | 6.680. − 4 | 1.145 | 318.8 | 571.6 | 3.675 | 4.921 | 1.065 | 0.502 |
| 210 | 0.3116 | 6.794. − 4 | 0.6370 | 329.1 | 576.5 | 3.707 | 4.885 | 1.071 | 0.544 |
| 220 | 0.5470 | 6.917. − 4 | 0.3772 | 339.7 | 581.2 | 3.756 | 4.854 | 1.080 | 0.577 |
| 230 | 0.9076 | 7.050. − 4 | 0.2352 | 350.6 | 585.9 | 3.804 | 4.828 | 1.091 | 0.603 |
| 240 | 1.4346 | 7.195. − 4 | 0.1532 | 361.7 | 590.5 | 3.852 | 4.805 | 1.105 | 0.626 |
| 250 | 2.174 | 7.351. − 4 | 0.1037 | 373.0 | 594.9 | 3.898 | 4.785 | 1.122 | 0.648 |
| 260 | 3.177 | 7.523. − 4 | 0.07237 | 384.5 | 599.0 | 3.942 | 4.768 | 1.143 | 0.673 |
| 270 | 4.497 | 7.733. − 4 | 0.05187 | 396.3 | 603.0 | 3.986 | 4.752 | 1.169 | 0.703 |
| 280 | 6.192 | 7.923. − 4 | 0.03803 | 408.2 | 606.6 | 4.029 | 4.738 | 1.193 | 0.741 |
| 290 | 8.324 | 8.158. − 4 | 0.02838 | 420.4 | 610.0 | 4.071 | 4.725 | 1.220 | 0.791 |
| 300 | 10.956 | 8.426. − 4 | 0.02148 | 432.7 | 612.8 | 4.113 | 4.713 | 1.257 | 0.854 |
| 310 | 14.17 | 8.734. − 4 | 0.01643 | 445.5 | 615.1 | 4.153 | 4.701 | 1.305 | 0.935 |
| 320 | 18.02 | 9.096. − 4 | 0.01265 | 458.6 | 616.7 | 4.194 | 4.688 | 1.372 | 1.036 |
| 330 | 22.61 | 9.535. − 4 | 9.753. − 3 | 472.4 | 617.3 | 4.235 | 4.674 | 1.460 | 1.159 |
| 340 | 28.03 | 1.010. − 3 | 7.479. − 3 | 487.2 | 616.5 | 4.278 | 4.658 | 1.573 | 1.308 |
| 350 | 34.41 | 1.086. − 3 | 5.613. − 3 | 503.7 | 613.3 | 4.324 | 4.637 | 1.718 | 1.486 |
| 360 | 41.86 | 1.212. − 3 | 4.036. − 3 | 523.7 | 605.5 | 4.378 | 4.605 | 1.897 | ∞ |
| 369.3[b] | 49.89 | 2.015. − 3 | 2.015. − 3 | 570.0 | 570.0 | 4.501 | 4.501 | ∞ | ∞ |

*(continued)*

# TABLE B.8 (continued)

Thermophysical Properties of Saturated Refrigerant 22

| T K | $\mu_f$ $10^{-4}$ Pa s | $\mu_g$ $10^{-4}$ Pa s | $k_f$ W/(m K) | $k_g$ W/(m K) | $v_{sf}$ m/s | $v_{sg}$ m/s | $Pr_f$ | $Pr_g$ | $\sigma$ N/m |
|---|---|---|---|---|---|---|---|---|---|
| 150 | | | 0.161 | | | | | | |
| 160 | | | 0.156 | | | | | | |
| 170 | 7.70 | | 0.151 | | | 142.6 | 5.39 | | |
| 180 | 6.47 | | 0.146 | | | 146.1 | 4.69 | | |
| 190 | 5.54 | | 0.141 | | | 149.4 | 4.16 | | |
| 200 | 4.81 | | 0.136 | | 1007 | 152.6 | 3.77 | | 0.024 |
| 210 | 4.24 | | 0.131 | | 957 | 155.2 | 3.47 | | 0.022 |
| 220 | 3.78 | | 0.126 | | 909 | 157.6 | 3.24 | | 0.021 |
| 230 | 3.40 | 0.100 | 0.121 | 0.0067 | 862 | 159.7 | 3.07 | 0.89 | 0.019 |
| 240 | 3.09 | 0.104 | 0.117 | 0.0073 | 814 | 161.3 | 2.92 | 0.89 | 0.017 |
| 250 | 2.82 | 0.109 | 0.112 | 0.0080 | 766 | 162.5 | 2.83 | 0.89 | 0.0155 |
| 260 | 2.60 | 0.114 | 0.107 | 0.0086 | 716 | 163.1 | 2.78 | 0.89 | 0.0138 |
| 270 | 2.41 | 0.118 | 0.102 | 0.0092 | 668 | 163.4 | 2.76 | 0.90 | 0.0121 |
| 280 | 2.25 | 0.123 | 0.097 | 0.0098 | 622 | 162.1 | 2.77 | 0.93 | 0.0104 |
| 290 | 2.11 | 0.129 | 0.092 | 0.0105 | 578 | 161.1 | 2.80 | 0.97 | 0.0087 |
| 300 | 1.98 | 0.135 | 0.087 | 0.0111 | 536 | 160.1 | 2.86 | 1.04 | 0.0071 |
| 310 | 1.86 | 0.141 | 0.082 | 0.0117 | 496 | 157.2 | 2.96 | 1.13 | 0.0055 |
| 320 | 1.76 | 0.148 | 0.077 | 0.0123 | 458 | 153.4 | 3.14 | 1.25 | 0.0040 |
| 330 | 1.67 | 0.157 | 0.072 | 0.0130 | 408 | 148.5 | 3.39 | 1.42 | 0.0026 |
| 340 | 1.51 | 0.171 | 0.067 | 0.0140 | 355 | 142.7 | 3.55 | 1.60 | 0.0014 |
| 350 | 1.30 | | 0.060 | | 290 | 135.9 | 3.72 | | 0.0008 |
| 360 | 1.06 | | | | | | | | |
| 369.3 | | | | | | | | | |

*Source:* Kakaç, S., Ed., *Boilers, Evaporators and Condensers,* Wiley, New York, pp. 812–813, 1991. With permission.

[a] 6.209. − 4 signifies 6.209 × $10^{-4}$.

[b] Critical point.

**TABLE B.9**

Thermophysical Properties of Refrigerant R22 at Atmospheric Pressure

| $T$ K | $v$ m³/kg | $h$ kJ/kg | $s$ kJ/(kg K) | $c_p$ kJ/(kg K) | $Z$ | $\bar{v}_s$ m/s | $\mu$ $10^{-6}$ Pa s | $k$ W/(m K) | $Pr$ |
|---|---|---|---|---|---|---|---|---|---|
| 232.3 | 0.2126 | 586.9 | 4.8230 | 0.608  | 0.9644 | 160.1 | 10.1 | 0.0067 | 0.893 |
| 240   | 0.2205 | 591.5 | 4.8673 | 0.6117 | 0.9682 | 163.0 | 10.4 | 0.0074 | 0.860 |
| 260   | 0.2408 | 604.0 | 4.8919 | 0.6255 | 0.9760 | 169.9 | 11.2 | 0.0084 | 0.838 |
| 280   | 0.2608 | 616.8 | 4.9389 | 0.6431 | 0.9815 | 176.2 | 12.0 | 0.0094 | 0.820 |
| 300   | 0.2806 | 630.0 | 4.9840 | 0.6619 | 0.9857 | 182.3 | 12.8 | 0.0106 | 0.804 |
| 320   | 0.3001 | 643.4 | 5.0274 | 0.6816 | 0.9883 | 188.0 | 13.7 | 0.0118 | 0.790 |
| 340   | 0.3196 | 657.3 | 5.0699 | 0.7017 | 0.9906 | 193.5 | 14.4 | 0.0130 | 0.777 |
| 360   | 0.3390 | 671.7 | 5.1111 | 0.7213 | 0.9923 | 198.9 | 15.1 | 0.0142 | 0.767 |
| 380   | 0.3583 | 686.5 | 5.1506 | 0.7406 | 0.9936 | 204.1 | 15.8 | 0.0154 | 0.760 |
| 400   | 0.3775 | 701.5 | 5.1892 | 0.7598 | 0.9945 | 209.1 | 16.5 | 0.0166 | 0.755 |
| 420   | 0.3967 | 717.0 | 5.2267 | 0.7786 | 0.9953 | 214.0 | 17.2 | 0.0178 | 0.753 |
| 440   | 0.4159 | 732.8 | 5.2635 | 0.7971 | 0.9961 | 218.8 | 17.9 | 0.0190 | 0.752 |
| 460   |        |       |        | 0.8150 |        | 223.5 | 18.6 | 0.0202 | 0.751 |
| 480   |        |       |        | 0.8326 |        | 227.9 | 19.3 | 0.0214 | 0.751 |
| 500   |        |       |        | 0.8502 |        |       | 19.9 | 0.0225 | 0.750 |

*Source:* Kakaç, S., Ed., *Boilers, Evaporators and Condensers*, Wiley, New York, p. 814, 1991. With permission.

**TABLE B.10**

Thermophysical Properties of Saturated Refrigerant R134a

| $T$ K | $P$ bar | $v_f$ m³/kg | $v_g$ m³/kg | $h_f$ kJ/kg | $h_g$ kJ/kg | $s_f$ kJ/(kg K) | $s_g$ kJ/(kg K) | $C_{pf}$ kJ/(kg K) | $C_{pg}$ kJ/(kg K) | $\mu_f$ 10⁻⁴ Pa s | $\mu_g$ 10⁻⁴ Pa s | $k_f$ W/(m K) | $k_g$ W/(m K) | $Pr_f$ | $Pr_g$ | $\sigma$ N/m |
|---|---|---|---|---|---|---|---|---|---|---|---|---|---|---|---|---|
| 200 | 0.070 | 0.000661 | 2.32 | −36.0 | 201.0 | −0.1691 | 1.0153 | | 0.732 | | | | | | | |
| 210 | 0.187 | 0.000674 | 0.906 | −26.5 | 208.1 | −0.1175 | 0.9941 | | | | | | | | | |
| 220 | 0.252 | 0.000687 | 0.698 | −15.3 | 214.5 | −0.0664 | 0.9758 | | | | | | | | | |
| 230 | 0.438 | 0.000701 | 0.416 | −3.7 | 220.8 | −0.0158 | 0.9602 | 1.113 | | | | | | | | |
| 240 | 0.728 | 0.000716 | 0.258 | 8.1 | 227.1 | 0.0343 | 0.9471 | 1.162 | 0.764 | 4.25 | 0.095 | 0.099 | 0.008 | 4.99 | 0.90 | |
| 250 | 1.159 | 0.000731 | 0.167 | 20.3 | 233.3 | 0.0840 | 0.9363 | 1.212 | 0.798 | 3.70 | 0.099 | 0.095 | 0.008 | 4.72 | 0.96 | 0.0145 |
| 260 | 1.765 | 0.000748 | 0.112 | 32.9 | 239.4 | 0.1331 | 0.9276 | 1.259 | 0.835 | 3.25 | 0.104 | 0.091 | 0.008 | 4.49 | 1.02 | 0.0131 |
| 270 | 2.607 | 0.000766 | 0.077 | 45.4 | 244.8 | 0.1817 | 0.9211 | 1.306 | 0.876 | 2.88 | 0.108 | 0.087 | 0.009 | 4.31 | 1.08 | 0.0117 |
| 280 | 3.721 | 0.000786 | 0.055 | 59.2 | 251.1 | 0.2299 | 0.9155 | 1.351 | 0.921 | 2.56 | 0.112 | 0.083 | 0.009 | 4.17 | 1.14 | 0.0103 |
| 290 | 5.175 | 0.000806 | 0.040 | 72.9 | 256.6 | 0.2775 | 0.9114 | 1.397 | 0.972 | 2.30 | 0.117 | 0.079 | 0.010 | 4.07 | 1.20 | 0.0090 |
| 300 | 7.02 | 0.000821 | 0.029 | 87.0 | 261.9 | 0.3248 | 0.9080 | 1.446 | 1.030 | 2.08 | 0.121 | 0.075 | 0.010 | 4.00 | 1.27 | |
| 310 | 9.33 | 0.000865 | 0.022 | 101.5 | 266.8 | 0.3718 | 0.9050 | 1.497 | 1.104 | 1.89 | 0.125 | 0.071 | 0.010 | 3.98 | 1.34 | |
| 320 | 12.16 | 0.000895 | 0.016 | 116.6 | 271.2 | 0.4189 | 0.9021 | 1.559 | 1.198 | 1.72 | 0.129 | 0.068 | 0.011 | 3.94 | 1.44 | |
| 330 | 15.59 | 0.000935 | 0.012 | 132.3 | 275.0 | 0.4663 | 0.8986 | 1.638 | 1.324 | 1.58 | 0.133 | 0.064 | 0.011 | 3.98 | 1.57 | |
| 340 | 19.71 | 0.000984 | 0.0094 | 148.9 | 277.8 | 0.5146 | 0.8937 | 1.750 | 1.520 | 1.45 | 0.137 | 0.060 | 0.012 | 4.23 | 1.74 | |
| 350 | 24.60 | 0.00105 | 0.0071 | 166.6 | 279.1 | 0.5649 | 0.8861 | 1.931 | 1.795 | 1.34 | 0.14 | 0.056 | 0.012 | 4.62 | 2.09 | |
| 360 | 30.40 | 0.00115 | 0.0051 | 186.5 | 277.7 | 0.6194 | 0.8721 | 2.304 | 2.610 | 1.20 | 0.16 | 0.054 | 0.013 | 5.16 | 3.21 | |
| 370 | 37.31 | 0.00139 | 0.0035 | 216.0 | 270.0 | 0.6910 | 0.8370 | | | 0.95 | 0.26 | | | | | |
| 374.3ᵃ | 40.67 | 0.00195 | 0.0020 | 248.0 | 248.0 | 0.7714 | 0.7714 | | | | | | | | | |

*Source:* Kakaç, S., Ed., *Boilers, Evaporators and Condensers*, Wiley, New York, p. 815, 1991. With permission.

ᵃ Critical point.

**TABLE B.11**

Thermophysical Properties of Refrigerant 134a
at Atmospheric Pressure

| $T$ K | $v$ m³/kg | $h$ kJ/kg | $s$ kJ/(kg K) | $c_p$ kJ/(kg K) | $Z$ | $\bar{v}_s$ m/s |
|---|---|---|---|---|---|---|
| 247 | 0.1901 | 231.5 | 0.940 | 0.787 | 0.957 | 145.9 |
| 260 | 0.2017 | 241.8 | 0.980 | 0.801 | 0.965 | 150.0 |
| 280 | 0.2193 | 258.1 | 1.041 | 0.827 | 0.974 | 156.3 |
| 300 | 0.2365 | 274.9 | 1.099 | 0.856 | 0.980 | 162.1 |
| 320 | 0.2532 | 292.3 | 1.155 | 0.885 | 0.984 | 167.6 |
| 340 | 0.2699 | 310.3 | 1.209 | 0.915 | 0.987 | 172.8 |
| 360 | 0.2866 | 328.8 | 1.263 | 0.945 | 0.990 | 177.6 |
| 380 | 0.3032 | 347.8 | 1.313 | 0.976 | 0.992 | 182.0 |
| 400 | 0.3198 | 367.2 | 1.361 | 1.006 | 0.994 | 186.0 |

*Source:* Kakaç, S., Ed., *Boilers, Evaporators and Condensers*, Wiley, New York, p. 816, 1991. With permission.

**TABLE B.12**

Thermophysical Properties of Unused Engine Oil

| $T$ K | $v_f$ m³/kg | $c_{pf}$ kJ/kg | $\mu_f$ Pa s | $k_f$ W/(m K) | $Pr_f$ | $\alpha_f$ m²/s |
|---|---|---|---|---|---|---|
| 250 | 1.093. – 3[a] | 1.72 | 32.20 | 0.151 | 367,000 | 9.60. – 8 |
| 260 | 1.101. – 3 | 1.76 | 12.23 | 0.149 | 144,500 | 9.32. – 8 |
| 270 | 1.109. – 3 | 1.79 | 4.99 | 0.148 | 60,400 | 9.17. – 8 |
| 280 | 1.116. – 3 | 1.83 | 2.17 | 0.146 | 27,200 | 8.90. – 8 |
| 290 | 1.124. – 3 | 1.87 | 1.00 | 0.145 | 12,900 | 8.72. – 8 |
| 300 | 1.131. – 3 | 1.91 | 0.486 | 0.144 | 6,450 | 8.53. – 8 |
| 310 | 1.139. – 3 | 1.95 | 0.253 | 0.143 | 3,450 | 8.35. – 8 |
| 320 | 1.147. – 3 | 1.99 | 0.141 | 0.141 | 1,990 | 8.13. – 8 |
| 330 | 1.155. – 3 | 2.04 | 0.084 | 0.140 | 1,225 | 7.93. – 8 |
| 340 | 1.163. – 3 | 2.08 | 0.053 | 0.139 | 795 | 7.77. – 8 |
| 350 | 1.171. – 3 | 2.12 | 0.036 | 0.138 | 550 | 7.62. – 8 |
| 360 | 1.179. – 3 | 2.16 | 0.025 | 0.137 | 395 | 7.48. – 8 |
| 370 | 1.188. – 3 | 2.20 | 0.019 | 0.136 | 305 | 7.34. – 8 |
| 380 | 1.196. – 3 | 2.25 | 0.014 | 0.136 | 230 | 7.23. – 8 |
| 390 | 1.205. – 3 | 2.29 | 0.011 | 0.135 | 185 | 7.10. – 8 |
| 400 | 1.214. – 3 | 2.34 | 0.009 | 0.134 | 155 | 6.95. – 8 |

*Source:* Kakaç, S., Ed., *Boilers, Evaporators and Condensers*, Wiley, New York, p. 826, 1991. With permission.

[a] The notation 1.093. – 3 signifies $1.093 \times 10^{-3}$.

# Reference

1. Kakaç, S. (Ed.), *Boilers, Evaporators and Condensers*, Wiley, New York, 1991.

# Appendix C: Bessel Functions

Second-order linear differential equations with variable coefficients of the form

$$x^2 \frac{d^2 y}{dx^2} + x \frac{dy}{dx} + (m^2 x^2 - v^2) y = 0 \tag{C.1}$$

are known as the *Bessel differential equations* [1–5], where $m$ is a parameter and $v$ is any real number. Since only the quantity $v^2$ appears in Equation C.1, we may also consider $v$ to be nonnegative without loss of generality. The general solution of Equation C.1 may be obtained by the use of the method of Frobenius and the result is

$$y(x) = C_1 J_v(mx) + C_2 Y_v(mx), \tag{C.2}$$

where the functions $J_v(mx)$ and $Y_v(mx)$ are known as the *Bessel functions of order $v$ of the first and second kind*, respectively. The function $J_v(mx)$ is defined, for all values of $v$, as

$$J_V(mx) = \sum_{k=0}^{\infty} (-1)^k \frac{(mx/2)^{2k+v}}{k! \Gamma(k+v-1)}. \tag{C.3}$$

If $v$ is not an integer or zero, then the function $Y_v(mx)$ is defined as

$$Y_V(mx) = \frac{(\cos v\pi) J_v(mx) - J_{-v}(mx)}{\sin v\pi}, \quad v \neq 0,1,2,3,\ldots, \tag{C.4}$$

and if $v$ is zero or an integer then it is given by

$$Y_v(mx) = \frac{2}{\pi} \left[ \ln \frac{mx}{2} + \gamma \right] J_v(mx) - \frac{1}{\pi} \sum_{k=0}^{v-1} \frac{(v-k-1)!}{k!} \left( \frac{mx}{2} \right)^{2k-v}$$

$$- \frac{1}{\pi} \sum_{k=0}^{\infty} (-1)^k \left[ \phi(k) + \phi(k+v) \right] \frac{(mx/2)^{2k+v}}{k!(v+k)!}, \quad v = 0,1,2,3,\ldots \tag{C.5}$$

The function $J_{-v}(mx)$ in Equation C.4 is obtained by replacing $v$ by $-v$ in Equation C.3, that is,

$$J_{-v}(mx) = \sum_{k=0}^{\infty} (-1)^k \frac{(mx/2)^{2k-v}}{k!\Gamma(k-v-1)}. \tag{C.6}$$

The *gamma junction* appearing in the denominator of Equations C.3 and C.6 is defined by

$$\Gamma(v) = \int_0^{\infty} \eta^{v-1} e^{-\eta} d\eta, \quad v > 0 \tag{C.7a}$$

from which it follows that

$$\Gamma(v+1) = v\Gamma(v), \quad v > 0. \tag{C.7b}$$

If $v$ is a positive integer, then this function has the following relation to the factorial function:

$$\Gamma(v+1) = v!, \quad \Gamma(1) = 1, \quad v = 1,2,3,.... \tag{C.8}$$

If $v$ is not an integer, then it can be shown that

$$\Gamma(v)\Gamma(1-v) = \frac{\pi}{\sin v\pi}. \tag{C.9}$$

In Equation C.5, the expression $\phi(k)$ is defined as

$$\phi(k) = \sum_{n-1}^{k} \frac{1}{n}, \quad \phi(0) = 0 \tag{C.10}$$

and $\gamma = 0.5772156\ldots$ is the *Euler constant*.
    If $v$ is an integer, then it can be shown that

$$J_{-v}(mx) = (-1)^v J_v(mx), \quad v = 1,2,3,... \tag{C.11}$$

If $v$ is not an integer, $J_v(mx)$ and $J_{-v}(mx)$ are linearly independent, and for this case, the general solution of Equation C.1 can also be written as

$$y(x) = D_1 J_v(mx) + D_2 J_{-v}(mx), \quad v \neq 0,1,2,3,... \tag{C.12}$$

    The behavior of the Bessel functions cannot easily be predicted from their series representations. For integer values of $v$, the general behavior of these functions is shown in Figure C.1a and b. Note that the Bessel functions of the second kind $Y_v$ are unbounded at $x = 0$.

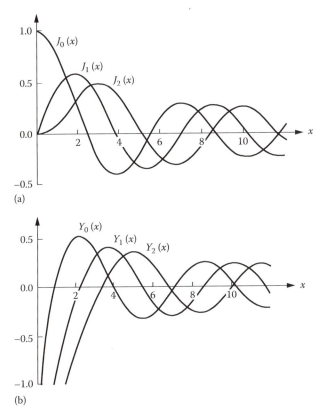

**FIGURE C.1**
Bessel functions of the (a) first kind and (b) second kind.

## C.1 Modified Bessel Functions

Second-order linear differential equations with variable coefficients of the form

$$x^2 \frac{d^2y}{dx^2} + x \frac{dy}{dx} - (m^2x^2 + v^2)y = 0 \tag{C.13}$$

are known as the *modified Bessel differential equations*. The general solution of Equation C.13 may be written as

$$y(x) = C_1 I_v(mx) + C_2 K_v(mx), \tag{C.14}$$

where the functions $I_v(mx)$ and $K_v(mx)$ are known as the *modified Bessel functions of order $v$ of the first and second kind*, respectively. The function $I_v(mx)$ is defined, for all values of $v$, as

$$I_v(mx) = (i)^{-v} I_v(imx), \quad i = \sqrt{-1}. \tag{C.15}$$

If $v$ is not an integer or zero, then the function $K_v(mx)$ is defined as

$$K_v(mx) = \frac{\pi}{2} \frac{I_{-v}(mx) - I_v(mx)}{\sin v\pi}, \quad v \neq 0,1,2,3,\ldots \tag{C.16}$$

and if $v$ is zero or an integer then it is given by

$$K_v(mx) = (-1)^{v+1} \left[ \ln \frac{mx}{2} + \gamma \right] I_v(mx)$$

$$+ \frac{1}{2} \sum_{k=0}^{v-1} (-1)^k \frac{(v-k-1)!}{k!} \left( \frac{mx}{2} \right)^{2k-v}$$

$$+ \frac{1}{2} (-1)^v \sum_{k=0}^{\infty} \left[ \phi(k) + \phi(k+v) \right] \frac{(mx/2)^{2k+v}}{k!(v+K)!}, \quad v = 0,1,2,3,\ldots \tag{C.17}$$

where the definitions of $\phi(k)$ and $\gamma$ are as defined before.

If $v$ is an integer, then

$$I_{-v}(mx) = I_v(mx), \quad v = 1,2,3,\ldots \tag{C.18}$$

If $v$ is not an integer, $I_{-v}(mx)$ and $I_v(mx)$ are linearly independent, and for this case, the general solution of Equation C.13 can also be written as

$$y(x) = D_1 I_v(mx) + D_2 I_{-v}(mx), \quad v \neq 0,1,2,3,\ldots \tag{C.19}$$

Graphs of the general behavior of these functions for integer values of $v$ are shown in Figure C.2a and b.

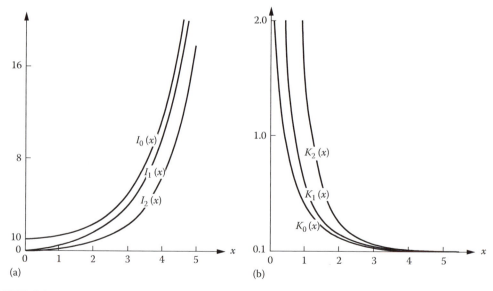

**FIGURE C.2**
Modified Bessel functions of the (a) first kind and (b) second kind.

## C.2  Asymptotic Formulas for the Bessel Functions

For large values of $x$, we have the following asymptotic formulas:

$$J_v(x) \cong \sqrt{\frac{2}{\pi x}} \cos\left(x - \frac{\pi}{4} - \frac{v\pi}{2}\right) \tag{C.20}$$

$$Y_v(x) \cong \sqrt{\frac{2}{\pi x}} \sin\left(x - \frac{\pi}{4} - \frac{v\pi}{2}\right) \tag{C.21}$$

$$I_v(x) \cong \frac{e^x}{\sqrt{2\pi x}} \tag{C.22}$$

$$K_v(x) \cong \sqrt{\frac{\pi}{2x}} e^{-x} \tag{C.23}$$

## C.3  Derivatives of the Bessel Functions

Some derivatives of the Bessel functions are as follows:

$$\frac{d}{dx}\left[W_v(mx)\right] = \begin{cases} mW_{v-1}(mx) - \left(\dfrac{v}{x}\right)W_v(mx), & W = J, Y, I \\[2ex] -mW_{v-1}(mx) - \left(\dfrac{v}{x}\right)W_v(mx), & W = K \end{cases} \tag{C.24}$$

$$\frac{d}{dx}\left[W_v(mx)\right] = \begin{cases} -mW_{v+1}(mx) + \left(\dfrac{v}{x}\right)W_v(mx), & W = J, Y, k \\[2ex] mW_{v+1}(mx) + \left(\dfrac{v}{x}\right)W_v(mx), & W = I \end{cases} \tag{C.25}$$

$$\frac{d}{dx}\left[x^v W_v(mx)\right] = \begin{cases} mx^v W_{v-1}(mx), & W = J, Y, I \\[2ex] -mx^v W_{v-1}(mx), & W = K \end{cases} \tag{C.26}$$

$$\frac{d}{dx}\left[x^{-v} W_v(mx)\right] = \begin{cases} -mx^{-v} W_{v+1}(mx), & W = J, Y, K \\[2ex] mx^{-v} W_{v+1}(mx), & W = I \end{cases} \tag{C.27}$$

## C.4    Equations Transformable into Bessel's Differential

The differential equation

$$x^2 \frac{d^2y}{dx^2} + (2k+1)x\frac{dy}{dx} + (\rho^2\alpha^2x^{2\rho} + k^2 - \rho^2 v^2)y(x) = 0, \qquad (C.28)$$

where $k$, $\alpha$, $\rho$, and $v$ are constants, has the following solution:

$$y(x) = x^{-k}\left[C_1 J_v(\alpha x^\rho) + C_2 Y_v(\alpha x^\rho)\right]$$

If $\alpha = 0$, Equation C.28 becomes a *Euler* or *Cauchy equation* and has the following solution:

$$y(x) = x^{-k}[C_1 x^{\rho v} + C_2 x^{-\rho v}] \qquad (C.29)$$

## References

1. Adams, J. A. and Rogers, D. F., *Computer-Aided Heat Transfer Analyses*, McGraw-Hill, New York, 1973.
2. Arpaci, V. S., *Conduction Heat Transfer*, Addison-Wesley, Reading, MA, 1966.
3. Özişik, M. N., *Heat Conduction*, Wiley, New York, 1980.
4. Hildebrand, F. B., *Advanced Calculus for Applications*, Prentice Hall, Englewood Cliffs, NJ, 1962.
5. Kakaç, S., Shah, R. K., and Aung, W. (Eds.), *Handbook of Single-Phase Convective Heat Transfer*, Wiley, New York, 1987.

# *Index*